T0203122

Lecture Notes in Computer Science 14434

Founding Editors

Gerhard Goos
Juris Hartmanis

The series Lecture Notes in Computer Science (LNCS), including its subseries Lecture Notes in Artificial Intelligence (LNAI) and Lecture Notes in Bioinformatics (LNBI), has established itself as a medium for the publication of new developments in computer science and information technology research, teaching, and education.

LNCS enjoys close cooperation with the computer science R & D community, the series counts many renowned academics among its volume editors and paper authors, and collaborates with prestigious societies. Its mission is to serve this international community by providing an invaluable service, mainly focused on the publication of conference and workshop proceedings and postproceedings. LNCS commenced publication in 1973.

Qingshan Liu · Hanzi Wang · Zhanyu Ma ·
Weishi Zheng · Hongbin Zha · Xilin Chen ·
Liang Wang · Rongrong Ji
Editors

Pattern Recognition and Computer Vision

6th Chinese Conference, PRCV 2023
Xiamen, China, October 13–15, 2023
Proceedings, Part X

 Springer

Editors
Qingshan Liu 🆔
Nanjing University of Information Science
and Technology
Nanjing, China

Zhanyu Ma 🆔
Beijing University of Posts
and Telecommunications
Beijing, China

Hongbin Zha 🆔
Peking University
Beijing, China

Liang Wang
Chinese Academy of Sciences
Beijing, China

Hanzi Wang 🆔
Xiamen University
Xiamen, China

Weishi Zheng 🆔
Sun Yat-sen University
Guangzhou, China

Xilin Chen 🆔
Chinese Academy of Sciences
Beijing, China

Rongrong Ji 🆔
Xiamen University
Xiamen, China

ISSN 0302-9743   ISSN 1611-3349   (electronic)
Lecture Notes in Computer Science
ISBN 978-981-99-8548-7   ISBN 978-981-99-8549-4   (eBook)
https://doi.org/10.1007/978-981-99-8549-4

This Springer imprint is published by the registered company Springer Nature Singapore Pte Ltd.
The registered company address is: 152 Beach Road, #21-01/04 Gateway East, Singapore 189721, Singapore

Paper in this product is recyclable.

Preface

Welcome to the proceedings of the Sixth Chinese Conference on Pattern Recognition and Computer Vision (PRCV 2023), held in Xiamen, China.

PRCV is formed from the combination of two distinguished conferences: CCPR (Chinese Conference on Pattern Recognition) and CCCV (Chinese Conference on Computer Vision). Both have consistently been the top-tier conference in the fields of pattern recognition and computer vision within China's academic field. Recognizing the intertwined nature of these disciplines and their overlapping communities, the union into PRCV aims to reinforce the prominence of the Chinese academic sector in these foundational areas of artificial intelligence and enhance academic exchanges. Accordingly, PRCV is jointly sponsored by China's leading academic institutions: the Chinese Association for Artificial Intelligence (CAAI), the China Computer Federation (CCF), the Chinese Association of Automation (CAA), and the China Society of Image and Graphics (CSIG).

PRCV's mission is to serve as a comprehensive platform for dialogues among researchers from both academia and industry. While its primary focus is to encourage academic exchange, it also places emphasis on fostering ties between academia and industry. With the objective of keeping abreast of leading academic innovations and showcasing the most recent research breakthroughs, pioneering thoughts, and advanced techniques in pattern recognition and computer vision, esteemed international and domestic experts have been invited to present keynote speeches, introducing the most recent developments in these fields.

PRCV 2023 was hosted by Xiamen University. From our call for papers, we received 1420 full submissions. Each paper underwent rigorous reviews by at least three experts, either from our dedicated Program Committee or from other qualified researchers in the field. After thorough evaluations, 522 papers were selected for the conference, comprising 32 oral presentations and 490 posters, giving an acceptance rate of 37.46%. The proceedings of PRCV 2023 are proudly published by Springer.

Our heartfelt gratitude goes out to our keynote speakers: Zongben Xu from Xi'an Jiaotong University, Yanning Zhang of Northwestern Polytechnical University, Shutao Li of Hunan University, Shi-Min Hu of Tsinghua University, and Tiejun Huang from Peking University.

We give sincere appreciation to all the authors of submitted papers, the members of the Program Committee, the reviewers, and the Organizing Committee. Their combined efforts have been instrumental in the success of this conference. A special acknowledgment goes to our sponsors and the organizers of various special forums; their support made the conference a success. We also express our thanks to Springer for taking on the publication and to the staff of Springer Asia for their meticulous coordination efforts.

We hope these proceedings will be both enlightening and enjoyable for all readers.

October 2023

Qingshan Liu
Hanzi Wang
Zhanyu Ma
Weishi Zheng
Hongbin Zha
Xilin Chen
Liang Wang
Rongrong Ji

Organization

General Chairs

Hongbin Zha	Peking University, China
Xilin Chen	Institute of Computing Technology, Chinese Academy of Sciences, China
Liang Wang	Institute of Automation, Chinese Academy of Sciences, China
Rongrong Ji	Xiamen University, China

Program Chairs

Qingshan Liu	Nanjing University of Information Science and Technology, China
Hanzi Wang	Xiamen University, China
Zhanyu Ma	Beijing University of Posts and Telecommunications, China
Weishi Zheng	Sun Yat-sen University, China

Organizing Committee Chairs

Mingming Cheng	Nankai University, China
Cheng Wang	Xiamen University, China
Yue Gao	Tsinghua University, China
Mingliang Xu	Zhengzhou University, China
Liujuan Cao	Xiamen University, China

Publicity Chairs

Yanyun Qu	Xiamen University, China
Wei Jia	Hefei University of Technology, China

Local Arrangement Chairs

Xiaoshuai Sun   Xiamen University, China
Yan Yan   Xiamen University, China
Longbiao Chen   Xiamen University, China

International Liaison Chairs

Jingyi Yu   ShanghaiTech University, China
Jiwen Lu   Tsinghua University, China

Tutorial Chairs

Xi Li   Zhejiang University, China
Wangmeng Zuo   Harbin Institute of Technology, China
Jie Chen   Peking University, China

Thematic Forum Chairs

Xiaopeng Hong   Harbin Institute of Technology, China
Zhaoxiang Zhang   Institute of Automation, Chinese Academy of
                         Sciences, China
Xinghao Ding   Xiamen University, China

Doctoral Forum Chairs

Shengping Zhang   Harbin Institute of Technology, China
Zhou Zhao   Zhejiang University, China

Publication Chair

Chenglu Wen   Xiamen University, China

Sponsorship Chair

Yiyi Zhou   Xiamen University, China

Exhibition Chairs

Bineng Zhong   Guangxi Normal University, China
Rushi Lan   Guilin University of Electronic Technology, China
Zhiming Luo   Xiamen University, China

Program Committee

Baiying Lei   Shenzhen University, China
Changxin Gao   Huazhong University of Science and Technology,
                               China
Chen Gong   Nanjing University of Science and Technology,
                               China
Chuanxian Ren   Sun Yat-Sen University, China
Dong Liu   University of Science and Technology of China,
                               China
Dong Wang   Dalian University of Technology, China
Haimiao Hu   Beihang University, China
Hang Su   Tsinghua University, China
Hui Yuan   School of Control Science and Engineering,
                               Shandong University, China
Jie Qin   Nanjing University of Aeronautics and
                               Astronautics, China
Jufeng Yang   Nankai University, China
Lifang Wu   Beijing University of Technology, China
Linlin Shen   Shenzhen University, China
Nannan Wang   Xidian University, China
Qianqian Xu   Key Laboratory of Intelligent Information
                               Processing, Institute of Computing
                               Technology, Chinese Academy of Sciences,
                               China
Quan Zhou   Nanjing University of Posts and
                               Telecommunications, China
Si Liu   Beihang University, China
Xi Li   Zhejiang University, China
Xiaojun Wu   Jiangnan University, China
Zhenyu He   Harbin Institute of Technology (Shenzhen), China
Zhonghong Ou   Beijing University of Posts and
                               Telecommunications, China

Contents – Part X

Neural Network and Deep Learning III

Dual-Stream Context-Aware Neural Network for Survival Prediction from Whole Slide Images

Junxiu Gao, Shan Jin, Ranran Wang, Mingkang Wang, Tong Wang, and Hongming Xu[✉]

School of Biomedical Engineering, Faculty of Medicine, Dalian University of Technology, Dalian 116024, China
mxu@dlut.edu.cn

Abstract. Whole slide images (WSI) encompass a wealth of information about the tumor micro-environment, which holds prognostic value for patients' survival. While significant progress has been made in predicting patients' survival risks from WSI, existing studies often overlook the importance of incorporating multi-resolution and multi-scale histological image features, as well as their interactions, in the prediction process. This paper introduces the dual-stream context-aware (DSCA) model, which aims to enhance survival risk prediction by leveraging multi-resolution histological images and multi-scale feature maps, along with their contextual information. The DSCA model comprises three prediction branches: two ResNet50 branches that learn features from multi-resolution images, and one feature fusion branch that aggregates multi-scale features by exploring their interactions. The feature fusion branch of the DSCA model incorporates a mixed attention module, which performs adaptive spatial fusion to enhance the multi-scale feature maps. Subsequently, the self-attention mechanism is developed to learn contextual and interactive information from the enhanced feature maps. The ordinal Cox loss is employed to optimize the model for generating patch-level predictions. Patient-level predictions are obtained by mean-pooling patch-level results. Experimental results conducted on colorectal cancer cohorts demonstrate that the proposed DSCA model achieves significant improvements over state-of-the-art methods in survival prognosis.

Keywords: Survival risk prediction · Whole slide images · Attention mechanism · Multi-scale features · Colorectal Cancer

1 Introduction

Survival analysis focuses on the time span from the beginning to the occurrence of an event, which is of great significance in determining the treatment strategy for cancer patients. Survival analysis from whole slide images (WSI) has gradually evolved from pathologists' subjective estimation to automatic feature extraction and indicative prediction using deep learning [4,6,16]. However,

Q. Liu et al. (Eds.): PRCV 2023, LNCS 14434, pp. 3–14, 2024.
https://doi.org/10.1007/978-981-99-8549-4_1

due to the large scale of histological slides and the high heterogeneity of tumor micro-environment (TME), building an automated survival prediction model via quantifying pathological features within and between tissues from WSI remains a challenging problem.

Recently, multiple instance learning (MIL) methods have gained significant popularity in performing survival prognosis from WSI. The MIL method commonly utilizes a feature extractor composed of convolutional neural networks (CNNs) to obtain instance-level features. These features are then aggregated using pooling layers to obtain the WSI-level feature representation. Finally, the MIL model is trained with WSI-level labels. Using the MIL pipeline, Yao et al. [26] proposed a deep attention multiple instance survival learning (DeepAttnMISL) model, which primarily relies on an attention-guided MIL method (termed as DeepMIL) [8] to predict patients' survival risks. Furthermore, Schirris et al. [18] developed an enhanced DeepMIL method called DeepSMILE by integrating attention-weighted variance pooling across distinct phenotypes and utilizing self-supervised pre-trained ShufflenetV2 as the feature extractor. To capture intricate relationships of instance-level input sequence by using the attention mechanism, Huang et al. [7] integrated feature embedding from pre-trained ResNet18 and position encoding with the Transformer encoder for survival outcome predictions. Jiang et al. [9] proposed an attention-based model (MHAttnSurv) which implemented the MIL method based on the multi-head attention mechanism. The above approaches utilize abundant attention mechanisms and feature extractors to capture meaningful relationships within the WSIs, enabling more accurate predictions of patient survival outcomes. While MIL methods are capable of capturing globally representative WSI features, their performance is often heavily reliant on the feature aggregator, which can sometimes overlook or neglect small pathological features during the feature aggregation process. Besides, due to the large scale of WSI and limitations in computer memory, instance-level features in MIL methods are typically computed via a pre-trained CNN model. However, there is no guarantee that these instance-level features are highly consistent with patient-level survival risks, which ultimately limits the performance of MIL methods.

In contrast to the aforementioned MIL methods, patch-based methods directly train the survival prediction model using histological image patches in an end-to-end fashion, allowing the model to learn survival-related histological information effectively. Mobadersany et al. [16] integrated patches from annotated ROIs and genomic data based on convolutional layers and fully connected layers in survival prognosis of Oligodendroglioma patients. Skrede et al. [20] developed a clinical prognostic model (DoMore-v1-CRC) using the MobileNetV2 allied to patches with high tumor content for the good or poor outcome prediction of colorectal cancer patients. Laleh et al. [10] predicted survival outcomes of gastrointestinal cancer patients by using a deep survival model (EE-Surv) that was trained with randomly selected histological image patches. These patch-based methods offer advantages in terms of end-to-end learning, but they do have some limitations. For example, patches at a specific resolution may have limited information and may not fully represent the entire WSI. Histological images at different resolutions can capture varying levels of lesion tissue information. Consequently, relying solely on patches at a single resolution may not capture the complete histological context present in the WSI, although it is challenging to achieve multi-scale context awareness when simultaneously learning multi-scale features from histological images at different resolutions.

To overcome the aforementioned limitations of previous studies, this paper introduces a novel end-to-end survival prediction model called the Dual-stream Context-aware (DSCA) model. The DSCA model is designed to efficiently learn and integrate multi-scale histological features extracted from multi-resolution images. Our contributions can be summarized as: (1) We design a dual-steam neural network that learns histological feature representations from multi-resolution images for survival risk prediction. (2) We adopt weighted histological feature map fusion across branches while improving channel and spatial attentions, and fuse feature representations using a multi-scale co-attention mechanism that incorporates context-aware learning. (3) The ordinal Cox loss is integrated to train our model, which further makes improvement over other comparative models in survival risk predictions.

2   Method

Figure 1 shows the flowchart of our DSCA model. As illustrated in Fig. 1, the DSCA is composed of two Resnet50 branches and one feature fusion branch. The fusion branch comprises eight Sequential Channel Spatial Attention (SCSA) modules, accompanied by weight units dedicated to intermediate features, an Efficient Spatial Feature Fusion (ESFF) module, and a Multi-scale Co-Attention (MSCA) module. Two Resnet50 branches extract features independently and generate prediction simultaneously in parallel. The fusion branch integrates multi-scale histological image features while incorporating context awareness, and then generates a prediction score. The prediction scores from three branches are concatenated together to make the multi-resolution patch-level survival risk prediction.

Data Pre-processing: By following previous studies [10,21], we divide each WSI into a set of non-overlapping patches of 512×512 pixels at 40× magnification, with background patches containing mainly white color pixels excluded. Then, 20× patches (see Fig. 1) are generated by utilizing the 40× patches as their centering regions, which means that neighboring patches at 20× magnification overlap by half. Considering the bias from staining and scanning of tissue samples, we generate stain-normalized patches by using the Macenko method [10,17]. The 40× stain-normalized patches are fed into the VGG19 model, which help to select patches belonging to categories of tumor (TUM), lymphocytes (LYM), and stroma (STR). Note that our VGG19 model was trained using a public dataset [27] that consists of 9 different tissue types in colorectal cancer slides, which provided over 96% accuracy during the independent testing. The TUM,

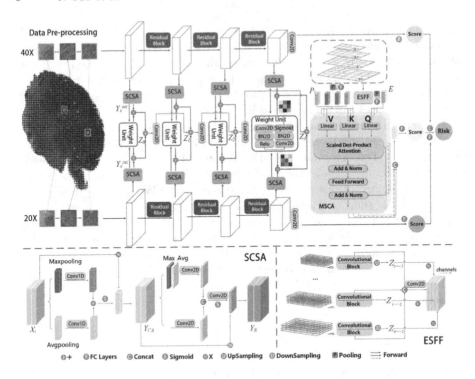

Fig. 1. The schematic diagram of our proposed DSCA model.

LYM and STR patches are selected, as they are considered informative for patients' survival prognosis. After patch selection, 400 to 1500 patches at $40\times$ magnification per WSI are derived.

Dual-Stream Feature Extractor with Multi-scale Feature Fusion:
Given an input $40\times$ patch and its corresponding $20\times$ patch, we feed them separately into two branches, resulting in multi-scale features fed into the fusion branch. To enhance representative power of multi-scale feature maps, the SCSA module (see Fig. 1) that explores channel and spatial attention is proposed. Specifically, given a feature map $X_i \in \mathbb{R}^{C_i \times H_i \times W_i}$, where $i \in \{0,1,2,3\}$, $C_i \in \{256,512,1024,2048\}$, the SCSA module adaptively adjusts the weights of different channels and spatial regions in a sequential manner, i.e.,

$$C_{CA} = \sigma\left(\text{Conv1D}(\text{GAP}(X_i)) + \text{Conv1D}(\text{GMP}(X_i))\right) \quad (1)$$
$$S_{SA} = \sigma\left(\text{Conv}(\text{Cat}(\text{Conv}(\text{Cat}(\text{AVG}(Y_{CA}), \text{MAX}(Y_{CA}))), \text{Conv}(Y_{CA})))\right) \quad (2)$$

where the GAP, GMP separately represent global average and max pooling, the MAX, AVG separately represent max and average pooling, σ means the sigmoid operation, and C_{CA}, S_{SA} separately refer to channel and spatial attentions. Our channel attention is improved from the efficient channel attention (ECA) as proposed in ECA-Net [24] by making use of both global max and average pooling for encoding global information among channels. The spatial attention extends the attention module as proposed in CBAM [25] by adding a convolutional operation branch to extract more details of local regions.

The enhanced feature map, i.e., $Y_s \in \mathbb{R}^{C_i \times H_i \times W_i}$, is generated from every SCSA module. Considering different scales of resolution-specific information, we design a novel feature fusion module which ensures the consistency across different stages of feature extraction. Inspired by the multi-task learning mechanism [13], our design is formulated as follows:

$$Z_i = WU^{20X}Y_s{}^{40X} + WU^{40X}Y_s{}^{20X} + \text{Conv}(Z_{i-1}) \tag{3}$$

where WU^{20X}, WU^{40X} represent the outputs of corresponding weight units from two branches. To ensure the continuous validity of the generated multi-scale features, we additionally implement a residual-style block across multiple stages to achieve skip connections following the weight-based addition operation. This approach ensures that the information flow remains uninterrupted and allows for effective integration of features among adjacent scales.

After generating feature maps $Z_i \in \mathbb{R}^{C_i \times H_i \times W_i}$, they are fed into our constructed ESFF module (see Fig. 1) for the initial spatial feature fusion. Inspired by the adaptively spatial feature fusion (ASFF) mechanism [14], the multi-scale feature maps Z_i are aggregated to generate the feature map Z, which contains both global and local information. The operations are defined as follows:

$$Z = \alpha^l Z_{0 \to l} + \beta^l Z_{1 \to l} + \gamma^l Z_{2 \to l} + \lambda^l Z_{3 \to l} \tag{4}$$

where $Z_{n \to l}$ denotes that Z_i is resized from level n to level l, α^l, β^l, γ^l, and λ^l refer to spatial importance weights of the feature maps from different levels to level l, which are adaptively learned by the network. It is empirically found that setting l to 2 can achieve the best performance. The feature map Z is shaped into the 2D-flattened token, i.e., $E \in \mathbb{R}^{1 \times dim}$, through pooling and multi-layer perceptron sequentially. Meanwhile, as illustrated in Fig. 1, the multi-scale feature maps Z_i, where $i \in \{0,1,2,3\}$, are mapped into the flattened tokens $P_i \in \mathbb{R}^{1 \times dim}$ with a unified dimension (e.g., $dim=64$) via the global average pooling and full connection operations.

Multi-scale Co-attention with Context-Aware Learning: Transformer [23] has the advantages in encoding global information of long-range sequences, which is incorporated to model communication and interaction between multi-resolution and multi-scale histological image features in this study. Specifically, the linearly projected features are fed into the scaled dot-product attention block, where the token $E \in \mathbb{R}^{1 \times dim}$ is utilized as the query (Q), and every token P_i is utilized as both the key (K) and value (V). As illustrated in Fig. 1, the calculation is performed as:

$$Attention(Q, K, V) = softmax(\frac{QW_q(KW_k)^T}{\sqrt{dim}})VW_v \tag{5}$$

$$O^i = Attention^i + FFD(K^i + Attention^i) \tag{6}$$

$$MSCA = Concat(O^1, O^2, ..., O^n)W_o \tag{7}$$

where W_q, W_k, W_v, W_o are trainable parameters, and n (i.e., $n = 4$) corresponds to the number of heads. FFD represents the block with linear projection and dropout layers. By incorporating relationship modeling across multi-scale feature representations, patch-level embedding contains more comprehensive information which is mapped to a score value via the multi-layer perceptron. Feature representations from both Resnet50 branches and one feature fusion branch are mapped to three score values which are finally fed into the output layer for survival risk prediction.

Ordinal Cox Model for Survival Prediction: Our DSCA survival prediction model is optimized via minimizing the ordinal Cox loss [19] function. Different from the Cox loss, the ordinal Cox loss considers survival analysis as a regression problem that focuses on the sequence of events. The key point of the ordinal Cox loss lies in the inclusion of a ranking regularization term by utilizing the ordinal survival information among different patients. Given the patient i observed with the event e_i at the time t_i, the ordinal Cox loss in the output layer is formulated as:

$$l(\theta, X_i) = \sum_{i:e_i=1} \left(\theta X_i - \log \sum_{j \in R(t_i)} \exp^{\theta X_i} \right.$$

$$\left. + \lambda \sum_{j \in R(t_i)} I(X_i, X_j) \max(0, 1 - r_{i,j}) \right) \tag{8}$$

where θ indicates the learnable parameters; X_i is the ith patch-level feature representation (i.e., concatenated prediction scores of three branches); $I(X_i, X_j) = 1$ if the ith patch belongs to a non-censored patient, whose recorded survival time t_i is shorter than that of t_j (i.e., the jth patch), otherwise $I(X_i, X_j) = 0$; $R(t_i)$ represents a risk set which includes patients with events that have not occurred at the time t_i; $r_{i,j}$ is the ordinal relationship of two patches computed from the ratio of two hazard functions, i.e., $r_{ij} = \exp(\theta(X_i - X_j))$. The regularization parameter λ takes 2.5 in this study. Note that our end-to-end DSCA model first generates patch-level survival risk predictions, and then mean-pooling is performed on patch-level predictions from each WSI to generate the patient-level survival risk score.

3   Experiments and Results

Dataset: Our experiments were conducted on two distinct cancer cohorts including Colon adenocarcinoma (COAD) and Rectum adenocarcinoma (READ) derived from The Cancer Genome Atlas (TCGA) project. TCGA project has collected WSIs and corresponding clinical information of cancer patients. Note that some TCGA patient slides with poor qualities or artifacts (e.g., pen markers) have been excluded for evaluation. Table 1 lists demographic and clinical information of two cohorts that are used in this study.

Table 1. Demographic and clinical information of study cohorts.

Cohort	No. of Patient	No. of WSI	No. of male	No. of Event	Avg OS (month)
TCGA-COAD	335	335	180	63	25.8
TCGA-READ	142	142	79	25	27.1

Experimental Settings: We implement and train our model with PyTorch by using two NVIDIA GeForce RTX3090 GPUs. We empirically set the minibatch size and learning rate as 64 and 4e-5. The Adam optimization algorithm is employed for training the proposed model. To ensure the reliability of our findings, we adopt the three-fold cross validation strategy where patients are partitioned into mutually exclusive subsets of equal size, with one fold serving as the test data in an iterative manner, while the remaining two folds constituted the training and validation datasets, ultimately computing the average result. For mitigating prediction bias towards the WSI with a large number of histological patches, 100 patches are randomly selected from three tissue types (TILs, TUM and STR) of each WSI in the training set, by following the same proportion of those patches in the corresponding WSI. All patches of three tissue types are included in validation and testing sets to ensure consistent prediction results. The widely-used concordance index (C-index) in survival analysis [26], i.e., CI $= \frac{1}{n} \sum_{i \in \{\delta_i = 1 | 1...N\}} \sum_{t_j > t_i} I(X_i\theta, X_j\theta)$, is utilized as the evaluation metric.

Comparisons with Other Prognosis Models: In this section, we compare the proposed DSCA model with a range of state-of-the-art models for survival prognosis. Apart from two-stage MIL (i.e., DeepAttnMISL [26], DeepSMILE [18]) and end-to-end methods (i.e., EE-Surv [10], DoMore-v1-CRC [20]) discussed above, we conduct comparative experiments with the following: (1) ResNet50+DeepMIL [8] and ViT+DeepMIL [5,8], both of which apply the self-supervised contrastive learning [3] to build feature extractors and the attention-based pooling to aggregate WSI-level representation; (2) MHAttnSurv [9] that applies the multi-head self-attention (MHSA) layer to aggregate different phenotypes for WSI representation; (3) DeepGraphSurv [12] and Patch-GCN [2], both of which leverage graph convolutional network (GCN) to model color texture and topology information; (4) Botnet [22] that employs Botnet50 as a replacement for ResNet50 in EE-Surv as the backbone while keeping the remaining settings unchanged.

Experimental results are presented in Table 2, where the CRC cohort includes all WSIs of TCGA-COAD and TCGA-READ patients. It is observed in Table 2 that the proposed DSCA-Ordinal Cox model provides the best performance, with

CI values of 0.685 and 0.679 on COAD and CRC cohorts, respectively. Among the WSI-level MIL methods, the ViT+DeepMIL [3, 26] and DeepSMILE [18] demonstrate overall good performance, with the CI values around 0.62. However, our proposed DSCA model surpasses these WSI-level MIL methods with a CI value over 6% higher, indicating its superior performance in survival prognosis tasks. Patch-GCN [2] provides the CI values of 0.644 and 0.611 on two cohorts, respectively, which is better than the DeepGraphSurv [12]. The end-to-end EE-Surv [10] model achieves the best performance among all existing methods, which provides the CI values of 0.660 and 0.652 on two cohorts, respectively. This indicates the advantage of end-to-end learning in survival prognosis tasks. Taken together, our proposed DSCA model offers substantial improvements compared with existing survival prediction models. It achieves this by leveraging multi-scale histological image features, capturing their interactions, and incorporating the ordinal Cox loss.

Table 2. The performance of different models on TCGA cohorts.

Methods	CI (TCGA-COAD)	CI (TCGA-CRC)
DeepAttnMISL [26]	0.589	0.608
ResNet50+DeepMIL [3, 26]	0.597	0.601
ViT+DeepMIL [3, 26]	0.623	0.618
MHAttnSurv [9]	0.604	0.619
DeepSMILE [18]	0.616	0.620
DeepGraphSurv [12]	0.588	0.576
Patch-GCN [2]	0.644	0.611
EE-Surv [10]	0.660	0.652
Botnet [22]	0.636	0.641
DoMore-v1-CRC [20]	0.617	0.622
Proposed DSCA-Cox	**0.673**	**0.672**
Proposed DSCA-Ordinal Cox	**0.685**	**0.679**

Comparison with Other Multi-scale Feature Fusion mechanisms: To verify the effectiveness of our designed multi-scale feature fusion approach, we compare the proposed DSCA model with several representative studies which also leverage the multi-scale feature fusion mechanisms. Specifically, the DSMIL [11] and MIST [1] use a pyramid concatenation strategy to generate integrated features and then utilize the patch-level attention to generate bag-level features. Different feature extractors such as Resnet and Swin Transformer are used in these two studies. Note that these two models were originally designed for solving classification tasks. We re-implement them by incorporating the Cox loss to make survival predictions for comparison. The MS-Trans [15] was developed to achieve multi-scale feature fusion by using the ViT architecture. Table 3 lists comparative performance of using different feature fusion mechanisms. As shown in Table 3, the proposed DSCA model outperforms comparative methods on both TCGA cohorts. In comparison to the concatenation operation used in DSMIL [11] and MIST [1] to combine multi-scale features, MS-Trans [15] excels in exploring the internal relationships within patch sequences. Our model focuses on modeling the sequential relationships of multi-scale features at each convolutional stage and achieves the best performance in our evaluation.

Table 3. Comparative performance of using different feature fusion mechanisms.

Methods	CI (TCGA-COAD)	CI (TCGA-CRC)
DSMIL [11]	0.645	0.650
MIST [1]	0.655	0.657
MS-Trans [15]	0.662	0.661
Proposed DSCA-Cox	**0.673**	**0.672**
Proposed DSCA-Ordinal Cox	**0.685**	**0.679**

Ablation Studies: Table 4 lists our ablation study results on COAD and CRC cohorts, respectively. We employ two Resnet50 models pre-trained on the ImageNet as the backbone, which separately accept input patches at $20\times$ and $40\times$ magnifications to jointly perform survival prognosis. Our proposed SCSA, ESFF, and MCSA modules are then adaptively integrated into the backbone model for improving survival risk prediction. Experimental results demonstrate that by including all our proposed modules can significantly improve the CI values compared with other ablation testings, which indicates the effectiveness of our proposed DSCA model.

Table 4. Ablation experiments.

Models	CI (TCGA-COAD)	CI (TCGA-CRC)
Backbone	0.665	0.654
Backbone+Weight Unit	0.667	0.660
Backbone+SCSA	0.670	0.662
Backbone+ESFF+MSCA	0.675	0.670
Ours	**0.685**	**0.679**

Visualization Results: Figure 2(a)(b) shows visual prediction results by our DSCA model on two patients' slides. Figure 2(c)(d) shows the Kaplan-Meier (KM) curves of patients in COAD and CRC cohorts, where the optimal cut-off risk scores are computed by using surv_cutpoint function of the R package survminer. As shown in Fig. 2(c)(d), colorectal cancer patients can be stratified into significantly different survival risk groups based on predicted risk scores by our proposed DSCA model with the log-rank p-value of less than 0.0001, which certifies the effective application of our proposed method in survival risk prognosis. Additional visualization results and source codes could be accessed from the supplementary material available at: https://github.com/gaogaojx/DSCA.

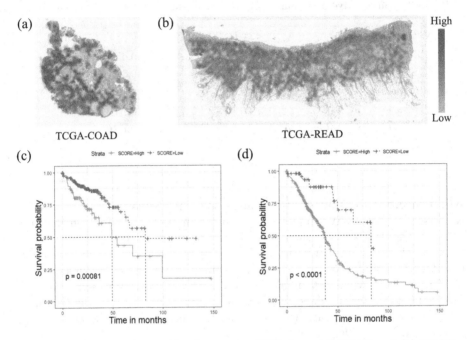

Fig. 2. Visualization of survival risk scores on WSIs from (a) COAD and (b) READ, and KM survival curves on (c) COAD cohort and (d) CRC cohort based on risk scores predicted by the proposed DSCA model.

4   Conclusion

In this paper, we propose an end-to-end survival risk prediction model termed as the DSCA, which fully integrates local contextual information at multi-resolution image patches. Our proposed feature fusion branch aggregates multi-scale feature maps with context-awareness using the self-attention mechanism, which enables the quantification of deep relationships within the same region. In addition, we propose the SCSA and ESFF modules for feature enhancement and aggregation, which help in more effectively learning features from local regions. Evaluations performed on colorectal cancer cohorts demonstrate that our DSCA model provides better performances in survival prognosis than comparative methods, which has the potential to be used in distinguishing high and low survival risk of colorectal cancer patients.

Acknowledgements. This work was supported by the National Natural Science Foundation of China (Grant No. 82102135), the Natural Science Foundation of Liaoning Province (Grant No. 2022-YGJC-36), and the Fundamental Research Funds for the Central Universities (Grant No. DUT22YG114, DUT21RC(3)038).

References

1. Cai, H., et al.: MIST: multiple instance learning network based on Swin transformer for whole slide image classification of colorectal adenomas. J. Pathol. **259**(2), 125–135 (2023)
2. Chen, R.J., et al.: Whole slide images are 2D point clouds: context-aware survival prediction using patch-based graph convolutional networks. In: de Bruijne, M., et al. (eds.) MICCAI 2021. LNCS, vol. 12908, pp. 339–349. Springer, Cham (2021). https://doi.org/10.1007/978-3-030-87237-3_33
3. Chen, T., Kornblith, S., Norouzi, M., Hinton, G.: A simple framework for contrastive learning of visual representations. In: International conference on machine learning, pp. 1597–1607. PMLR (2020)
4. Coudray, N., et al.: Classification and mutation prediction from non-small cell lung cancer histopathology images using deep learning. Nat. Med. **24**(10), 1559–1567 (2018)
5. Dosovitskiy, A., et al.: An image is worth 16×16 words: transformers for image recognition at scale. arXiv preprint arXiv:2010.11929 (2020)
6. Geessink, O.G.F., et al.: Computer aided quantification of intratumoral stroma yields an independent prognosticator in rectal cancer. Cell. Oncol. **42**(3), 331–341 (2019). https://doi.org/10.1007/s13402-019-00429-z
7. Huang, Z., Chai, H., Wang, R., Wang, H., Yang, Y., Wu, H.: Integration of patch features through self-supervised learning and transformer for survival analysis on whole slide images. In: de Bruijne, M., et al. (eds.) MICCAI 2021. LNCS, vol. 12908, pp. 561–570. Springer, Cham (2021). https://doi.org/10.1007/978-3-030-87237-3_54
8. Ilse, M., Tomczak, J., Welling, M.: Attention-based deep multiple instance learning. In: International conference on machine learning, pp. 2127–2136. PMLR (2018)
9. Jiang, S., Suriawinata, A.A., Hassanpour, S.: MHAttnSurv: multi-head attention for survival prediction using whole-slide pathology images. Comput. Biol. Med. **158**, 106883 (2023)
10. Laleh, N.G., Echle, A., Muti, H.S., Hewitt, K.J., Schulz, V., Kather, J.N.: Deep learning for interpretable end-to-end survival prediction in gastrointestinal cancer histopathology. In: COMPAY 2021: The Third MICCAI Workshop on Computational Pathology (2021)
11. Li, B., Li, Y., Eliceiri, K.W.: Dual-stream multiple instance learning network for whole slide image classification with self-supervised contrastive learning. In: Proceedings of the IEEE/CVF Conference on Computer Vision and Pattern Recognition, pp. 14318–14328 (2021)

12. Li, R., Yao, J., Zhu, X., Li, Y., Huang, J.: Graph CNN for survival analysis on whole slide pathological images. In: Frangi, A.F., Schnabel, J.A., Davatzikos, C., Alberola-López, C., Fichtinger, G. (eds.) MICCAI 2018. LNCS, vol. 11071, pp. 174–182. Springer, Cham (2018). https://doi.org/10.1007/978-3-030-00934-2_20

13. Liu, S., Johns, E., Davison, A.J.: End-to-end multi-task learning with attention. In: Proceedings of the IEEE/CVF Conference on Computer Vision and Pattern Recognition, pp. 1871–1880 (2019)

14. Liu, S., Huang, D., Wang, Y.: Learning spatial fusion for single-shot object detection. arXiv preprint arXiv:1911.09516 (2019)

15. Lv, Z., Lin, Y., Yan, R., Wang, Y., Zhang, F.: TransSurv: transformer-based survival analysis model integrating histopathological images and genomic data for colorectal cancer. IEEE/ACM Transactions on Computational Biology and Bioinformatics (2022)

16. Mobadersany, P., et al.: Predicting cancer outcomes from histology and genomics using convolutional networks. Proc. Nat. Acad. Sci. **115**(13), E2970–E2979 (2018)

17. Roy, S., kumar Jain, A., Lal, S., Kini, J.: A study about color normalization methods for histopathology images. Micron **114**, 42–61 (2018)

18. Schirris, Y., Gavves, E., Nederlof, I., Horlings, H.M., Teuwen, J.: DeepSMILE: contrastive self-supervised pre-training benefits MSI and HRD classification directly from H&E whole-slide images in colorectal and breast cancer. Med. Image Anal. **79**, 102464 (2022)

19. Shao, W., Wang, T., Huang, Z., Han, Z., Zhang, J., Huang, K.: Weakly supervised deep ordinal cox model for survival prediction from whole-slide pathological images. IEEE Trans. Med. Imaging **40**(12), 3739–3747 (2021)

20. Skrede, O.J., et al.: Deep learning for prediction of colorectal cancer outcome: a discovery and validation study. Lancet **395**(10221), 350–360 (2020)

21. Srinidhi, C.L., Ciga, O., Martel, A.L.: Deep neural network models for computational histopathology: a survey. Med. Image Anal. **67**, 101813 (2021)

22. Srinivas, A., Lin, T.Y., Parmar, N., Shlens, J., Abbeel, P., Vaswani, A.: Bottleneck transformers for visual recognition. In: Proceedings of the IEEE/CVF Conference on Computer Vision and Pattern Recognition, pp. 16519–16529 (2021)

23. Vaswani, A., et al.: Attention is all you need. Advances in neural information processing systems, vol. 30 (2017)

24. Wang, Q., Wu, B., Zhu, P., Li, P., Zuo, W., Hu, Q.: ECA-Net: efficient channel attention for deep convolutional neural networks. In: Proceedings of the IEEE/CVF Conference on Computer Vision and Pattern Recognition, pp. 11534–11542 (2020)

25. Woo, S., Park, J., Lee, J.-Y., Kweon, I.S.: CBAM: convolutional block attention module. In: Ferrari, V., Hebert, M., Sminchisescu, C., Weiss, Y. (eds.) ECCV 2018. LNCS, vol. 11211, pp. 3–19. Springer, Cham (2018). https://doi.org/10.1007/978-3-030-01234-2_1

26. Yao, J., Zhu, X., Jonnagaddala, J., Hawkins, N., Huang, J.: Whole slide images based cancer survival prediction using attention guided deep multiple instance learning networks. Med. Image Anal. **65**, 101789 (2020)

27. Zhao, K., et al.: Artificial intelligence quantified tumour-stroma ratio is an independent predictor for overall survival in resectable colorectal cancer. EBioMedicine **61**, 103054 (2020)

A Multi-label Image Recognition Algorithm Based on Spatial and Semantic Correlation Interaction

Jing Cheng, Genlin Ji[(✉)], Qinkai Yang, and Junzhao Hao

School of Computer and Electronic Information/School of Artificial Intelligence, Nanjing Normal University, Nanjing, China
glji@njnu.edu.cn

Abstract. Multi-Label Image Recognition (MLIR) approaches usually exploit label correlations to achieve good performance. Two types of label correlations principally studied, i.e., the spatial and semantic correlations. However, most of the existing algorithms for multi-label image recognition consider semantic correlations and spatial correlations respectively, and often require additional information support. Although some algorithms simultaneously capture the semantic and spatial correlations of labels, they ignore the intrinsic relationship between the two. Specifically, only considering spatial correlations will misidentify some difficult objects in the image. For example, different categories of objects with similar appearance and close distance are mistaken for the same category, and semantic correlations can constrain the error caused by spatial correlations. In this work, we propose a multi-label image recognition algorithm based on transformer, named Spatial and Semantic Correlation Interaction (SSCI). Transformer is used to model the internal relationship between spatial correlations and semantic correlations to improve the recognition ability of the model for difficult objects. Experiments on the public datasets MS-COCO, VOC2007 and VOC2012 show that the mAP values reach 84.1%, 95.0% and 95.4%, respectively. Compared with other MLIR algorithms, the proposed algorithm can significantly improve the recognition performance of multi-label images.

Keywords: multi-label image recognition · transformer · semantic correlations · spatial correlations

1 Introduction

With the development of network and information technology, people can share plenty of images on the social media at any time through mobile smart devices. In the face of the generation of a large number of visual data, how to mine the useful information in these visual data is one of the current hot research issues. As a practical and challenging computer vision task, multi-label image recognition aims to identify the objects presented by an image and assign corresponding labels. Compared with single-label image recognition, the key and

Q. Liu et al. (Eds.): PRCV 2023, LNCS 14434, pp. 15–27, 2024.
https://doi.org/10.1007/978-981-99-8549-4_2

challenge of multi-label image recognition is to identify objects with different categories, scales and spatial locations in the image. Multi-label image recognition has many application scenarios in real life, such as scene understanding [13], medical image recognition [1], human attribute recognition [11]. Considering the combination of these different labels is huge and challenging for deep neural networks. Therefore, it is reasonable to learn the rich information of labels and model their correlation to improve the recognition performance of multi-label images.

Existing multi-label image recognition algorithms mainly model the spatial correlations or semantic correlations of labels. For modeling spatial correlations, the most commonly used network architecture is the convolution-based attention module. However, the CNN-based algorithm is difficult to capture long-term spatial correlations because the receptive field of the convolution kernel is limited to the kernel size. Chen et al. [3] use the transformer to model the correlation between the features of all spatial locations so as not to consider the short- and long-range spatial correlations. For modeling semantic correlations, previous algorithms [2,4,5] mainly chose the structure of the graph to explicitly define semantic correlations. These algorithms [2,4,17] mostly establish semantic correlations between labels by manually defining graph structures. In order to define the graph, these algorithms are also associated with the statistical results of the training set. In addition, these algorithms also draw support from additional information, such as word-embeddings from well-learned language models. Most of the existing algorithms consider semantic correlations and spatial correlations respectively, and often require additional information support. Spatial correlations mainly consider the correlations between object pixels in different spatial positions of the image, but only considering spatial correlations will cause different categories of objects with similar appearance and close spatial positions in the image to be misjudged as the same category, resulting in the lack of some difficult object categories. Semantic correlations mainly capture the co-existence of labels, but it depends on the statistical results of the training set often cause over-fitting problem, that is, predicting the categories that do not exist in image.

To solve the above problems, we propose a multi-label image recognition algorithm, named Spatial and Semantic Correlation Interaction (SSCI), to improve the recognition ability of multi-label image recognition for difficult objects. Different from the previous algorithms, we consider both spatial correlations and semantic correlations, and fully explore the intrinsic relationship between spatial and semantic correlations through the transformer structure [14], and our algorithm does not require additional information support. Specifically, we fully model the spatial correlations by fusing the extracted multi-scale image features, and we capture the semantic correlations of the image by building the Semantic Correlation Module(SCM). The spatial and semantic correlation information obtained above is input into the encoder to model the intrinsic relationship between them. Because the features belonging to the same category are bound to have higher correlation, under this constraint condition, the correlation degree between different categories of features previously misallocated to the same category can be weakened, so that they can be distinguished. The details of the algorithm will be described in detail in the third section. In summary, the main contributions of this paper are as follows:

(1) A simple Semantic Correlation Module (SCM) is designed to comprehensively capture the semantic correlations of labels without additional information support;
(2) A multi-label image recognition algorithm based on transformer is proposed, named Spatial and Semantic Correlation Interaction (SSCI), to model the internal relationship between spatial correlations and semantic correlations to improve the recognition ability of the model for difficult objects.

2 Related Work

In recent years, the representative algorithms of multi-label image recognition can be divided into three types , i.e., the correlation-agnostic, semantic correlation and spatial correlation algorithms.

2.1 Correlation-Agnostic Algorithms

Previous pioneering work tends to roughly locate multiple objects to identify [8,10,16]. For example, Wei et al. [16] proposed the HCP to generate hundreds of object proposals for each image and aggregate prediction scores of these proposals to obtain the final prediction. Gao et al. [8] proposed the MCAR to minimize the number of candidate regions and maintain the diversity of these regions as high as possible. Inspired by the great success of Transformer in natural language, Dosovitskiy et al. [6] proposed the ViT algorithm, which directly applies Transformer on non-overlapping image patches to achieve image classification. C-Tran proposed by Lanchantin et al. [10] only exploits the transformer encoder to explore the complex dependence between image visual features and labels. However, these algorithms fail to learn the correlation between labels explicitly.

2.2 Spatial Correlation Algorithms

Spatial correlations consider the correlations between object pixels in different spatial locations in multi-label images. It is closely related to image feature description and mainly describes the intra-class dependencies of features. In recent years, there are works to capture label correlations between different spatial positions in multi-label images. For example, Zhu et al. [21] developed a spatial regularization network to focus on objectiveness regions, and further learned the label correlation of these regions through self-attention. Guo et al. [9] proposed visual attention consistency to constrain the spatial attention regions and ensure spatial attention regions can maintain consistent on different augmentations. Although the spatial correlations algorithm has achieved good results, these cnn-based algorithms cannot capture the long-term spatial correlations. Chen et al. [3] proposed SST algorithm to apply a transformer to the feature sequence to ensure that the correlation between the features of all spatial locations is considered and established, so as not to consider the short- and long-range spatial correlation. However, only considering the spatial correlations will cause different categories of objects with similar appearance and close spatial distance in the image to be misclassified as the same category.

2.3   Semantic Correlation Algorithms

Semantic correlations in multi-label image recognition can be understood as the co-occurrence relationship of labels, which mainly describes the label dependence between different objects. Some recent works use graph structure to model label correlations. The ML-GCN proposed by Chen et al. [4] construct a label correlation graph based on conditional probability to help predict the labels of images. The SSGRL proposed by Chen et al. [2] learns the label correlations by using the recurrent networks to capture the correlation between graph nodes. Ye et al. [18] proposed ADD-GCN model the label correlation from global to local. Chen et al. [5] decomposed the visual representation of an image into a set of label-aware features and proposed P-GCN to encode such features into interdependent image-level prediction scores. However, if only the label co-occurrence relationship is emphasized, some discriminative features in the image may be ignored, resulting in overfitting of the model. Xu et al. [17] adopted two elaborated complementary task decomposition strategies to learn the joint pattern and specific category pattern of labels. However, the construction of the labels co-occurrence relationship not only requires additional auxiliary information, but also produces overfitting.

3   Methodology

In this section, we elaborate on the proposed a multi-label image recognition algorithm of Spatial and Semantic Correlation Interaction(SSCI). The overall framework of the algorithm is shown in Fig. 1. Firstly, we give the definition of multi-label image recognition. Then, we describe each part of the proposed algorithm framework in detail. Finally, we give the loss function and algorithm description of SSCI.

3.1   Definition of Multi-label Image Recognition

Here, we give an introduction to the notations used in the paper. We denote the training set as $D = \{(\boldsymbol{I}^1, \boldsymbol{y}^1), ..., (\boldsymbol{I}^N, \boldsymbol{y}^N)\}$, in which N is the number of training samples. Let $\boldsymbol{I}^n$ denote an input image. $\boldsymbol{y}^n = \{y_1^n, ..., y_C^n\} \in \{0,1\}^C$ is the label vector of the n-th sample and C is the label number. y_c^n is assigned to 1 if label c exists in the n-th image, assigned to 0 if it does not exist. The goal of multi-label image recognition is to construct a model f and train the model with the training set D, so that the model can predict the labels $\hat{\boldsymbol{y}}$ corresponding to an new multi-label image $\hat{\boldsymbol{I}}$: $\hat{\boldsymbol{y}} = f(\hat{\boldsymbol{I}})$.

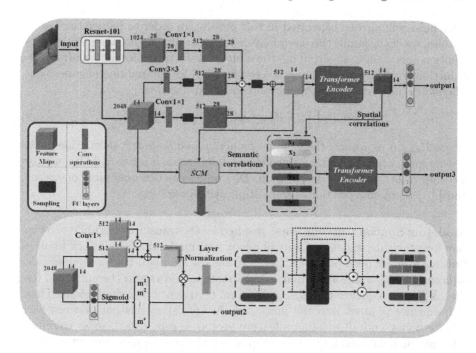

Fig. 1. The overall framework of our proposed SSCI algorithm.

3.2   The Framework of SSCI

Image Feature Extraction: Given an image I^n, in theory, we can use any
CNN based models to learn the features of the image I^n. In our experiment, following [2,4,5,8,18,21], we use ResNet-101 as the backbone for a fair comparison.
Because the high-level feature semantic information is rich, but the object location is rough ; the low-level feature semantic information is less, but the object
position is more accurate. So we output the feature map of different layers. It
can be formulated as :

$$X_i^n = f_{cnn}(I^n, \theta_{cnn}), i = 1, 2 \tag{1}$$

where f_{cnn} is the CNN-based backbone and θ_{cnn} denotes parameters of the backbone CNN. $X_1^n \in \mathbb{R}^{d_1 \times h_1 \times w_1}$ is the output of the third layer of Resnet101,the
corresponding dimension of $d_1 \times h_1 \times w_1$ is $1024 \times 28 \times 28$. $X_2^n \in \mathbb{R}^{d_2 \times h_2 \times w_2}$ as
the output of the fourth layer, the corresponding dimension is $2048 \times 14 \times 14$.

Multi-scale Feature Fusion: Subsequently, we use different convolution kernels for the latter two layers to obtain feature maps with different receptive
field sizes, e.g., 1×1 and 3×3 convolution. By upsampling features of different scales to the same size, we can use position-wise multiplication operation to
perform information fusion to extract common positions information. However,

some noise will also be generated in this process. In order to reduce the impact of noise, we downsample the fused feature information, and use the position-wise addition operation to superimpose the multi-scale feature information after noise reduction on the deep features extracted by CNN for information enhancement. Therefore, this serialization operation can be expressed as :

$$\boldsymbol{X}_{en}^n = D(\prod(U(R_{dim}(\boldsymbol{X}_1^n, \boldsymbol{X}_2^n)))) + R_{dim}(\boldsymbol{X}_2^n) \tag{2}$$

where $R_{dim}(\cdot)$ represents the channel reduction and scale transformation of the feature map, where the transformed channel dimension is set to 512 dimensions ; $U(\cdot)$ represents upsampling operation, $D(\cdot)$ represents downsampling operation, and $\prod(\cdot)$ represents position-wise multiplication operation.

Modeling Spatial Correlations: Inspired by the transformer's ability to capture long-term dependencies, we will adopt it to model the correlations between features from all pixel positions. Here, we use the standard transformer encoder structure as the transformer unit transformation, and regard the enhanced feature $\boldsymbol{X}_{en}^n$ as a sequence containing d feature vectors of $h \times w$ dimensions. Then, the sequence is input into the unit, and the spatial correlation between each feature vector and all remaining feature vectors is established, so as to obtain a feature map with comprehensive spatial correlations. Finally, we restore these feature vectors to the original spatial size and represent them as spatial correlation features.

$$\boldsymbol{X}_{spat}^n = R_{re}(f_{trans}(R_{hw}(\boldsymbol{X}_{en}^n); \theta_{trans})) \tag{3}$$

where $R_{hw}(\cdot)$ denotes that $\boldsymbol{X}_{en}^n \in \mathbb{R}^{d \times h \times w}$ is flattened as $\boldsymbol{X}_{en}^n \in \mathbb{R}^{d \times (h \times w)}$. f_{trans} denotes the transformer unit, θ_{trans} denotes the relevant parameters of the transformer unit.$R_{re}(\cdot)$ denotes that these feature vectors are restored to the original spatial size, and $\boldsymbol{X}_{spat}^n \in \mathbb{R}^{d \times h \times w}$ denotes a feature with spatial correlations.

Modeling Semantic Correlations: In order to achieve end-to-end training and avoid the overfitting problem caused by traditional manual graphs, we propose the SCM module to construct semantic correlations. Firstly, we need to introduce a category-specific semantic learning module to learn the semantic representation of category c. There are different algorithms [2,18] to implement this module. In this paper, we use semantic activation mapping [18]. Firstly, we fuse deep features and enhanced features to obtain feature maps with more semantic information. Then, we generate category-specific activation maps through semantic attention, and use these maps to convert the transformed feature map into the content-aware category representations.

$$\boldsymbol{V}_C^n = \boldsymbol{M}_C^n(R_{dim}(\boldsymbol{X}_2^n) + \boldsymbol{X}_{en}^n R_{dim}(\boldsymbol{X}_2^n)) \tag{4}$$

here, $m_c^n \in \mathbb{R}^{d \times c}$ from $\boldsymbol{M}_C^n$ is the weights of the c-th activation map. Subsequently, we input the semantic representations $\boldsymbol{V}_C^n$ into the multi-head self-attention module, and obtain the overall semantic correlations representations

V_G^n by calculating the correlation A_C^n between the semantic representations of all categories. Then we carry out position-level multiplication on the global semantic correlation information and the initial semantic information of specific categories to further enhance the semantic correlations. We use V_G^n to represent the global label semantic correlations.

$$A_C^n = softmax(\frac{(W_Q^n V_C^n)^{\mathrm{T}}(W_K^n V_C^n)}{\sqrt{d}}) \tag{5}$$

$$V_G^n = (A_C^n V_C^n W_C^n)V_C^n \tag{6}$$

where $V_G^n \in \mathbb{R}^{c \times d}$, c represents the number of categories, and d is 512 dimensions. It is worth mentioning that the SCM is specific to each image, which not only eliminates the complex manual design of capturing semantic correlations, but also avoids the overfitting problem caused by the pre-established label correlation based on all images in the training set.

Spatial Correlations and Semantic Correlations Interaction: In order to model the interaction between spatial and semantic correlations, we use the transformer encoder to capture the dependencies between them. Our formula allows us to easily input semantic correlation information and spatial correlation information jointly into a transformer encoder. And transformer encoders are appropriate because they are order invariant and allow learning any type of dependencies between all spatial and semantic information. Specially, we can represent each vector $x_i \in \mathbb{R}^d$ from X_{spat}^n, with i ranging from 1 to $h \times w$. We express the semantic correlation information as $V_G^n = [v_1, ..., v_c]$, $v_i \in \mathbb{R}^d$. Let $E = \{x_1, ..., x_{h \times w}, v_1, ..., v_c\}$ be the embedding set is input to the transformer encoder.

$$E' = f_{trans}(E, \theta_{trans}) \tag{7}$$

where we denote the final output of the transformer unit after L layers as $E' = \{x_1', ..., x_{h \times w}', v_1', ..., v_c'\}$. Finally, after modeling the spatial and semantic dependencies through the transformer unit, the classifier will perform the final label prediction.

3.3 Loss Function

As shown in Fig. 1, we can aggregate all the output vectors to get the final prediction score $\hat{\mathbf{y}} = [\hat{y}_1, \hat{y}_2, ..., \hat{y}_c]$ in the algorithm to predict more reliable results.

$$\hat{\mathbf{y}} = \lambda_1 \mathbf{output1} + \lambda_2 \mathbf{output2} + \lambda_3 \mathbf{outputs3}, \lambda_1 + \lambda_2 + \lambda_3 = 1 \tag{8}$$

where the value of $\lambda_1, \lambda_2, \lambda_3$ is the proportion of different output weights. We supervise the final score $\hat{\mathbf{y}}$ and train the whole SSCI with the traditional multi-label classification loss as follows:

$$L(\mathbf{y}, \hat{\mathbf{y}}) = \sum_{c=1}^{C} y_c \log(\sigma(\hat{y}_c)) + (1 - y_c \log(1 - \sigma(\hat{y}_c))) \tag{9}$$

where $\sigma(\cdot)$ is the sigmoid function. The algorithm description of SSCI specific training process is given in Algorithm 1.

Algorithm 1. Algorithm SSCI

Input: The multi-label image training set $D = \{(\boldsymbol{I}^n, \boldsymbol{y}^n)\}_{n=1}^{N}$, the number of epoch e, the size of image i, the learning rate lr, the batch size b ;

Output: The predictive label vector $\hat{\boldsymbol{y}}^n$

1: Initialize the model parameters .
2: **for** $i = 1$ to e **do**
3:    Divide the training set D into b batchs.
4:    **for** $k = 1$ to N/b **do**
5:      Extract the image features $\boldsymbol{X}_1, \boldsymbol{X}_2$ by Eq. (1);
6:      Multi-scale feature fusion $\boldsymbol{X}_{en}$ by Eq. (2) ;
7:      Model spatial correlations $\boldsymbol{X}_{spat}$ by Eq. (3) ;
8:      Model semantic correlations $\boldsymbol{V}_G$ by Eq. (4), Eq. (5) and Eq. (6) ;
9:      Interacte spatial correlations and semantic correlations $\boldsymbol{E}'$ by Eq. (7) ;
10:      Compute the final label confidence $\hat{\boldsymbol{y}}$ by Eq.(8) ;
11:      Compute loss $L(\boldsymbol{y}, \hat{\boldsymbol{y}})$ by Eq.(9) ;
12:      Backward $L(\boldsymbol{y}, \hat{\boldsymbol{y}})$ and update parameters ;
13:    **end for**
14: **end for**

---

4 Experiments

In this section, we first introduce the evaluation metrics and our implementation details. Then, we compare SSCI with other existing mainstream algorithms on three public multi-label image recognition datasets, namely MS-COCO [12], Pascal VOC2007 [7] and Pascal VOC2012 [7]. Finally, we conduct ablation study and give corresponding experimental results.

4.1 Evaluation Metrics

In order to fairly compare with other existing algorithms, we follow most of the current work [4,18,21] to adopt the average of overall/per-class precision (OP/CP), overall/per class recall (OR/CR), overall/per-class F1-score (OF1/CF1) and the mean Average Precision (mAP) as evaluation metrics. When measuring precision/recall/F1-score, the label is considered as positive if its confident score is great than 0.5. Generally, the OF1, CF1 and mAP are more important than other metrics.

Table 1. The performance comparison on MS-COCO 2014 dataset. The best results are marked as bold.

Imgsize	Algorithm	mAP	ALL					
			CP	CR	CF1	OP	OR	OF1
448 × 448	Resnet101(GMP)	81.5	84.6	69.1	76.1	85.9	73.0	78.9
	DSDL [19]	81.7	**88.1**	62.9	73.4	**89.6**	65.3	75.6
	CPCL [20]	82.8	85.6	71.1	77.6	86.1	74.5	79.9
	ML-GCN [4]	83.0	85.1	72.0	78.0	85.8	75.4	80.3
	F-GCN [15]	83.2	85.4	72.4	78.3	86.0	75.7	80.5
	P-GCN [5]	83.2	84.9	72.7	78.3	85.0	76.4	80.5
	SST [3]	**84.2**	86.1	72.1	78.5	87.2	75.4	80.8
	ours	84.1	85.3	**73.6**	**79.0**	85.8	**77.1**	**81.2**
448 × 576	ADD-GCN [18]	85.2	84.7	**75.9**	**80.1**	84.9	**79.4**	82.0
	ours	**85.4**	**87.9**	72.2	79.3	**89.4**	75.9	**82.1**
576 × 576	SSGRL [2]	83.8	**89.9**	68.5	76.8	**91.3**	70.8	79.7
	C-Tran [10]	85.1	86.3	74.3	79.9	87.7	76.5	81.7
	ours	**85.5**	86.0	**75.1**	**80.2**	86.0	**78.9**	**82.3**

4.2 Implementation Details

In our experiment, we use pre-trained ResNet-101 as the backbone of our proposed model and set a same input size to train and evaluate models for fair comparison. We use 4 attention heads and 3 layers shared transformer encoder to capture the correlation information, the dimension of the attention head and feed-forward head is set to 512 and 2048, respectively. The probability of all dropout in transformers is set to 0.1. For the MS-COCO dataset and the Pascal VOC dataset, we set the learning rates of 0.03 and 0.05, respectively. Standard stochastic gradient descent with momentum of 0.9 is chosen as the model optimizer, and the weight decay is set to 10^{-4}. The batch size of each GPU is 32. We train our model for 60 epoch in total and the learning rate is reduced by a factor of 0.1 at 30 and 50 epoch, respectively. We apply different weight coefficients to the final prediction output. For the MS-COCO dataset, we set the weights of 0.3,0.3,0.4, and for Pascal VOC, we set the weights of 0.4,0.2,0.4 to obtain the best performance. The experiment is carried out on GeForce RTX 3090 GPU.

4.3 Comparison with Other Mainstream Algorithms

Performance of the MS-COCO Dataset: Table 1 shows the performance comparison results of our model with other mainstream models on MS-COCO. We find that our proposed algorithm outperforms the traditional multi-label recognition algorithms by a considerable margin. Specifically, our algorithm obtains a second best mAP of 84.1%, slightly lower than that of the best performing SST by 0.1%. In addition, for SSGRL [2], C-Tran [10] and ADD-GCN [18], we also carried out experimental comparisons at the same input size. The experimental mAP is 1.7% higher than the mAP of SSGRL [2], 0.4% higher than the mAP of C-Tran [10], and 0.2% slightly higher than the mAP of ADD-GCN [18].

Table 2. The performance comparison on VOC2007 dataset. The best results are marked as bold.

Algorithm	aero	bike	bird	boat	bottle	bus	car	cat	chair	cow	table	dog	horse	motor	person	plant	sheep	sofa	train	tv	mAP
ResNet101(GMP)	99.2	97.9	97.0	96.1	76.3	93.5	97.4	97.5	80.2	93.3	81.6	97.0	96.7	95.1	98.8	79.7	95.5	82.3	97.4	92.9	92.3
SSGRL [2]	99.5	97.1	97.6	97.8	82.6	94.5	96.7	98.1	78.0	**97.0**	85.6	97.8	98.3	96.4	98.1	84.9	96.5	79.8	98.4	92.8	93.4
ADD-GCN [18]	99.8	98.2	97.5	98.3	79.8	93.9	96.5	97.8	80.1	93.3	84.5	98.6	98.6	**96.9**	98.6	83.2	95.1	82.7	98.3	94.4	93.3
ML-GCN [4]	99.5	98.5	98.6	98.1	80.8	94.6	97.2	98.2	82.3	95.7	86.4	98.2	98.4	96.7	99.0	84.7	96.7	84.3	98.9	93.7	94.0
F-GCN [15]	99.5	98.5	98.7	98.2	80.9	94.8	97.3	98.3	82.5	95.7	86.6	98.2	98.4	96.7	99.0	84.8	96.7	84.4	99.0	93.7	94.1
P-GCN [5]	99.6	98.6	98.4	98.7	81.5	94.8	97.6	98.2	83.1	96.0	87.1	98.3	98.5	96.3	99.1	**87.3**	95.5	85.4	98.9	93.6	94.3
DSDL [19]	99.8	98.7	98.4	97.9	81.9	95.4	97.6	98.3	83.3	95.0	**88.6**	98.0	97.9	95.8	99.0	86.6	95.9	86.4	98.6	94.4	94.4
CPCL [20]	99.6	98.6	98.5	98.8	81.9	95.1	97.8	98.2	83.0	95.5	85.5	98.4	98.5	97.9	99.0	86.6	97.0	84.9	99.1	94.3	94.4
SST [3]	99.8	98.6	**98.9**	98.4	**85.5**	94.7	97.9	**98.6**	83.0	96.8	85.7	**98.8**	**98.9**	95.7	99.1	85.4	96.2	84.3	99.1	**95.0**	94.5
ours	**99.9**	**99.1**	98.6	**98.8**	83.9	**96.2**	**98.1**	98.0	**85.2**	96.3	86.0	98.6	98.0	96.5	**99.2**	87.1	**98.3**	**87.2**	**99.2**	94.9	**95.0**

Table 3. The performance comparison on VOC2012 dataset. The best results are marked as bold.

Algorithm	aero	bike	bird	boat	bottle	bus	car	cat	chair	cow	table	dog	horse	motor	person	plant	sheep	sofa	train	tv	mAP
ResNet101(GMP)	99.3	97.3	97.8	96.3	77.9	95.5	97.9	97.3	80.7	82.9	97.6	84.2	97.6	95.9	98.5	80.6	95.7	82.3	98.2	93.2	93.2
DSDL [19]	99.4	95.3	97.6	95.7	83.5	94.8	93.9	98.5	85.7	94.5	83.8	98.4	97.7	95.9	98.5	80.6	95.7	82.3	98.2	93.2	93.2
SSGRL [2]	99.5	95.1	97.4	96.4	**85.8**	94.5	93.7	**98.9**	86.7	96.3	84.6	**98.9**	98.6	96.2	98.7	82.2	98.2	84.2	98.1	93.5	93.9
ADD-GCN [18]	99.7	96.1	98.2	96.4	**85.8**	95.9	95.0	**98.9**	86.5	96.8	84.6	98.6	97.5	96.5	98.8	82.5	97.1	84.0	98.7	93.5	94.1
SST [3]	99.7	98.7	98.6	98.9	84.4	**96.5**	98.3	98.5	83.3	**97.8**	89.2	98.5	98.6	**98.2**	**99.1**	86.9	98.4	86.3	**99.6**	94.7	95.2
ours	**99.9**	**98.8**	**98.6**	**99.3**	82.6	96.3	**98.6**	98.3	**85.9**	97.3	88.5	98.7	98.3	97.9	98.9	86.2	**98.9**	**89.4**	99.3	**95.8**	**95.4**

It is worth noting that our algorithm model does not require any pre-trained object detection model to generate a large number of recommendations to locate object regions, nor does it require any statistical label co-occurrence information to capture semantic dependencies.

Performance of Pascal VOC Dataset: Table 2 and Table 3 show the performance comparison results of our model with other mainstream models on VOC 2007 and VOC 2012. Our algorithm achieves the best mAP of 95.0% and 95.4%, respectively. It is worth noting that in order to make a fair comparison with SSGRL [2] and ADD-GCN [18], we quote the indicators given in the newly published literature [3]. It can be seen from the table that our algorithm has a good effect on most of the object recognition in the dataset. Especially for difficult objects such as ' chair ' and ' sofa ', which are easy to be occluded, our algorithm shows superior recognition results. This also shows that the interaction between spatial correlations and semantic correlations can better identify such difficult objects that are often obscured.

4.4   Evaluation of the SSCI Effectiveness

In this part, we experimented on the contribution of each part of the SSCI. First, we use the standard ResNet101 backbone network as a benchmark. Then, the experimental results under different combinations of spatial correlations, semantic information and semantic correlations are given. Among them, the semantic relevance is constructed through the proposed SCM module. From Table 4, we can see that after adding the spatial correlation module and the semantic correlation module respectively, the mAP of the data set has been improved. Then

Table 4. Research on the ablation of spatial and semantic interaction.

	Spa-cor	Sem	Sem-cor	MS-COCO	VOC 2007	VOC2012
Baseline				81.5	92.3	93.3
ours	✓			**82.6**	**94.5**	**93.8**
ours	✓	✓		**83.3**	**94.8**	**94.9**
ours			✓	**83.6**	**94.3**	**94.9**
ours	✓		✓	**84.1**	**95.0**	**95.4**

we interact spatial correlations with semantic information, and we can see that its experimental results are higher than those using spatial correlations alone, but lower than those using semantic correlations alone. It can be seen that the semantic information has a certain guiding role in the construction of spatial correlations, but because these semantic information is not established between the dependence, the guidance of semantic information is not so perfect. Finally, we let the semantic correlations and spatial correlations interact, and the experimental results are further improved than before. Experimental results show the effectiveness of the proposed method.

5   Conclusion

In this work, we propose a multi-label image recognition algorithm named SSCI. SSCI first models the spatial correlations and semantic correlations of the labels respectively, and then maps the spatial correlations and semantic correlations to the same space and inputs them into the transformer to model the correlation between them. Extensive experiments conducted on public benchmarks demonstrate the effectiveness and rationality of our SSCI algorithm.

References

1. Alam, M.U., Baldvinsson, J.R., Wang, Y.: Exploring LRP and Grad-CAM visualization to interpret multi-label-multi-class pathology prediction using chest radiography. In: 2022 IEEE 35th International Symposium on Computer-Based Medical Systems (CBMS), pp. 258–263 (2022)
2. Chen, T., Xu, M., Hui, X., Wu, H., Lin, L.: Learning semantic-specific graph representation for multi-label image recognition. In: 2019 IEEE/CVF International Conference on Computer Vision (ICCV), pp. 522–531 (2019)

3. Chen, Z.M., Cui, Q., Zhao, B., Song, R., Zhang, X., Yoshie, O.: SST: spatial and semantic transformers for multi-label image recognition. IEEE Trans. Image Process. **31**, 2570–2583 (2022)
4. Chen, Z.M., Wei, X.S., Wang, P., Guo, Y.: Multi-label image recognition with graph convolutional networks. In: 2019 IEEE/CVF Conference on Computer Vision and Pattern Recognition (CVPR), pp. 5172–5181 (2019)
5. Chen, Z.M., Wei, X.S., Wang, P., Guo, Y.: Learning graph convolutional networks for multi-label recognition and applications. IEEE Trans. Pattern Anal. Mach. Intell. **45**(6), 6969–6983 (2023)
6. Dosovitskiy, A., et al.: An image is worth 16×16 words: transformers for image recognition at scale. In: International Conference on Learning Representations (2021)
7. Everingham, M., Gool, L.V., Williams, C.K.I., Winn, J.M., Zisserman, A.: The pascal visual object classes (VOC) challenge. Int. J. Comput. Vision **88**, 303–338 (2010)
8. Gao, B.B., Zhou, H.Y.: Learning to discover multi-class attentional regions for multi-label image recognition. IEEE Trans. Image Process. **30**, 5920–5932 (2021)
9. Guo, H., Zheng, K., Fan, X., Yu, H., Wang, S.: Visual attention consistency under image transforms for multi-label image classification. In: 2019 IEEE/CVF Conference on Computer Vision and Pattern Recognition (CVPR), pp. 729–739 (2019)
10. Lanchantin, J., Wang, T., Ordonez, V., Qi, Y.: General multi-label image classification with transformers. In: 2021 IEEE/CVF Conference on Computer Vision and Pattern Recognition (CVPR), pp. 16473–16483 (2021)
11. Li, Yining, Huang, Chen, Loy, Chen Change, Tang, Xiaoou: Human attribute recognition by deep hierarchical contexts. In: Leibe, Bastian, Matas, Jiri, Sebe, Nicu, Welling, Max (eds.) ECCV 2016. LNCS, vol. 9910, pp. 684–700. Springer, Cham (2016). https://doi.org/10.1007/978-3-319-46466-4_41
12. Lin, T.-Y., et al.: Microsoft COCO: common objects in context. In: Fleet, David, Pajdla, Tomas, Schiele, Bernt, Tuytelaars, Tinne (eds.) ECCV 2014. LNCS, vol. 8693, pp. 740–755. Springer, Cham (2014). https://doi.org/10.1007/978-3-319-10602-1_48
13. Shao, J., Kang, K., Loy, C.C., Wang, X.: Deeply learned attributes for crowded scene understanding. In: 2015 IEEE Conference on Computer Vision and Pattern Recognition (CVPR), pp. 4657–4666 (2015)
14. Vaswani, A., et al.: Attention is all you need. In: Guyon, I., Luxburg, U.V., Bengio, S., Wallach, H., Fergus, R., Vishwanathan, S., Garnett, R. (eds.) Advances in Neural Information Processing Systems, vol. 30. Curran Associates, Inc. (2017)
15. Wang, Y., Xie, Y., Liu, Y., Zhou, K., Li, X.: Fast graph convolution network based multi-label image recognition via cross-modal fusion, pp. 1575–1584. Association for Computing Machinery, New York, NY, USA (2020)
16. Wei, Y., et al.: HCP: a flexible CNN framework for multi-label image classification. IEEE Trans. Pattern Anal. Mach. Intell. **38**(9), 1901–1907 (2016)
17. Xu, J., Huang, S., Zhou, F., Huangfu, L., Zeng, D., Liu, B.: Boosting multi-label image classification with complementary parallel self-distillation. In: Proceedings of the Thirty-First International Joint Conference on Artificial Intelligence, IJCAI, pp. 1495–1501 (2022)
18. Ye, Jin, He, Junjun, Peng, Xiaojiang, Wu, Wenhao, Qiao, Yu.: Attention-driven dynamic graph convolutional network for multi-label image recognition. In: Vedaldi, Andrea, Bischof, Horst, Brox, Thomas, Frahm, Jan-Michael. (eds.) ECCV 2020. LNCS, vol. 12366, pp. 649–665. Springer, Cham (2020). https://doi.org/10.1007/978-3-030-58589-1_39

19. Zhou, F., Huang, S., Xing, Y.: Deep semantic dictionary learning for multi-label image classification. ArXiv abs arXiv:2012.12509 (2020)
20. Zhou, F., Huang, S., Liu, B., Yang, D.: Multi-label image classification via category prototype compositional learning. IEEE Trans. Circuits Syst. Video Technol. **32**(7), 4513–4525 (2022)
21. Zhu, F., Li, H., Ouyang, W., Yu, N., Wang, X.: Learning spatial regularization with image-level supervisions for multi-label image classification. In: 2017 IEEE Conference on Computer Vision and Pattern Recognition (CVPR), pp. 2027–2036 (2017). https://doi.org/10.1109/CVPR.2017.219

Hierarchical Spatial-Temporal Network for Skeleton-Based Temporal Action Segmentation

Chenwei Tan[1], Tao Sun[2(✉)], Talas Fu[1], Yuhan Wang[3], Minjie Xu[1],
and Shenglan Liu[2]

[1] Faculty of Electronic Information and Electrical Engineering, Dalian University of
Technology, Dalian 116024, Liaoning, China
{tcw2000,oyontalas}@mail.dlut.edu.cn, 704215404@qq.com
[2] School of Innovation and Entrepreneurship, Dalian University of Technology,
Dalian 116024, Liaoning, China
{dlutst,liusl}@mail.dlut.edu.cn
[3] School of Computer Science, Fudan University, Shanghai 200433, China
yuhanwang22@m.fudan.edu.cn

Abstract. Skeleton-based Temporal Action Segmentation (TAS) plays
an important role in analyzing long videos of motion-centered human
actions. Recent approaches perform spatial and temporal information
modeling simultaneously in the spatial-temporal topological graph, leading to high computational costs due to the large graph magnitude. Additionally, multi-modal skeleton data has sufficient semantic information,
which has not been fully explored. This paper proposes a Hierarchical
Spatial-Temporal Network (HSTN) for skeleton-based TAS. In HSTN,
the Multi-Branch Transfer Fusion (MBTF) module utilizes a multi-
branch graph convolution structure with an attention mechanism to capture spatial dependencies in multi-modal skeleton data. In addition, the
Multi-Scale Temporal Convolution (MSTC) module aggregates spatial
information and performs multi-scale temporal information modeling to
capture long-range dependencies. Extensive experiments on two challenging datasets are performed and our proposed method outperforms
the State-of-the-Art (SOTA) methods.

Keywords: Temporal action segmentation · Multi-modal fusion ·
Graph convolution

1 Introduction

The TAS task has significant practical applications in video surveillance [3],
activity analysis [1], and human-computer interaction [13]. It aims to assign
frame-level labels for long untrimmed videos. This task is particularly challenging
as it requires the precise classification of actions at a fine temporal granularity.

Efficiently modeling long sequences of skeleton data in both spatial and temporal dimensions is essential for skeleton-based TAS. Benjamin et al. [6] proposed MS-GCN to address this challenge. In the initial stage, MS-GCN uses

© The Author(s), under exclusive license to Springer Nature Singapore Pte Ltd. 2024
Q. Liu et al. (Eds.): PRCV 2023, LNCS 14434, pp. 28–39, 2024.
https://doi.org/10.1007/978-981-99-8549-4_3

spatial graph convolutions and dilated temporal convolutions to exploit the spatial configuration of the joints and their long-term temporal dynamics. This approach allows MS-GCN to extract topological information from a spatial-temporal graph. However, skeleton-based TAS usually involves videos with long durations, resulting in the creation of large spatial-temporal graphs formed by skeleton sequences. Processing such large graphs leads to high computational complexity. Besides, MS-GCN adopts an early fusion approach for multi-modal skeleton data, which concatenates different modality skeleton data in the channel dimension. Nevertheless, concatenating multi-modal data without feature extraction is not the best approach for feature fusion [18].

In this paper, we introduce a hierarchical model that effectively captures the spatial-temporal correlations of skeleton sequences while maintaining computational efficiency. In contrast to MS-GCN, our network realizes spatial and temporal modeling through two sequential modules: the Multi-Branch Transfer Fusion (MBTF) module and the Multi-Scale Temporal Convolution (MSTC) module. By using such a hierarchical network structure, we can extract high-level spatial information more comprehensively in MBTF. This is because it is more effective to model the correlations of joints within the same frame rather than modeling the correlations of all joints across all frames [26]. Afterwards, the MSTC module incorporates multiple dilated convolution layers with different dilation rates in parallel to capture temporal information at varying scales. As the MSTC module no longer needs to model spatial information, we aggregate high-level spatial semantic information to reduce the dimensionality of the feature data, which will help decrease the computational complexity of the model. In the MBTF module, we adopt different approaches for each branch to generate three distinct forms of feature representations: joint, velocity and bone features. To fuse these multi-modal features, we employ the attention mechanism to conduct mid-fusion. Our contributions are three-fold: (1) We introduce a model that is implemented using a novel hierarchical approach. Through this approach, the model can reduce computational complexity with pooling while ensuring discriminative feature extraction. (2) We propose the MBTF module to capture the enriched spatial semantic information from multi-modal skeleton data. Additionally, the MSTC module is proposed to capture temporal information across different scales. (3) Our method outperforms the state-of-the-art methods on two challenging datasets.

2 Related Work

2.1 Temporal Action Segmentation

Previous approaches identify action segments utilizing the sliding window and remove superfluous hypotheses with non-maximum suppression [11,19]. Recent advanced works involve Temporal Convolutional Networks (TCNs) to enlarge the temporal receptive field for modeling long sequences. Lea et al. [14] proposed an encoder-decoder TCN that incorporates pooling and upsampling in the temporal dimension to capture long-distance dependencies. MS-TCN [5] introduced a cascaded architecture for refining the prediction through multiple stages. Instead of using temporal convolution, Yi et al. [25] attempted to model temporal sequence information using the Transformer architecture, leveraging its ability to model relations within sequence data. To address the over-segmentation problem, Wang et al. [23] and Ishikawa et al. [8] respectively proposed boundary prediction approaches in BCN and ASRF for obtaining boundaries. Wang et al. introduced local barrier pooling to aggregate local predictions using boundaries, while Ishikawa et al. used boundaries to rectify errors within action segments. For the skeleton-based TAS task, Benjamin et al. [6] introduced MS-GCN to address the inadequacy of the TCN model's initial stage in capturing spatial hierarchies between human joints.

2.2 Skeleton-Based Action Recognition

Early skeleton-based action recognition methods can be classified into two categories: RNN and CNN. RNN approaches [4,9,21] treat the skeleton data as a temporal sequence and leverage the modeling capability of RNNs to extract temporal features for action recognition. CNN-based methods [12,15] transform the skeleton data into grid-like images and directly employ 2D CNNs for high-level feature extraction. However, these methods have a limitation. They do not effectively learn the spatial relations between skeleton joints. This is due to the lack of utilization of the spatial topologies of skeleton data. Recent studies have focused on utilizing Graph Convolutional Networks (GCNs) to better capture the spatial information of skeleton data. ST-GCN [24], proposed by Yan el al., constructed a spatial-temporal graph and employed graph convolutions to learn the spatial-temporal dependencies of skeleton data. However, the heuristically predefined graph lacks the ability to fully represent dynamic human actions. To address this issue, 2 s-AGCN [20] proposed a novel adaptive GCN that learns the spatial adjacency graph from the data. Further, CTR-GCN [2] introduced a more refined graph learning method that learns not only a shared graph but also channel-specific graphs. Moreover, EfficientGCN [22] suggested an early fusion multi-input branch structure to extract sufficient semantic features and designed an efficient network with the bottleneck structure.

3 Method

3.1 Network Architecture

We propose a network composed of three components in sequence: the Multi-Branch Transfer Fusion (MBTF) module, the Multi-Scale Temporal Convolution (MSTC) module, and the refinement module. Figure 2 depicts the overall structure of our proposed Hierarchical Spatial-Temporal Network (HSTN). Firstly, the MBTF utilizes three branches to extract body semantic information from three modalities of skeleton data and fuses the features of each branch using the attention mechanism. Secondly, the MSTC aggregates features at different temporal scales and captures long-range dependencies. Lastly, the refinement module applies a two-stage TCN [5] to refine the initial prediction for improved continuity. The structure of TCN is illustrated in Fig. 1(b). The MBTF and MSTC together constitute the feature extractor of the network, which models the spatial and temporal information of the skeleton sequences.

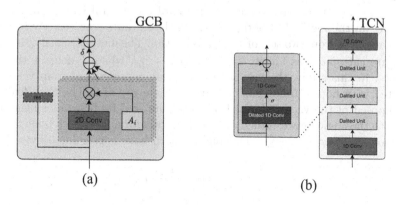

Fig. 1. (a) Graph convolution block. δ denotes the BatchNorm. $\oplus$ demotes the elementwise summation. $\otimes$ denotes the matrix multiplication. A_i is i-th normalized adjacency matrix, (b) Single stage temporal convolution network and Dilated Unit. σ denotes the ReLU activation.

3.2 Multi-Branch Transfer Fusion Module

The Multi-Branch Transfer Fusion (MBTF) module is a frame-level module designed to capture the semantic information of the spatial graph in a single frame. Previous research [20,22] has demonstrated that fusing multi-modal skeleton data can enhance the performance of a model. To incorporate multimodal skeleton data, we utilize Graph Convolution Block (GCB) branches to extract spatial semantic information from different modalities. The structure of the GCB is displayed in Fig. 1(a). Each branch undergoes batch normalization to standardize the data. Since simple concatenation of the output of each branch, as performed in [22], is insufficient for fusing the information of multi-modal features, we introduce the multi-modal squeeze and excitation attention mechanism [10] at the output position of the branches to facilitate mid-level feature fusion. By leveraging the enhanced representation power of the network resulting from the attention mechanism, this module enables the more comprehensive fusion of features from the outputs of different branches. Finally, we employ a GCB to extract high-level features. In this work, we utilize three modalities of skeleton data [22]: (1) joint features, (2) velocity features and (3) bone features.

Suppose that the original 3D coordinate set of a skeleton sequence is $\mathcal{X} = \{x \in \mathbb{R}^{C \times T \times V \times M}\}$, where C, T, V and M denote the coordinate dimensions, sequence length, number of joints and number of individuals, respectively. In this study, we focus on single-person motion analysis where $M = 1$, hence simplifying the skeleton sequence to $\mathcal{X} = \{x \in \mathbb{R}^{C \times T \times V}\}$. In the following sections, we introduce $x_{t,i} \in \mathbb{R}^C$ to denote the coordinates of point i at time t.

The joint features are composed of absolute and relative coordinates. Relative coordinates refer to the position of all joints relative to the central joint of human body. The relative coordinates, denoted as $r_{t,i}$, are calculated as follow: $r_{t,i} = x_{t,i} - x_{t,c}$. Here, c represents the index of the central joint, which varies depending on the method of obtaining the skeleton data. The velocity features are concatenated from two sets of motion velocities. The motion velocity, denoted as $z_{t,i}$, are calculated as follow: $z_{t,i}^{(k)} = x_{t+k,i} - x_{t,i}$, where k denotes the gap of frames. In this study, the velocity features are concatenated from velocities calculated with both a frame gap of 1 and a frame gap of 2. The bone features are composed of the vectors formed by each joint and its first-order neighbors along with the angle between them. The calculation formula of each bone feature is as follow: $b_{t,i} = x_{t,i} - x_{t,i_{adj}}$. The angle of each bone is calculated by:

$$a_{t,i,w} = \arccos(\frac{b_{t,i,w}}{\|b_{t,i}\|}) \tag{1}$$

Here, i_{adj} represents the index of the neighboring joint for the i-th joint, while w represents the index of a particular dimension in 3-dimensional coordinates.

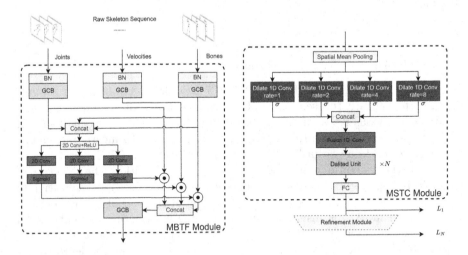

Fig. 2. The overall architecture of proposed HSTN. $\odot$ denotes the elementwise multiplication. σ denotes the ReLU activation.

3.3   Multi-Scale Temporal Convolution Module

The Multi-Scale Temporal Convolution (MSTC) module is designed to exploit long-term temporal relations at different scales. Processing the complete skeleton sequences leads to significant computational demands. Therefore, we introduce a spatial mean pooling layer to merge spatial information within each frame, thus reducing the computational burden while retaining spatial information. The first layer of this module utilizes convolutions with varying dilation rates to capture temporal patterns across diverse temporal scales. We employ a 1×1 convolution for feature fusion. The receptive field is gradually expanded by stacking 9 layers of Dilated Units, where dilation rates increase with layer number. Figure 1(b) depicts the structure of the Dilated Unit. Finally, we apply a fully connected layer to obtain the initial prediction.

3.4   Loss Function

The loss function consists of a frame-level cross-entropy loss function L_{cls} for classification and a smoothing loss function L_{smo} [5] that penalizes discontinuity in predictions.

$$L = L_{cls} + \lambda L_{smo} = \frac{1}{T} \sum_t - \log\left(\hat{y}_{t,c}\right) + \frac{\lambda}{TC} \sum_{t,c} \left(\log\left(\hat{y}_{t-1,c}\right) - \log\left(\hat{y}_{t,c}\right)\right)^2 \quad (2)$$

T is the number of video frames. C is the number of action classes. $\hat{y}_{t,c}$ denotes the network's predicted probability of label c at time t. λ is the hyperparameter that controls the weight of L_{smo}. It should be noted that this model has multiple stages, and the final loss is obtained by combining the losses of all stages.

4   Experiments

4.1   Setup

Datasets. Our work evaluates the model's performance on two benchmark skeleton-based TAS datasets, namely Peking University - Continuous Multi-Modal Human Action Understanding (PKU-MMD) [16] and Motion-Centered Figure Skating (MCFS) [17]. PKU-MMD is a 3D human action understanding dataset with 1009 videos of 52 distinct actions captured from three camera perspectives on 13 subjects. The videos contain 25 body joints and have a frame rate of 30 frames per second. MCFS is a motion-centered dataset comprising 271 single-person figure skating videos of varying lengths between 2 to 72 s. It utilizes classification criteria at three levels, where a sub-dataset with 22 classes (MCFS22) and another sub-dataset with 130 classes (MCFS130) were used for this study. During training and testing, features with full video frame rates were used for both datasets. To assess model performance, we employed the fixed test/train partition for PKU-MMD and the five-fold cross-validation for MCFS.

Metrics. The effectiveness is assessed using several metrics, including frame-level accuracy, normalized edit distance score, and F1 scores for segment overlap at thresholds of 10%, 25% and 50%.

Implementation Details. The branch and trunk GCBs of the MBTF module have 6/32 and 96/64 input/output channels, respectively. During training, the loss function employs $\lambda = 0.15$, while using the Adam optimizer with a learning rate of 0.0005. For all experiments, the model was trained for 100 epochs with a batch size of 4 on a single NVIDIA GeForce RTX 2080 GPU.

4.2 Effect of Hierarchical Model

In HSTN, we employ a hierarchical approach to model the spatial-temporal information in the skeleton sequences. To validate its effectiveness, we perform a comparative analysis between HSTN and MS-GCN (a non-hierarchical model) at the initial stage. The results are presented in Table 1 and Table 2 below. Benefiting from the hierarchical structure, the MSTC module no longer needs to maintain the spatial dimension of the skeleton data, thus enabling the aggregation of spatial features by the pooling layer. This allows HSTN to require fewer FLOPs and less memory compared to MS-GCN. Specifically, on the MCFS22 dataset, HSTN demands approximately 1/5 of the FLOPs and 1/2 of the memory required by MS-GCN. Moreover, this hierarchical framework facilitates a more effective capture of spatial dependencies. At the initial stage, both HSTN and MS-GCN employ an equal number of temporal convolution layers. However, despite employing fewer graph convolutions, HSTN achieves higher accuracy results, surpassing MS-GCN by 4.1%. Nevertheless, while HSTN falls behind MS-GCN in terms of segment-level evaluation metrics, these deficiencies can be addressed through the subsequent refinement module.

Table 1. Comparison of hierarchical model and non-hierarchical model on MCFS22. * denotes the initial stage.

	Acc	Edit	F1@{10,25,50}		
MS-GCN*	74.2	**17.2**	**22.5**	**19.7**	**15.2**
Ours*	**78.3**	12.0	15.4	14.1	11.6

Table 2. FLOPs, Params and GPU memory cost of MS-GCN and HSTN on MCFS22 (batch size 1).

	FLOPs	Params	Memory
MS-GCN*	17.6G	255.0K	6.6G
Ours*	3.4G	258.8K	2.8G

4.3 Effect of Multiple Modalties

Multi-modal skeleton data is of significance as it provides rich semantic information. In this section, we conduct ablation experiments on models with varying numbers of branches and modalities to assess their impact on performance. The results are shown in Table 3. The findings indicate that the use of multi-modal data can effectively improve the performance of the model. Specifically, the model that employs multi-modal data outperforms the one that uses uni-modal data by 3.7% in accuracy and 3.4% in F1@50, respectively. Among models with the same number of branches, those employing the velocity modality perform better. This suggests that compared to the static information of joint and bone features, the dynamic information of velocity features can better represent the characteristics of actions. However, for the PKU-MMD dataset, using all modalities does not result in the best performance. On the contrary, the performance even slightly decreases compared to the best-performing 2-branch model. Hence, for the PKU-MMD dataset, we adopt the 2-branch model.

Table 3. Ablation study on the number of branches and modalities on PKU-MMD.

Input	Acc	Edit	F1@{10,25,50}		
Joint	68.0	69.3	68.8	63.5	50.5
Velocity	70.2	70.7	70.8	65.6	53.4
Bone	69.0	69.4	69.1	63.6	50.0
Joint+Velocity	72.7	71.6	73.1	68.6	55.9
Joint+Bone	70.4	70.1	70.1	65.1	52.4
Velocity+Bone	**73.9**	**73.5**	**74.0**	**69.3**	**56.8**
All	73.4	72.6	73.3	68.4	56.0

4.4 Effect of Multi-modal Fusion Methods

To verify the effectiveness of our proposed MBTF module in feature fusion, we compare its performance against the Multi-Input Branches (MIB) module, which directly concatenates features from different branches. The experimental results on the PKU-MMD dataset using models with different numbers of branches are reported in Table 4. The results indicate that across all metrics, MBTF outperforms MIB when using 1-branch or 2-branch models. In the case of a 1-branch model, the attention structure in MBTF degrades to the SE attention [7]. Despite this, MBTF still demonstrates an improvement compared to MIB as the attention mechanism enhances the model's expressive power. Meanwhile, with a 2-branch model, MBTF shows the best performance improvement compared to MIB by up to 1.7% and 1.4% in accuracy and F1@50 respectively. This implies that MBTF can more effectively fuse multi-modal data. However, when using a 3-branch model, the performance of the two modules is similar. This indicates that on the PKU-MMD dataset, the attention mechanism in MBTF causes overfitting when there are enough modalities, resulting in a performance decline. Overall, the results show that MBTF can facilitate the fusion of multi-modal data.

4.5 Effect of Multi-Scale Temporal Convolution

In order to investigate the impact of multi-scale temporal information on model performance, we conduct a comparative experiment between the MSTC module

Table 4. Comparison of MBTF and MIB on PKU-MMD.

Input	Acc	Edit	F1@{10,25,50}		
Velocity(MIB)	69.3	70.1	70.4	65.2	52.4
Velocity(MBTF)	70.2	70.7	70.8	65.6	53.4
Velocity+Bone (MIB)	72.2	72.3	73.1	68.0	55.4
Velocity+Bone(MBTF)	**73.9**	**73.5**	**74.0**	**69.3**	**56.8**
All (MIB)	72.4	73.0	73.7	68.4	55.6
All(MBTF)	73.4	72.6	73.3	68.4	56.0

and the Single-Scale Temporal Convolution (SSTC) module. In the MSTC module, temporal convolutions with different dilation rates are used in the first layer. This helps the model capture local information at different temporal scales. In contrast, the SSTC module uses the Dilated Unit at the first layer, while the other parts remain consistent with the MSTC module. In the SSTC module, the dilation rate of the first Dilated Unit is set to 1, which means that this module can only capture information from adjacent three frames in its first layer. The results in Table 5 demonstrate that the use of multi-scale temporal information leads to a comprehensive improvement in the model's performance, with increases of 1% and 1.6% in accuracy and F1@50 respectively.

Table 5. Comparing single-scale and multi-scale temporal convolution on PKU-MMD.

	Acc	Edit	F1@{10,25,50}		
single-scale	72.9	71.9	73.1	68.3	55.2
multi-scale	**73.9**	**73.5**	**74.0**	**69.3**	**56.8**

4.6   Comparision with State-of-the-Art

In this section, we compare our proposed HSTN with classical TAS methods and the SOTA skeleton-based TAS methods on the PKU-MMD and MCFS datasets. As depicted in Table 6 and Table 7, our model outperforms other methods in all metrics on the two datasets, with improvements of up to 3.8% and 4.2% in edit distance and F1 scores on the MCFS22 dataset. We visualize the prediction as shown in Fig. 4. Due to the large GPU memory consumption of the MS-GCN model on the MCFS dataset, we reduce its batch size to 1.

Additionally, as illustrated in Fig. 3, the stacked bar chart shows the improvements in evaluation metrics brought by different stages of the model. It is evident that our method achieves higher accuracy in the initial stage than MS-GCN, indicating that the features extracted from the initial stage of our model are more discriminative. The initial stage contributes to the improvement in accuracy,

Table 6. Comparison with the state-of-the-art methods on PKU-MMD, MCFS22. Underlined denotes that the result was obtained using a batch size of 1.

Dataset	PKU-MMD					MCFS22				
Method	Acc	Edit	F1@{10,25,50}			Acc	Edit	F1@{10,25,50}		
MS-TCN [5]	67.4	68.2	68.5	63.3	49.0	72.6	63.0	63.4	58.4	47.5
ASFormer [25]	68.1	69.5	69.4	64.0	50.7	70.4	63.6	64.0	59.3	47.6
MS-GCN [6]	72.5	71.7	71.0	65.3	52.9	<u>74.1</u>	<u>66.4</u>	<u>67.3</u>	<u>63.3</u>	<u>53.9</u>
Ours	**73.9**	**73.5**	**74.0**	**69.3**	**56.8**	**76.8**	**70.2**	**71.5**	**67.4**	**57.7**

Table 7. Comparison with the state-of-the-art methods on MCFS130.

Dataset	MCFS130				
Method	Acc	Edit	F1@{10,25,50}		
MS-TCN [5]	64.8	55.7	56.0	51.3	41.3
ASFormer [25]	63.7	53.5	53.9	49.2	38.2
MS-GCN [6]	<u>64.4</u>	<u>52.1</u>	<u>52.7</u>	49.2	<u>39.7</u>
Ours	**68.1**	**60.0**	**61.7**	**57.5**	**48.1**

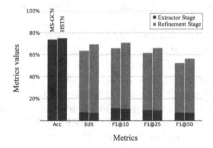

Fig. 3. Impact of different stages of MS-GCN and HSTN on metrics on MCFS22.

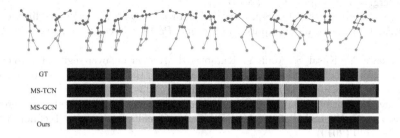

Fig. 4. The visualization of qualitative results on MCFS22 for different methods.

while the refinement stage refines the initial prediction to enhance continuity. The improvement in continuity is based on the accuracy of the initial prediction.

5   Conclusion

This paper proposes a network with a novel hierarchical structure for skeleton-based TAS, called Hierarchical Spatial-Temporal Network (HSTN). This network models the spatial-temporal information of skeleton data through two sequential modules. Firstly, a frame-level module is proposed to model spatial information for topological graphs. This module handles multi-modal skeleton data by utilizing a multi-branch graph convolution structure with an attention mechanism. We also propose a temporal module to capture long-range dependencies in the skeleton sequences at different scales. Our proposed method achieves state-of-the-art results on two benchmark datasets.

References

1. Chen, H.T., Chen, H.S., Lee, S.Y.: Physics-based ball tracking in volleyball videos with its applications to set type recognition and action detection. In: 2007 IEEE International Conference on Acoustics, Speech and Signal Processing (ICASSP 2007), vol. 1, pp. I–1097. IEEE (2007)
2. Chen, Y., Zhang, Z., Yuan, C., Li, B., Deng, Y., Hu, W.: Channel-wise topology refinement graph convolution for skeleton-based action recognition. In: Proceedings of the IEEE/CVF International Conference on Computer Vision, pp. 13359–13368 (2021)
3. Collins, R.T., Lipton, A.J., Kanade, T.: Introduction to the special section on video surveillance. IEEE Trans. Pattern Anal. Mach. Intell. **22**(8), 745–746 (2000)
4. Du, Y., Wang, W., Wang, L.: Hierarchical recurrent neural network for skeleton based action recognition. In: Proceedings of the IEEE Conference on Computer Vision and Pattern Recognition, pp. 1110–1118 (2015)
5. Farha, Y.A., Gall, J.: MS-TCN: multi-stage temporal convolutional network for action segmentation. In: Proceedings of the IEEE/CVF Conference on Computer Vision and Pattern Recognition, pp. 3575–3584 (2019)
6. Filtjens, B., Vanrumste, B., Slaets, P.: Skeleton-based action segmentation with multi-stage spatial-temporal graph convolutional neural networks. IEEE Trans. Emerg. Top. Comput. 1–11 (2022)
7. Hu, J., Shen, L., Sun, G.: Squeeze-and-excitation networks. In: Proceedings of the IEEE Conference on Computer Vision and Pattern Recognition, pp. 7132–7141 (2018)
8. Ishikawa, Y., Kasai, S., Aoki, Y., Kataoka, H.: Alleviating over-segmentation errors by detecting action boundaries. In: Proceedings of the IEEE/CVF Winter Conference on Applications of Computer Vision, pp. 2322–2331 (2021)
9. Jiang, X., Xu, K., Sun, T.: Action recognition scheme based on skeleton representation with DS-LSTM network. IEEE Trans. Circuits Syst. Video Technol. **30**(7), 2129–2140 (2019)
10. Joze, H.R.V., Shaban, A., Iuzzolino, M.L., Koishida, K.: MMTM: multimodal transfer module for CNN fusion. In: Proceedings of the IEEE/CVF Conference on Computer Vision and Pattern Recognition, pp. 13289–13299 (2020)
11. Karaman, S., Seidenari, L., Del Bimbo, A.: Fast saliency based pooling of fisher encoded dense trajectories. In: ECCV THUMOS Workshop, p. 5 (2014)

12. Kim, T.S., Reiter, A.: Interpretable 3d human action analysis with temporal convolutional networks. In: 2017 IEEE Conference on Computer Vision and Pattern Recognition Workshops (CVPRW), pp. 1623–1631. IEEE (2017)
13. Krüger, V., Kragic, D., Ude, A., Geib, C.: The meaning of action: a review on action recognition and mapping. Adv. Robot. **21**(13), 1473–1501 (2007)
14. Lea, C., Flynn, M.D., Vidal, R., Reiter, A., Hager, G.D.: Temporal convolutional networks for action segmentation and detection. In: proceedings of the IEEE Conference on Computer Vision and Pattern Recognition, pp. 156–165 (2017)
15. Li, C., Zhong, Q., Xie, D., Pu, S.: Skeleton-based action recognition with convolutional neural networks. In: 2017 IEEE International Conference on Multimedia and Expo Workshops (ICMEW), pp. 597–600. IEEE (2017)
16. Liu, C., Hu, Y., Li, Y., Song, S., Liu, J.: PKU-MMD: a large scale benchmark for continuous multi-modal human action understanding. arXiv preprint arXiv:1703.07475 (2017)
17. Liu, S., et al.:: Temporal segmentation of fine-gained semantic action: a motion-centered figure skating dataset. In: Proceedings of the AAAI Conference on Artificial Intelligence, pp. 2163–2171 (2021)
18. Ramachandram, D., Taylor, G.W.: Deep multimodal learning: a survey on recent advances and trends. IEEE Signal Process. Mag. **34**(6), 96–108 (2017)
19. Rohrbach, M., Amin, S., Andriluka, M., Schiele, B.: A database for fine grained activity detection of cooking activities. In: 2012 IEEE Conference on Computer Vision and Pattern Recognition, pp. 1194–1201. IEEE (2012)
20. Shi, L., Zhang, Y., Cheng, J., Lu, H.: Two-stream adaptive graph convolutional networks for skeleton-based action recognition. In: Proceedings of the IEEE/CVF Conference on Computer Vision and Pattern Recognition, pp. 12026–12035 (2019)
21. Song, S., Lan, C., Xing, J., Zeng, W., Liu, J.: An end-to-end spatio-temporal attention model for human action recognition from skeleton data. In: Proceedings of the AAAI Conference on Artificial Intelligence (2017)
22. Song, Y.F., Zhang, Z., Shan, C., Wang, L.: Stronger, faster and more explainable: a graph convolutional baseline for skeleton-based action recognition. In: Proceedings of the 28th ACM International Conference on Multimedia, pp. 1625–1633 (2020)
23. Wang, Z., Gao, Z., Wang, L., Li, Z., Wu, G.: Boundary-aware cascade networks for temporal action segmentation. In: Vedaldi, A., Bischof, H., Brox, T., Frahm, J.-M. (eds.) ECCV 2020. LNCS, vol. 12370, pp. 34–51. Springer, Cham (2020). https://doi.org/10.1007/978-3-030-58595-2_3
24. Yan, S., Xiong, Y., Lin, D.: Spatial temporal graph convolutional networks for skeleton-based action recognition. In: Proceedings of the AAAI Conference on Artificial Intelligence, pp. 7444–7452 (2018)
25. Yi, F., Wen, H., Jiang, T.: Asformer: transformer for action segmentation, p. 236 (2021)
26. Zhang, P., Lan, C., Zeng, W., Xing, J., Xue, J., Zheng, N.: Semantics-guided neural networks for efficient skeleton-based human action recognition. In: Proceedings of the IEEE/CVF Conference on Computer Vision and Pattern Recognition, pp. 1112–1121 (2020)

Multi-behavior Enhanced Graph Neural Networks for Social Recommendation

Xinglong Wu[1] , Anfeng Huang[1] , Hongwei Yang[1] , Hui He[1(✉)] ,
Yu Tai[1] , and Weizhe Zhang[1,2]

[1] Harbin Institute of Technology, Harbin 150001, China
{xlwu,1190300425}@stu.hit.edu.cn,
{yanghongwei,hehui,taiyu,wzzhang}@hit.edu.cn
[2] Pengcheng Laboratory, Shenzhen 518055, China

Abstract. Social recommendation has gained more and more attention by utilizing the social relationships among users, alleviating the data sparsity problem in collaborative filtering. Most existing social recommendation approaches treat the preference propagation process coarse-grained, ignoring the different diffusion patterns targeting corresponding interaction behaviors. However, this may be inappropriate because of the interplay between multi-behavior and social relations. Therefore, in this paper, we propose a novel framework, **MB-Soc**, for Multi-**B**ehavior Enhanced **Soc**ial Recommender, to model the mutual effect between users' multiple behaviors and social connections. In MB-Soc, we first devise a single behavior-based social diffusion module to depict behavioral trust propagation. Moreover, to support behavior integration, we propose an intent embedding to ensure behavior independency. In addition, we design a Self-Supervised Learning-based behavior integration module to capture the correlations among multiple behaviors. Extensive experiments conducted on two real-world datasets demonstrate the effectiveness of our model.

Keywords: Social Recommendation · Multi-behavior
Recommendation · Graph Neural Networks · Self-Supervised Learning

1 Introduction

Recommender System (RS) has played an essential role in our daily life due to data explosion and information overload, focusing on mining the latent user-item distribution patterns and exploring user preference diffusion paradigm based on historical interactions. Collaborative Filtering (CF) [19] has become the mainstream recommendation solution that collaboratively factorizes interaction records between users and items into a latent representation space and then utilizes these latent embeddings to predict the potential interactions. However, data sparsity and cold-start problems remain the main challenges in CF scenarios that seriously impede the recommendation performance. To address such challenges, social recommendation has garnered significant attention in recent years and has become a mainstream recommendation branch.

Q. Liu et al. (Eds.): PRCV 2023, LNCS 14434, pp. 40–52, 2024.
https://doi.org/10.1007/978-981-99-8549-4_4

Social recommendation utilizes the trust regularization or preference propagation between users provided by Social Networking Services (SNSs) to enhance the recommendation performance. The core idea lies in social influence theory, which proposes that users tend to share similar items with their friends through social connections [11]. Based on such theory, social recommendation methods can be generally divided into two categories: (1) relationship regularization-based CF models and (2) trust propagation-based CF models. Specifically, relationship regularization-based methods [8,10,22] utilize the social relationships as auxiliary restrictions to enhance the representations of users. Trust propagation-based methods [4,14,18] explicitly model the preference diffusion process via learning representations obtained from social connections.

Although these methods have shown promising results, a prominent drawback is that *they only model the trust diffusion process coarse-grained and ignore the microscopic propagation under multi-behavior in social recommendations*. We reckon ignoring the heterogeneous social influence diffusion patterns under multiple behaviors may result in a suboptimal recommendation result. The multi-typed behavior of users may have different quantities and scopes of influence for socially connected users. If we can leverage such differentiated propagation information between multiple behaviors, it is possible to depict the comprehensive trust diffusion process and thus improve the performance of social recommendation. This is also explained in the following example.

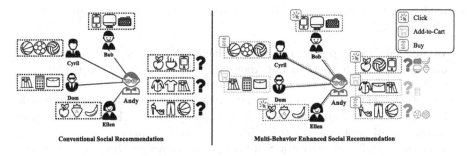

Fig. 1. A toy example: Multi-behavior enhanced social RS generates precise recommendations with microscopic behavior calculated. E.g., Andy's next click would be electronic products or fruits, influenced by Bob or Ellen with click behavior (best viewed in color).

A Motivating Example. We illustrate the abovementioned phenomenon and explicate our motivation in a toy example, as shown in Fig. 1. Each user possesses multiple interaction behaviors and distinct preference types (e.g., *Bob* prefers electronic products; *Cyril* prefers sports; *Dom* prefers official accessories; *Ellen* prefers fruits), with different behavior propagating through users individually. Conventional social RSs overlook microscopic multi-typed behaviors, and users' preference is propagated integrally, blind to the influence of different behaviors. Taking *Andy* as the ego user, we can see that conventional social RSs cannot differentiate between various social connections, and the recommendation to *Andy* might be inaccurate since his interaction sequence contains multiple types of items. However, the behavior-aware RS can generate more precise recommendations according to preference diffusion and behavior influence with multi-behavioral information included. Specifically, with behavior information calculated, we can infer that *Andy*'s next potential click would be electronic products or fruit, influenced by *Bob* and *Ellen* through the same behavior type—*'click'*.

As illustrated in the above example, behavior information would promote the preference diffusion process in the trust network and thus improve the performance of social recommendations. However, traditional social recommenders only depict the users as integral units without differentiating the trust propagation process concerning interaction behavior. Thus, the novel multi-behavior social recommendation architecture faces a new challenge that has not been addressed by current literature, i.e., **CH1**: *'how to propagate behavioral preferences and trust information among users uniformly?'*

Moreover, to distinguish different facets of multi-behavior and effectively extract the main characteristics of entities in multi-behavior recommendations, the latent correlation exploration between multiple behaviors is essential. Hence, the second challenge of our work is **CH2**: *'how to model the heterogeneity in multi-behavior recommendations and dig into their relevance?'*

Our Approach and Contributions. Facing the above challenges, we propose a multi-behavior-enhanced social recommendation model to explore the fine-grained trust propagation in social recommendations. Our main contributions are summarized below:

- We highlight the importance of multi-behavior enhancement of social recommendations. To the best of our knowledge, we are the first to explore the multi-behavior social recommendation problem.
- Targeting **CH1**, we propose a Graph Neural Network (GNN)-based multi-behavior social recommendation architecture, dubbed MB-Soc, to combine the propagation of behavioral feedback and social connections.
- Targeting **CH2**, we devise a Self-Supervised Learning (SSL)-based multi-behavior integration framework to ensure the relevance and variety of multiple behaviors simultaneously and explore the latent intent of users.
- Extensive experiments conducted on two real-world datasets demonstrate the superiority and effectiveness of our MB-Soc model.

2   Related Work

In this section, we review the related literature in two main categories: (1) social recommendation and (2) multi-behavior recommendation.

Social Recommendation: Social recommendation [1] utilizes social connections as side information to enhance recommendation and alleviate the effect of data sparsity problems. Based on the integration paradigm of social relationships, we generally categorize social recommenders into two main groups: (1) social regularization methods and (2) trust propagation methods.

Social Regularization Methods: Early social RSs regularize the representation of socially connected users (friends or social neighbors) on the basis of CF, considering the social constraints as an auxiliary optimization objective. User representations are socially enhanced with different learning paradigms, including (1) shared representation-based factorization (e.g., TrustMF [22] and LOCABAL [15]); (2) representation reconstruction (e.g., SoReg [10] and SocialMF [8]).

Trust Propagation Methods: DNN has been prosperous owing to the significant power of learning representation. Thus explicit modeling of trust diffusion process has gained much attention and become the state-of-the-art method. Different learning paradigms have been absorbed to learn various diffusion properties of trust propagation. We roughly review the related literature in the following 3 categories: (1) GNN-based methods integrate neighbor embeddings on social graphs (including DiffNet [18] and GraphRec [4]). (2) Multi-Layer Perceptron (MLP)-based methods utilize the DNN architecture to explore the latent relationships in social connections (e.g., DGRec [13]). (3) Recurrent Neural Network (RNN)-based methods utilize the chronological order to dig into the dependency relationships (e.g., ARSE [14]). However, these methods ignore the heterogeneous behavior distribution.

Multi-behavior Recommendation: Multi-behavior recommendation is flourishing in recommendation scenarios, aiming to explore the relevance between different types of interactions. The multi-behavior information is mainly utilized as auxiliary knowledge for different recommendation tasks. For example, GNMR [20] augments CF by exploiting multi-behavior patterns. MB-STR [23] introduces the multi-behavior sequences into sequential recommendations. Similarly, KHGT [21] applies it to Knowledge-Graph-enhanced recommendations. However, integrating behavior information into social RSs remains unexplored.

Integrating multi-behavior CF into interdisciplinary scenarios also holds great promise. ECF-S/W [9] demonstrates impressive performance in lung cancer treatment, thereby advancing the application of CF in medical science. Calculating multi-behaviors can further enhance applying CF methods in practical applications.

3   Preliminaries

We begin with the introduction of key notations and represent the typical recommendation scenario with user set U ($U = \{u_1, u_2, \ldots, u_M\}$) and item set I ($I = \{i_1, i_2, \ldots, i_N\}$), where $M = |U|$ and $N = |I|$ denote the numbers of users and items. Moreover, we further define the multi-behavior interaction graph as:

Multi-behavior Interaction Graph: G_I. With the awareness of user-item interaction under different behaviors, we construct the multi-behavior interaction graph as $G_I = (U, I, E_I)$, where $E_I \in U \times I$ contains total B types of behaviors (e.g., *add-to-cart*, *add-to-favorite*, *buy*, etc.). For each $e_{ui}^b \in E_I$, $e_{ui}^b = 1$ denotes that user u has interacted with the item i under the behavior type b, where $b = 1, 2, \cdots, B$; otherwise, $e_{ui}^b = 0$ indicates no interactions between u and i under the behavior type b.

Social Trust Graph: G_S. Social trust graph is constructed to incorporate the trust connection information remaining in social relationships: $G_S = (U, E_S)$, where $E_S \in U \times U$ denotes the trust connections. For each $e_{uv}^s \in E_S$, $e_{uv}^s = 1$ denotes that the user u and v are social neighbors (friends) in the social graph; otherwise, if $e_{uv}^s = 0$, they are not socially connected.

Problem Statement: On the basis of the above-defined notations, we give the formal definition of the multi-behavior social recommendation as follows:

Given a set of users U, a set of items I, multi-behavior interaction graph G_I between users and items, and social trust graph G_S between users, the goal of multi-behavior social recommendation is to predict the probability y_{ui} of the interaction between the user u and the item i under the behavior of type b.

4   Methodology

In this section, we introduce our MB-Soc model that incorporates fine-grained multi-behavior information into social recommendation architecture as shown in Fig. 2. The MB-Soc framework consists of four main parts, i.e., (1) Embedding Layer, (2) Propagation Layer, (3) Multi-Behavior Integration Layer, and (4) Prediction Layer. We provide detailed explanations of each component's implementation in the following parts.

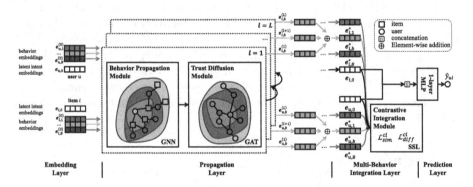

Fig. 2. Architecture of MB-Soc. Embedding Layer generates behavior and intent embeddings; Propagation Layer performs single-behavior propagation and trust diffusion; Multi-Behavior Integration Layer integrates multi-behavior embeddings of each entity; Prediction Layer predicts pair-wise interaction scores.

4.1   Embedding Layer

Embedding Layer takes the user and item IDs as the input and generates free ID embeddings for further calculation. We represent the ID embedding of user u and item i under the behavior of type $b \in \{1, \cdots, B\}$ with $\mathbf{e}_{u,b}^{(0)} \in \mathbb{R}^d$, and $\mathbf{e}_{i,b}^{(0)} \in \mathbb{R}^d$, respectively, where d is the embedding size. Notably, we represent different behavior type $b \in \{1, \cdots, B\}$ with distinct behavior embeddings. Moreover, for each user u and item i, we initialize a latent intent embedding to model the integral intention of the entity under different behaviors, which we denote as $\mathbf{e}_{u,0}$ and $\mathbf{e}_{i,0}$, respectively. Since it is not our primary focus, we uniformly represent all embeddings with d-dimensional vectors for computation convenience.

4.2   Propagation Layer

We employ Propagation Layer for modeling the embedding propagation under the single behavior type b. The propagation process is subdivided into Behavior Propagation Module and Trust Diffusion Module, taking charge of user-item preference inferring and social influence formulation.

Behavior Propagation Module. Behavior Propagation Module is responsible for disseminating the single-behavior embeddings initialized in the Embedding Layer. Regarding the graph-structured interaction data, we naturally resort to GNNs to portray the diffusion process. Specifically, in the Behavior Propagation Module, we mainly depict the intra-behavior preference diffusion phenomenon under user-item interactions within a single behavior type b. To do so, we first split the multi-behavior graph into B single-behavior graphs, namely $G_I^1, G_I^2, \cdots, G_I^B$. Formally, for any behavior type b, we define the single behavior graph as $G_I^b = \{U, I, E_I^b\}$, where $E_I^b = \{e_{ui}^{b'} | e_{ui}^{b'} = 1 \wedge b' = b\}$.

On the single behavior interaction graph G_I^b, for any adjacent tuplet (u, i), we define the message propagation as:

$$\mathbf{e}_{u,b}^{(l+1)} \leftarrow \sum_{i \in \mathcal{N}_u^b} \frac{1}{\sqrt{|\mathcal{N}_u^b|} \cdot \sqrt{|\mathcal{N}_i^b|}} \mathbf{e}_{i,b}^{(l)}, \qquad \mathbf{e}_{i,b}^{(l+1)} \leftarrow \sum_{u \in \mathcal{N}_i^b} \frac{1}{\sqrt{|\mathcal{N}_i^b|} \cdot \sqrt{|\mathcal{N}_u^b|}} \mathbf{e}_{u,b}^{(l)}, \quad (1)$$

where $\mathcal{N}_u^b$ and $\mathcal{N}_i^b$ denote the adjacent neighbors of user u and item i under behavior type b. We employ a symmetric normalization to avoid the imbalance caused by different numbers of neighbors.

Trust Diffusion Module. We further proceed with the trust propagation process in Trust Diffusion Module to calculate the interplay of social users. To assess the pairwise influence between social users, we design a Graph Attention Network (GAT) for trust diffusion calculation.

Specifically, we design a two-layer MLP network to calculate the pairwise attention coefficient:

$$\beta_{uv}^b = \mathbf{W}_2^\top \cdot \sigma \left(\mathbf{W}_1 \cdot \left(\mathbf{e}_{u,b}^{(l)} || \mathbf{e}_{v,b}^{(l)} \right) + \mathbf{b}_1 \right) + b_2, \qquad (2)$$

where $\mathbf{W}_1 \in \mathbb{R}^{d \times 2d}$, $\mathbf{b}_1 \in \mathbb{R}^d$, $\mathbf{W}_2 \in \mathbb{R}^d$, and $b_2 \in \mathbb{R}$ are trainable parameters. We opt for the sigmoid function for $\sigma(\cdot)$. We additionally apply a softmax operation on the attention score β_{uv}^b to obtain the formal attention score α_{uv}^b as follows:

$$\alpha_{uv}^b = \frac{\exp\left(\beta_{uv}^b\right)}{\sum_{v' \in \mathcal{N}_u^S} \exp\left(\beta_{uv'}^b\right)}, \tag{3}$$

where $\mathcal{N}_u^S$ denotes the adjacent neighbors of user u on graph G_S. After acquiring the attention coefficient between any social tuplet, we formulate the graph attention diffusion process as:

$$\mathbf{h}_{u,b}^{(l+1)} = \sum_{v \in \mathcal{N}_u^S} \alpha_{uv}^b \mathbf{e}_{v,b}^{(l)}. \tag{4}$$

All the adjacent social users transmit the social influence attentively with the attention weight α_{uv}^b under the single behavior type b. Furthermore, to overcome the gradient calculation problem, we apply a skip-connection to the social representations and utilize the Mean Pooling operation for $\mathbf{h}_{u,b}^{(l)}$ and $\mathbf{e}_{u,b}^{(l)}$:

$$\mathbf{e}_{u,b}^{(l+1)} \leftarrow \text{Mean} - \text{Pooling}\left(\mathbf{h}_{u,b}^{(l+1)}, \mathbf{e}_{u,b}^{(l)}\right). \tag{5}$$

Multi-hop Propagation. Stacking multiple GNN layers can expand the receptive field [6,17], capturing the correlations in broader scopes. We generalize the 1-hop propagation in Eqn. (1) and Trust Diffusion Module to multi-hops, and obtain the final embeddings of user u (item i) with the propagation depth $l \in \{1, 2, \cdots, L\}$, i.e., $\mathbf{e}_{u,b}^{(l)}$ ($\mathbf{e}_{i,b}^{(l)}$), where L denotes the maximum propagation depth. The overall single behavior representations $\mathbf{e}_{u,b}^*$ and $\mathbf{e}_{i,b}^*$ are obtained with mean-pooling, representing the equalized behavior characteristic:
$\mathbf{e}_{u,b}^* = \frac{1}{L+1} \sum_{l=0}^{L} \mathbf{e}_{u,b}^{(l)}, \mathbf{e}_{i,b}^* = \frac{1}{L+1} \sum_{l=0}^{L} \mathbf{e}_{i,b}^{(l)}$.

4.3   Multi-behavior Integration Layer

In this section, we integrate entity embeddings under different behaviors. For each entity, different behavior embeddings represent distinct facets of entity characteristics. The goal is to explore the distribution patterns of behavior embeddings and the relationships between them. Thus a natural solution is deploying an SSL architecture to excavate the correlation of different behaviors. Specifically, we resort to Contrastive Learning (CL) for automatic variety representation.

We deploy a latent intent embedding to model the integral intention of user behaviors. For notation convenience, we denote p, q for any entities (including users and items). The learning objective is to minimize the distance between the latent intent embedding $(\mathbf{e}_{p,0})$ and the behavior embedding $(\mathbf{e}_{p,b}^*)$ of the same entity and maximize the distances between any different behavior embeddings:

$$\mathcal{L}_{sim}^{cl} = -\sum_{b=1}^{B} \sum_{p \in U \cup I} \log\left(\frac{\exp\left(sim\left(\mathbf{e}_{p,0}, \mathbf{e}_{p,b}^*\right)/\tau\right)}{\sum_{q \in U \cup I} \exp\left(sim\left(\mathbf{e}_{p,0}, \mathbf{e}_{q,b}^*\right)/\tau\right)}\right), \tag{6}$$

$$\mathcal{L}_{diff}^{cl} = -\sum_{b=1}^{B} \sum_{b' \in [B] \backslash b} \sum_{p \in U \cup I} \log \left(\frac{\exp \left(sim \left(\mathbf{e}_{p,b}^*, \mathbf{e}_{p,b'}^* \right) / \tau \right)}{\sum_{q \in U \cup I} \exp \left(sim \left(\mathbf{e}_{p,b}^*, \mathbf{e}_{q,b'}^* \right) / \tau \right)} \right), \quad (7)$$

where $sim(\cdot, \cdot) : \mathbb{R}^d \times \mathbb{R}^d \to \mathbb{R}$ is the discriminator function for similarity calculation, which is implemented by the simple but effective inner product operation; τ is the temperature parameter to control the speed of convergence.

We utilize the term $\mathcal{L}_{sim}^{cl}$ as the similarity loss term for contrastive learning, minimizing the distance between latent intent embedding and different behavior embeddings. The $\mathcal{L}_{sim}^{cl}$ ensures that the latent intent embedding can reveal comprehensive aspects of entity behaviors. On the other hand, the term $\mathcal{L}_{diff}^{cl}$ is utilized to measure the distance between any two behavior embeddings. The less $\mathcal{L}_{diff}^{cl}$, the more similar that different behavior embeddings $\mathbf{e}_{p,b}$ and $\mathbf{e}_{p,b'}^*$ are. We aim to model the variety and diversity of different embedding behaviors. Thus the inverse of the term $\mathcal{L}_{diff}^{cl}$ is included in the total loss.

4.4 Prediction Layer

After CL-based Multi-Behavior Integration Layer, the latent intent embedding and the overall behavior are consistent in the latent representation space. Thus, we utilize the latent intent to represent the overall characteristic of each entity. We obtain the complete prediction utilizing a 1-layer linear MLP architecture to endow the model with more expressing ability [7]:

$$\hat{y}_{ui} = \mathbf{W}^\top \left(\mathbf{e}_{u,0} \| \mathbf{e}_{i,0} \right) + b, \quad (8)$$

where $\mathbf{W} \in \mathbb{R}^{2d}$ and $b \in \mathbb{R}$ are the trainable parameters.

4.5 Model Training

We formulate the main learning loss with the Mean Squared Error (MSELoss), and the contrastive loss term is included to enhance the expressiveness of the latent intent embedding and diversity of entity characteristics:

$$\mathcal{L}^{main} = \frac{1}{N} \sum_{u=0}^{M} \sum_{i=1}^{N} \left(y_{ui} - \hat{y}_{ui} \right)^2. \quad (9)$$

$$\mathcal{L}^{cl} = \mathcal{L}_{sim}^{cl} - \mathcal{L}_{diff}^{cl}. \quad (10)$$

We additionally include the L_2 penalty term for preventing the overfitting problem. Thus, the total objective function is formulated as follows:

$$\mathcal{L} = \mathcal{L}^{main} + \eta_1 \mathcal{L}^{cl} + \eta_2 \mathcal{L}^{Reg}, \quad (11)$$

where η_1 and η_2 are weight decay factors.

We adopt the pre-training & fine-tuning learning strategy to promote the recommendation capacity. Specifically, in the pre-training phase, we randomly initialize the model, utilizing the $\mathcal{L}^{cl}$ loss as the objective term. While in the fine-tuning phase, we initialize embeddings with the pre-trained values and optimize the model with all loss terms, i.e., $\mathcal{L}$.

5   Experiments

We conduct experiments on two real-world datasets to answer the following research questions:

RQ1. How does our MB-Soc model compare with other baseline models?

RQ2. How do key components of MB-Soc model impact the performance?

RQ3. How do key hyper-parameters impact the performance of MB-Soc model?

In the following parts, we first introduce the experimental settings, followed by answers to the above research questions.

5.1   Experimental Settings

Datasets and Evaluation Metrics. To evaluate the performance of our MB-Soc model, we conduct experiments on two real-world datasets, i.e., Ciao[1] [5] and Epinions[2] [12], which are widely adopted in social recommendations [4]. We adopt the same data processing procedure to make a fair comparison—we split them into multi-behaviors following conventional studies [20,23]: *dislike* ($1 \leq r < 3$), *neutral* ($r = 3$), and *like* ($r > 3$).

We utilize the widely-adopted MAE and $RMSE$ for the comparison. Notably, a smaller value of MAE and $RMSE$ represents better performance, and prior literature has verified that even minor improvements of MAE and $RMSE$ can significantly improve the recommendation quality [15].

Table 1. Overall Performance Comparison.

Dataset	Training	Metrics	NCF	STAR	SocialMF	SoReg	TrustMF	SAMN	EATNN	DiffNet	SNGCF	DSL	GraphRec	MB-Soc
Ciao	80%	RMSE	1.0617	1.0295	1.0657	1.0782	1.0518	1.0543	1.0313	1.0369	1.0306	1.0418	0.9794	**0.8968**
		MAE	0.8062	0.7687	0.8321	0.8593	0.8113	0.7890	0.7667	0.7723	0.7781	0.7879	0.7387	**0.6848**
	60%	RMSE	1.0824	1.0461	1.0714	1.0855	1.0678	1.0924	1.0742	1.0585	1.0445	1.0775	1.0093	**0.9194**
		MAE	0.8251	0.7932	0.8378	0.8420	0.8262	0.8116	0.7973	0.7935	0.7877	0.8145	0.7540	**0.6955**
Epinions	80%	RMSE	1.1476	1.0946	1.1494	1.1576	1.1314	1.1366	1.1187	1.1095	1.1022	1.1665	1.0631	**1.0438**
		MAE	0.9072	0.8582	0.8730	0.8797	0.8642	0.8671	0.8545	0.8438	0.8650	0.8952	0.8168	**0.8008**
	60%	RMSE	1.1645	1.1183	1.1692	1.1789	1.1553	1.1899	1.1385	1.1241	1.1173	1.2081	1.0878	**1.0561**
		MAE	0.9097	0.8843	0.8973	0.9184	0.8757	0.8995	0.8663	0.8534	0.8758	0.9263	0.8441	**0.8122**

Compared Methods and Parameter Settings. To verify the superiority and effectiveness of our model, we compare MB-Soc with 11 baseline models, including state-of-the-art non-socialization CF methods, social regularization methods, and trust propagation methods. **NCF** [7] and **STAR** [24] utilize GNNs to collaboratively factorize the interactions. **SocialMF** [8], **SoReg** [10], and **TrustMF** [22] utilize the social regularization to enhance CF performance. **SAMN** [3], **EATNN** [2], **DiffNet** [18], **SNGCF** [17] (socialized NGCF model), **DSL** [16], and **GraphRec** [4] explicitly model the social propagation process in RSs. The detailed parameter settings of MB-Soc are shown in Section A.

---

[1] http://www.cse.msu.edu/~tangjili/trust.html.

[2] http://www.trustlet.org/downloaded_epinions.html.

5.2 Performance Comparison (RQ1)

We demonstrate the overall performance comparisons in Table 1. Below are our observations:

- *Our MB-Soc model consistently outperforms other state-of-the-art models by a large gap.* In specific, MB-Soc outperforms the best-performed baseline by at least 7.30%, 7.76%, 1.82%, 2.91% on Ciao 80%, Ciao 60%, Epinions 80%, and Epinions 60%, respectively. This indicates the superiority of our proposed multi-behavior-enhanced social recommendation.
- *In general, DNN-based trust-propagation methods outperform social regularized methods.* This verifies the effectiveness of explicit modeling of the embedding propagation of the trust relationships and further confirms the efficacy of the GNN-based Propagation Layer.
- *MB-Soc achieves a better performance improvement on Ciao compared with Epinions.* This may be because of the denser social connections in Ciao. We argue MB-Soc would perform even better on a more complicated social graph.

5.3 Ablation Study (RQ2)

We perform ablation studies to explore the separate effects of each single module. We denote our model without the multi-behavior module, social diffusion module, and both modules as MB-Soc w/o MB, MB-Soc w/o Soc, and MB-Soc w/o MB & Soc, respectively. We demonstrate the performance comparisons of ablation experiments on Ciao 80% in Fig. 3 due to space limitation. Results on the other datasets exhibit the same conclusion.

We can observe that (1) MB-Soc consistently achieves the best performance against other variants. This verifies the necessity of each component of MB-Soc, a delicate combination of all components yielding the best performance. (2) MB-Soc w/o Soc exhibits a significant performance degradation compared with MB-Soc. This verifies the importance of the modeling of social propagation. (3) MB-Soc w/o MB exhibits a relatively minor degradation compared with MB-Soc. This may be because the multi-behavior enhancement is focused on the social propagation process. More detailed ablation experiments are shown in Section B.1.

5.4 Parameter Analysis (RQ3)

In this section, we perform analysis experiments w.r.t. key parameters to ensure the optimal assignment, including L, d, and τ. We demonstrate the results of analysis experiments on Ciao 60% considering space limitation.

Regarding the Propagation Depth L of the consecutive GNN and GAT modules, we denote the model with the maximum depth L as MB-Soc-L and demonstrate them in Fig. 4. Here are our observations: (1) With the increase of L, the performance of MB-Soc-L becomes better when $L \leq 2$. This indicates that the expressiveness of the model is getting better with a bigger receptive field.

This indicates that with the increase of GNN depth, the capacity of MB-Soc is correspondingly increasing. This indicates that the model's expressiveness is improving with a bigger receptive field. (2) However, MB-Soc-L tends to degrade as $L \geq 2$ and exhibits an unstable performance. This may be because the over-smoothness problem influences the embedding capacity of expression.

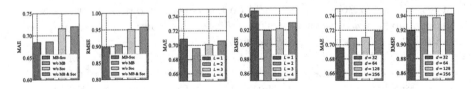

Fig. 3. Ablation Studies.   **Fig. 4.** Impact of L.   **Fig. 5.** Impact of d.

We uniformly set the dimension of all model embeddings as d to save the computation cost. To investigate the optimal embedding dimension, we conduct experiments w.r.t. d and denote the variant with different embedding size as MB-Soc-d, performance comparisons shown in Fig. 5. Below are our observations: (1) MB-Soc-d exhibits a consistent performance degradation with the increase of d. This might be because that MB-Soc-d suffers from overfitting with a great dimension d. Since our model represents multiple behaviors with independent embeddings, a relatively small value of dimension d is conceivable. (2) Jointly analyzing Fig. 5 with Table 1, we can see that even MB-Soc-256 outperforms most of the baselines. This reveals the effectiveness of our MB-Soc model apart from the influence of different embedding dimensions.

Denoting model variants with different temperature τ as MB-Soc-τ, MB-Soc-τ consistently achieves the best performance with $\tau = 0.1$. Further analyses are shown in Section B.2. Moreover, we denote MB-Soc without normalization operation in the graph as MB-Soc w/o norm. MB-Soc consistently outperforms the variant without normalization with varying graph depths, verifying the effectiveness of our design of propagation in the graph. More detailed analyses are shown in Section B.3. We release the appendix at https://github.com/WuXinglong-HIT/MB-Soc.

6   Conclusion and Future Work

In this paper, we propose a multi-behavior enhanced social recommendation framework, dubbed MB-Soc, which distinguishes the fine-grained multi-behavioral trust diffusion process in social recommendations. The main innovations in MB-Soc are behavioral trust propagation architecture and contrastive behavior integration. Specifically, regarding behavioral trust propagation, we propose a heterogeneous graph neural network-based single behavioral trust diffusion architecture. Moreover, regarding behavior integration, we propose a Self-Supervised Learning-based multi-behavior integration paradigm that aligns with user intention and highlights the differences between each behavior. Finally, extensive experiments demonstrate the superiority of our model.

In the future, we will further fine-tune our model to fully realize the potential of our model and investigate the specific efficacy of each behavior.

Acknowledgements. This work was supported in part by the Joint Funds of the National Natural Science Foundation of China (Grant No. U22A2036), the National Key Research and Development Program of China (2020YFB1406902), the Key-Area Research and Development Program of Guangdong Province (2020B0101360001), and the GHfund C (20220203, ghfund202202033706).

References

1. Camacho, L.A.G., et al.: Social network data to alleviate cold-start in recommender system: a systematic review. Inf. Process. Manag. **54**(4), 529–544 (2018)
2. Chen, C., Zhang, M., Wang, C., Ma, W., et al.: An efficient adaptive transfer neural network for social-aware recommendation. In: SIGIR, pp. 225–234 (2019)
3. Chen, C., Zhang, M., et al.: Social attentional memory network: modeling aspect- and friend-level differences in recommendation. In: WSDM, pp. 177–185 (2019)
4. Fan, W., et al.: Graph neural networks for social recommendation. In: WWW, pp. 417–426 (2019)
5. Guo, G., Zhang, J., Thalmann, D., Yorke-Smith, N.: ETAF: an extended trust antecedents framework for trust prediction. In: ASONAM, pp. 540–547 (2014)
6. He, X., Deng, K., Wang, X., Li, Y., et al.: LightGCN: simplifying and powering graph convolution network for recommendation. In: SIGIR, pp. 639–648 (2020)
7. He, X., Liao, L., Zhang, H., Nie, L., Hu, X., Chua, T.: Neural collaborative filtering. In: WWW, pp. 173–182 (2017)
8. Jamali, M., Ester, M.: A matrix factorization technique with trust propagation for recommendation in social networks. In: RecSys, pp. 135–142 (2010)
9. Luo, S., Xu, J., et al.: Artificial intelligence-based collaborative filtering method with ensemble learning for personalized lung cancer medicine without genetic sequencing. Pharmacol. Res. **160** (2020)
10. Ma, H., Zhou, D., Liu, C., Lyu, M.R., King, I.: Recommender systems with social regularization. In: WSDM, pp. 287–296 (2011)
11. Marsden, P.V., Friedkin, N.E.: Network studies of social influence. Sociol. Methods Res. **22**(1), 127–151 (1993)
12. Massa, P., Avesani, P.: Trust-aware recommender systems. In: RecSys, pp. 17–24 (2007)
13. Song, W., Xiao, Z., Wang, Y., Charlin, L., et al.: Session-based social recommendation via dynamic graph attention networks. In: WSDM, pp. 555–563 (2019)
14. Sun, P., Wu, L., Wang, M.: Attentive recurrent social recommendation. In: SIGIR, pp. 185–194 (2018)
15. Tang, J., Hu, X., Gao, H., Liu, H.: Exploiting local and global social context for recommendation. In: IJCAI, pp. 2712–2718 (2013)
16. Wang, T., Xia, L., Huang, C.: Denoised self-augmented learning for social recommendation. In: IJCAI, pp. 2324–2331 (2023)

17. Wang, X., He, X., Wang, M., Feng, F., Chua, T.: Neural graph collaborative filtering. In: SIGIR, pp. 165–174 (2019)
18. Wu, L., Sun, P., Fu, Y., Hong, R., Wang, X., Wang, M.: A neural influence diffusion model for social recommendation. In: SIGIR, pp. 235–244 (2019)
19. Wu, Xinglong, He, Hui, Yang, Hongwei, Tai, Yu., Wang, Zejun, Zhang, Weizhe: PDA-GNN: propagation-depth-aware graph neural networks for recommendation. World Wide Web **26**(5), 3585–3606 (2023)
20. Xia, L., Huang, C., Xu, Y., et al.: Multi-behavior enhanced recommendation with cross-interaction collaborative relation modeling. In: ICDE, pp. 1931–1936 (2021)
21. Xia, L., Huang, C., et al.: Knowledge-enhanced hierarchical graph transformer network for multi-behavior recommendation. In: AAAI, pp. 4486–4493 (2021)
22. Yang, B., Lei, Y., Liu, J., Li, W.: Social collaborative filtering by trust. IEEE Trans. Pattern Anal. Mach. Intell. **39**(8), 1633–1647 (2017)
23. Yuan, E., Guo, W., He, Z., Guo, H., Liu, C., Tang, R.: Multi-behavior sequential transformer recommender. In: SIGIR, pp. 1642–1652 (2022)
24. Zhang, J., Shi, X., Zhao, S., King, I.: STAR-GCN: stacked and reconstructed graph convolutional networks for recommender systems. In: IJCAI, pp. 4264–4270 (2019)

A Complex-Valued Neural Network Based Robust Image Compression

Can Luo[1], Youneng Bao[1,2], Wen Tan[1,2], Chao Li[1,2], Fanyang Meng[2], and Yongsheng Liang[1(✉)]

[1] Harbin Institute of Technology, Shenzhen, China
{21s152071,ynbao}@stu.hit.edu.cn, liangys@hit.edu.cn
[2] Peng Cheng Laboratory, Shenzhen, China
mengfy@pcl.ac.cn

Abstract. Recent works on learned image compression (LIC) based on convolutional neural networks (CNNs) have achieved great improvement with superior rate-distortion performance. However, the robustness of LIC has received little investigation. In this paper, we proposes a complex-valued learned image compression model based on complex-valued convolutional neural networks (CVCNNs) to enhance its robustness. Firstly, we design a complex-valued neural image compression framework, which realizes compression with complex-valued feature maps. Secondly, we build a module named modSigmoid to implement a complex-valued nonlinear transform and a split-complex entropy model to compress complex-valued latent. The experiment results show that the proposed model performs comparable compression performance with a large parameter drop. Moreover, we adopt the adversarial attack method to examine robustness, and the proposed model shows better robustness to adversarial input compared with its real-valued counterpart.

Keywords: Deep learning · image compression · complex-valued convolutional neural networks · robustness

1 Introduction

Image compression is an essential research to compress images into binary data, which reduces the overhead of image storage and transmission. Traditional lossy image compression methods, such as JPEG, JPEG2000, and BPG [4], adopt a hybrid coding framework consisting of transformation, quantization, and entropy coding. On the one hand, these traditional methods are difficult to further improve in compression performance after decades of development. On the other hand, traditional methods are limited by hand-crafted designs and are hard to optimize jointly.

C. Luo and Y. Bao—These authors contributed to the work equally and should be regarded as co-first authors.

© The Author(s), under exclusive license to Springer Nature Singapore Pte Ltd. 2024
Q. Liu et al. (Eds.): PRCV 2023, LNCS 14434, pp. 53–64, 2024.
https://doi.org/10.1007/978-981-99-8549-4_5

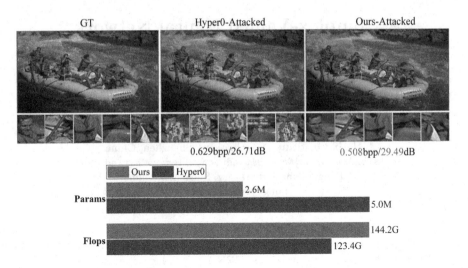

Fig. 1. Adversarial reconstructed images (noise level = 1e-4) and computational complexity of Hyper0 and Ours method.

In recent years, learned image compression models have shown great potential to improve compression performance [11,18]. All the existing models are built upon real-valued convolutional neural networks and commonly consist of modules like transformation, quantization, and entropy coding. Some of these models even show better compression performance compared with traditional methods.

However, these learned models improve compression performance while neglecting robustness. Previous works have investigated the robustness of image compression systems where imperceptible perturbations of input images can precipitate a significant increase in the bitrate of their compressed latent and severe reconstructed distortion. To improve robustness, [12] proposes a model incorporating an attention module and factorized entropy model and [5] applies the iterative adversarial finetuning to refine pre-trained models. These methods can improve the robustness of the model to a certain extent, but there are still problems like complex schemes and low flexibility.

Complex-valued neural networks (CVNNs) are artificial neural networks (ANNs) that process inputs with complex-valued parameters. They possess more benefits over their real-valued counterparts, such as larger representational capacity [14] and more robust embedding [19]. A solution using CVNNs to explore their capabilities in image denoising has achieved good results in PSNR and MS-SSIM scores [15], which proved the complex field's potential and inspired us.

In this paper, we aim to build a more robust complex-valued neural image compression model based on the advantages of complex-valued embedding. Contributions of this paper are summarized as follows:

i.) We propose a complex-valued image compression model based CVCNNs. To the best of our knowledge, this is the first work to explore the application of CVCNNs in image compression.

ii.) We build a channel attention module and a split-complex entropy model for image compression based CVCNNs, which improves rate-distortion performance.

iii.) Extensive experiments show that the proposed compression model achieves better robustness with negligible performance drop and fewer parameters compared with the real-valued counterpart.

2   Related Works

2.1   Neural Image Compression

The development of previously learned compression models has spanned several years and has many related works. In 2017, Balle et al. [1] proposed an end-to-end optimized image compression framework based on CNNs with GDN non-linearity embedded in analysis and synthesis transformation. To solve the problem of quantization underivability and realize end-to-end joint optimization, [1] replace the true quantization with additive uniform noise. Some works directly pass the gradient through the quantization layer without correction in the back-propagation process. When the network is trainable, in order to achieve better rate-distortion performance, some works adopt residual blocks, non-local modules, multi-scale networks and invertible structures to further reduce spatial and channel redundancy in the compression models. Furthermore, the entropy model is also a critical module in compression models, which estimates the marginal distribution of latent representation to save more transmission bit rates. Recent studies on entropy model, such as hyperprior model [2], context model [13] and Gaussian mixture model [6], have greatly improved the entropy estimation and rate-distortion performance. To conclude, benefiting from the strong modeling capacity of neural networks, the learned compression models show better performance than classical methods, and the gap is further widening. However, few studies have explored model robustness, though it is crucial for practical deployment.

2.2   Adversarial Attack

The adversarial attack is a method to generate adversarial examples formed by applying imperceptible perturbations to inputs from the dataset so that perturbed inputs make the model output wrong results, which is widely used to evaluate the robustness of the deep learning models. In 2014, Szegedy et al. [17] suggest that the machine learning models are vulnerable to adversarial examples and propose an attack method based on optimization to generate adversarial perturbation. Goodfellow et al. [10] argue that the vulnerability of neural network models to adversarial samples is caused by their linear characteristics. Based on

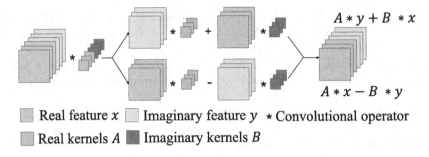

$$A * y + B * x$$

$$A * x - B * y$$

- 🟩 Real feature x ⬜ Imaginary feature y * Convolutional operator
- 🟩 Real kernels A ⬛ Imaginary kernels B

Fig. 2. Details of Complex convolutional operator

this, they proposed a fast gradient sign method (FGSM) for generating adversarial examples.

Recent studies have combined neural image compression (NIC) with the adversarial attack to explore the robustness of NIC models to adversarial distributions. Tong et al. [5] take reconstructed image as the attack target to generate adversarial examples, and then adopt the iterative adversarial finetuning algorithm to refine the pre-trained model. Liu et al. [12] attacked the bit rate of the model, aiming to make the model consume more resources to realize denial-of-service attack. They propose a model consisting of an attention module and factorized entropy model to improve robustness. Although these methods can improve the robustness of the model to a certain extent, there are still problems, such as complex schemes and low flexibility.

2.3 Complex-Valued Convolutional Neural Networks

Many works suggest that applying complex values in ANNs could enhance the representational capacity [14] and possess more benefits such as better generalization [16] and noise-robust memory mechanisms [8]. In computer vision, Trabelsi et al. [7] redesigned convolution, batch normalization and nonlinear activation functions in the complex domain, which could realize operations such as convolution, activation and batch normalization for complex-valued inputs. Quan et al. [15] use CVCNNs to explore the capabilities of the complex domains in image denoising, and the proposed model has achieved good results in terms of robustness and performance.

In addition, complex-valued transformation is closely related to biological visual perception. The human visual cortex is controlled by certain complex number of cells [9], which are selective for direction and frequency. As also known in computer vision, the phase information of the image provides the shape, edge, and orientation information, and the phase information is sufficient to recover most of the information encoded in the amplitude. The benefits listed above of complex-valued transformation motivated us to use CVCNNs to construct a light and robust LIC model.

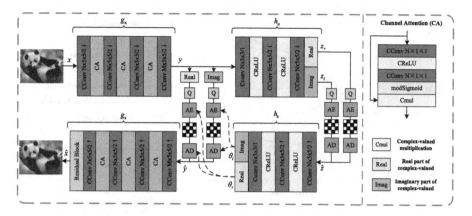

Fig. 3. Network architecture of our method (CHyper0). Convolution parameters are denoted as: the number of filters × kernel height × kernel width / down- or upsampling stride, where ↑ indicates upsampling and ↓ downsampling. $N = 64$ and $M = 96$ are chosen for the 4 lower λ values, and $N = 96$ and $M = 160$ for the 2 higher values. Q denotes quantization, and AE, AD represents arithmetic encoder and decoder.

3   Proposed Method

3.1   Overall Framework

Figure 3 provides a high-level overview of our generalized compression model named CHyper0-CA, which contains three main improvements compared with Hyper0 [2]. Firstly, we replace the real convolution layer in the baseline model with the complex-valued convolution layer (CConv) Fig. 2, and each input and output of the complex convolution layer contain two feature maps (i.e. the real and imaginary parts), whereas they could be viewed as one channel, so we can halve the number of parameters in complex convolution layer [7,15]. Secondly, in the nonlinear transformation part, we use a channel attention module incorporating the proposed modSigmoid to replace the original GDN to reduce the redundancy between complex channels. Finally, in order to fuse the information between the real and imaginary parts of the three-channel complex image generated by the decoder, we concatenate them in the channel dimension and adopt a real-valued residual block.

In addition, we model the real part and imaginary part as zero-mean Gaussian distribution with their own standard deviation (i.e. θ_r and θ_i) respectively, where the standard deviations are from the complex parameters predicted by hyper network:

$$\theta = h_s(\mathrm{Q}(h_a(\mathbf{y}))),$$
$$\theta_r = \mathrm{Real}(\theta), \ \theta_i = \mathrm{Imag}(\theta), \tag{1}$$

where h_a and h_s denote hyper codec, Real and Imag denote the real part and imaginary part of complex values respectively, Q represents the quantization process.

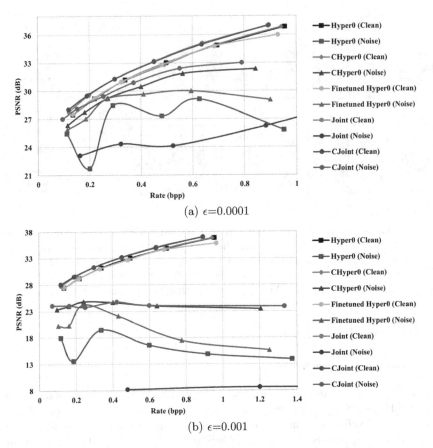

Fig. 4. Rate-distortion curves of PSNR attack on Kodak dataset with $\epsilon = 0.0001$ (top) and $\epsilon = 0.001$ (bottom). CHyper0 and CJoint are the proposed methods, Hyper0 is [2], Finetuned Hyper0 is [5], Joint is [13].

After designing the structure of our model, it needs to be trained in an end-to-end manner. Similar to most image compression problems, two loss terms rate R and distortion D need to be optimized:

$$
\begin{aligned}
Loss &= R + \lambda \cdot D \\
&= \mathbb{E}\left[-\log_2 p(\hat{z}_r))\right] + \mathbb{E}\left[-\log_2 p(\hat{z}_i)\right] + \mathbb{E}\left[-\log_2 p(\hat{y}_r|\hat{z})\right] \\
&\quad + \mathbb{E}\left[-\log_2 p(\hat{y}_i|\hat{z})\right] + \lambda \cdot \mathbb{E}||x - \hat{x}||_2^2,
\end{aligned}
\tag{2}
$$

where $p(\cdot)$ denotes the probability of feature maps at the bottleneck layer and $p(\cdot|\hat{z})$ denotes the conditional probability of latent representations, λ is hyper parameter to control the trade-off between rate and distortion.

Figure 1 shows the complexity of the whole model, including the number of parameters and FLOPs. It can be seen that compared with Hyper0, our model can greatly reduce parameters though the computational complexity is little higher than Hyper0.

3.2 Nonlinear Transform

Most real-valued LIC models apply GDN to realize nonlinear transform in main codec. We first try to improve the GDN, in which we apply separate GDNs on both of the real and the imaginary part. We call the improved transformation as SGDN (Split-GDN). The expression of SGDN is as follows:

$$\mathbf{y} = \text{GDN}(\text{Real}(\mathbf{x})) + i \cdot \text{GDN}(\text{Imag}(\mathbf{x})) \tag{3}$$

where $\mathbf{x}$ and $\mathbf{y}$ denote complex-valued input and output. However, it is found that SGDN is not compatible with complex feature maps, which shows performance drop in compression. To tackle this problem, we propose a channel attention module detailed in Fig. 3 to substitute complex GDN. A new complex activation function named modSigmoid is proposed in this module. The modSigmoid is defined as:

$$\text{modSigmoid}(\mathbf{h}) = \text{Sigmoid}(|\mathbf{h}| + b) \exp i\theta_{\mathbf{h}}, \tag{4}$$

where $\mathbf{h} \in \mathbb{C}$, $\theta_{\mathbf{h}}$ is the phase of $\mathbf{h}$, and $b \in \mathbb{R}$ is a learnable bias. The intuition behind modSigmoid is to normalize magnitude without destroying phase information. With this activation, the channel attention module could learn the redundancy in channel dimension and generate importance weights scaled in the range of 0–1 based on the amplitude of features, so as to improve the representational capacity.

Table 1. The rate-distortion performance of original images and adversarial examples tested on Kodak for different models with quality = 6 and $\epsilon = 0.001$.

Model	Original		Adversarial		PSNR Drop Rate↓
	PSNR↑	Bpp↓	PSNR↑	Bpp↓	
Hyper0 [2]	36.8450	0.9561	13.9159	1.3731	0.6223
Finetuned Hyper0 [5]	35.8027	0.9657	15.6110	1.2521	0.5639
CHyper0-CA	36.8229	0.9424	23.4368	1.2063	0.3635
Joint [13]	36.9540	0.8920	9.2346	5.7112	0.7499
CJoint	37.0158	0.8946	23.8821	1.3325	0.3548

In hyper codecs, we use CReLU [7] to introduce nonlinearity on the complex-valued vector. The CReLU applies ReLU on the real part and imaginary part respectively:

$$\text{CReLU}(\mathbf{x}) = \text{ReLU}(\text{Real}(\mathbf{x})) + i \cdot \text{ReLU}(\text{Imag}(\mathbf{x})) \tag{5}$$

60   C. Luo et al.

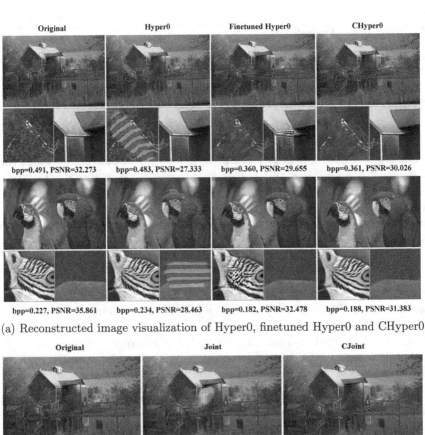

bpp=0.491, PSNR=32.273   bpp=0.483, PSNR=27.333   bpp=0.360, PSNR=29.655   bpp=0.361, PSNR=30.026

bpp=0.227, PSNR=35.861   bpp=0.234, PSNR=28.463   bpp=0.182, PSNR=32.478   bpp=0.188, PSNR=31.383

(a) Reconstructed image visualization of Hyper0, finetuned Hyper0 and CHyper0

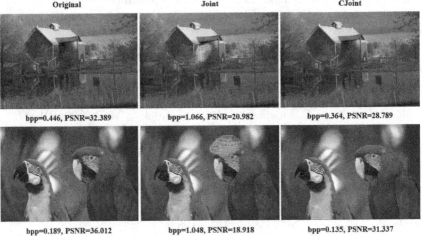

bpp=0.446, PSNR=32.389   bpp=1.066, PSNR=20.982   bpp=0.364, PSNR=28.789

bpp=0.189, PSNR=36.012   bpp=1.048, PSNR=18.918   bpp=0.135, PSNR=31.337

(b) Reconstructed image visualization of Joint and CJoint

Fig. 5. Visualization of original and adversarial reconstructed images for different models with $\epsilon = 0.0001$ and quality = 4.

4 Experiment Results

4.1 Experiment Setup

Datasets. The CLIC dataset[1] is used for training and we augment it by randomly cropping the images into 256×256 patches. To test the performance of our model, we utilize Kodak dataset[2] which contains 24 images with size 768×512 to evaluate rate-distortion performance and robustness.

Training Setting. We adopt CompressAI [3] to implement our models and retrain the Hyper0 model, where we followed most of their training settings. All models are trained for 100 epochs with six different λ values.

Adversarial Examples Generating. We adopt the white-attack method to generate adversarial samples. While generating adversarial examples, we need to cause the greatest distortion under limited perturbation. Therefore, the attack method can be realized by optimizing the formula in [5]. In adversarial training, we followed the training settings in [5]. Specifically, we run 10000 steps with learning rate $1e - 3$ and different ϵ to generate adversarial examples for our models and Hyper0.

Evaluation Setting. We measure the quality of reconstructed images with PSNR. In the process of adversarial attack, we focus on the PSNR drop averaged on the Kodak dataset.

$$\text{PSNR drop} = \frac{\text{PSNR}(x, \hat{x}) - \text{PSNR}(x, \hat{x}^{'})}{\text{PSNR}(x, \hat{x})}. \tag{6}$$

To ensure fairness, models are all trained and tested on the same machine (NVIDIA Tesla V100).

4.2 Results and Comparison

We adopt the iterative adversarial finetuning algorithm proposed in [5] to update the pre-trained Hyper0 and we follow the same settings except that we train and finetune Hyper0 with PSNR. To verify effectiveness of the proposed complex-valued modules, we also utilized CVCNNs and employed the similar improvements as in Hyper0 to obtain a complex-valued joint autoregressive model (i.e. CJoint) [13]. After that, We use two noise levels (i.e. $\epsilon = 0.001$, 0.0001) to evaluate and compare the robustness of our models and the real-valued counterpart models.

---

[1] http://compression.cc/.
[2] http://r0k.us/graphics/kodak/.

Figure 4 shows the rate-distortion curves of different models after and before the adversarial attack. As shown in this figure, our model shows better robustness for different qualities with both two noise levels. Table 1 provides specific value of Fig. 4. We compared different models under the conditions of q = 6 and noise level = 0.001. The "Original" column represents the rate-distortion performance before the attack, while the "Adversarial" column represents the rate-distortion performance after the attack. The "PSNR DROP" column indicates the change in distortion before and after the attack, where a smaller value indicates better robustness of the model.

To vividly show the performance of different models, we visualize the reconstructed images of adversarial examples in Fig. 5(a), where the first column represents reconstructed output of original clean image in Hyper0 and the last three columns denote reconstructed images of adversarial examples in Hyper0, finetuned Hyper0 and CHyper0-CA. Similarly, we visualize the reconstructed images of joint autoregressive model and complex-valued joint autoregressive model in Fig. 5(b). To make the comparison fair, same quality, noise level and images are chosen. The adversarial examples generated from same images are with different noise perturbations with respect to models.

From Fig. 5, we can conclude that the images reconstructed by real-valued models namely Hyper0 and Joint after attack show severe distortion in PSNR. Especially for the Joint model, its context entropy model can lead to a higher vulnerability to slight perturbations, making it less robust compared to Hyper0 under the same noise level. Anti-attack real-valued model like finetuned Hyper0 improve robustness with better visual quality. However, there is still evident distortion in the adversarial reconstruction. Additionally, complex-valued models like CHyper0 and CJoint can maintain a relatively good reconstruction quality under the same attack conditions, without exhibiting significant distortion in local regions. This proves that the proposed complex-valued models can resist noise attacks more effectively.

4.3   Ablation Study

We call CHyper0 with SGDN nonlinearity CHyper0-SGDN. We train CHyper0-CA, CHyper0-SGDN and Hyper0 models with same training settings. The rate-distortion performance curves are shown in Fig. 6. We can see that CHyper0-CA can achieve similar perforamce with Hyper0 and CHyper0-SGDN shows worse results than the other two models particularly in high bitrate. This result demonstrates the CA can reduce more redundancy in complex-valued channels.

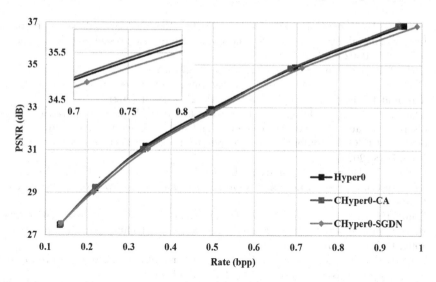

Fig. 6. Rate-distortion performance comparison averaged on Kodak dataset for different models.

5   Conclusions

In this paper, we use CVCNNs to refine the existing LIC model and propose a more robust complex-valued image compression model, where we propose a complex activation layer to implement channel attention to realize nonlinear transform in real-valued image compression networks and a split-complex entropy model to encode and decode complex-valued latent representation. After designing our model, we compare performance in rate-distortion and adopt white-box attack method to evaluate robustness of the proposed model and baseline model. The experiment results show that our complex-valued model has stronger anti-attack capacity than the real-valued network model when the rate-distortion performance is similar to that of the real-valued counterpart.

Acknowledgment. This research was supported by the National Natural Science Foundation of China (Grant No. 62031013), the Guangdong Province Key Construction Discipline Scientific Research Capacity Improvement Project (Grant No. 2022ZDJS117), and the project of Peng Cheng Laboratory. The computing resources of Pengcheng Cloudbrain are used in this research.

References

1. Ballé, J., Laparra, V., Simoncelli, E.P.: End-to-end optimized image compression. arXiv preprint arXiv:1611.01704 (2016)
2. Ballé, J., Minnen, D., Singh, S., Hwang, S.J., Johnston, N.: Variational image compression with a scale hyperprior. arXiv preprint arXiv:1802.01436 (2018)

3. Bégaint, J., Racapé, F., Feltman, S., Pushparaja, A.: Compressai: a pytorch library and evaluation platform for end-to-end compression research. arXiv preprint arXiv:2011.03029 (2020)
4. Bellard, F.: BPG image format (2014). 1, 2 (2016)
5. Chen, T., Ma, Z.: Towards robust neural image compression: Adversarial attack and model finetuning. arXiv preprint arXiv:2112.08691 (2021)
6. Cheng, Z., Sun, H., Takeuchi, M., Katto, J.: Learned image compression with discretized gaussian mixture likelihoods and attention modules. In: Proceedings of the IEEE/CVF Conference on Computer Vision and Pattern Recognition, pp. 7939–7948 (2020)
7. Chiheb, T., Bilaniuk, O., Serdyuk, D., et al.: Deep complex networks. In: International Conference on Learning Representations (2017). https://openreview.net/forum
8. Danihelka, I., Wayne, G., Uria, B., Kalchbrenner, N., Graves, A.: Associative long short-term memory. In: International Conference on Machine Learning, pp. 1986–1994. PMLR (2016)
9. Dow, B.M.: Functional classes of cells and their laminar distribution in monkey visual cortex. J. Neurophysiol. **37**(5), 927–946 (1974)
10. Goodfellow, I.J., Shlens, J., Szegedy, C.: Explaining and harnessing adversarial examples. arXiv preprint arXiv:1412.6572 (2014)
11. Li, B., Xin, Y., Li, C., Bao, Y., Meng, F., Liang, Y.: Adderic: towards low computation cost image compression. In: ICASSP 2022–2022 IEEE International Conference on Acoustics, Speech and Signal Processing (ICASSP), pp. 2030–2034 (2022). https://doi.org/10.1109/ICASSP43922.2022.9747652
12. Liu, K., Wu, D., Wang, Y., Feng, D., Tan, B., Garg, S.: Denial-of-service attacks on learned image compression. arXiv preprint arXiv:2205.13253 (2022)
13. Minnen, D., Ballé, J., Toderici, G.D.: Joint autoregressive and hierarchical priors for learned image compression. Adv. Neural Inf. Process. Syst. **31** (2018)
14. Nitta, T.: The computational power of complex-valued neuron. In: Kaynak, O., Alpaydin, E., Oja, E., Xu, L. (eds.) Artificial Neural Networks and Neural Information Processing — ICANN/ICONIP 2003, pp. 993–1000. Springer, Heidelberg (2003). https://doi.org/10.1007/3-540-44989-2_118
15. Quan, Y., Chen, Y., Shao, Y., Teng, H., Xu, Y., Ji, H.: Image denoising using complex-valued deep CNN. Pattern Recogn. **111**, 107639 (2021)
16. Singhal, U., Xing, Y., Yu, S.X.: Co-domain symmetry for complex-valued deep learning. In: Proceedings of the IEEE/CVF Conference on Computer Vision and Pattern Recognition, pp. 681–690 (2022)
17. Szegedy, C., et al.: Intriguing properties of neural networks. arXiv preprint arXiv:1312.6199 (2013)
18. Yin, S., Li, C., Bao, Y., Liang, Y., Meng, F., Liu, W.: Universal efficient variable-rate neural image compression. In: ICASSP 2022–2022 IEEE International Conference on Acoustics, Speech and Signal Processing (ICASSP), pp. 2025–2029 (2022). https://doi.org/10.1109/ICASSP43922.2022.9747854
19. Yu, S.: Angular embedding: a robust quadratic criterion. IEEE Trans. Pattern Anal. Mach. Intell. **34**(1), 158–173 (2011)

Binarizing Super-Resolution Neural Network Without Batch Normalization

Xunchao Li and Fei Chao[✉]

School of Informatics, Xiamen University, Fujian 361005, People's Republic of China
lixunchao@stu.xmu.edu.cn, fchao@xmu.edu.cn

Abstract. In this paper, our objective is to propose a model binarization method aimed at addressing the challenges posed by over-parameterized super-resolution (SR) models. Our analysis reveals that binary SR models experience significant performance degradation, primarily attributed to their sensitivity towards weight/activation distributions, particularly when devoid of Batch Normalization (BN) layers. Consequently, we undertake the following endeavors in this study: First, we conduct a comprehensive analysis to examine the impact of BN layers on SR models based on Binary Neural Networks (BNNs). Second, we propose an asymmetric binarizer that can be reparameterized to adaptively adjust the transition point for activation binarization. Third, we introduce a progressive gradient estimator that modifies weight smoothness and controls weight flipping to stabilize the training procedure in the absence of BN layers. Through extensive experiments, we demonstrate that our proposed method exhibits significant performance improvements. For instance, when binarizing EDSR and scaling up input images by a factor of $\times 4$, our approach achieves a PSNR decrease of less than 0.4dB on the Urban100 benchmark.

Keywords: Super-Resolution · Model compression and Acceleration · Binary neural network

1 Introduction

Single image super-resolution (SISR) is a fundamental task in computer vision that aims to reconstruct a high-resolution (HR) image from a low-resolution (LR) input [28, 29, 39]. SISR plays a critical role in enhancing the resolution of photographs taken by users, especially for portable devices with limited storage and computational resources. Significant efforts have been devoted to solving this challenging ill-posed problem over the past few decades. Furthermore, the rapid advancement of convolutional neural networks (CNNs) [7, 11, 27] has led to the development of state-of-the-art SISR models [4, 5, 14, 19].

However, the high memory usage and computational complexity present a significant challenge for portable devices such as smartphones and wearable gadgets. Consequently, there has been considerable attention in the visual community towards learning methods that can compress and accelerate CNN-based SISR models [2, 6, 9, 38].

Supplementary Information The online version contains supplementary material available at https://doi.org/10.1007/978-981-99-8549-4_6.

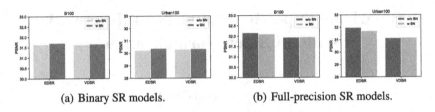

(a) Binary SR models.   (b) Full-precision SR models.

Fig. 1. Influence of BN layers on SR models including EDSR [19] and VDSR [14] (×2 upscaling). We test the performance on two example benchmarks BSD100 [24] and Urban100 [12], and observe (a) better binary SR models with BN layers while (b) better full-precision SR models without BN layers.

One promising approach for achieving lightweight and efficient models is network quantization, which discretizes the full-precision weights and activations [9,10,38]. In extreme cases, binary neural networks (BNNs) represent the full-precision variables in 1-bit format, taking on values of either −1 or +1. Compared to the original convolutional operations, BNNs offer substantial reductions in network size (by a factor of 32) and computational speedups (by a factor of 58) through the use of xnor and bit-count logics [26]. Consequently, several BNN-based SR approaches have been proposed in recent years. For instance, Ma et al. [23] introduced the first weight-binarized SR model, while Xin et al. [33] developed a bit accumulation mechanism to approximate full-precision convolutions when binarizing both weights and activations. Jiang et al. [13] subsequently invented a binary training mechanism based on feature distribution to replace batch normalization (BN) layers in SR models. More recently, Li et al. [18] proposed dynamically adjusting the binarization threshold based on the feature's local mean for improved performance.

However, a noticeable performance gap exists between BNN-based SR models and their full-precision counterparts. Most existing methods for binarizing SR models are developed based on studies focused on high-level vision tasks, emphasizing the importance of batch normalization (BN) layers in stabilizing BNN training and enhancing model capacity [3,22,26]. This observation is consistent with the results shown in Fig. 1(a), where the performance of BNN-based SR models improves when a BN layer is inserted after binarized convolution. Two main factors contribute to this improvement. First, BN mitigates weight flipping, thereby stabilizing the training process [8,20,36]. Second, BN layers promote more uniform per-channel activation distributions, reducing the challenges associated with activation binarization.

Regrettably, it is widely acknowledged that removing BN layers enhances the performance of full-precision SR models, as also observed in Fig. 1(b). This discrepancy arises because low-level vision tasks, such as SR, are sensitive to the scale information of images, while BN restricts the flexibility of activation ranges [38]. This conflict motivates us to seek an SR-specific method for improved binarization, rather than simply incorporating BN layers.

Based on the aforementioned analysis, instead of directly introducing additional BN layers in BNN-based SR models, we propose several techniques to stabilize training and preserve network flexibility. To address the challenge of binarizing diverse channel-wise distributions, we employ a channel-wise asymmetric binarizer for activations inspired by AdaBin [30] in our SR model. This binarizer allows reparameterization of additional parameters during the inference phase to maintain the efficiency of BNNs. Furthermore, we conduct a comprehensive analysis of BN layers and discover that BN layers contribute to the stabilization of BNN-based SR model training from a gradient perspective. In order to stabilize training and retain model flexibility, we design a progressive gradient estimator that gradually adjusts weight smoothness and controls the weight flipping ratio. By combining these approaches, we achieve significant performance improvements on various benchmark datasets. For example, our methods achieve a PSNR of 27.91dB on VDSR [14] and 28.41dB on EDSR [19] when performing ×4 upscaling on Set14 [15].

2 Related Work

BNN (binary neural network) is a low-bit model that represents most full-precision variables in a 1-bit format, enabling efficient matrix multiplication through bit-wise xnor and bit-count logics. While originating from high-level vision tasks such as classification [20,22,25,30,36] and object detection [32,34,35], BNN models oriented towards super-resolution (SR) have also gained attention in recent years. The pioneering work by Me et al. [23] introduced weight binarization for SR models. Subsequently, Xin et al. [33] proposed a bit-accumulation mechanism (BAM) that accumulates full-precision values from previous layers before binarization, thereby improving the approximation of full-precision convolutions even when both weights and activations are quantized. In [13], a bit training mechanism (BTM) was introduced to effectively reduce the performance gap between BNN models and their full-precision counterparts in the context of SR. Recently, Li et al. [18] presented a local means scheme that preserves more detailed information in feature maps using dynamic thresholds. Additionally, they proposed a gradient approximator to adaptively optimize gradients. These advancements highlight the ongoing exploration of BNN models in the SR domain, leveraging techniques such as weight binarization, bit-accumulation mechanisms, bit training, and dynamic thresholding with gradient approximation.

3 Method

3.1 Batch Normalization in SR Models

In the case where no batch normalization (BN) layers are used, as depicted in Fig. 3, the input $\mathbf{X}^r$ in super-resolution (SR) models exhibits diverse per-channel distributions with asymmetry. This differs from models designed for high-level tasks, where BN layers are employed to unify per-channel activation distributions.

On the other hand, BN layers contribute to the stability of BNNs, as supported by previous work [21], which analyzes the weight flipping ratio in each layer. Specifically, the weight flipping ratio measures the proportion of weights that change signs from training iteration t to $t+1$ and can be formulated as:

$$R_{t \to t+1} = \frac{\sum_{w^r \in \mathbf{X}^r} \mathcal{I}\big(\text{sign}(w_{t+1}^r) \neq \text{sign}(w_t^r)\big)}{\text{len}(\mathbf{X}^r)}, \tag{1}$$

where $\mathcal{I}(\cdot)$ returns 1 if the input is true and 0 otherwise, and $\text{len}(\cdot)$ returns the number of elements in the input. The weight flipping ratio is computed at the last step of each epoch. Experimental results on EDSR, as shown in the left Fig. 4(a), reveal that inserting BN layers leads to a smaller weight flipping ratio during training, indicating better stability compared to models without BN layers.

To analyze this phenomenon, we consider the weight updating process using stochastic gradient descent, given by:

$$w_{t+1}^r = w_t^r - \eta \frac{\partial \mathcal{L}}{\partial w_t^r}, \tag{2}$$

where η denotes the learning rate.

It is evident that the difference in weight flipping with and without BN layers mainly stems from the discrepancy in gradients. To delve deeper into the analysis, the BN layer processes the convolutional output $\mathbf{Y}^r$ as follows[1]:

$$\tilde{\mathbf{Y}}^r = \left(\frac{\mathbf{Y}^r - \mu}{\sqrt{\sigma^2 + \epsilon_{BN}}} \right) \gamma + \beta, \tag{3}$$

where μ and σ are statistical parameters computed over an incoming data batch, and γ and β are learnable parameters. Here, ϵ_{BN} is a small constant introduced for numerical stability.

Hence, for an SR model with BN layers, the gradient $\frac{\partial \mathcal{L}}{\partial w_t^r}$ can be derived as:

$$\frac{\partial \mathcal{L}}{\partial w_t^r} \stackrel{\text{STE}}{\approx} \frac{\partial \mathcal{L}}{\partial \tilde{y}_t^r} \frac{\partial \tilde{y}_t^r}{\partial y_t^r} \underbrace{\frac{\gamma}{\sqrt{\sigma^2 + \epsilon_{BN}}}} \frac{\partial y_t^r}{\partial w_t^b}, \tag{4}$$

where $\frac{\partial \tilde{y}_t^r}{\partial y_t^r} = \frac{\gamma}{\sqrt{\sigma^2 + \epsilon_{BN}}}$ is an additional term compared to the BN-free case. It is observed that $\frac{\partial \tilde{y}_t^r}{\partial y_t^r}$ gradually decreases during network training and eventually becomes less than 1, as shown in the right Fig. 4(a). Consequently, BN layers reduce the weight flipping ratio by diminishing the gradient $\frac{\partial \mathcal{L}}{\partial w_t^r}$, thus stabilizing the network training process.

However, as discussed in Sect. 1, SR models rely on the scale information of images, and most existing models remove BN layers to achieve better performance. To binarize models without BN layers, we introduce an asymmetric binarizer and a smoothness-controlled estimator to achieve comparable efficacy to BN layers without actually including them.

---

[1] BN layers are inserted within each channel. For simplicity, we omit the channel-wise index here.

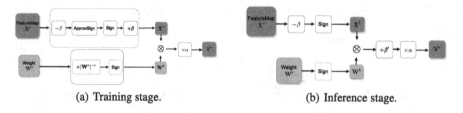

(a) Training stage.   (b) Inference stage.

Fig. 2. Our binarization framework. Noting that the extra parameter β are fused in a unify vector after binary convolution.

3.2 Channel-Wise Asymmetric Binarizer for Activations

Inspired by Si-BNN [31] and AdaBin [30], we propose a channel-wise asymmetric binarizer that retains more information from activations by applying a channel-wise shifting factor β to replace the input $\mathbf{X}^r$:

Fig. 3. Per-channel input distribution without/with BN layers of the first residual block of EDSR [19].

$$\hat{\mathbf{X}}^r = \text{sign}(\mathbf{X}^r - \beta) + \beta = \mathbf{X}^b + \beta. \quad (5)$$

Thus, the symmetric center in each channel adapts to the introduced shifting factor β. In this paper, we make β learnable for better adaptive fitting. Consequently, the convolutional output can be approximated as:

$$\mathbf{Y}^r \approx \boldsymbol{\alpha} \cdot (\mathbf{W}^b \otimes \hat{\mathbf{X}}^r) = \boldsymbol{\alpha} \cdot (\mathbf{W}^b \otimes (\mathbf{X}^b + \beta)) = \boldsymbol{\alpha} \cdot (\mathbf{W}^b \otimes \mathbf{X}^b + \mathbf{W}^b \otimes \beta),$$

where $\boldsymbol{\alpha}$ is a channel-wise scaling factor, and "·" denotes channel-wise multiplication. The convolution operation $\mathbf{W}^b \otimes \mathbf{X}^b$ can be efficiently implemented using xnor and bitcount logics, which have been observed to provide at least 58× speedups [26]. After training the network, $\mathbf{W}^b \otimes \beta$ can be pre-computed offline for all inputs, incurring negligible computation cost. The difference between training and inference can be observed in Fig. 2.

In the backward propagation, the straight-through estimator (STE) [26] is adopted to solve the non-differentiable issue of sign function, leading to gradient of any weight $w^r \in \mathbf{W}^r$:

$$\frac{\partial \mathcal{L}}{\partial w^r} = \frac{\partial \mathcal{L}}{\partial w^b} \frac{\partial w^b}{\partial w^r} \approx \frac{\partial \mathcal{L}}{\partial w^b}. \quad (6)$$

As for input $x^r \in \mathbf{X}^r$, we consider ApproxSign [22] as an alternative of sign function to compute the gradient as:

$$\frac{\partial \mathcal{L}}{\partial x^r} = \frac{\partial \mathcal{L}}{\partial x^b} \cdot \frac{\partial x^b}{\partial x^r} = \frac{\partial \mathcal{L}}{\partial x^b} \frac{\partial \text{ApproxSign}(x^r)}{\partial x^r}, \quad (7)$$

$$\frac{\partial \text{ApproxSign}(x^r)}{\partial x^r} = \begin{cases} 2 + 2x^r, & \text{if } -1 \le x^r < 0, \\ 2 - 2x^r, & \text{if } 0 \le x^r < 1, \\ 0, & \text{otherwise.} \end{cases} \quad (8)$$

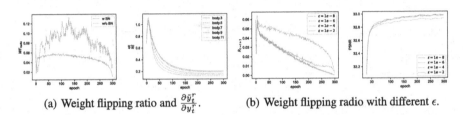

(a) Weight flipping ratio and $\frac{\partial \tilde{y}_t^r}{\partial y_t^r}$. (b) Weight flipping radio with different ϵ.

Fig. 4. We test EDSR [19] on Set14 [16] dataset to analyze: (a) Weight flipping ratio with/without BN layers and the gradient from BN layers. (b) Weight flipping ratio and PNSR performance with different values of ϵ.

3.3   Smoothness-Controlled Estimator

To address the absence of batch normalization (BN) layers, we propose a technique to mitigate their impact by smoothing the magnitude of weights before applying the sign function. This is achieved through a progressive smoothness-controlled estimator, denoted as $F(w^r)$, which is defined as follows:

$$\hat{w}^r = F(w^r) = \frac{w^r}{|w^r|_{.detach()}^{\tau}}, \tag{9}$$

where, $|\cdot|$ computes the absolute value of its input, and detach() prevents the gradient from being backpropagated through this operation. The smoothing factor τ is gradually increased from 0.1 to 1 during training epochs, following the annealing schedule given by Eq. (10):

$$\tau_i = \frac{\tau_{max} - \tau_{min}}{e - 1}e^{i/I} + \frac{e \cdot \tau_{min} - \tau_{max}}{e - 1}, \tag{10}$$

In this equation, τ_{min} is set to 0.1 and τ_{max} is set to 1. The variable I represents the total number of training epochs, and i represents the current training epoch. The binarized weight w^b is obtained by applying the sign function to $\hat{w}^r$:

$$w^b = \text{sign}(\hat{w}^r). \tag{11}$$

This progressive smoothness-controlled estimator ensures a more uniform weight distribution in the early stages of training when τ_i is small. As training progresses, the weight magnitudes gradually shift from values close to 0, which are generally considered uncertain, to values closer to 1, which provide more stability during network training. The left figure in Fig. 4(b) illustrates the weight flipping curve over different training epochs, demonstrating a similar trend to the case with BN layers depicted in the left figure of Fig. 4(a). Therefore, our smoothness-controlled estimator serves as a viable alternative to BN layers in super-resolution (SR) models.

To handle the gradient from the smoothness-controlled estimator F, which is $|w^r|^{-\tau}$ and significantly amplifies weights close to zero, we slightly modify Eq. (9) as follows:

$$\hat{w}^r = F(w^r) = \frac{w^r}{\text{clamp}(|w^r|, \min = \epsilon)^{\tau}}, \tag{12}$$

Table 1. Quantitative evaluation on VDSR [14].

method	bit(W/A)	Scale	Set5 [1] PSNR/SSIM	Set14 [16] PSNR/SSIM	BSD100 [24] PSNR/SSIM	Urban100 [12] PSNR/SSIM
full-precision	32/32	×2	37.64/0.959	33.22/0.914	31.94/0.896	31.14/0.918
Bicubic	-/-	×2	33.66/0.930	30.24/0.869	29.56/0.843	26.88/0.840
BAM [33]	1/1	×2	36.60/0.953	32.41/0.905	31.69/0.893	30.48/0.910
IBTM [13]	1/1	×2	37.06/0.956	32.72/0.908	31.53/0.889	29.96/0.902
LMBN [18]	1/1	×2	37.21/0.956	32.85/0.910	31.66/0.892	30.31/0.908
ours	1/1	×2	37.30/0.957	32.93/0.911	31.71/0.893	30.55/0.911
full-precision	32/32	×4	31.60/0.888	28.24/0.773	27.33/0.728	25.45/0.764
Bicubic	-/-	×4	28.42/0.810	26.00/0.703	25.96/0.668	23.14/0.658
BAM [33]	1/1	×4	30.31/0.860	27.48/0.754	26.87/0.708	24.45/0.720
IBTM [13]	1/1	×4	30.83/0.873	27.76/0.761	27.03/0.717	24.73/0.736
LMBN [18]	1/1	×4	30.96/0.875	27.88/0.764	27.09/0.720	24.84/0.742
ours	1/1	×4	31.08/0.878	27.91/0.764	27.12/0.720	24.92/0.743

where ϵ is a clipping threshold that prevents the gradient from becoming excessively large. The right figure in Fig. 4(b) illustrates the influence of ϵ, with a value of $\epsilon = 1e-8$ used in this study.

4 Experimentation

4.1 Experiment Setup

We extensively trained our models from scratch using the DIV2K dataset, consisting of 800 images [19], over 300 epochs. Subsequently, we assessed the models' performance on four widely recognized benchmark datasets, namely Set5 [1], Set14 [16], BSD100 [24], and Urban100 [12]. Throughout the training process, we maintained a fixed batch size of 16 and employed the Adam optimizer with hyperparameters $\beta_1 = 0.99$ and $\beta_2 = 0.999$. The initial learning rate was set to 2×10^{-4} and was halved at the 200th epoch. Consistent with previous studies on super-resolution using Binary Neural Networks (BNN) [13,18,33], we binarized all weights and activations of the feature extraction modules, while the remaining modules were kept in full-precision mode. To evaluate the effectiveness and versatility of our approach, we conducted comprehensive experiments on two well-established super-resolution models, namely VDSR [14] and EDSR [19]. We generated low-resolution images using the bicubic algorithm and utilized the Peak Signal-to-Noise Ratio (PSNR) and Structural Similarity Index (SSIM) on the luminance channel Y of the YCbCr color space as evaluation metrics to assess performance.

Experiment on VDSR

VDSR [14] is an interpolation-based super-resolution (SR) model that does not utilize Batch Normalization (BN) layers. In line with the methodologies employed in

Table 2. Quantitative evaluation on EDSR-standard [19].

method	bit(W/A)	Scale	Set5 [1] PSNR/SSIM	Set14 [16] PSNR/SSIM	BSD100 [24] PSNR/SSIM	Urban100 [12] PSNR/SSIM
full-precision	32/32	×2	38.11/0.960	33.92/0.920	32.32/0.901	32.93/0.935
Bicubic	-/-	×2	33.66/0.930	30.24/0.869	29.56/0.843	26.88/0.840
IBTM [13]	1/1	×2	37.80/0.960	33.38/0.916	32.04/0.898	31.49/0.922
LMBN [18]	1/1	×2	37.73/0.959	33.25/0.915	31.97/0.897	31.33/0.920
ours	1/1	×2	37.81/0.960	33.38/0.916	32.06/0.898	31.61/0.923
full-precision	32/32	×4	32.46/0.897	28.80/0.788	27.71/0.742	26.64/0.803
Bicubic	-/-	×4	28.42/0.810	26.00/0.703	25.96/0.668	23.14/0.658
IBTM [13]	1/1	×4	31.84/0.890	28.33/0.777	27.42/0.732	25.54/0.769
LMBN [18]	1/1	×4	31.72/0.888	28.31/0.774	27.39/0.729	25.54/0.767
ours	1/1	×4	31.89/0.891	28.41/0.777	27.46/0.732	25.71/0.774

BAM [33] and BTM [13], we divided the middle 18 convolution layers of VDSR into nine blocks, each incorporating a short connection.

Our experimental results, as illustrated in Table 1, demonstrate the substantial superiority of our approach compared to previous works focused on SR. For instance, we achieved a PSNR of 30.55dB on the Urban100 dataset with ×2 upsampling, outperforming LMBN's performance by 0.24dB.

Experiments on EDSR

EDSR is a classical SR model that also does not employ BN layers. Several quantization methods [17,37] have demonstrated the effectiveness of quantization on a 16-block variant with 64 channels per convolution. However, most binarization methods for SR models [13,18] evaluate their performance on a standard implementation with 32 convolution blocks and 256 channels per convolution. The latter model has a larger number of parameters, making it more amenable to binarization with a smaller performance gap.

Following previous works [13,18], we applied our methods to the 32-block version and referred to it as "EDSR-standard". The experimental results for EDSR-standard are presented in Table 2. We compared our approach with two state-of-the-art methods, namely IBTM [13] and LMBN [18]. Our approach exhibited the best performance, surpassing the results of these methods by a significant margin in the case of ×4 upsampling.

4.2 Ablation Study

Impact of SCE and AsymmBin

Our proposed method introduces a smoothness-controlled estimator (SCE) as a plug-in module for super-resolution (SR) models. To assess its effectiveness, we conducted ablation experiments on several models, as presented in Table 3. The results clearly demonstrate that incorporating SCE into the models leads to a significant improvement

Table 3. Ablation of SCE and AsymmBin.

AsymmBin	SCE	Set5 PSNR	Set14 PSNR	BSD100 PSNR	Urban100 PSNR
No	No	37.39	32.94	31.68	30.29
No	Yes	37.48	33.00	31.78	30.64
layer-wise	No	37.44	32.97	31.74	30.50
channel-wise	No	37.44	32.98	31.75	30.52
channel-wise	Yes	37.49	33.02	31.77	30.58

Table 4. Comparison of with and without BN.

Methods	Set5 PSNR	Set14 PSNR	BSD100 PSNR	Urban100 PSNR
w/o BN	37.39	32.93	31.68	30.29
w BN	37.47	32.96	31.72	30.41
w/o BN, ours	37.49	33.02	31.77	30.58

in performance. While our experiments indicate that AsymmBin also enhances performance, its effect is not as pronounced as SCE. Furthermore, the combination of SCE and AsymmBin achieves even better performance compared to the model utilizing SCE alone.

Comparison with Additional BN

As depicted in Fig. 1(a), the inclusion of extra batch normalization (BN) layers in binary super-resolution (SR) models can yield a substantial performance boost. To establish our method's superiority, we conducted a comparative experiment between our approach and a model augmented with BN layers following convolutions. The results of this experiment are presented in Table 4. The first row of the table represents the performance of a BN-free model that replaces our gradient estimator with a straight-through estimator (STE). The second row corresponds to the model incorporating BN layers and utilizing STE. It is observed that the model with BN layers achieves better performance across the benchmarks. Moreover, our method surpasses the BN-augmented model, particularly on benchmarks characterized by higher complexity, such as BSD100 and Urban100.

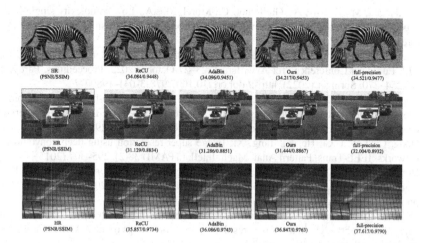

Fig. 5. Visualization of $\times 2$ upscaling on three LR images.

4.3 Visualization

We carefully selected one low-resolution (LR) image from each of the Set14, BSD100, and Urban100 benchmarks to visualize and compare the performance of our method with two other conventional binarization methods, namely ReCU and AdaBin. The results are presented in Fig. 5, where the three LR images are arranged in three rows. Our method achieves higher Peak Signal-to-Noise Ratio (PSNR) and Structural Similarity Index (SSIM) values compared to ReCU and AdaBin. Additionally, the visual quality of our method closely resembles that of the full-precision model, which attains the highest PSNR and SSIM values.

5   Conclusion

In this paper, we conducted a comprehensive analysis of the impact of batch normalization (BN) layers on binary neural network (BNN)-based super-resolution (SR) models. Our findings indicate that while BN layers can reduce the gradient of latent weights in BNNs and improve their performance, they may not be suitable for SR models, as previously suggested in works such as [19,38]. To address this issue, we proposed a general binarization method that eliminates the reliance on BN layers by employing the channel-wise asymmetric binarizer(AsymmBin) and the smoothness-controlled estimator (SCE). Our method demonstrates superior performance compared to models incorporating BN layers. We evaluated the effectiveness of our method on three classic SR models, including VDSR [14] and EDSR [19], and surpassed the state-of-the-art methods [13,18,33].

References

1. Bevilacqua, M., Roumy, A., Guillemot, C., Morel, M.L.A.: Low-complexity single-image super-resolution based on nonnegative neighbor embedding. In: British Machine Vision Conference (BMVC) (2012)
2. Chen, B., et al.: Arm: any-time super-resolution method. In: Computer Vision-ECCV 2022: 17th European Conference, Tel Aviv, Israel, 23–27 October 2022, Proceedings, Part XIX, pp. 254–270. Springer (2022). https://doi.org/10.1007/978-3-031-19800-7
3. Chen, T., Zhang, Z., Ouyang, X., Liu, Z., Shen, Z., Wang, Z.: "bnn-bn=?": training binary neural networks without batch normalization. In: Proceedings of the IEEE/CVF conference on Computer Vision and Pattern Recognition (CVPR), pp. 4619–4629 (2021)
4. Dong, C., Loy, C.C., He, K., Tang, X.: Image super-resolution using deep convolutional networks. IEEE Trans. Pattern Anal. Mach. Intell. (TPAMI) **38**(2), 295–307 (2015)
5. Dong, C., Loy, C.C., Tang, X.: Accelerating the super-resolution convolutional neural network. In: Leibe, B., Matas, J., Sebe, N., Welling, M. (eds.) ECCV 2016. LNCS, vol. 9906, pp. 391–407. Springer, Cham (2016). https://doi.org/10.1007/978-3-319-46475-6_25
6. Gao, G., Li, W., Li, J., Wu, F., Lu, H., Yu, Y.: Feature distillation interaction weighting network for lightweight image super-resolution. In: Proceedings of the AAAI conference on artificial intelligence (AAAI), vol. 36, pp. 661–669 (2022)
7. He, K., Zhang, X., Ren, S., Sun, J.: Deep residual learning for image recognition. In: Proceedings of the IEEE Conference on Computer Vision and Pattern Recognition (CVPR), pp. 770–778 (2016)

8. Helwegen, K., Widdicombe, J., Geiger, L., Liu, Z., Cheng, K.T., Nusselder, R.: Latent weights do not exist: Rethinking binarized neural network optimization. In: Advances in Neural Information Processing Systems (NeurIPS) 32 (2019)

9. Hong, C., Baik, S., Kim, H., Nah, S., Lee, K.M.: Cadyq: content-aware dynamic quantization for image super-resolution. In: Computer Vision-ECCV 2022: 17th European Conference, Tel Aviv, Israel, 23–27 October 2022, Proceedings, Part VII, pp. 367–383. Springer (2022). https://doi.org/10.1007/978-3-031-20071-7_22

10. Hong, C., Kim, H., Baik, S., Oh, J., Lee, K.M.: Daq: channel-wise distribution-aware quantization for deep image super-resolution networks. In: Proceedings of the IEEE/CVF Winter Conference on Applications of Computer Vision (WACV), pp. 2675–2684 (2022)

11. Huang, G., Liu, Z., Van Der Maaten, L., Weinberger, K.Q.: Densely connected convolutional networks. In: Proceedings of the IEEE Conference on Computer Vision and Pattern Recognition (CVPR), pp. 4700–4708 (2017)

12. Huang, J.B., Singh, A., Ahuja, N.: Single image super-resolution from transformed self-exemplars. In: Proceedings of the IEEE/CVF Conference on Computer Vision and Pattern Recognition (CVPR), pp. 5197–5206 (2015)

13. Jiang, X., Wang, N., Xin, J., Li, K., Yang, X., Gao, X.: Training binary neural network without batch normalization for image super-resolution. In: Proceedings of the AAAI Conference on Artificial Intelligence (AAAI), pp. 1700–1707 (2021)

14. Kim, J., Lee, J.K., Lee, K.M.: Accurate image super-resolution using very deep convolutional networks. In: Proceedings of the IEEE conference on Computer Vision and Pattern Recognition (CVPR), pp. 1646–1654 (2016)

15. Ledig, C., et al.: Photo-realistic single image super-resolution using a generative adversarial network. In: Proceedings of the IEEE conference on Computer Vision and Pattern Recognition (CVPR), pp. 4681–4690 (2017)

16. Ledig, C., et al.: Photo-realistic single image super-resolution using a generative adversarial network. In: Proceedings of the IEEE/CVF Conference on Computer Vision and Pattern Recognition (CVPR), pp. 4681–4690 (2017)

17. Li, H., et al.: PAMS: quantized super-resolution via parameterized max scale. In: Vedaldi, A., Bischof, H., Brox, T., Frahm, J.-M. (eds.) ECCV 2020. LNCS, vol. 12370, pp. 564–580. Springer, Cham (2020). https://doi.org/10.1007/978-3-030-58595-2_34

18. Li, K., et al.: Local means binary networks for image super-resolution. IEEE Trans. Neural Netw. Learn. Syst. (TNNLS) (2022)

19. Lim, B., Son, S., Kim, H., Nah, S., Mu Lee, K.: Enhanced deep residual networks for single image super-resolution. In: Proceedings of the IEEE/CVF conference on Computer Vision and Pattern Recognition Workshops (CVPRW), pp. 136–144 (2017)

20. Lin, M., et al.: Rotated binary neural network. Adv. Neural Inform. Process. Syst. (NeurIPS) **33**, 7474–7485 (2020)

21. Liu, Z., Shen, Z., Li, S., Helwegen, K., Huang, D., Cheng, K.T.: How do adam and training strategies help bnns optimization. In: International Conference on Machine Learning (ICML), pp. 6936–6946. PMLR (2021)

22. Liu, Z., et al.: Bi-Real Net: enhancing the performance of 1-Bit CNNs with improved representational capability and advanced training algorithm. In: Ferrari, V., Hebert, M., Sminchisescu, C., Weiss, Y. (eds.) ECCV 2018. LNCS, vol. 11219, pp. 747–763. Springer, Cham (2018). https://doi.org/10.1007/978-3-030-01267-0_44

23. Ma, Y., Xiong, H., Hu, Z., Ma, L.: Efficient super resolution using binarized neural network. In: Proceedings of the IEEE/CVF Conference on Computer Vision and Pattern Recognition Workshops (CVPRW), (2019)

24. Martin, D., Fowlkes, C., Tal, D., Malik, J.: A database of human segmented natural images and its application to evaluating segmentation algorithms and measuring ecological statistics. In: Proceedings of the IEEE/CVF International Conference on Computer Vision (ICCV), pp. 416–423 (2001)

25. Qin, H., et al.: Forward and backward information retention for accurate binary neural networks. In: Proceedings of the IEEE/CVF conference on Computer Vision and Pattern Recognition (CVPR), pp. 2250–2259 (2020)

26. Rastegari, M., Ordonez, V., Redmon, J., Farhadi, A.: XNOR-Net: imagenet classification using binary convolutional neural networks. In: Leibe, B., Matas, J., Sebe, N., Welling, M. (eds.) ECCV 2016. LNCS, vol. 9908, pp. 525–542. Springer, Cham (2016). https://doi.org/10.1007/978-3-319-46493-0_32

27. Simonyan, K., Zisserman, A.: Very deep convolutional networks for large-scale image recognition. arXiv preprint arXiv:1409.1556 (2014)

28. Timofte, R., De Smet, V., Van Gool, L.: Anchored neighborhood regression for fast example-based super-resolution. In: Proceedings of the IEEE International Conference on Computer Vision (ICCV), pp. 1920–1927 (2013)

29. Tipping, M., Bishop, C.: Bayesian image super-resolution. In: Advances in Neural Information Processing Systems (NeurIPS) 15 (2002)

30. Tu, Z., Chen, X., Ren, P., Wang, Y.: Adabin: improving binary neural networks with adaptive binary sets. In: Computer Vision-ECCV 2022: 17th European Conference, Tel Aviv, Israel, 23–27 October 2022, Proceedings, Part XI. pp. 379–395. Springer (2022). https://doi.org/10.1007/978-3-031-20083-0_23

31. Wang, P., He, X., Cheng, J.: Toward accurate binarized neural networks with sparsity for mobile application. IEEE Trans. Neural Netw. Learn. Syst. (TNNLS) (2022)

32. Wang, Z., Wu, Z., Lu, J., Zhou, J.: Bidet: an efficient binarized object detector. In: Proceedings of the IEEE/CVF conference on Computer Vision and Pattern Recognition (CVPR), pp. 2049–2058 (2020)

33. Xin, J., Wang, N., Jiang, X., Li, J., Huang, H., Gao, X.: Binarized neural network for single image super resolution. In: Vedaldi, A., Bischof, H., Brox, T., Frahm, J.-M. (eds.) ECCV 2020. LNCS, vol. 12349, pp. 91–107. Springer, Cham (2020). https://doi.org/10.1007/978-3-030-58548-8_6

34. Xu, S., et al.: Ida-det: an information discrepancy-aware distillation for 1-bit detectors. In: Computer Vision-ECCV 2022: 17th European Conference, Tel Aviv, Israel, 23–27 October 2022, Proceedings, Part XI, pp. 346–361. Springer (2022). https://doi.org/10.1007/978-3-031-20083-0_21

35. Xu, S., Zhao, J., Lu, J., Zhang, B., Han, S., Doermann, D.: Layer-wise searching for 1-bit detectors. In: Proceedings of the IEEE/CVF Conference on Computer Vision and Pattern Recognition (CVPR), pp. 5682–5691 (2021)

36. Xu, Z., et al.: Recu: reviving the dead weights in binary neural networks. In: Proceedings of the IEEE/CVF International Conference on Computer Vision (ICCV), pp. 5198–5208 (2021)

37. Zhang, Y., Tian, Y., Kong, Y., Zhong, B., Fu, Y.: Residual dense network for image super-resolution. In: Proceedings of the IEEE Conference on Computer Vision and Pattern Recognition (CVPR), pp. 2472–2481 (2018)

38. Zhong, Y., et al.: Dynamic dual trainable bounds for ultra-low precision super-resolution networks. In: Computer Vision-ECCV 2022: 17th European Conference, Tel Aviv, Israel, 23–27 October 2022, Proceedings, Part XVIII, pp. 1–18. Springer (2022). https://doi.org/10.1007/978-3-031-19797-0_1

39. Zhu, Y., Zhang, Y., Yuille, A.L.: Single image super-resolution using deformable patches. In: Proceedings of the IEEE Conference on Computer Vision and Pattern Recognition (CVPR), pp. 2917–2924 (2014)

Infrared and Visible Image Fusion via Test-Time Training

Guoqing Zheng[1], Zhenqi Fu[1], Xiaopeng Lin[1], Xueye Chu[1], Yue Huang[1,2],
and Xinghao Ding[1,2(✉)]

[1] School of Informatics, Xiamen University, Xiamen, China
`dxh@xmu.edu.cn`
[2] Institute of Artificial Intelligence, Xiamen University, Xiamen, China

Abstract. Infrared and visible image fusion (IVIF) is a widely used technique in instrument-related fields. It aims at extracting contrast information from the infrared image and texture details from the visible image and combining these two kinds of information into a single image. Most auto-encoder-based methods train the network on natural images, such as MS-COCO, and test the model on IVIF datasets. This kind of method suffers from domain shift issues and cannot generalize well in real-world scenarios. To this end, we propose a self-supervised test-time training (TTT) approach to facilitate learning a better fusion result. Specifically, a new self-supervised loss is developed to evaluate the quality of the fusion result. This loss function directs the network to improve the fusion quality by optimizing model parameters with a small number of iterations in the test time. Besides, instead of manually designing fusion strategies, we leverage a fusion adapter to automatically learn fusion rules. Experimental comparisons on two public IVIF datasets validate that the proposed method outperforms existing methods subjectively and objectively.

Keywords: Infrared image · Image fusion · Domain shift · Test-time training · Deep learning

1 Introduction

Infrared images (IR) contain thermal radiation information, and visible images (VI) have rich texture details. Infrared and visible image fusion (IVIF) aims to combine meaningful and complete information from images captured by visible and infrared sensors. As a result, the generated image contains richer information and is more favorable for subsequent computer vision tasks. IVIF techniques have been widely applied in object tracking and detection, urban security, and vehicle navigation [2,32].

In the past decades, many methods have been developed to fuse IR-VI images. Typical traditional methods are multi-scale transform-based (MST) methods and representation learning-based (RL) approaches. MST methods [25] use a specific transformation model to extract multi-scale features and manually design rules to fuse images. RL approaches include sparse representation (SR) [29], joint

Q. Liu et al. (Eds.): PRCV 2023, LNCS 14434, pp. 77–88, 2024.
https://doi.org/10.1007/978-981-99-8549-4_7

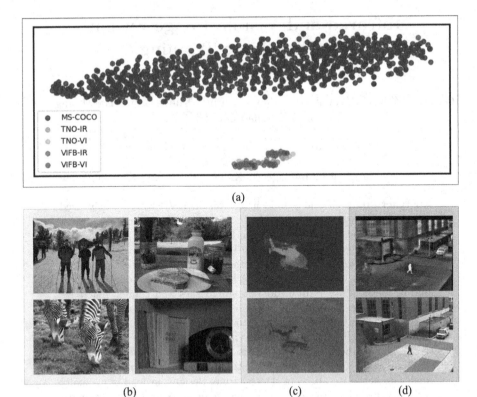

Fig. 1. (a) Visualization of image features using t-SNE [18]. Image features are generated by ResNet18 pre-trained on ImageNet [3]. These features are clustered in different centers, indicating an apparent discrepancy between the distribution of MS-COCO [11] and IVIF datasets. (b) The examples in MS-COCO. (c) and (d) are examples of IR and VI images in the TNO [24] dataset and VIFB [33] dataset, respectively.

sparse representation (JSR) [31], low-rank representation (LRR) [14], and latent low-rank representation (LatLRR) [13]. Similar to MST methods, the fusion rules in RL approaches often require manual design, which may degrade the fusion performance because source images are complex and diverse.

Recently, various deep learning-based solutions for IVIF have been presented, such as auto-encoder-based (AE) methods [6–8,26], convolutional neural network-based (CNN) methods [10], generative adversarial network-based (GAN) approaches [5], and transformer-based [23] solutions. Deep learning-based methods leverage the powerful nonlinear fitting ability of deep neural networks to make the fused images have the desired distribution. As a result, deep learning-based methods provide promising results against traditional approaches. However, deep learning-based approaches still struggle with domain shift problems, especially for AE-based solutions that train the fusion network on MS-COCO and test the model on IR-VI images. The MS-COCO and IR-VI datasets have different characters, and the domain distribution difference between them is significant. For demonstration, we visualize the domain shift in Fig. 1.

Test-time training (TTT) techniques [22] are proposed to deal with the domain shift problem and promote the model performance for each test instance. TTT updates the network parameters before predicting. The optimization is based on specific self-supervised loss or tasks. For example, Liu et al. [15] proposed a TTT strategy that employs an auxiliary network to help the dehazing model better adapt to the domain of interest. In [4], masked auto-encoders were explored to address the one-shot learning problem, and this method improved generalization performance in various vision benchmarks.

In this paper, we propose a self-supervised TTT method that updates model parameters during the testing phase to improve the AE-based methods for IVIF. Concretely, we propose a new self-supervised loss based on a mutual attention mechanism to guide the network optimization in the test time. Moreover, we also propose a fusion adapter to automatically learn fusion rules instead of manually designing fusion strategies. Our contributions can be summarized as follows:

(1) We apply TTT strategies to improve the generalization performance of AE-based IVIF methods. A mutual attention mechanism loss is designed for self-supervised optimization.
(2) We propose a fusion adapter to learn and fuse features from different source images adaptively.
(3) Extensive experiments on two public IVIF datasets demonstrate that with a small number of iterations, the proposed method outperforms the state-of-the-art methods.

2   Method

2.1   Overall Framework

As illustrated in Fig. 2, the AE-based algorithms learn feature representations with RGB images in a self-supervised manner. Handcrafted fusion strategies are adopted in the test time to fuse IR-VI images. Note that both the encoder and the decoder are fixed during the test time. Obviously, AE-based solutions have poor generation performance because significant domain gaps exist between training and testing data. In contrast, the proposed self-supervised TTT method updates network parameters for each test sample to generate better fusion images. Besides, rather than manually design fusion strategies, we design an adapter to fuse two images. The whole network, including the encoder, the decoder, and the adapter, is optimized end-to-end during the test-time training with few iterations. Note that this paper does not focus on designing sophisticated network structures. The encoder and decoder can be any existing well-designed models.

2.2   Training and Testing

Training-Time Training. Assuming that we have a collection of large-scale dataset with training instance $X_1, X_2, ..., X_n$ drawn i.i.d from a distribution P.

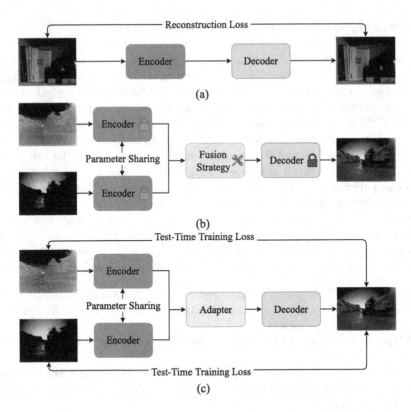

Fig. 2. Illustration of the difference between the existing AE-based IVIF methods and the proposed TTT method. (a) The AE-based solutions train the fusion network on MS-COCO and (b) test the network on IR-VI images by involving a handcrafted fusion strategy. Instead of performing (b), we propose (c) a self-supervision test-time training loss for updating the model parameters, and use a fusion adapter module to extract and fuse features of source images adaptively.

We aim to train an encoder f_θ and decoder g_θ to learn effective feature representations. Specifically, we train the encoder and decoder using a self-supervised method by minimizing the following commonly used loss function $\mathcal{L}_{tr}$:

$$g'_\theta, f'_\theta = \arg\min_{f_\theta, g_\theta} \frac{1}{n} \sum_{i=1}^{n} \mathcal{L}_{tr}(X_i, g_\theta(f_\theta(X_i))), \tag{1}$$

$$\mathcal{L}_{tr} = \mathcal{L}_{pixel} + \mathcal{L}_{SSIM}, \tag{2}$$

$$\mathcal{L}_{pixel} = \left\| X_i - \hat{X}_i \right\|_F^2, \tag{3}$$

$$\mathcal{L}_{SSIM} = 1 - SSIM(X_i, \hat{X}_i), \tag{4}$$

$$\hat{X}_i = g_\theta(f_\theta(X_i))), \tag{5}$$

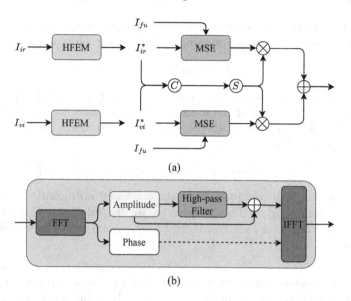

(a)

(b)

Fig. 3. (a) Overview of the proposed loss function. (b) The high-frequency enhanced module (HFEM). C refers to the concatenation operation, and S denotes the softmax function.

where $\mathcal{L}_{pixel}$ indicates the pixel level loss, $\mathcal{L}_{SSIM}$ denotes the structure similarity loss. $\|\cdot\|_F$ indicates the Frobenius norm. X_i and $\hat{X}_i$ denote the input and reconstructed images, respectively. SSIM [27] represents the structure similarity measurement.

Test-Time Training. Let I refers to the IR-VI paired dataset with test samples $(I_{ir}^1, I_{vi}^1), ..., (I_{ir}^m, I_{vi}^m)$ drawn i.i.d from a distribution Q. Since the distribution Q may significantly different with P, we develop a self-supervised loss function to reduce the domain gap and promote the fusion performance. On each test input (I_{ir}, I_{vi}), we perform test-time training to minimize the following loss:

$$g_\theta^*, h_\theta^*, f_\theta^* = \arg \min_{g_\theta', h_\theta', f_\theta'} \mathcal{L}_{TTT}(I_{ir}, I_{vi}, I_{fu}), \qquad (6)$$

where $g_\theta^*, h_\theta^*, f_\theta^*$ denotes the optimized network that will be used to generate the final result. I_{fu} is the initial fusion result that obtained via the pre-trained encoder f_θ', decoder g_θ', and an initial adapter h_θ:

$$I_{fu} = g_\theta'(h_\theta(f_\theta'(I_{ir}), f_\theta'(I_{vi}))), \qquad (7)$$

To combine meaningful information of source images, the self-supervised reconstruction loss L_{TTT} is calculated based on enhanced version of source images. As shown in Fig. 3, original IR and VI images are first inputted to a high-frequency enhanced module (HFEM) to calculate enhanced versions I_{ir}^*

and I_{vi}^*. Then, we propose a mutual attention mechanism that concatenates the I_{ir}^* and I_{vi}^* to generate two attention masks, which can be expressed as:

$$m_1, m_2 = Softmax(Concate(I_{ir}^*, I_{vi}^*)), \tag{8}$$

These two attention masks force the network to pay more attention to the high-frequency information of the source images, and thus the fusion result will contain rich texture details. With the guidance of attention masks, our L_{TTT} can be defined as follows:

$$\mathcal{L}_{TTT}(I_{ir}, I_{vi}, I_{fu}) = m_1 \cdot \|I_{fu} - I_{ir}^*\|_F^2 + m_2 \cdot \|I_{fu} - I_{vi}^*\|_F^2, \tag{9}$$

Note that, for each test pairs, the gradient-based optimization for Eq. (6) always starts from f_θ', g_θ', and h_θ. Same as [4], we discard g_θ^*, h_θ^*, f_θ^* after making a reconstructed result on each test input (I_{ir}, I_{vi}), and reset the weights to f_θ', g_θ', and h_θ.

Adapter. Instead of manually designing fusion rules, the proposed method develops a learnable adapter to fuse features of two source images. This is a benefit for our TTT framework, that allows the network to update parameters in the test time. The calculation of the proposed adapter h_θ is defined as follow:

$$H = Conv(ReLU(Conv(I_{co}))), \tag{10}$$

$$I_{co} = Concat(f_\theta'(I_{ir}), f_\theta'(I_{vi})))), \tag{11}$$

where H is the output feature of the adapter, I_{co} is the concatenate of features generated by the encoder, $Conv(\cdot)$ is a $2D$ convolutional layer, $ReLU(\cdot)$ is the nonlinear activation function. Our adapter has two convolutional layers and one activation function layer. During the TTT process, the adapter parameters will be updated along with the encoder and decoder to further adapt to the data distribution of the test samples.

3 Experiments

3.1 Experiment Configuration

Baseline Model. In this work, we focus on the problem of domain shifts in IVIF, especially for AE-based solutions. Therefore, we select several representation AE-based methods as our baseline models. Specifically, we choose DenseFuse [6] as our default baseline model. We also conduct experiments on other AE-based solutions [7,26] to verify the effectiveness of our proposed method in the ablation study.

Training Dataset. During the training-time training step, similar to existing AE-based networks, the encoder and the decoder are pre-trained on MS-COCO [11] dataset, which consists of more than 80,000 daily RGB images. All images are converted into grayscale versions and resized to 256×256.

Testing Datasets. This study validates the proposed method on two publicly IVIF datasets. A total of 42 pairs of images are utilized. In the TNO test dataset [24], there are 21 infrared and visible image pairs of military scenes. The remaining 21 pairs are from the VIFB [33] dataset with diverse environments, including indoor, outdoor, low-light, and over-exposure scenes.

Implementation Details. During the training-time training process, we use the same epoch, batch size, learning rate, and optimizer as in the DenseFuse [6]. For the test-time training, Adam is used as an optimizer with a learning rate of 1e-4. We update the network with five iterations for each test sample during test-time training. All the experiments are conducted on the Pytorch platform, using a GeForce RTX 3090 GPU with 24GB memory.

Evelution Metrics. For quantitative analysis, we adopt five metrics to evaluate our method and other comparative approaches, including mutual information (MI) [19], information entropy (EN) [20], sum of the correlations of differences (SCD) [1], multiscale structural similarity (MS-SSIM) [28], and visual information fidelity (VIFF) [21].

Comparison Methods. To verify the effectiveness of our proposed method, we compared the proposed with nine existing state-of-the-art algorithms, including two traditional methods: gradient transfer fusion (GTF) [16], MDLatLRR [9], and seven deep-learning-based methods: DeepFuse [8], DenseFuse [6], Nest-Fuse [7], U2Fusion [30], DDcGAN [17], Res2Fusion [26] and TLCM [12].

3.2 Performance Comparison on TNO

The qualitative results on the TNO dataset are shown in Fig. 4. All methods can fuse the infrared image's thermal radiation information and the visible image's structure information, and our proposed method achieves the best visual quality with sharper texture details. Visually, the details of GTF, MDLatLLR, Deep-Fuse, U2Fusion, and TLCM are not clear enough, and the texture details of the landing gear are blurred. The fused image of DDcGAN is ambiguous and suffers from distortion. Although DenseFuse, NestFuse, and Res2Fuse can fuse more thermal radiation information, textures and details are degraded. Our method introduces TTT on the AE-based method, which preserves clear texture information and can better balance the two kinds of information of the source images. The quantitative results on the TNO dataset are shown in Table 1. As can be seen, our method achieves promising performance according to five measurements. Specifically, the comparison between our method and DenseFuse demonstrates that the proposed TTT can significantly improve the fusion performance. Besides, our method achieves the best MS-SSIM and SCD, demonstrating the superiority of the proposed method.

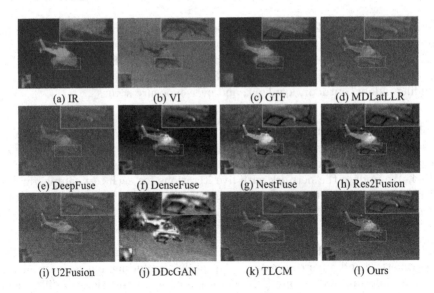

(a) IR   (b) VI   (c) GTF   (d) MDLatLLR

(e) DeepFuse   (f) DenseFuse   (g) NestFuse   (h) Res2Fusion

(i) U2Fusion   (j) DDcGAN   (k) TLCM   (l) Ours

Fig. 4. Qualitative comparison of the proposed method with the state-of-the-art methods on "helicopter" on the TNO dataset. The red and green boxes show the tail and landing gear of the helicopter, respectively. (Color figure online)

Table 1. Comparisons of different methods on the TNO and VIFB dataset. Red indicates the best result, and blue represents the second best result.

Methods	TNO					VIFB				
	EN	MI	MS-SSIM	SCD	VIFF	EN	MI	MS-SSIM	SCD	VIFF
GTF	6.590	13.181	0.812	0.908	0.595	6.545	13.091	0.809	0.802	0.314
DeepFuse	6.438	12.876	0.881	1.473	0.681	6.694	13.388	0.887	1.312	0.399
DenseFuse	6.451	12.902	0.869	1.484	0.686	6.982	13.965	0.919	1.497	0.520
MDLatLLR	6.489	12.979	0.904	1.489	0.710	6.759	13.518	0.913	1.329	0.775
NestFuse	6.930	13.961	0.858	1.578	0.926	6.921	13.841	0.893	1.455	0.861
U2Fusion	6.468	12.936	0.899	1.459	0.681	7.143	14.285	0.907	1.461	0.832
DDcGAN	7.479	14.957	0.777	1.450	0.714	7.484	14.968	0.789	1.291	0.725
Res2Fusion	6.764	13.529	0.866	1.727	0.818	6.851	13.702	0.862	1.372	0.852
TLCM	6.957	13.914	0.917	1.643	0.785	6.947	13.895	0.917	1.458	0.480
Ours	6.796	13.593	0.936	1.887	0.783	7.046	14.092	0.946	1.692	0.825

3.3 Performance Comparison on VIFB

We further evaluate our method on the VIFB dataset. Qualitative results are shown in Fig. 5. As can be observed, for the brightness of the pedestrian in the red box and the detail of the car in the green box, our method has rich thermal radiation information and texture details. In addition, our method can effectively balance the global brightness of the image. The quantitative results on the VIFB dataset are shown in Table 1. The proposed method achieves the best MS-SSIM

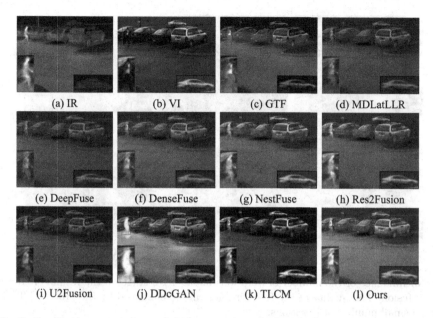

(a) IR   (b) VI   (c) GTF   (d) MDLatLLR

(e) DeepFuse   (f) DenseFuse   (g) NestFuse   (h) Res2Fusion

(i) U2Fusion   (j) DDcGAN   (k) TLCM   (l) Ours

Fig. 5. Qualitative comparison of the proposed method with the state-of-the-art methods on "walking2" on the VIFB dataset. The red and green boxes show the pedestrian and car, respectively. (Color figure online)

Table 2. Ablation studies on the impact of the TTT and adapter on the TNO dataset. The best is marked in bold.

Model	TTT	Adapter	EN	MI	MS-SSIM	SCD	VIFF
DenseFuse	×	×	6.571	13.143	0.788	1.592	**0.940**
	✓	×	6.787	13.574	0.929	1.846	0.782
	✓	✓	**6.797**	**13.593**	**0.936**	**1.887**	0.783
NestFuse	×	×	**6.892**	**13.785**	0.879	1.758	**0.924**
	✓	×	6.796	13.592	**0.929**	1.849	0.787
	✓	✓	6.730	13.459	**0.929**	**1.881**	0.786
Res2Fusion	×	×	6.764	13.529	0.866	1.727	**0.878**
	✓	×	6.770	13.540	0.924	1.842	0.779
	✓	✓	**6.790**	**13.580**	**0.925**	**1.849**	0.790

and SCD scores. Our method can largely improve the performance of DenseFuse with the proposed TTT strategy.

3.4 Ablation Study

The Effectiveness of TTT. The proposed method can be added to most existing AE-based IVIF methods. Here, we choose DenseFuse, NestFuse, and

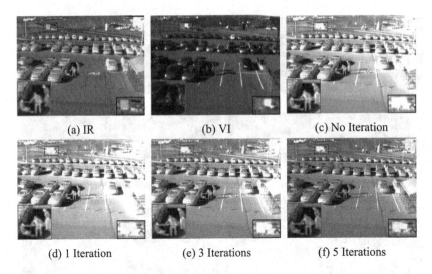

(a) IR   (b) VI   (c) No Iteration

(d) 1 Iteration   (e) 3 Iterations   (f) 5 Iterations

Fig. 6. Influence of the number of iterations in the TTT stage on the results of IR-VI image fusion. The quality of the fused images can be enhanced by the proposed TTT after a small number of iterations.

Res2Fusion as the baseline methods to verify the proposed method. During the training-time training phase, MS-COCO is employed to train the fusion model, and the hyperparameter settings are consistent with each original baseline method. The experiment results of different baseline methods with and without TTT are listed in Table 2. From the table, one can observe that TTT can greatly improve performance.

The Effectiveness of the Fusion Adapter. To understand the role of the adapter, we replace the adapter with feature averaging, denoted by without adapter. Test results are reported in Table 2, which indicates that the proposed adapter is better than handcrafted fusion rules. This is because our adapter is learnable and can handle diverse and complex real-world scenarios.

Ablation Studies of the Iteration. We investigate the impact of the number of iterations in the TTT stage. As shown in Fig. 6, with the iterations increase, the visual quality of the fused image gradually improves. Note that, in our experiments, continually increasing the iterations can obtain different quality outputs. But we found five updates have already achieved promising results. By comprehensively considering the run-time and performance, we set the default iteration times as five.

4 Conclusion

In this paper, we focus on developing test-time training methods to improve the generalization performance of AE-based networks. We propose a self-supervision loss to drive the model parameters updating during the test time. This loss function is based on a mutual attention mechanism. Additionally, a fusion adapter module is proposed to adaptively fuse features of two source images. Extensive experiments and ablation studies strongly support the proposed method.

Acknowledgements. The work was supported in part by the National Natural Science Foundation of China under Grant 82172033, U19B2031, 61971369, 52105126, 82272071, 62271430, and the Fundamental Research Funds for the Central Universities 20720230104.

References

1. Aslantas, V., Bendes, E.: A new image quality metric for image fusion: the sum of the correlations of differences. Aeu-Inter. J. Electr. Commun. **69**(12), 1890–1896 (2015)
2. Das, S., Zhang, Y.: Color night vision for navigation and surveillance. Transp. Res. Rec. **1708**(1), 40–46 (2000)
3. Deng, J., Dong, W., Socher, R., Li, L.J., Li, K., Fei-Fei, L.: Imagenet: a large-scale hierarchical image database. In: 2009 IEEE Conference on Computer Vision and Pattern Recognition, pp. 248–255. Ieee (2009)
4. Gandelsman, Y., Sun, Y., Chen, X., Efros, A.: Test-time training with masked autoencoders. Adv. Neural. Inf. Process. Syst. **35**, 29374–29385 (2022)
5. Gao, Y., Ma, S., Liu, J.: Dcdr-gan: a densely connected disentangled representation generative adversarial network for infrared and visible image fusion. IEEE Trans. Circ. Syst. Video Technol. (2022)
6. Li, H., Wu, X.J.: Densefuse: a fusion approach to infrared and visible images. IEEE Trans. Image Process. **28**(5), 2614–2623 (2018)
7. Li, H., Wu, X.J., Durrani, T.: Nestfuse: an infrared and visible image fusion architecture based on nest connection and spatial/channel attention models. IEEE Trans. Instrum. Meas. **69**(12), 9645–9656 (2020)
8. Li, H., Wu, X.J., Kittler, J.: Infrared and visible image fusion using a deep learning framework. In: 2018 24th International Conference On Pattern Recognition (ICPR), pp. 2705–2710. IEEE (2018)
9. Li, H., Wu, X.J., Kittler, J.: Mdlatlrr: a novel decomposition method for infrared and visible image fusion. IEEE Trans. Image Process. **29**, 4733–4746 (2020)
10. Li, Q., et al.: A multilevel hybrid transmission network for infrared and visible image fusion. IEEE Trans. Instrum. Meas. **71**, 1–14 (2022)
11. Lin, T.-Y., et al.: Microsoft COCO: common objects in context. In: Fleet, D., Pajdla, T., Schiele, B., Tuytelaars, T. (eds.) ECCV 2014. LNCS, vol. 8693, pp. 740–755. Springer, Cham (2014). https://doi.org/10.1007/978-3-319-10602-1_48
12. Lin, X., Zhou, G., Tu, X., Huang, Y., Ding, X.: Two-level consistency metric for infrared and visible image fusion. IEEE Trans. Instrum. Meas. **71**, 1–13 (2022)
13. Liu, G., Lin, Z., Yan, S., Sun, J., Yu, Y., Ma, Y.: Robust recovery of subspace structures by low-rank representation. IEEE Trans. Pattern Anal. Mach. Intell. **35**(1), 171–184 (2012)
14. Liu, G., Lin, Z., Yu, Y.: Robust subspace segmentation by low-rank representation. In: Proceedings of the 27th International Conference on Machine Learning (ICML 2010), pp. 663–670 (2010)

15. Liu, H., Wu, Z., Li, L., Salehkalaibar, S., Chen, J., Wang, K.: Towards multi-domain single image dehazing via test-time training. In: Proceedings of the IEEE/CVF Conference on Computer Vision and Pattern Recognition, pp. 5831–5840 (2022)
16. Ma, J., Chen, C., Li, C., Huang, J.: Infrared and visible image fusion via gradient transfer and total variation minimization. Inform. Fusion **31**, 100–109 (2016)
17. Ma, J., Xu, H., Jiang, J., Mei, X., Zhang, X.P.: Ddcgan: a dual-discriminator conditional generative adversarial network for multi-resolution image fusion. IEEE Trans. Image Process. **29**, 4980–4995 (2020)
18. Van der Maaten, L., Hinton, G.: Visualizing data using t-sne. J. Mach. Learn. Res. **9**(11) (2008)
19. Piella, G.: A general framework for multiresolution image fusion: from pixels to regions. Inform. Fusion **4**(4), 259–280 (2003)
20. Roberts, J.W., Van Aardt, J.A., Ahmed, F.B.: Assessment of image fusion procedures using entropy, image quality, and multispectral classification. J. Appl. Remote Sens. **2**(1), 023522 (2008)
21. Sheikh, H.R., Bovik, A.C.: Image information and visual quality. IEEE Trans. Image Process. **15**(2), 430–444 (2006)
22. Sun, Y., Wang, X., Liu, Z., Miller, J., Efros, A., Hardt, M.: Test-time training with self-supervision for generalization under distribution shifts. In: International Conference on Machine Learning, pp. 9229–9248. PMLR (2020)
23. Tang, W., He, F., Liu, Y.: Ydtr: infrared and visible image fusion via y-shape dynamic transformer. IEEE Trans. Multimedia (2022)
24. Toet, A.: The tno multiband image data collection. Data Brief **15**, 249–251 (2017)
25. Vishwakarma, A.: Image fusion using adjustable non-subsampled shearlet transform. IEEE Trans. Instrum. Meas. **68**(9), 3367–3378 (2018)
26. Wang, Z., Wu, Y., Wang, J., Xu, J., Shao, W.: Res2fusion: infrared and visible image fusion based on dense res2net and double nonlocal attention models. IEEE Trans. Instrum. Meas. **71**, 1–12 (2022)
27. Wang, Z., Bovik, A.C., Sheikh, H.R., Simoncelli, E.P.: Image quality assessment: from error visibility to structural similarity. IEEE Trans. Image Process. **13**(4), 600–612 (2004)
28. Wang, Z., Simoncelli, E.P., Bovik, A.C.: Multiscale structural similarity for image quality assessment. In: The Thrity-Seventh Asilomar Conference on Signals, Systems & Computers 2003, vol. 2, pp. 1398–1402. IEEE (2003)
29. Wright, J., Yang, A.Y., Ganesh, A., Sastry, S.S., Ma, Y.: Robust face recognition via sparse representation. IEEE Trans. Pattern Anal. Mach. Intell. **31**(2), 210–227 (2008)
30. Xu, H., Ma, J., Jiang, J., Guo, X., Ling, H.: U2fusion: a unified unsupervised image fusion network. IEEE Trans. Pattern Anal. Mach. Intell. **44**(1), 502–518 (2020)
31. Zhang, Q., Fu, Y., Li, H., Zou, J.: Dictionary learning method for joint sparse representation-based image fusion. Opt. Eng. **52**(5), 057006–057006 (2013)
32. Zhang, X., Demiris, Y.: Visible and infrared image fusion using deep learning. IEEE Trans. Pattern Anal. Mach. Intell. (2023)
33. Zhang, X., Ye, P., Xiao, G.: Vifb: a visible and infrared image fusion benchmark. In: Proceedings of the IEEE/CVF Conference on Computer Vision and Pattern Recognition Workshops, pp. 104–105 (2020)

Graph-Based Dependency-Aware Non-Intrusive Load Monitoring

Guoqing Zheng[1], Yuming Hu[2], Zhenlong Xiao[1,2]([✉]), and Xinghao Ding[1,2]

[1] School of Informatics, Xiamen University, Xiamen, China
zlxiao@xmu.edu.cn
[2] Institute of Artificial Intelligence, Xiamen University, Xiamen, China

Abstract. Non-intrusive load monitoring (NILM) is able to analyze and predict users' power consumption behaviors for further improving the power consumption efficiency of the grid. Neural network-based techniques have been developed for NILM. However, the dependencies of multiple appliances working simultaneously were ignored or implicitly characterized in their models for disaggregation. To improve the performance of NILM, we employ a graph structure to explicitly characterize the temporal dependencies among different appliances. Specially, we consider the prior temporal knowledge between the appliances in the working state, construct a weighted adjacency matrix to represent their dependencies. We also introduce hard dependencies of each appliance to prevent the sparsity of the weighted adjacency matrix. Furthermore, the non-sequential dependencies are learned among appliances using a graph attention network based on the weighted adjacency matrix. An encoder-decoder architecture based on dilated convolutions is developed for power estimation and state detection at the same time. We demonstrate the proposed model on the UKDALE dataset, which outperforms several state-of-the-art results for NILM.

Keywords: Non-intrusive load monitoring · Graph structure learning · Graph neural network · Time series

1 Introduction

With the development of clean energy such as wind power and photovoltaic, power consumption capacity has attracted lots of interest in literature, where load monitoring is an important technique since it is able to provide reasonable consumption suggestions to improve power consumption efficiency by analysing and predicting users' power consumption behaviours. Intrusive load monitoring requires a large number of sensors to collect data from various appliances, leading to very expensive hardware costs in real applications. Non-intrusive load monitoring [12] can identify and extract the power consumption of a single appliance through the aggregated total power consumption of the household, showing advantages of low cost and users' privacy preservation [1].

As shown in Fig. 1, NILM is a computational approach to achieve power estimation and state detection of individual appliances from a single aggregated

© The Author(s), under exclusive license to Springer Nature Singapore Pte Ltd. 2024
Q. Liu et al. (Eds.): PRCV 2023, LNCS 14434, pp. 89–100, 2024.
https://doi.org/10.1007/978-981-99-8549-4_8

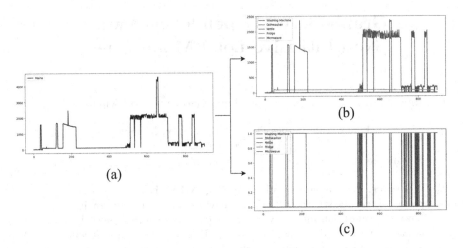

Fig. 1. The purpose of NILM is to separate the power demand of individual appliances from the aggregated signal and achieve state estimation. (a) The aggregated power signal, (b) the power of individual appliances for power estimation, and (c) the state of individual appliances for state detection.

meter in a smart grid system. Note that the energy consumption behaviours of various appliances are not often independent of each other. Many of them even have similar power consumption characteristics. Moreover, the aggregated signal may contain complicated noise, and appliances may be arbitrarily switched ON/OFF. These all bring difficulties in NILM [7].

HMM-based [16,17] and optimization-based [6,18] methods were developed for the NILM problem. However, they strongly rely on domain knowledge, and manual extraction may be required [11]. Deep neural networks (DNNs) were applied to learn features for NILM automatically, e.g., convolutional neural networks (CNNs) were studied in [13,21,23,24]. To further investigate the dependencies of power consumption sequence, long short-term memory networks (LSTMs) [5,19] were employed. Recently, graph neural networks (GNNs) [2,15] has been applied to NILM to capture the correlation information between different states. However, these methods only focus on a single appliance, where each appliance trains a separate DNN model. This approach is computationally expensive and fails to consider the dependencies among multiple appliances. To address the above issue, multi-task-multi-appliances NILM problem have been considered in [9], power estimation and state detection are conducted simultaneously on multiple appliances. In [4], causal convolutions are introduced to improve the performance further. However, they are implicitly characterized in the feature space.

Due to the superior ability of graph representations learning, graph structure learning has received extensive attention and been widely used in multiple time series forecasting [20], traffic forecasting [25], anomaly detection [3], etc. GNNs have powerful feature learning abilities based on the graph structure. Therefore,

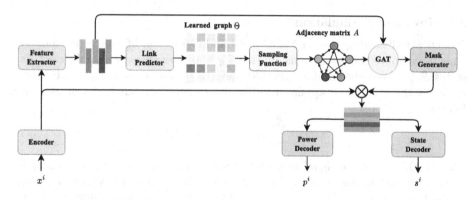

Fig. 2. Architecture of GRAD-NILM for non-intrusive load monitoring.

we consider the dependencies between appliances working simultaneously as the relationship among graph nodes, and propose the GRAph-based Dependency-aware method for NILM (GRAD-NILM). Especially, We characterize the temporal dependencies between appliances as a directed weighted adjacency matrix, i.e., the co-occurrence probability graph, where the dependencies of appliances are explicitly constructed. Then, graph attention network (GAT) [22] is exploited to extract the non-sequential dependencies between appliances. The features learned by GAT are used to generate masks for different appliances to reconstruct the aggregated signals. Based on the shared masks, both the power estimation and state detection can be performed by the proposed GRAD-NILM simultaneously. The main contributions of this paper are summarized as:

(1) The co-occurrence probability graph is explicitly constructed to learn the temporal dependencies of appliances, and the non-sequential dependencies are learned via GAT, which can facilitate the NILM for multiple appliances.
(2) Appliances are represented by different masks, and the mask feature spaces are shared between different appliances in the GRAD-NILM model to improve the power disaggregation performance.
(3) Compared with the state-of-the-art multi-tasks-multi-appliances methods, it does not need to perform quantile regression processing [9] on the dataset for Grad-NILM, but performs better.

The rest of this paper is organized as follows. Section 2 presents the main results, and numerical studies are discussed in Sect. 3. Finally, Sect. 4 concludes this paper.

2 Proposed Method

In the following, we first introduce the problem formulation of the multi-task-multi-appliances NILM task. Then, we present the proposed GRAD-NILM, which consists of graph structure learning, graph attention network-based representation learning, and the encoder-decoder module for NILM tasks. An overview of the proposed architecture is shown in Fig. 2.

2.1   Problem Formulation

The energy disaggregation in NILM is a blind separation of source problem. It aims to estimate the power consumption $\mathcal{P}_n(t)$ of an appliance n and detect its states $\mathcal{S}_n(t)$ from the total aggregated meter $\mathcal{X}(t)$ for $n = 1, 2, ...N$, where N indicates the number of appliances. The NILM problem can be formulated as follows:

$$\mathcal{X}(t) = \sum_{n=1}^{N} \mathcal{P}_n(t) \cdot \mathcal{S}_n(t) + \varepsilon(t), \tag{1}$$

where $\varepsilon(t)$ is the noise signal at time t, $t = 1, 2, ..., T$, and T is the total length of data. By setting a sliding window with a length of L and a step size of β, the original data $\mathcal{X} \in \mathbb{R}^{1 \times T}$, $\mathcal{P}_n \in \mathbb{R}^{1 \times T}$, and $\mathcal{S}_n \in \mathbb{R}^{1 \times T}$ are segmented to obtain X, S_n and P_n with a total number of M, where $X \in \mathbb{R}^{M \times L} = \{x^1, ..., x^M\}$, $P_n \in \mathbb{R}^{M \times L} = \{p_n^1, ..., p_n^M\}$ and $S_n \in \mathbb{R}^{M \times L} = \{s_n^1, ..., s_n^M\}$. Notably, $x^\ell \in \mathbb{R}^{1 \times L}$, $p_n^\ell \in \mathbb{R}^{1 \times L}$ and $s_n^\ell \in \mathbb{R}^{1 \times L}$. For model predicting, given a input data $x^\ell(t)$, the goal is to calculate the power signal $\hat{p}^\ell(t) = \{\hat{p}_1^\ell(t), ..., \hat{p}_N^\ell(t)\}$ and the corresponding states $\hat{s}^\ell(t) = \{\hat{s}_1^\ell(t), ..., \hat{s}_N^\ell(t)\}$.

2.2   Co-occurrence Probability Graph

Notations. A directed graph without self-loop edges is defined as $G = (V, E)$, where V is the node set of G, and E is the edge set. $A \in \mathbb{R}^{N \times N}$ denotes the weighted adjacency matrix of G, and the (i, j)-th entry A_{ij} indicates the dependency between node i and node j. The size N of the adjacency matrix equates to the number of appliances because of the dependencies among them, and every node in G corresponds a single appliance.

The prior knowledge of appliances' working states $s^\ell = \{s_1^\ell(t), ..., s_N^\ell(t)\}$ will be applied to construct the co-occurrence probability graph for characterizing the temporal dependencies of appliances. The co-occurrence probability graph is defined by a weighted adjacency matrix $\Theta^* \in \mathbb{R}^{N \times N}$ with its entries denoted by

$$\theta_{ij}^* = \begin{cases} \frac{\Gamma_i^e - \Gamma_j^b}{L}, & i \neq j, \\ 0, & i = j, \end{cases} \tag{2}$$

where L is the length of the time series window, $\Gamma_i^e \in \mathbb{R}^{1 \times L}$ represents the end of the working time instance of the i-th appliance, and $\Gamma_j^b \in \mathbb{R}^{1 \times L}$ is the beginning time of the j-th appliance. Since the co-occurrence probability graph generated by Eq. (2) is sparse, the hard dependency between appliances i and j is considered, we introduce a new adjacency matrix $\Theta' \in \mathbb{R}^{N \times N}$ with its entries defined as

$$\theta_{ij}' = \begin{cases} 1, if \ \ \Gamma_i^e > 0 \ \ and \ \ i \neq j, \\ 0, \qquad\quad i = j. \end{cases} \tag{3}$$

Then, the ultimate co-occurrence probability graph $\hat{\Theta} \in \mathbb{R}^{N \times N}$ can be obtained by the weighted adjacency matrix as

$$\hat{\Theta} = R(Q(\Theta') + Q(\Theta^*)), \tag{4}$$

where $Q(\cdot)$ is the quantile threshold function used to filter out the abnormal instantaneous impulse response, and $R(r) = sigmoid(r-1)$ is a rescaling trick function. The co-occurrence probability graph $\hat{\Theta}$ characterizes the relationships between the co-working appliances. Such a representation of dependencies would be able to facilitate the graph neural network-based NILM.

2.3 Graph Structure Learning

As shown in Fig. 2, the adjacency matrix A can be obtained by sampling over the learned graph Θ, where the sampling function is a differentiable function to output discrete values between 0 and 1. Matrix A can be randomly parameterized by Bernoulli distribution, that is, for each entry A_{ij} is independent, we have $A_{ij} \sim Ber(\Theta_{ij})$. However, the gradient calculated by Bernoulli sampling may make the back propagation-based end-to-end training unavailable. Hence, the Gumbel reparameterization strategy [14] is further employed by defining the parametrization of A as

$$A_{ij} = \sigma(log(\frac{\Theta_{ij}}{1-\Theta_{ij}}) + \frac{g_{ij}^1 - g_{ij}^2}{\tau}), \tag{5}$$

where $\sigma(\cdot)$ is the sigmoid function, $g_{ij}^1, g_{ij}^2 \sim Gumbel(0,1)$ for all i, j. τ is a temperature coefficient. When $\tau \to 0$, $A_{ij} = 1$ holds with probability Θ_{ij}, otherwise, $A_{ij} = 0$ with the other probabilities.

To calculate graph Θ, we employ a feature extractor consisting of 1D convolution layers to generate a node feature matrix $H \in \mathbb{R}^{N \times D}$, $H = \{h_1, ..., h_N\}$ where h_i denotes the feature vector of appliance i with feature dimension D. Then, a predictor composed of two fully connected layers is applied to generate a link probability Θ_{ij} using the pair of (h_i, h_j) .

2.4 Graph Attention Neural Network

Driven by graph neural networks and attention mechanisms, GAT can assign different weights to the adjacent nodes and would thus be more effective for directed graph learning. In this section, GAT-based representation learning is built to extract the dependencies between different appliances.

In this paper, the attention score between appliances v_i and v_j in GAT can be computed by

$$\alpha_{ij} = \frac{exp(\phi(a^T[Wh_i\|Wh_j]))}{\sum_{k \in \mathcal{N}_i} exp(\phi(a^T[Wh_i\|Wh_k]))}, \tag{6}$$

where $h_i \in H$ is the feature vector of appliance v_i, $\|$ is the concatenation operation, and $\phi(\cdot)$ is the LeakyReLU nonlinear function. $\mathcal{N}_i$ indicates the neighborhood set of appliance v_i in graph G, W is a linear weight matrix, and a is the parameter of the fully connected layer to calculate the dependency between appliances v_i and v_j.

The coefficient h'_i can be obtained by computing the node-level average of the K-head GAT, i.e.,

$$h'_i = \varphi(\frac{1}{K}\sum_{k=1}^{K}\sum_{j\in\mathcal{N}_i}\alpha_{ij}^k \boldsymbol{W}^k h_j), \qquad (7)$$

where $\varphi(\cdot)$ is the RELU nonlinear function, and $\boldsymbol{W}^k$ is the k-th weight matrix. h'_i aggregates the feature information of neighboring nodes through GAT. As a result, the operation of Eq. (7) is performed on each node in the graph to obtain the graph-level feature representation H'. Such non-sequential dependencies of different appliances may contribute to the model generalization for NILM.

2.5 Encoder-Decoder Module

In order to embed the original signal into the feature space, we use an encoder to generate the feature $F \in \mathbb{R}^{L \times D}$ from the aggregated signal $x^i \in \mathbb{R}^{1 \times L}$. The sequential dependencies of different appliances are characterized by enlarging the receptive field using dilated convolution layers in the encoder. The encoder is expressed as

$$F = \phi(x^i \cdot E^\top), \qquad (8)$$

where $E \in \mathbb{R}^{D \times 1}$ is the dilated convolution layers parameter of the encoder, $(\cdot)^\top$ is the transpose operation, and $\phi(\cdot)$ is the PReLU nonlinear function. A mask generator composed of 1D convolution layers is used to estimate the N masks $m_i \in \mathbb{R}^{L \times D}$, $i = 1, ..., N$, where the main purpose is to extract each appliance in the mixed feature from the GAT output H'. By applying the masks m_i to the mixture representation F, the reconstructed representation for appliances $z_i \in \mathbb{R}^{L \times D}$ can be defined as

$$z_i = F \odot m_i, \qquad (9)$$

where $\odot$ denotes the element-wise multiplication. z_i can be decoded using two different parameter-sharing decoders as

$$\hat{s^i} = z_i \cdot \Omega, \qquad (10)$$

$$\hat{p^i} = z_i \cdot \Lambda, \qquad (11)$$

where $\Lambda, \Omega \in \mathbb{R}^{N \times L}$ are the deconvolution filters of the decoder based on dilated convolution layers, $\hat{p^i} = \{\hat{p_1^i}, ..., \hat{p_N^i}\}$, and $\hat{s^i} = \{\hat{s_1^i}, ..., \hat{s_N^i}\}$. Therefore, $\hat{p^i} \in \mathbb{R}^{N \times L}$ and $\hat{s^i} \in \mathbb{R}^{N \times L}$ are the final results of power estimation and state detection from the input aggregated power signal x^i, respectively.

3 Numerical Studies and Discussions

3.1 Dataset and Experiment Setup

Dataset. The proposed method is evaluated on the public UKDALE [10] dataset. The UKDALE dataset comprises sub-metered power consumption readings with a resolution of 1/6 Hz and aggregate power consumption readings with a resolution of 1 Hz that were gathered from five residential buildings in the UK. In this experiment, we use the dataset of House 1 from January to March 2015, select kettle (KT), fridge (FRZ), dishwasher (DW), washing machine (WM) and microwave (MW) for disaggregation. The data preprocessing steps are the same as the configuration in [9], the total aggregated data is obtained by adding the sum of the selected appliances with several other additional appliances.

Experimental Setup. We set the total duration time T from January to March 2015, the input sequence length $L = 128$ of the sliding window, and the sliding step $\beta = 32$ for the UKDALE dataset. In graph representation learning, the number of graph nodes $N = 5$ indicates the dependencies among five appliances, the feature dimension of each node $D = 15$, and the multi-head attention $K = 2$ for GAT. We train the model with 80 epochs, use Adam as the optimizer, and set the initial learning rate to be 1e-3. The loss function for GRAD-NILM is defined by

$$\mathcal{L} = \mathcal{L}_{MSE}(p^i, \hat{p^i}) + \lambda * \mathcal{L}_{CE}(s^i, \hat{s^i}) + \mu * \mathcal{L}_{CE}(\Theta, \hat{\Theta}), \tag{12}$$

$$\mathcal{L}_{MSE}(p^i, \hat{p^i}) = \frac{1}{LN} \sum_{l=1}^{L} \sum_{n=1}^{N} (p_n^i(l) - \hat{p_n^i}(l))^2, \tag{13}$$

$$\mathcal{L}_{CE}(s^i, \hat{s^i}) = -\frac{1}{LN} \sum_{l=1}^{L} \sum_{n=1}^{N} s_n^i(l) \cdot log \frac{exp(\hat{s_n^i}(l))}{\sum_k^2 exp(\hat{s_n^i}(l))}, \tag{14}$$

$$\mathcal{L}_{CE}(\Theta, \hat{\Theta}) = - \sum_{n=1}^{N \times N} (\Theta_n log\hat{\Theta}_n + (1 - \Theta_n)log(1 - \hat{\Theta}_n)), \tag{15}$$

where $\lambda = 0.1$, $\mu = 0.1$, and each appliance has $k \in \{0, 1\}$ meaning the state ON and OFF.

Comparison Method. We compare our model to several state-of-the-art methods that were also developed for multi-task-multi-appliance NILM disaggregation on the same dataset: (1) 1D-CNN. The model based on the CNN architecture presented in [8] consists of a fourstage CNN layer and three MLP layers. (2) Unet-NILM [9]. A one-dimensional CNN based on the U-Net architecture, which consider dependencies among multiple appliances. (3) Conv-Tas-NILM [4]. A fully convolutional framework for end-to-end NILM. It is a causal model for multi appliance source separation.

Table 1. Comparison with the state-of-the-art methods on the UKDALE dataset. ↑ stands for the higher the better, while ↓ stands for the lower the better. Best results are shown in bold.

Metric	Model	WM	DW	KT	FRZ	MW	Average
MAE ↓	1D-CNN	14.34	9.45	7.73	7.31	11.94	9.55
	Unet-NILM	13.01	10.90	6.21	11.94	5.05	9.42
	Conv-NILM-Net	10.08	**2.58**	5.51	6.11	5.96	6.05
	GRAD-NILM(Ours)	**9.99**	3.10	**1.59**	**5.73**	**2.74**	**4.63**
NDE ↓	1D-CNN	0.08	0.08	0.20	0.11	0.51	0.20
	Unet-NILM	0.07	0.06	0.15	0.23	0.23	0.15
	Conv-NILM-Net	**0.05**	**0.02**	0.16	**0.10**	0.36	0.09
	GRAD-NILM (Ours)	0.06	0.04	**0.03**	**0.10**	**0.14**	**0.07**
EAC ↑	1D-CNN	0.86	0.89	0.84	0.90	0.65	0.83
	Unet-NILM	0.87	0.87	0.87	0.84	0.80	0.85
	Conv-NILM-Net	**0.90**	**0.97**	0.88	**0.92**	0.76	0.89
	GRAD-NILM (Ours)	**0.90**	0.96	**0.97**	**0.92**	**0.89**	**0.93**
F1-Score ↑	1D-CNN	**0.90**	0.83	0.92	0.94	0.70	0.86
	Unet-NILM	**0.90**	0.76	0.94	0.91	0.87	0.88
	Conv-NILM-Net	**0.90**	**0.97**	0.83	0.94	0.78	0.88
	GRAD-NILM (Ours)	**0.90**	0.81	**0.98**	**0.95**	**0.91**	**0.91**

3.2 Metrics and Comparisons

Metrics. In this work, we evaluate the effectiveness of the proposed model quantitatively using both regression and classification metrics. Three standard metrics were used for regression evaluation, including the mean absolute error (MAE), estimated accuracy (EAC), and normalized disaggregation error (NDE). MAE shows how well models perform power consumption estimation in NILM and calculates it as

$$\mathcal{L}_{MSE}(p^i, \hat{p^i}) = \frac{1}{LN} \sum_{l=1}^{L} \sum_{n=1}^{N} \left| p_n^i(l) - \hat{p_n^i}(l) \right|, \tag{16}$$

EAC is a common metric for evaluationg disaggregated power and gives the total estimated accuracy defined as

$$EAC = 1 - \frac{\sum_{l=1}^{L} \sum_{n=1}^{N} \left| p_n^i(l) - \hat{p_n^i}(l) \right|}{2 \sum_{l=1}^{L} \sum_{n=1}^{N} p_n^i(l)}, \tag{17}$$

The NDE metric quantifies the normalized error between the predicted value and ground truth by calculating the squared difference, which defined as

$$NDE = \frac{\sum_{l=1}^{L} \sum_{n=1}^{N} (p_n^i(l) - \hat{p_n^i}(l))^2}{\sum_{l=1}^{L} \sum_{n=1}^{N} p_i^n(l)^2}, \tag{18}$$

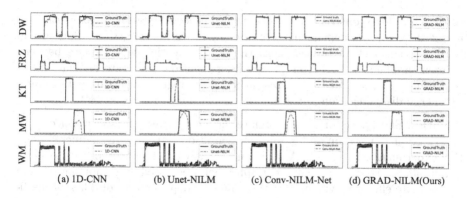

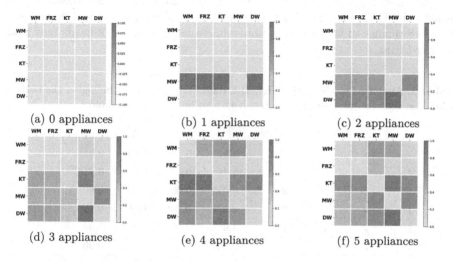

Fig. 3. Examples of disaggregation results one the UKDALE dataset.

(a) 0 appliances

(b) 1 appliances

(c) 2 appliances

(d) 3 appliances

(e) 4 appliances

(f) 5 appliances

Fig. 4. Visualization of the learned graph Θ.

Since the state detection is a label classification task, we use F1-Score to evaluate. The F1-Score averages over all labels and is defined as

$$F1 = \frac{1}{N} \sum_{n=1}^{N} \frac{2 \cdot TP^i}{2 \cdot TP^i + FP^i + FN^i}, \tag{19}$$

where TP^i is true positive, FP^i is false positive and FN^i is false negative. High F1-Score indicates high performance on classification task.

Comparisons. Table 1 presents the results of the comparison between GRAD-NILM and other previous works. As observed, GRAD-NILM outperforms the other methods on the average metrics and has the best performance in most cases. Figure 4 shows the learned structure Θ of six different cases from the test set, i.e., there are 0, 1, 2, 3, 4 and 5 appliances working simultaneously. Our method is able to construct the co-occurrence probability graph to explicitly represent the temporal dependencies of appliances working together, which would facilitate the GAT to further learn the non-sequential dependencies between different appliances for better power disaggregation and state estimation. As shown in Fig. 3, our method shows much better regression fitting performance for energy disaggregation, especially on the datasets of KT and MW.

Ablation Study. To further investigate the effectiveness of co-occurrence probability graph generation and graph representation learning, we conducted ablation experiments. In Table 2, A, B, C and D show different methods to generate graph labels. We can see that the graph labels generated by combining Θ^* and Θ' are more effective, where Θ^* is the temporal dependencies of different appliances based on relative time relationships, and Θ' takes the hard dependencies into account. D, E, F and G show that an appropriate number of heads would benefit the GAT learning. That is, a smaller number of heads will deteriorate the performance of feature extraction, while a larger number of heads may lead to over-fitting.

Table 2. Abalation study on the UKDALE dataset. ↑ stands for the higher the better, while ↓ stands for the lower the better. Best results are shown in bold.

Config	w/ Θ^*	w/ Θ'	K-Head	MAE ↓	NDE ↓	EAC ↑	F1-Score ↑
A	×	×	2	5.57	0.10	0.91	0.91
B	×	✓	2	4.78	0.08	0.93	0.91
C	✓	×	2	5.03	0.09	0.92	0.90
D	✓	✓	2	**4.63**	**0.03**	**0.93**	**0.91**
E	✓	✓	1	4.98	0.08	0.92	0.90
F	✓	✓	4	4.84	0.08	0.92	0.90
G	✓	✓	8	5.01	0.08	0.92	0.90

4   Conclusion

In this paper, we propose the GRAD-NILM by utilizing a graph structure to represent the dependencies between multiple appliances and employing an encoder-decoder architecture for power disaggregation. The co-occurrence probability graph can explicitly model the temporal dependencies of the appliances, and the non-sequential dependencies are further characterized by GAT. Such dependencies of different appliances would help to improve the performance of multi-task-multi-appliance NILM based on the encoder-decoder architecture. Comparative and ablation experiments are presented to demonstrate the efficiency of the proposed GRAD-NILM model.

Acknowledgements. The work was supported in part by the National Natural Science Foundation of China under Grant 62271430, 82172033, U19B2031, 61971369, 52105126, 82272071, and the Fundamental Research Funds for the Central Universities 20720230104.

References

1. Angelis, G.F., Timplalexis, C., Krinidis, S., Ioannidis, D., Tzovaras, D.: NILM applications: literature review of learning approaches, recent developments and challenges. Energy Buildings **261**, 111951 (2022)
2. Athanasoulias, S., Sykiotis, S., Kaselimi, M., Protopapadakis, E., Ipiotis, N.: A first approach using graph neural networks on non-intrusive-load-monitoring. In: Proceedings of the 15th International Conference on PErvasive Technologies Related to Assistive Environments, pp. 601–607 (2022)
3. Dai, E., Chen, J.: Graph-augmented normalizing flows for anomaly detection of multiple time series. arXiv preprint arXiv:2202.07857 (2022)
4. Decock, J., Kaddah, R., Read, J., et al.: Conv-NILM-net, a causal and multi-appliance model for energy source separation. arXiv preprint arXiv:2208.02173 (2022)
5. de Diego-Otón, L., Fuentes-Jimenez, D., Hernández, Á., Nieto, R.: Recurrent LSTM architecture for appliance identification in non-intrusive load monitoring. In: 2021 IEEE International Instrumentation and Measurement Technology Conference (I2MTC), pp. 1–6. IEEE (2021)
6. Elhamifar, E., Sastry, S.: Energy disaggregation via learning powerlets and sparse coding. In: Twenty-Ninth AAAI Conference on Artificial Intelligence (2015)
7. Faustine, A., Mvungi, N.H., Kaijage, S., Michael, K.: A survey on non-intrusive load monitoring methodies and techniques for energy disaggregation problem. arXiv preprint arXiv:1703.00785 (2017)
8. Faustine, A., Pereira, L.: Multi-label learning for appliance recognition in NILM using fryze-current decomposition and convolutional neural network. Energies **13**(16), 4154 (2020)
9. Faustine, A., Pereira, L., Bousbiat, H., Kulkarni, S.: UNet-NILM: a deep neural network for multi-tasks appliances state detection and power estimation in nilm. In: Proceedings of the 5th International Workshop on Non-Intrusive Load Monitoring, pp. 84–88 (2020)
10. Figueiredo, M., Ribeiro, B., de Almeida, A.: Electrical signal source separation via nonnegative tensor factorization using on site measurements in a smart home. IEEE Trans. Instrum. Meas. **63**(2), 364–373 (2013)
11. Gopinath, R., Kumar, M., Joshua, C.P.C., Srinivas, K.: Energy management using non-intrusive load monitoring techniques-state-of-the-art and future research directions. Sustain. Urban Areas **62**, 102411 (2020)
12. Hart, G.W.: Nonintrusive appliance load monitoring. Proc. IEEE **80**(12), 1870–1891 (1992)
13. He, J., et al.: MSDC: exploiting multi-state power consumption in non-intrusive load monitoring based on a dual-CNN model. arXiv preprint arXiv:2302.05565 (2023)
14. Jang, E., Gu, S., Poole, B.: Categorical reparameterization with gumbel-softmax. arXiv preprint arXiv:1611.01144 (2016)

15. Jiao, X., Chen, G., Liu, J.: A non-intrusive load monitoring model based on graph neural networks. In: 2023 IEEE 2nd International Conference on Electrical Engineering, Big Data and Algorithms (EEBDA), pp. 245–250. IEEE (2023)
16. Kim, H., Marwah, M., Arlitt, M., Lyon, G., Han, J.: Unsupervised disaggregation of low frequency power measurements. In: Proceedings of the 2011 SIAM International Conference on Data Mining, pp. 747–758. SIAM (2011)
17. Kolter, J.Z., Jaakkola, T.: Approximate inference in additive factorial HMMs with application to energy disaggregation. In: Artificial intelligence and statistics, pp. 1472–1482. PMLR (2012)
18. Lin, Y.H., Tsai, M.S., Chen, C.S.: Applications of fuzzy classification with fuzzy c-means clustering and optimization strategies for load identification in NILM systems. In: 2011 IEEE International Conference on Fuzzy Systems (FUZZ-IEEE 2011), pp. 859–866. IEEE (2011)
19. Mauch, L., Yang, B.: A new approach for supervised power disaggregation by using a deep recurrent LSTM network. In: 2015 IEEE Global Conference on Signal and Information Processing (GlobalSIP), pp. 63–67. IEEE (2015)
20. Shang, C., Chen, J., Bi, J.: Discrete graph structure learning for forecasting multiple time series. arXiv preprint arXiv:2101.06861 (2021)
21. Shin, C., Joo, S., Yim, J., Lee, H., Moon, T., Rhee, W.: Subtask gated networks for non-intrusive load monitoring. In: Proceedings of the AAAI Conference on Artificial Intelligence, vol. 33, pp. 1150–1157 (2019)
22. Veličković, P., Cucurull, G., Casanova, A., Romero, A., Lio, P., Bengio, Y.: Graph attention networks. arXiv preprint arXiv:1710.10903 (2017)
23. Yu, M., Wang, B., Lu, L., Bao, Z., Qi, D.: Non-intrusive adaptive load identification based on siamese network. IEEE Access 10, 11564–11573 (2022)
24. Zhang, C., Zhong, M., Wang, Z., Goddard, N., Sutton, C.: Sequence-to-point learning with neural networks for non-intrusive load monitoring. In: Proceedings of the AAAI Conference on Artificial Intelligence, vol. 32 (2018)
25. Zhang, Q., Chang, J., Meng, G., Xiang, S., Pan, C.: Spatio-temporal graph structure learning for traffic forecasting. In: Proceedings of the AAAI Conference on Artificial Intelligence, vol. 34, pp. 1177–1185 (2020)

Few-Shot Object Detection via Classify-Free RPN

Songlin Yu, Zhiyu Yang, Shengchuan Zhang$^{(\boxtimes)}$, and Liujuan Cao

School of Informatics, Xiamen University, Fujian 361005, People's Republic of China
{31520211154004,yangzhiyu}@stu.xmu.edu.cn,
{zsc_2016,caoliujuan}@xmu.edu.cn

Abstract. The research community has shown great interest in few-shot object detection, which focuses on detecting novel objects with only a small number of annotated examples. Most of the works are based on the Faster R-CNN framework. However, due to the absence of annotated data for novel instances, models are prone to base class bias, which can result in misclassifying novel instances as background or base instances. Our analysis reveals that although the RPN is class-agnostic in form, the binary classification loss possesses class-awareness capabilities, which can lead to the base class bias issue. Therefore, we propose a simple yet effective classify-free RPN. We replace the binary classification loss of the RPN with Smooth L1 loss and adjust the ratio of positive and negative samples for computing the loss. This avoids treating anchors matched with novel instances as negative samples in loss calculation, thereby mitigating the base class bias issue. Without any additional computational cost or parameters, our method achieves significant improvements compared to other methods on the PASCAL VOC and MS-COCO benchmarks, establishing state-of-the-art performance.

Keywords: Object detection · Few-shot object detection · RPN

1 Introduction

In recent years, with the rapid growth of data volume and computational power, fully supervised deep neural networks have made significant advances in tasks such as image classification [13], object detection [3,10,22,23], and instance segmentation [12]. However, fully supervised learning requires a large amount of labeled training data, and obtaining labeled data is often an expensive and time-consuming process. For certain tasks, it may be challenging to acquire a sufficient amount of labeled data. Human learning of new knowledge typically occurs through cognitive processes such as perception, thinking, reasoning, and experience, and it often requires only a small number of examples to be completed successfully. To address these challenges, the paradigm of few-shot learning [6] provides an effective method that allows achieving good performance with a limited number of labeled samples.

A plethora of research has emerged to address the problem of few-shot object detection(FSOD) [4,8,11,15,18,20,29]. These works can be mainly categorized

© The Author(s), under exclusive license to Springer Nature Singapore Pte Ltd. 2024
Q. Liu et al. (Eds.): PRCV 2023, LNCS 14434, pp. 101–112, 2024.
https://doi.org/10.1007/978-981-99-8549-4_9

into two major approaches: meta-learning based methods [4,11,29] and transfer-learning based methods [8,15,20]. Meta-learning based approaches aim to learn from the knowledge gained from few-shot learning tasks to adapt to new tasks. This approach involves designing a meta-learning model that can quickly learn from a small number of samples and adapt to novel classes. Transfer-learning based approaches aim to improve model performance in FSOD by leveraging the knowledge learned by a pre-trained model on large-scale base datasets. This approach involves using the weights of a pre-trained model on base datasets as initial parameters and fine-tuning them on the novel datasets, allowing the model to better adapt to the novel classes. Our work is based on the transfer-learning paradigm.

Faster R-CNN [23] is a classic object detection algorithm. The introduction of Region Proposal Networks (RPN) allows the network to automatically generate candidate boxes, eliminating the need for manually designed box generation steps in traditional methods. As a typical two-stage object detector, Faster R-CNN has been widely applied in the field of FSOD due to the strong generalization ability of its RPN. However, due to the lack of novel class annotations during the base pre-training phase, the network may exhibit a bias towards the base classes, treating the novel classes as background during the base pre-training. This directly leads to the difficulty of detecting novel class instances during the novel fine-tuning phase. To alleviate the issue of base class bias in RPN, we propose replacing the binary classification loss of RPN with a regression loss, and adjusting the ratio of positive and negative samples for loss computation. Experimental results demonstrate that the proposed classify-free RPN achieves favorable performance. Our contributions are summarized as follows:

- We re-evaluate the issue of base class bias in FSOD from the perspective of RPN and found that the binary classification loss of RPN possesses class-awareness capabilities, which leads to a severe base class bias during the base pre-training phase.
- This paper introduces the classify-free RPN, which replaces the binary classification loss of RPN with a regression loss and adjusts the ratio of positive and negative samples for loss computation. This approach alleviates the problem of RPN treating novel class instances as background.
- Comprehensive experimental results on PASCAL VOC and MS-COCO show that our approach outperforms state-of-the-art on FSOD tasks.

2   Related Work

2.1   Object Detection

Object detection [3,10,22,23] is a fundamental task in computer vision. Current algorithms for object detection can be divided into three categories: two-stage detectors, one-stage detectors and query-based detectors. Two-stage detectors [1,23] first proposed a set of Region of Interests (RoIs) and then predict the class probabilities and bounding box offsets for each RoI. Unlike two-stage detectors, one-stage detectors [21] do not require proposing RoIs. Instead, they directly predict class probabilities and bounding box offsets on the CNN features. DETR [3] pioneered the concept of query-based detectors. Query-based detectors directly employ cross-attention between queries and image features to decode class probabilities and bounding boxes. Building upon DETR, numerous improvements have emerged [30], resulting in significant progress in terms of convergence speed and accuracy.

2.2 Few-Shot Learning

Few-shot learning (FSL) [7,9,19,25,27] is a machine learning task that aims to learn novel classes using an extremely limited number of training samples. According to the different techniques of FSL methods, existing approaches can be divided into the following three categories: 1. Metric based methods. Matching Network [27] utilizes class prototypes and attention mechanisms for one-shot learning. Prototypical Network [25] models the similarity between classes by introducing prototype vectors and transform the classification task into computing the distances between samples and prototype vectors. 2. Optimization Based methods. MAML [7] optimizes the initial parameters of a model to enable it to quickly converge to good parameters on a small support set of samples. MixupDG [19] transforms the model selection problem into an optimization problem. 3. Parameter generation based methods. DFSL [9] introduces an attention-based classification weight generator that can infer the weight vector of novel classes.

2.3 Few-Shot Object Detection

FSOD [4,8,15,20,28] aims to identify and localize novel objects with only a limited number of samples provided. Based on the model architecture, existing methods can be categorized into the following two classes: 1. Meta-learning based methods. FR-FSDet [15] enhances the discriminative power for novel classes by reweighting the features. Attention RPN [4] applies a spatial attention module on the feature map produced by the RPN. This module can adaptively learn which regions are more important and enhance the feature representation of those regions. 2. Transfer-learning based methods. TFA [28] freezes most of the network parameters and utilizes a small number of samples to fine-tune the classification and regression heads of the detector for learning new classes. DefRCN [20] extends Faster R-CNN by introducing a gradient decoupled layer for multistage decoupling and a prototypical calibration block for multi-task decoupling. DCFS [8] tackles the problem from the perspective of missing labels. Building upon DCFS, we observed a significant base class bias issue in the binary classification head of the RPN. It tends to incorrectly treat novel instances as background. To address this, we propose a classify-free RPN, which mitigates the base class bias problem in the RPN.

3   Methodology

In this section, we first introduce the problem formulation of FSOD. Then, based on DCFS, we conducted a series of preliminary experiments to validate our hypothesis: RPN has class-awareness and suffers from a base class bias issue during the base pre-training stage, treating novel instances as background. Finally, we will provide a detailed explanation of the proposed Classify-free RPN.

3.1   Problem Setting

Followed by previous works [20,28], we adopt the standard problem formulation of FSOD. Specifically, let the entire training set be denoted as D, and let C represent all the classes. The dataset D is divided into two sets: D_{base} and D_{novel}, with the corresponding classes divided into C_{base} and C_{novel}. Where $D = D_{base} \cup D_{novel}$, $C = C_{base} \cup C_{novel}$, and $C_{base} \cap C_{novel} = \emptyset$. D_{base} contains a large number of annotated instances for each class, while D_{novel} has only K (usually less than 10) annotated instances for each class. Given a test sample (x, y), where x represents an input image with N objects, represented as $y = \{(c_i, b_i), i = 1, \ldots, N\}$. Here, c_i represents the class of the object, with $c_i \in C_{base} \cup C_{novel}$, and b_i represents the bounding box of the object. Our goal is to train a robust detector F_{final} based on dataset D, which can achieve good performance on the test set D_{test}. The classes in D_{test} are represented by C_{test}, where $C_{test} \in C_{base} \cup C_{novel}$. Typically, following the conventional transfer learning process, the entire training process consists of two stages. The base pre-training stage involves training an initial model F_{init} using D_{base} to create a base class detector F_{base}. Then, In the novel fine-tuning stage, F_{base} is fine-tuned using D_{novel} (or a combination of a D_{base}'s subset and D_{novel}) to obtain the final model F_{final}. F_{final} is capable of detecting both C_{base} and C_{novel} simultaneously.

3.2   Analysis of the Base Class Bias Issue in RPN

Table 1. RPN's recall of base class and novel class in base pre-training. **Threshold** represents different IoU thresholds, while **bAR@100** and **nAR@100** represent the recall for base classes and novel classes, respectively.

Threshold	0.5	0.6	0.7	0.8	0.9
bAR@100	90.5	86.9	76.5	40.2	3.9
nAR@100	89.5	84.1	71.5	34.6	2.9

Is RPN truly class-agnostic? Unfortunately, our answer is no. During the base pre-training stage, due to the lack of annotated information for novel classes, anchors that match a potential novel class instance will be assigned as negative

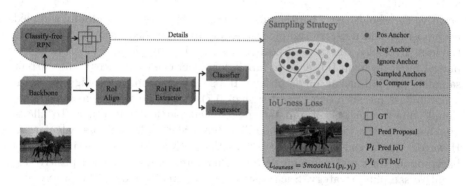

Fig. 1. The framework of our approach. The **Sampling Strategy** and **IoU-ness Loss** modules provide a detailed demonstration of the improvements made to the RPN.

samples. If such an anchor is involved in loss computation, the binary classification loss of RPN may mistakenly classify that anchor as background, resulting in a bias towards base classes in the network. In other words, the RPN tends to treat novel class instances as background while focusing on detecting base class instances. We compare the recall of the RPN after base pre-training using the PASCAL VOC dataset with the split-1 configuration. Table 1 demonstrates the difference in recall of the RPN for base class and novel class at different IoU thresholds. In the experiment, we consider the top 100 proposals generated by the RPN for computation. The results indicate that when the IoU threshold is set to 0.5, there is no significant difference in recall between base class and novel class. However, as the IoU threshold increases, the recall for base class becomes noticeably higher than novel class. This confirms the presence of base class bias in the RPN.

3.3   Classify-Free RPN

Figure 1 illustrates the overall framework. Our approach primarily focuses on the RPN structure. Specifically, we made modifications in two parts: sampling strategy and loss formulation.

Sampling Strategy. We first review the traditional process of RPN label assignment. RPN initially generates multiple anchors at each position on the feature map. Each anchor typically has multiple scales and aspect ratios to cover objects of different sizes and shapes. Let $A = \{a_i, i = 1 \ldots N\}$ represents all the anchors, and $G = \{g_i, i = 1 \ldots M\}$ represents all the ground truth boxes in the image. RPN calculates the IoU between each anchor and all the ground truth boxes, and assigns one of the following labels to them: 1. Positive Label: If an anchor has an IoU exceeds a certain threshold (e.g., 0.7). 2. Negative Label: If an anchor has an IoU lower than a certain threshold (e.g., 0.3) with all ground truth boxes. 3. Ignore Label: If an anchor has an IoU between the positive and negative thresholds (e.g., between 0.3 and 0.7). Anchors are randomly sampled from positive and negative samples according to a specified ratio for calculating the loss. Typically, the ratio of positive to negative samples is 1 : 1. Due to the partial label problem (absence of labels for novel class instances), the current sampling method has an inherent flaw. If a selected anchor has a high IoU with a novel class instance (e.g., greater than 0.7), it will be treated as a negative sample due to the lack of annotations for that particular instance. To alleviate this issue, the ideal is only sampling positive samples for loss calculation. However, this would significantly reduce the RPN's ability to distinguish the background (refer to Table 5 for detail). Therefore, we propose a 9:1 positive-to-negative sampling strategy. Such a strategy can greatly reduce the probability of including anchors that match potential novel class instances in the loss calculation, while maintaining the RPN's ability to differentiate the background. The **Sampling Strategy** section in Fig. 1 provides a detailed demonstration of our improvement.

IoU-ness Loss. The typical RPN loss consists of two components: $\mathcal{L}_{cls}$ and $\mathcal{L}_{reg}$, which represent the classification loss and regression loss, respectively. The $\mathcal{L}_{cls}$ represents the binary classification loss and is calculated using the following formula:

$$\mathcal{L}_{cls} = -\frac{1}{N} \sum_{i=1}^{N} [t_i \log(p_i) + (1 - t_i) \log(1 - p_i)] \tag{1}$$

Here, N represents the number of anchors involved in the loss calculation, p_i represents the predicted class score, and t_i represents the assigned label for the anchor. When $t_i = 1$, it indicates a positive sample, while $t_i = 0$ represents a negative sample. Physically, this loss represents the probability of an anchor containing an object. Although $\mathcal{L}_{cls}$ is formally class-agnostic, [16] suggests that using binary classification loss in RPN leads to a severe bias towards the seen class. To address this issue, we propose replacing the binary classification loss with IoU-ness loss. Specifically, in binary classification loss, the ground truth is represented discretely as either 0 for negative samples or 1 for positive samples. However, in IoU-ness loss, the ground truth is represented as a continuous value ranging from 0 to 1, indicating the IoU between the anchor and the ground truth box. Let p_i represent the predicted IoU-ness and y_i represent the IoU between the anchor and the matched ground truth box. The $\mathcal{L}_{iouness}$ is computed using the smooth $\mathcal{L}_1$ loss, and it can be represented by the following formula:

$$\mathcal{L}_{iouness} = -\frac{1}{N} \sum_{i=1}^{N} \begin{cases} 0.5 \times (p_i - y_i), & if |p_i - y_i| < 1 \\ |p_i - y_i| - 0.5, & otherwise \end{cases} \tag{2}$$

The **IoU-ness Loss** section in Fig. 1 provides a detailed demonstration of our method.

4   Experiments

In this section, we first introduce the experimental setup in Sect. 4.1, then compare our method with previous state-of-the-art on multiple benchmarks in Sect. 4.2, and finally provide comprehensive ablation studies in Sect. 4.3.

4.1   Experimental Setup

Datasets. We have tested our approach on two widely used benchmark datasets in computer vision, namely PASCAL VOC and MS-COCO. To ensure a fair comparison with previous studies [28], we use the same data splits and annotations provided by TFA [28].

PASCAL VOC includes 20 categories divided into three splits. Each split consists of 15 base classes and 5 novel classes. We randomly select K objects for each novel class, where K can be 1, 2, 3, 5, or 10. And the VOC2007 test set is used to evaluate model's performance. We report the Average Precision (IoU = 0.5) of the novel classes.

MS-COCO consists of 80 categories. We select the 20 categories that are present in the PASCAL VOC dataset as novel classes, while the remaining 60 categories are used as base classes. We train the model based on K (10, 30) instances for each novel class and evaluate on the MS-COCO validation set. We report the Average Precision (IoU=0.5:0.95) on the novel classes.

Experimental Details. We conduct our experiments using Detectron2 on 4 NVIDIA GPU 3090 with CUDA 11.3. Our approach uses the Faster R-CNN [23] framework for object detection, with a ResNet-101 [13] backbone that has been pre-trained on ImageNet [24]. To train our network, we adopt a two-stage transfer-learning method. Firstly, we initialize the network with parameters pre-trained on ImageNet to train the base classes. Then, we fine-tune the network on K-shots for each class. We optimize our network using SGD with a mini-batch size of 16, momentum of 0.9, and weight decay of $5e^{-5}$ on 4 GPUs. The learning rate for base pre-training and novel fine-tuning is set to 0.02 and 0.01, respectively.

4.2   Comparison with the State-of-the-Art

PASCAL VOC. Table 2 presents the results of our method on the PASCAL VOC dataset with three different splits, considering different numbers of shots: 1, 2, 3, 5, and 10. It can be observed that our method achieves state-of-the-art performance in most settings, validating the effectiveness of our approach. It is worth noting that there is a significant performance gap among different splits, which can be attributed to the varying label quantities of different categories in

Table 2. FSOD performance (AP_{50}) for novel classes on PASCAL VOC. The top-performing results and the runner-up are colored in red and blue, respectively.

Method/Shots		Novel Set 1					Novel Set 2					Novel Set 3				
		1	2	3	5	10	1	2	3	5	10	1	2	3	5	10
Meta Learning	FSRW [15]	14.8	15.5	26.7	33.9	47.2	15.7	15.2	22.7	30.1	40.5	21.3	25.6	28.4	42.8	45.9
	MetaDet [29]	18.9	20.6	30.2	36.8	49.6	21.8	23.1	27.8	31.7	43.0	20.6	23.9	29.4	43.9	44.1
	DCNet [14]	33.9	37.4	43.7	51.1	59.6	23.2	24.8	30.6	36.7	46.6	32.3	34.9	39.7	42.6	50.7
	CME [18]	41.5	47.5	50.4	58.2	60.9	27.2	30.2	41.4	42.5	46.8	34.3	39.6	45.1	48.3	51.5
	TIP [17]	27.7	36.5	43.3	50.2	59.6	22.7	30.1	33.8	40.9	46.9	21.7	30.6	38.1	44.5	40.9
	FCT [11]	38.5	49.6	53.5	59.8	64.3	25.9	34.2	40.1	44.9	47.4	34.7	43.9	49.3	53.1	56.3
Transfer Learning	TFA [28]	39.8	36.1	44.7	55.7	56.0	23.5	26.9	34.1	35.1	39.1	30.8	34.8	42.8	49.5	49.8
	GFSD [5]	42.4	45.8	45.9	53.7	56.1	21.7	27.8	35.2	37.0	40.3	30.2	37.6	43.0	49.7	50.1
	FSCE [26]	44.2	43.8	51.4	61.9	63.4	27.3	29.5	43.5	44.2	50.2	37.2	41.9	47.5	54.6	58.5
	FADI [2]	50.3	54.8	54.2	59.3	63.2	30.6	35.0	40.3	42.8	48.0	45.7	49.7	49.1	55.0	59.6
	DeFRCN [20]	53.6	57.5	61.5	64.1	60.8	30.1	38.1	47.0	53.3	47.9	48.4	50.9	52.3	54.9	57.4
	DCFS [8]	56.6	59.6	62.9	65.6	62.5	29.7	38.7	46.2	48.9	48.1	47.9	51.9	53.3	56.1	59.4
	Ours	54.7	59.6	62.7	66.5	66.3	33.1	42.0	48.0	51.2	52.8	49.9	52.8	55.3	59.7	61.6

Fig. 2. Visualization of our method and the DCFS on MS-COCO validation images. Best viewed in color and zoom in.

the PASCAL VOC dataset. Our method achieves the best performance under the split3 setting for all shots, showing a significant improvement compared to the second best. We speculate that this improvement is due to the fact that our method is more class-agnostic, thus being less affected by different class divisions. This strongly confirms the robustness of our approach.

MS-COCO. Table 3 presents the results on the MS-COCO dataset. Compared to the PASCAL VOC dataset, our method does not show a significant improve-

Table 3. FSOD performance (AP) for novel classes on MS-COCO. The top-performing results and the runner-up are colored in red and blue, respectively.

Method/Shots		10 shot	30 shot
Meta Learning	FSRW [15]	5.6	9.1
	MetaDet [29]	7.1	11.3
	DCNet [14]	12.8	18.6
	CME [18]	15.1	16.9
	TIP [17]	16.3	18.3
	FCT [11]	15.3	20.2
Transfer Learning	TFA [28]	10.0	13.7
	GFSD [5]	10.5	13.8
	FSCE [26]	11.1	15.3
	FADI [2]	12.2	16.1
	DeFRCN [20]	18.5	22.6
	DCFS [8]	19.5	22.7
	Ours	18.6	23.0

ment on the MS-COCO, but it is still comparable to the SOTA methods. MS-COCO dataset is known to be more challenging as it contains a larger number of instances per image. Achieving a large margin performance improvement on such a dataset is indeed a difficult task. However, this motivates us to continue exploring and pushing the boundaries in our future research.

Visualization. Figure 2 presents the visual results of our approach. Compared to DCFS, our method exhibits fewer false detections. In the first row, second column, DCFS misclassifies the clothing of the boy as a zebra, while in the second row, second column, DCFS misidentifies a stone as an elephant. However, there are also some bad cases in our approach, such as in the second row, third column, where our method misses the detection of the smaller bear.

4.3 Ablation Study

Table 4. RPN's recall for base classes and novel classes in base pre-training stage. $\triangle$ represents thre difference between bAR and nAR.

	Threshold	0.5	0.6	0.7	0.8	0.9	avg
DCFS	bAR@100	90.5	86.9	76.5	40.2	3.9	-
	nAR@100	89.5	84.1	71.5	34.6	2.9	-
	$\triangle$	1.0	2.8	5.0	5.6	1.0	3.1
Ours	bAR@100	89.8	87.5	75.8	40.1	3.6	-
	nAR@100	89.7	86.7	74.9	38.5	3.2	-
	$\triangle$	0.1	0.8	0.9	1.6	0.4	0.8

Table 5. Effectiveness of the pos/neg sample ratio and the IoU-ness loss in RPN. The **pos/neg** indicates the proportion of positive and negative samples used to compute the loss in RPN.

pos/neg	loss type		AP_{50}
	bin-cls	IoU-ness	
0.50 : 0.50	✓	✗	65.6
0.50 : 0.50	✗	✓	66.5
0.75 : 0.25	✗	✓	66.9
0.90 : 0.10	✗	✓	67.8
1.00 : 0.00	✗	✓	60.1

Recall of RPN. We conduct ablation experiments on the PASCAL VOC dataset, specifically using the split-1 configuration, to validate the alleviating effect of our method on the RPN base class bias issue. Table 4 shows the recall difference of RPN between base class instances and novel class instances during the base pre-training stage for both DCFS and our method. It can be observed that our method achieves comparable recall to DCFS for base class instances, but significantly improves the recall for novel class instances. Under the IoU threshold of 0.8, our method achieves a recall of 38.5% for novel class instances, while DCFS only achieves 34.6%. The recall difference between the base class and novel class in our method is significantly smaller than DCFS. On average, across five threshold settings, our method is 0.8%, while DCFS is 3.1%. This validates the effectiveness of our method.

Effectiveness of Classify-Free RPN. In this section, we utilize the split-1 configuration of the PASCAL VOC dataset. Table 5 presents the performance of 5-shot novel AP_{50} under different configurations of RPN. Comparing the first and second rows, when we replace the binary classification loss of the original RPN with the IoU-ness loss, AP_{50} has improved by 0.9%. This validates the effectiveness of our IoU-ness loss. Comparing the second to fourth rows, we observe that as the proportion of positive samples in the loss calculation increases, there is a corresponding increase in AP_{50}. This validates our insight that negative samples might match potential novel class instances, treating them as background. It is worth noting that if only positive samples are included in the loss calculation, it will result in a decrease in the RPN's ability to distinguish the background, ultimately leading to a decline.

5   Conclusion

In this paper, we address the issue of base class bias in FSOD, and propose a novel approach to tackle this problem from the perspective of RPN for the first time. Specifically, we begin by replacing the binary classification loss of the RPN with the Smooth L1 loss based on the IoU between anchors and the corresponding ground truthes. Additionally, We modify the sampling strategy of the RPN. This alleviates treating certain anchors that match potential novel instances as negative samples when calculating loss. Extensive experiments have shown that our method improves the performance of FSOD. One drawback of our method is that the performance improvement on MS-COCO is not significant. We hope this work provide a new perspective for addressing the base class bias issue in FSOD within the research community.

References

1. Cai, Z., Vasconcelos, N.: Cascade R-CNN: delving into high quality object detection. In: Proceedings of the IEEE Conference on Computer Vision and Pattern Recognition, pp. 6154–6162 (2018)
2. Cao, Y., Wang, J., Jin, Y., Wu, T., Chen, K., Liu, Z., Lin, D.: Few-shot object detection via association and discrimination. In: Advance in Neural Information Processing System, vol. 34, pp. 16570–16581 (2021)
3. Carion, N., Massa, F., Synnaeve, G., Usunier, N., Kirillov, A., Zagoruyko, S.: End-to-end object detection with transformers. In: Vedaldi, A., Bischof, H., Brox, T., Frahm, J.-M. (eds.) ECCV 2020. LNCS, vol. 12346, pp. 213–229. Springer, Cham (2020). https://doi.org/10.1007/978-3-030-58452-8_13
4. Fan, Q., Zhuo, W., Tang, C.K., Tai, Y.W.: Few-shot object detection with attention-RPN and multi-relation detector. In: Proceedings of the IEEE/CVF Conference on Computer Vision and Pattern Recognition, pp. 4013–4022 (2020)
5. Fan, Z., Ma, Y., Li, Z., Sun, J.: Generalized few-shot object detection without forgetting. In: Proceedings of the IEEE/CVF Conference on Computer Vision and Pattern Recognition, pp. 4527–4536 (2021)
6. Fink, M.: Object classification from a single example utilizing class relevance metrics. In: Advances in Neural Information Processing Systems, vol. 17 (2004)
7. Finn, C., Abbeel, P., Levine, S.: Model-agnostic meta-learning for fast adaptation of deep networks. In: International Conference on Machine Learning, pp. 1126–1135. PMLR (2017)
8. Gao, B.B., et al.: Decoupling classifier for boosting few-shot object detection and instance segmentation. In: Advances in Neural Information Processing Systems (2022)
9. Gidaris, S., Komodakis, N.: Dynamic few-shot visual learning without forgetting. In: Proceedings of the IEEE Conference on Computer Vision and Pattern Recognition, pp. 4367–4375 (2018)
10. Girshick, R.: Fast R-CNN. In: Proceedings of the IEEE International Conference on Computer Vision, pp. 1440–1448 (2015)
11. Han, G., Ma, J., Huang, S., Chen, L., Chang, S.F.: Few-shot object detection with fully cross-transformer. In: Proceedings of the IEEE/CVF Conference on Computer Vision and Pattern Recognition, pp. 5321–5330 (2022)
12. He, K., Gkioxari, G., Dollár, P., Girshick, R.: Mask R-CNN. In: Proceedings of the IEEE International Conference on Computer Vision, pp. 2961–2969 (2017)
13. He, K., Zhang, X., Ren, S., Sun, J.: Deep residual learning for image recognition. In: Proceedings of the IEEE Conference on Computer Vision and Pattern Recognition, pp. 770–778 (2016)

14. Hu, H., Bai, S., Li, A., Cui, J., Wang, L.: Dense relation distillation with context-aware aggregation for few-shot object detection. In: Proceedings of the IEEE/CVF Conference on Computer Vision and Pattern Recognition, pp. 10185–10194 (2021)
15. Kang, B., Liu, Z., Wang, X., Yu, F., Feng, J., Darrell, T.: Few-shot object detection via feature reweighting. In: Proceedings of the IEEE/CVF International Conference on Computer Vision, pp. 8420–8429 (2019)
16. Kim, D., Lin, T.Y., Angelova, A., Kweon, I.S., Kuo, W.: Learning open-world object proposals without learning to classify. IEEE Robot. Autom. Lett. **7**(2), 5453–5460 (2022)
17. Li, A., Li, Z.: Transformation invariant few-shot object detection. In: Proceedings of the IEEE/CVF Conference on Computer Vision and Pattern Recognition, pp. 3094–3102 (2021)
18. Li, B., Yang, B., Liu, C., Liu, F., Ji, R., Ye, Q.: Beyond max-margin: class margin equilibrium for few-shot object detection. In: Proceedings of the IEEE/CVF Conference on Computer Vision and Pattern Recognition, pp. 7363–7372 (2021)
19. Lu, W., Wang, J., Wang, Y., Ren, K., Chen, Y., Xie, X.: Towards optimization and model selection for domain generalization: a mixup-guided solution. arXiv preprint arXiv:2209.00652 (2022)
20. Qiao, L., Zhao, Y., Li, Z., Qiu, X., Wu, J., Zhang, C.: Defrcn: decoupled faster R-CNN for few-shot object detection. In: Proceedings of the IEEE/CVF International Conference on Computer Vision, pp. 8681–8690 (2021)
21. Redmon, J., Divvala, S., Girshick, R., Farhadi, A.: You only look once: unified, real-time object detection. In: Proceedings of the IEEE Conference on Computer Vision and Pattern Recognition, pp. 779–788 (2016)
22. Redmon, J., Farhadi, A.: Yolo9000: better, faster, stronger. In: Proceedings of the IEEE Conference on Computer Vision and Pattern Recognition, pp. 7263–7271 (2017)
23. Ren, S., He, K., Girshick, R., Sun, J.: Faster R-CNN: towards real-time object detection with region proposal networks. In: Advances in Neural Information Processing Systems, vol. 28 (2015)
24. Russakovsky, O., et al.: ImageNet large scale visual recognition challenge. Int. J. Comput. Vis. **115**, 211–252 (2015)
25. Snell, J., Swersky, K., Zemel, R.: Prototypical networks for few-shot learning. In: Advances in Neural Information Processing Systems, vol. 30 (2017)
26. Sun, B., Li, B., Cai, S., Yuan, Y., Zhang, C.: FSCE: few-shot object detection via contrastive proposal encoding. In: Proceedings of the IEEE/CVF Conference on Computer Vision and Pattern Recognition, pp. 7352–7362 (2021)
27. Vinyals, O., Blundell, C., Lillicrap, T., Wierstra, D., et al.: Matching networks for one shot learning. In: Advances in Neural Information Processing Systems, vol. 29 (2016)
28. Wang, X., Huang, T.E., Darrell, T., Gonzalez, J.E., Yu, F.: Frustratingly simple few-shot object detection. arXiv preprint arXiv:2003.06957 (2020)
29. Wang, Y.X., Ramanan, D., Hebert, M.: Meta-learning to detect rare objects. In: Proceedings of the IEEE/CVF International Conference on Computer Vision, pp. 9925–9934 (2019)
30. Zhu, X., Su, W., Lu, L., Li, B., Wang, X., Dai, J.: Deformable DETR: deformable transformers for end-to-end object detection. arXiv preprint arXiv:2010.04159 (2020)

IPFR: Identity-Preserving Face Reenactment with Enhanced Domain Adversarial Training and Multi-level Identity Priors

Lei Zhu, Ge Li[(✉)], Yuanqi Chen, and Thomas H. Li

School of Electronic and Computer Engineering, Shenzhen Graduate School,
Peking University, Beijing, China
geli@ece.pku.edu.cn

Abstract. In the face reenactment task, identity preservation is challenging due to the leakage of driving identity and the complexity of source identity. In this paper, we propose an Identity-Preserving Face Reenactment (IPFR) framework with impressive expression and pose transfer. To address the leakage of driving identity, we develop an enhanced domain discriminator to eliminate the undesirable identity in the generated image. Considering the complexity of source identity, we inject multi-level source identity priors to keep the identity domain of generated image close to that of source. In detail, firstly, we utilize a 3D geometric prior from the 3D morphable face model (3DMM) to control face shape and reduce artifacts caused by occlusion in the module of generating motion field; secondly, we use an identity texture prior extracted by face recognition network to supervise the final stage, aiming to make the identity domain of the generated close to that of source. Extensive experiments demonstrate that our method significantly outperforms state-of-the-art methods on image quality and identity preservation. Ablation studies are also conducted to further validate the effectiveness of our individual components.

Keywords: Enhanced domain adversarial training · 3D shape-aware identity · Attention mechanism

1 Introduction

Face reenactment aims to synthesize realistic images of a source identity with head poses and facial expressions from a driving face. Such technology has great potential in the fields of virtual reality and film production. However, preserving identity is extremely challenging especially when a huge discrepancy of identity, head poses, and facial expressions exists in source and driving face. The reenacted face should capture the unique facial characteristics and details specific to the source individual while incorporating the desired head poses and expressions from the driving face.

Rapid progress has been achieved on face reenactment [15,19,21,24]. A series of works [12,14] mainly decouple facial motion from identity information in

Q. Liu et al. (Eds.): PRCV 2023, LNCS 14434, pp. 113–124, 2024.
https://doi.org/10.1007/978-981-99-8549-4_10

the driving image. A serious problem behind is that the generated image's identity may be easily affected by the driving motion. The dominant motion descriptor is divided into two categories: 1) 3DMM parameters [5,14], which contains disentangled identity, expression, pose and other attributes; 2) landmark image [12,15]. It is easier to cause identity mismatch problems when using landmark, [12] proposes a landmark transformer to transform driving landmark images to the source identity. Although 3DMM motion parameters [5,14] do not contain identity information unlike landmarks [12,15]. However, in the process of injecting 3DMM motion, many methods [2,14,16,21] overfit some of the driving's identity information and cause the leakage of the driving's identity to cater to motion or restore the occlusion area. [5] use an image discriminator and [20] use an identity coefficient loss to enforce the reenacted face to have the same identity as the source. However, these methods are unable to eliminate the leakage of driving identity information fundamentally. In this paper, we use domain adversarial training [6] with data augmentation to make the latent vector encoded from 3DMM motion parameters without any driving identity. The key design is a domain discriminator with a strong ability to distinguish identity domains. We use the latent vector encoded from primary and reorganized driving 3DMM motion parameters and identity domains of driving as training samples and labels. In training, we train the discriminator to improve its ability to distinguish identity domains. Meanwhile, we introduce the domain loss as an adversarial loss into the optimization process of the latent space encoding network through a negative coefficient. When the ability of the domain discriminator is strong enough, if the domain loss is still large, we can assume the latent vector contains almost no driving identity.

In addition to the leakage of driving identity, there exists the problem that the source identity is not perfect in many methods [14,15,21]. Due to the sensitivity of the human eye to identity, after removing the driving identity information, we need to perfect the source identity information. Actually, identity information is multi-layered. Common identity representations can be divided into: 1) identity information in disentangled 3D face descriptors; 2) feature embedding encoded from the face image. Identity information in disentangled 3D face descriptor [17,22] learns to acquire the geometry shape of a 3D face built upon the prior knowledge of 3DMM [1], while identity feature extracted by face recognition network [3] focuses more on texture and is insensitive to the geometric structure [11]. In this paper, to obtain accurate facial geometric-level and texture-level identity simultaneously, we inject the multi-level priors in different stages.

Our framework is shown in Fig. 1 and the reenactment can be described in the following three steps. Firstly, it maps 3DMM descriptor [4] to produce a latent vector, and we use enhanced domain adversarial training to keep the latent space away from driving identity domain. Secondly, it predicts an optical flow under the guidelines of the latent vector. In the process of modeling the movement, we inject a 3D face shape prior to this stage by using a cross-attention mechanism, which helps to model a more accurate facial shape. Then, we fuse the identity feature prior extracted by the face recognition network and the latent vector to supervise the final stage. We evaluate our framework qualitatively and quantitatively on the public dataset: VoxCeleb [13]. Experimental results show the framework can facilitate the generation of accurate identity information with impressive expression and pose transfer. Our main contributions can be summarized as follows:

- We propose a novel face reenactment framework, which can generate identity-preserving results with impressive expression and pose transfer. Our method significantly outperforms state-of-the-art methods on image quality and preserving identity both in facial shape and texture.
- We utilize the domain adversarial training with data augmentation to filter driving identity information introduced in the generated result due to network overfitting.
- We use a shape-guided high-frequency attention block to inject facial geometric-level identity prior in the warping module, and we inject texture-level identity feature prior extracted by a face recognition network into the final refining module to preserve the identity perfectly.

2   Methods

Let $\mathbf{I}_s$ be a source image and $\mathbf{I}_d$ a driving image respectively. We aim to generate an image $\mathbf{I}_g$ with the expressions and poses of $\mathbf{I}_d$ and other attributes of $\mathbf{I}_s$ such as identity, illumination and background. The proposed model is illustrated in Fig. 1, and we will discuss each module in detail.

2.1   Target Motion Encoder and 3D Shape Encoder

The 3DMM provides us with disentangled parameters of face shape, expression and head pose. We use a pre-trained face reconstruction model provided by [4] to regress 3DMM parameters $\mathbf{v} = (\boldsymbol{\alpha}, \boldsymbol{\beta}, \boldsymbol{p})$ from a face image, where $\boldsymbol{\alpha} \in \mathbb{R}^{80}$, $\boldsymbol{\beta} \in \mathbb{R}^{64}$, $\boldsymbol{p} \in \mathbb{R}^{6}$ describe the identity, expression, and pose respectively. With parameter $\mathbf{m}_i \equiv \{\boldsymbol{\beta}_i, \boldsymbol{p}_i\}$, the target motions of face i can be clearly expressed, and $\boldsymbol{\alpha} \in \mathbb{R}^{80}$ denotes the geometric structure of source image. Following the same mechanism as PIRenderer [14], the motion parameter $\mathbf{m}_i$ of face i is replaced by window parameters with continuous frames centered on the face i. Ultimately, $\mathbf{m} \equiv \mathbf{m}_{i-k:i+k} \equiv \{\boldsymbol{\beta}_i, \boldsymbol{p}_i, \dots, \boldsymbol{\beta}_{i\pm k}, \boldsymbol{p}_{i\pm k}\}$ is defined as target motion descriptor, the facial shape parameters in continuous frames are defined as source facial shape descriptor. The target motion descriptor $\mathbf{m}$ and source facial shape descriptor is then encoded into latent vector $\mathbf{z}_m \in \mathcal{Z}_m$, $\mathbf{z}_{id} \in \mathcal{Z}_{id}$ by target motion encoder and 3D shape encoder separately.

2.2   3D Shape-Aware Warping Module

We employ a 3D shape-aware warping module $\mathcal{F}_s$ to model the motion field. Here, we define these motions as the coordinate offset Θ and learn them from the source image $\mathbf{I}_s$, the motion latent vector $\mathbf{z}_m$ and the facial shape latent vector $\mathbf{z}_{id}$.

$$\Theta = \mathcal{F}_s\left(\mathbf{I}_s, \mathbf{z}_m, \mathbf{z}_{id}\right) \tag{1}$$

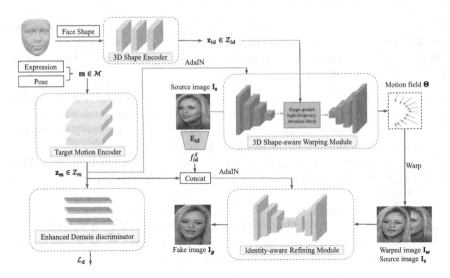

Fig. 1. Overview of our proposed model. The reenactment can be described in the following three steps. Firstly, the 3DMM pose and expression descriptor [4] is mapped to produce a latent vector, and we utilize enhanced domain adversarial training to keep the latent space away from driving identity domain. Secondly, an optical flow is predicted under the guidelines of the latent vector. Meanwhile, we inject a 3D face shape prior to this stage by using a high-frequency cross-attention block, which helps to model a more accurate facial shape. Then, we fuse the identity feature prior extracted by a face recognition network and the latent vector to supervise the final stage.

The module consists of two submodules: an offset-estimating auto-encoder module and a shape-guided high-frequency attention block. To model the motion field, we inject the motion latent vector z_m with AdaIN [9] operation after each convolution layer in the auto-encoder. In order to repair 3D geometric shape of high-frequency occluded area caused by a large movement of the head, we use a cross-attention block with a high-frequency attention mask in the lower decoder layer to inject face geometric outline information that does not exist in the source image into the high-frequency area of feature maps. Face geometric outline information is the facial shape latent vector z_{id} derived from 3D face shape parameters of continuous frames. The attention mechanism helps to reduce the negative impact of occlusion and maintain identity information in the geometric level. The **shape-guided high-frequency attention (SHA)** block is shown in Fig. 2 which is similar to [18]. Finally, the warping module outputs the offsets Θ, we use warping equation $\mathbf{I}_w = \Theta(\mathbf{I}_s)$ to obtain a preliminary reenactment result.

2.3 Identity-Aware Refining Module

The warping module is only efficient at spatially transforming the source images, to add source image texture and eliminate artifacts we employ a refining network

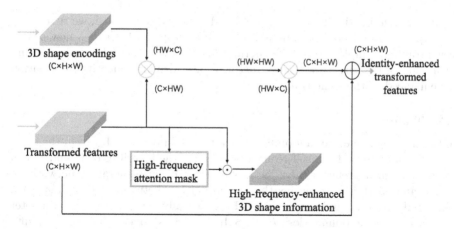

Fig. 2. The framework of the proposed SHA block, which takes the transformed features and 3D shape latent vector as inputs and learns high-frequency source facial shape.

$\mathcal{F}_r$ designed with an encoder-decoder architecture to transform from the source image $\mathbf{I}_s$ and warped image $\mathbf{I}_w$ into the generated image $\mathbf{I}_g$. However, the network is insensitive to identity texture, so we concatenate the latent vector $\mathbf{z}_m$ and the face feature vector $\mathbf{f}_{id}^s$ extracted by a face recognition network [3], and then inject it with AdaIN operation after each convolution layer to supplement texture-level identity information:

$$\mathbf{I}_g = \mathcal{F}_r \left(\mathbf{I}_s, \mathbf{I}_w, \mathbf{z}_m, \mathbf{f}_{id}^s \right) \tag{2}$$

2.4 Enhanced Domain Discriminator

To ensure that the encoded motion latent space $\mathcal{Z}_m$ does not contain the driving's identity information, domain adversarial training is utilized to optimize the target motion encoder. The core is using an adversarial loss to erase driving's identity due to network overfitting. Specifically, we design a domain discriminator as a 3-layer multilayer perceptron (MLP). It takes the latent vector $\mathbf{z}_m$ as input and infers its identity label as:

$$\hat{K} = D(\mathbf{z}_m) \tag{3}$$

where $D(\cdot)$ denotes the MLP, $\hat{K}$ denotes the predicted identity label, we optimize the MLP with cross-entropy loss as:

$$\mathcal{L}_d = H(K, \hat{K}) \tag{4}$$

where $H(\cdot)$ denotes cross-entropy loss, and K denotes the one-hot coding of driving's identity label, which is assumed as a vector with the length of id numbers in the training set.

It is worth mentioning that since 3DMM is fully disentangled, we reorganize the expression and pose parameters to do **data augmentation** as new adversarial training samples, so that the performance of the domain discriminator and the ability of the target motion extractor to filter driving's identity information are enhanced at the same time.

2.5   Training

In the training stage, we randomly select a pair of images $(\mathbf{I}_s, \mathbf{I}_d)$ from the same person's video to perform self-reenactment, while the identities of $\mathbf{I}_s$ and $\mathbf{I}_d$ can be different in inference stage. The loss function of our generation network $\mathcal{L}_g$ is the same as [14]. $\mathcal{L}_g$ contains the perceptual loss between warped image $\mathbf{I}_w$, generated image $\mathbf{I}_g$ and target image $\mathbf{I}_d$. Especially, we use a hyperparameter λ to introduce the domain loss $\mathcal{L}_d$ in Subsect. 2.4 as adversarial loss $\mathcal{L}_{adv}$ into the generation network to keep the latent space away from driving's identity domain, so the final loss function is defined:

$$\mathcal{L} = \mathcal{L}_g - \lambda \mathcal{L}_d \tag{5}$$

We maximize this adversarial loss while training the target motion encoder and optimize the domain discriminator alternately. When the domain discriminator is powerful enough with data augmentation, if the adversarial loss is still large, we assume the latent space contains almost no identity information of the driving. Here we set $\lambda = 0.2$. The identity domain's changing process is illustrated in Fig. 3. Id_d, Id_s denote the center of driving and source identity domain individually.

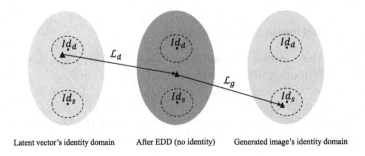

Fig. 3. Illustration of the identity domain change, EDD denotes the enhanced domain discriminator.

3   Experiment

3.1   Experimental Setup

In this subsection, we first introduce the dataset used for model training and the details of the training process. Then, we will proceed to explain the evaluation metrics used during testing.

Dataset and Training Details. We use VoxCeleb dataset for training and testing. Following the same pre-processing method described in [15], a total of 17193 and 514 videos with frames length varying from 64 to 1024 are obtained with a resolution of 256 for the training set and testing set separately. Our model is trained in two stages. The warping module with its affiliate networks is pre-trained 100 epochs. Then we train the whole network for 100 epochs. It is worth mentioning that the enhanced domain discriminator is trained from start to finish. The reason is identity leakage caused by overfitting happens through two stages. Our all experiments are conducted on four RTX 3090 GPUs with a batch size of 20. ADAM [10] optimizer is used for training with an initial learning rate of 1×10^{-4} and a decreased learning rate of 1×10^{-5} after 150 epochs.

Evaluation Metrics. We evaluate our model using the following quality metrics: Learned Perceptual Image Patch Similarity [23] (**LPIPS**) is utilized to estimate the reconstruction error, which computes the perceptual distance between the generated images and target images. Fréchet Inception Distance [7] (**FID**) is used to estimate the difference between the distributions of the generated and real images. Especially, we calculate the cosine similarity of identity (**CSIM**) feature extracted by the pretrained face recognition model ArcFace [3], which is utilized to evaluate the degree of identity preservation. What's more, we use the Average Expression Distance (**AED**) and Average Pose Distance (**APD**) to evaluate the effect of imitation of head poses and expressions following the method in [14].

3.2 Comparisons

We evaluate the performance of our model in two reenactment tasks: (1) same-identity reenactment, where the source and driving image are from the same person (2) cross-identity reenactment, where the source and driving image are from two persons. We compare our approach with the following methods: FOMM [15], PIRenderer [14], DaGAN [8]. The evaluation results are summarized in Table 1. And an analysis was conducted on the reasons behind the results demonstrated by each method.

Same-Identity Reenactment. The quantitative results of the same-identity reenactment task are shown in the left of Table 1. Our method achieves the best results on CSIM, FID, LPIPS and impressive results on AED and APD. This shows that the results obtained by our method can better ensure the authenticity of identity information and generate more realistic images under the condition of successful motion transfer.

Cross-Identity Reenactment. In practice, cross-identity reenactment has a wider range of applications and is more difficult to implement. The quantitative results are in the right of Table 1. Our approach is much better than others at maintaining identity information, which can be inferred from the highest CSIM with source image and the lowest CSIM with driving image (CSIM(dr)). In addition, our framework produces more realistic results with a higher FID. In

120   L. Zhu et al.

order to show the effect of our model intuitively, we provide qualitative results in Fig. 4. Compared with other methods, we generate identity-preserving results with an accurate facial shape.

Results Analysis. Allow us to elucidate the rationale behind the supremacy of our proposed approach in contrast to several other state-of-the-art methods. Firstly, we conduct a thorough analysis of the limitations inherent in various state-of-the-art techniques. Subsequently, we expound on how our method effectively mitigates these shortcomings. While FOMM endeavors to estimate motion fields through sparse key points, it might not excel in tasks demanding quantitative reenactment due to pronounced dissimilarities in facial structure between the source and driving images. Additionally, facial expressions are not well-imitated. PIRenderer is good at transferring expressions, but when the identity of the source and driving images is quite different, the generated image's identity remains poor, and the driving identity information leaks into the result. This is because PIRenderer injects the mapped driving motion latent into the generation process, which can overfit driving identity information during training. Our method addresses these challenges comprehensively. To begin, we employ an enhanced domain discriminator to counteract the leakage of driving identity information, distinguishing us from PIRenderer. Furthermore, we employ a shape-guided high-frequency attention block to rectify the incompleteness in source map face geometry caused by head movement, resulting in a more accurate facial structure than that achieved by all the contrasting methods. Additionally, we incorporate the identity feature vector obtained from a face recognition network into the final generation stage, facilitating the amalgamation of transposed

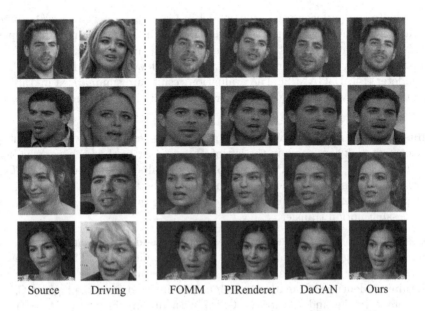

Source   Driving   FOMM   PIRenderer   DaGAN   Ours

Fig. 4. Qualitative comparisons of cross-reenactment on the VoxCeleb dataset.

expressions and human-perceptible identity information. The confluence of these advancements leads to visually credible outcomes.

Table 1. Quantitative comparisons on the VoxCeleb dataset.

	Self-reenactment					Cross-reenactment				
	CSIM ↑	FID ↓	LPIPS ↓	AED ↓	APD ↓	CSIM ↑	CSIM(dr) ↓	FID ↓	AED ↓	APD ↓
FOMM	0.821	13.12	0.112	0.192	0.048	0.502	0.257	42.41	0.281	0.069
PIRenderer	0.827	12.49	0.115	**0.173**	0.041	0.535	0.261	36.25	**0.268**	0.060
DaGAN	0.829	11.56	0.113	0.175	**0.039**	0.496	0.259	49.37	0.279	0.066
Ours	**0.834**	**10.23**	**0.106**	0.175	0.041	**0.609**	**0.228**	**32.14**	0.270	**0.060**

Comparison of Algorithm Complexity. We train our framework on the VoxCeleb dataset for 200 epochs using four RTX 3090 GPUs. The batch size is set to 20 in our experiments (Be consistent with the baseline). The overall consumption time is about 3 days. For the computational complexity, we test the flops, model params, and forward time per image and the results are shown in Table 2. This result shows that our method is almost optimal in terms of computation, model parameters, and inference time.

Table 2. Comparison of algorithm complexity.

	Flops (G)	Params (M)	Time (ms)
FOMM	**56.14**	59.56	40.19
PIRenderer	65.49	22.76	24.86
DaGAN	64.18	46.08	56.03
Ours	65.50	**22.54**	**23.28**

3.3 Ablation Study

We perform ablation experiments on the testing set to investigate the effectiveness of our method. We add proposed multiple sub-modules to the base model overlayed and give quantitative results in Table 3, qualitative results in Fig. 5. After using the enhanced domain discriminator to filter driving identity introduced due to network overfitting, the generated results (e.g. the fourth column in Fig. 5) have significantly less driving identity and more source identity clearly. Furthermore, we inject identity feature into the generation stage, which makes the results full of source texture (e.g. the fifth column in Fig. 5), especially in lips and chin, while we can find some inaccurate facial shape, particularly the contours of the face that are occluded in the source image. Finally, we add a shape-guided high-frequency attention block to refine the facial geometric shape, obviously the shape of the final generated face is more accurate. (e.g. the sixth column in Fig. 5)

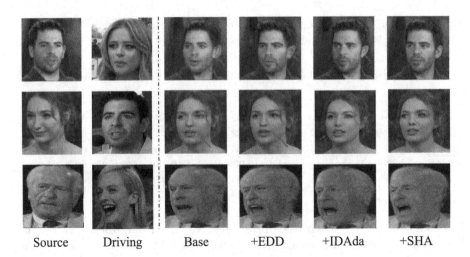

| Source | Driving | Base | +EDD | +IDAda | +SHA |

Fig. 5. Qualitative results of the ablation study, red arrows point to areas that are still not accurate enough. (Color figure online)

Table 3. Quantitative ablation results of cross-reenactment on VoxCeleb. Enhanced domain discriminator (EDD), ArcFace identity feature (IDAda), shape-guided high-frequency attention (SHA) are added to base model in turn.

	CSIM ↑	CSIM(dr) ↓	FID ↓	AED ↓	APD ↓
Base	0.535	0.263	36.25	**0.268**	0.060
+ EDD.	0.601	0.233	32.51	0.271	0.061
+ IDAda.	0.603	0.231	32.81	0.269	0.062
+ SHA.	**0.609**	**0.228**	**32.14**	0.270	**0.060**

4 Limitation

Although our method surpasses several state-of-the-art approaches, it does not perform optimally on more extreme task scenarios, such as when there is a huge discrepancy between the source and driver images in terms of pose and expression, or when there is severe occlusion in the source face. These limitations will be further explored in our future work.

5 Conclusion

In this paper, we propose a novel identity-preserving face reenactment framework with impressive pose and expression transfer. We use a domain adversarial loss to keep the motion latent vector away from the identity domain of driving. Meanwhile, we utilize 3D geometric-level and texture-level identity prior in two stages to close to source identity domain. Experimental results show the effectiveness of our proposed framework.

Ackownlegement. This work is supported by Shenzhen Fundamental Research Program (GXWD20201231165807007-20200806163656003) and National Natural Science Foundation of China (No. 62172021). We thank all reviewers for their valuable comments.

References

1. Blanz, V., Vetter, T.: A morphable model for the synthesis of 3d faces. In: SIGGRAPH (1999)
2. Burkov, E., Pasechnik, I., Grigorev, A., Lempitsky, V.: Neural head reenactment with latent pose descriptors. In: CVPR (2020)
3. Deng, J., Guo, J., Xue, N., Zafeiriou, S.: ArcFace: additive angular margin loss for deep face recognition. In: CVPR (2019)
4. Deng, Y., Yang, J., Xu, S., Chen, D., Jia, Y., Tong, X.: Accurate 3d face reconstruction with weakly-supervised learning: from single image to image set. Cornell University (2019)
5. Doukas, M.C., Zafeiriou, S., Sharmanska, V.: HeadGAN: one-shot neural head synthesis and editing. In: ICCV (2021)
6. Ganin, Y., et al.: Domain-adversarial training of neural networks. J. Mach. Learn. Res. **17**(1), 2030–2063 (2016)
7. Heusel, M., Ramsauer, H., Unterthiner, T., Nessler, B., Hochreiter, S.: GANs trained by a two time-scale update rule converge to a local nash equilibrium. In: Advances in Neural Information Processing Systems, vol. 30 (2017)
8. Hong, F.T., Zhang, L., Shen, L., Xu, D.: Depth-aware generative adversarial network for talking head video generation. In: CVPR (2022)
9. Huang, X., Belongie, S.: Arbitrary style transfer in real-time with adaptive instance normalization. In: ICCV (2017)
10. Kingma, D.P., Ba, J.: Adam: a method for stochastic optimization. arXiv preprint arXiv:1412.6980 (2014)
11. Liu, J., et al.: Identity preserving generative adversarial network for cross-domain person re-identification. IEEE Access **7**, 114021–114032 (2019)
12. Liu, J., et al.: Li-Net: large-pose identity-preserving face reenactment network. In: 2021 IEEE International Conference on Multimedia and Expo (ICME). IEEE (2021)
13. Nagrani, A., Chung, J.S., Zisserman, A.: VoxCeleb: a large-scale speaker identification dataset. arXiv preprint arXiv:1706.08612 (2017)
14. Ren, Y., Li, G., Chen, Y., Li, T.H., Liu, S.: PIRenderer: controllable portrait image generation via semantic neural rendering. In: ICCV (2021)
15. Siarohin, A., Lathuilière, S., Tulyakov, S., Ricci, E., Sebe, N.: First order motion model for image animation. In: NeurIPS (2019)
16. Siarohin, A., Woodford, O.J., Ren, J., Chai, M., Tulyakov, S.: Motion representations for articulated animation. In: CVPR (2021)
17. Wang, Y., et al.: HifiFace: 3d shape and semantic prior guided high fidelity face swapping. IJCAI (2021)
18. Xu, M., Chen, Y., Liu, S., Li, T.H., Li, G.: Structure-transformed texture-enhanced network for person image synthesis. In: ICCV (2021)
19. Yang, K., Chen, K., Guo, D., Zhang, S.H., Guo, Y.C., Zhang, W.: Face2face ρ: real-time high-resolution one-shot face reenactment. In: Avidan, S., Brostow, G., Cissé, M., Farinella, G.M., Hassner, T. (eds.) ECCV 2022. LNCS, vol. 13673, pp. 55–71. Springer, Cham (2022). https://doi.org/10.1007/978-3-031-19778-9_4

20. Yao, G., Yuan, Y., Shao, T., Zhou, K.: Mesh guided one-shot face reenactment using graph convolutional networks. In: ACMMM (2020)
21. Yin, F., et al.: StyleHEAT: one-shot high-resolution editable talking face generation via pre-trained styleGAN. In: Avidan, S., Brostow, G., Cissé, M., Farinella, G.M., Hassner, T. (eds.) ECCV 2022. LNCS, vol. 13677, pp. 85–101. Springer, Cham (2022). https://doi.org/10.1007/978-3-031-19790-1_6
22. Zhang, H., Ren, Y., Chen, Y., Li, G., Li, T.H.: Exploiting multiple guidance from 3dmm for face reenactment. In: AAAI Workshop (2023)
23. Zhang, R., Isola, P., Efros, A.A., Shechtman, E., Wang, O.: The unreasonable effectiveness of deep features as a perceptual metric. In: CVPR (2018)
24. Zheng, M., Karanam, S., Chen, T., Radke, R.J., Wu, Z.: HifiHead: one-shot high fidelity neural head synthesis with 3d control. In: IJCAI (2022)

L2MNet: Enhancing Continual Semantic Segmentation with Mask Matching

Wenbo Zhang, Bocen Li, and Yifan Wang[✉]

Dalian University of Technology, Dalian 116024, China
wyfan@dlut.edu.cn

Abstract. Continual semantic segmentation (CSS) aims to continuously learn a semantic segmentation model that incorporates new categories while avoiding forgetting the previously seen categories. However, CSS faces a significant challenge known as weight shift, which leads to the network mistakenly predicting masks belonging to new categories instead of their actual categories. To mitigate this phenomenon, we propose a novel module named mask matching module, which transfers pixel-level prediction task into a mask-level feature matching task by computing the similarity between mask features and prototypes. Further, we introduce a new paradigm and a network called **Learn-to-Match** (L2M) Net, which alleviates weight shift and gains remarkable improvements on long settings by leveraging mask-level feature matching. Our method can be easily integrated into various network architectures without extra memory and data cost. Experiments conducted on the Pascal-VOC 2012 and ADE20K datasets demonstrate that, particularly on long settings where CSS encounters more challenging settings, our method achieves a remarkable 10.6% improvement in terms of all mean Intersection over Union (mIoU) and establishes a new state-of-the-art performance in the demanding CSS settings.

Keywords: Semantic segmentation · Continual semantic segmentation · Class-incremental semantic segmentation

1 Introduction

Semantic segmentation is an important task in computer vision, which aims to assign each the corresponding semantic category to every pixel in an image. This makes semantic segmentation widely applicable in many practical applications, such as autonomous driving, medical image analysis, and smart homes. However, in real-world, the assumption of independent and identically distributed (i.i.d.) [7] does not hold for many settings. As a consequence, deep neural networks are prone to experiencing *catastrophic forgetting* [15,33]. In another word, the acquisition of new knowledge will significantly lower the performance on the previously learned knowledge. In recent years, there has been a growing focus on addressing this issue within the framework of CSS [11,26,27].

© The Author(s), under exclusive license to Springer Nature Singapore Pte Ltd. 2024
Q. Liu et al. (Eds.): PRCV 2023, LNCS 14434, pp. 125–136, 2024.
https://doi.org/10.1007/978-981-99-8549-4_11

Beyond that, recent works [12,27] show that weight shift plays an important role in catastrophic forgetting, which means that the network exhibiting a higher tendency to incorrectly classify an object as a newly learned category, rather than its actual category. Some methods [12,27] establish relationships among classes or normalize classifier to solve this problem. While they have made significant strides in CSS, they remain unsatisfactory on long settings due to a lack of comprehensive understanding of the relationships between different categories.

Most CSS methods adopt the paradigm that involves feature extraction and pixel-level mask prediction, which we call learn-to-predict. However, this paradigm is prone to suffer from weight shift due to the training of the classifier, leading to potential issues in model stability. In this paper, we introduce a novel paradigm and network called the **Learn-to-Match** (L2M) Net to address the issue of weight shift in CSS. Our proposed method transfers a pixel-level prediction task into a mask-level feature matching task, which aims to learn the relationships between features and different categories.

Specifically, we replace classifier to a novel module called the Mask Matching module, which calculates the similarity between mask features and category prototypes. By leveraging this module, our method surpasses the CSS approaches of directly predicting mask categories, enabling the network to develop a more comprehensive understanding of the relationships between features and different categories. Experimental results demonstrate the effectiveness of our learn-to-match paradigm in mitigating weight shift phenomenon and achieving remarkable performance gains. Specifically, our approach achieves a significant improvement of 10.6% in terms of overall mean Intersection over Union (mIoU) when facing longer setup, demonstrating its superiority in handling continual learning scenarios.

The main contributions can be summarized as follows:

(1) We propose an efficient CSS paradigm and a network called L2MNet. This paradigm treats CSS as a mask-level feature matching task. Comparing with pixel-level prediction method, it effectively alleviates weight shift issues on previously learned categories while simultaneously improving the performance on newly introduced categories.
(2) We propose a mask matching module, which aims to learn the relation ships between mask-level features and prototypes. It can be easily utilized with various CSS approaches.
(3) We experimentally show that our method significantly improves recent CSS approaches on several CSS settings and sets a new state-of-the-art under long settings.

2   Related Work

Semantic Segmentation. With the development of deep learning technique, semantic segmentation methods based on CNN and transformer have achieved impressive results. Encoder-decoder [10,37], multi-scale pooling [5], larger kernels [36] and multi-branch [31] become the most common structure for extracting and fusing feature information. Thanks to the global context modeling ability of transformer [34], several works [6,16,39] design the variants of self-attention mechanisms and achieve great success.

Continual Learning. Continual Learning studies the problem of learning from an infinite stream of data, with the goal of gradually extending acquired knowledge and avoiding forgetting. According to the division of incremental batches and the availability of task identities, the typical continual learning scenarios can be divided into task-incremental learning, class-incremental learning and domain-incremental learning [7,18,35]. Our work studies class-incremental learning for semantic segmentation. Rehearsal-based methods [4,8,20,28,29] prevent catastrophic forgetting by replaying a subset of exemplars stored in limited memory. iCaRL [28] builds class exemplar sets and trains model by self-distillation loss. SER [20] selects what experience memory will be stored for better distribution matching. Regularization-based methods [21–23,40] introduce an extra regularization term in the loss function. PODNet [13] applies an efficient spatial-based distillation-loss. Model structure based methods [30,32,42] change or expand model structure for parameter isolation or task-specific training.

Continual Semantic Segmentation. Continuous semantic segmentation is a sub problem of continuous learning, extending existing continual learning method [23] for semantic segmentation. Most methods are improved based on continual learning methods. PLOP [11] proposes a multi-scale pooling distillation scheme Local POD to preserve spatial information. [26] presents some latent representation shaping techniques to prevent forgetting. REMINDER [27] focuses on weight shift and uses semantic similarity between classes as a prior. RCIL [41] dynamically expands the network with a representation compensation module and proposes new distillation mechanism. Some methods [3,12,19,24,38] use exemplar-memory or extra models for current step learning. [19,24] generate extra samples from the class space of past learning steps.

3   Method

3.1   Preliminaries and Revisiting

Preliminaries. CSS aims to train the model f in $t = 0...T$ steps, where t refers to the number of times that new labels are provided. Let $\mathcal{D} = \{X, Y\}$ denote the training dataset, where X, Y are images and labels, respectively. At step t, given a model f_{t-1} trained on $\{\mathcal{D}_0, \mathcal{D}_1, ..., \mathcal{D}_{t-1}\}$ with $\{\mathcal{C}_0, \mathcal{C}_1, ..., \mathcal{C}_{t-1}\}$ classes continually, the model is supposed to learn the discrimination for all seen classes $\mathcal{C}_{0:t}$ when only $\mathcal{D}_t$ with newly added $\mathcal{C}_t$ classes is accessible. Notably, when training on $\mathcal{D}_t$, labels Y_t only consists of current classes $\mathcal{C}_t$, which means all other classes (i.e., old classes $\mathcal{C}_{0:t-1}$ and future classes $\mathcal{C}_{t+1:T}$) are labeled as background class c_b. Hence, pixels labeled as background class gradually become the foreground classes during future incremental steps.

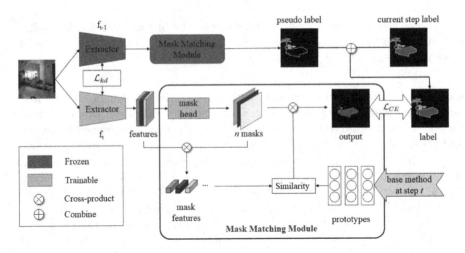

Fig. 1. Overview of the proposed L2M framework. At step t, we first get prototypes by *base method*. Then, given the input image, it is sent to the previous step model f_{t-1} for pseudo-labeling. Meanwhile, the current model f_t extracts features and predicts masks throughout mask head. Then, we get mask features from the above with cross-product, and compute the similarity between mask features and prototypes. At last we calculate the cross-production between similarity and predicted masks for final output.

3.2 Proposed Learn-to-Match Framework

Figure 1 illustrates the framework of our proposed method. The input image is processed by a feature extractor, composed of a ResNet-101 and an ASPP module. The extracted features are then fed into our mask matching module, which combines them with prototypes to produce final segmentation results. At step t, the frozen model f_{t-1} generates pseudo label of old classes and combines it with current step label. We train the model f_t using cross-entropy loss with combined labels and distillation loss between old model f_{t-1} and current model f_t.

Prototype Computing. Prototypes (i.e., class-centroids) are vectors that are representation of each category that appears in the dataset [26], and they contribute in forming the latent prototypical feature space. To generate the prototypes, a simple solution is to mine the features through clustering, extracted by the feature extractor with pixel-level labels. Mined features are computed with a running average updated at each training batch. Given a batch of extracted features $\mathcal{F} \in \mathbb{R}^{B \times H \times W \times C}$, we compute an in-batch average of $\mathcal{F}_i$, where $i = 1, ..., BHW$. The prototype of class c can be written as

$$p_c = \frac{\sum_{i=1}^{BHW} \mathcal{F}_i \mathbb{1}[y_i = c]}{|\{i : y_i = c\}|}, \tag{1}$$

where $\mathbb{1}[y_i = c] = 1$ if the label of pixel i is c, else 0 otherwise. $|\cdot|$ denotes cardinality and $p_c \in \mathbb{R}^{1 \times C}$. Significantly, we compute prototypes by any CSS method, called the *base method*, such as PLOP and REMINDER. Specifically, at step t, we first train a model by *base method* and get prototypes as Eq. 1, then we train L2M model with prototypes. To obtain accurate prototypes, we compute them at the end of *base method*'s each step and only update those prototypes when the ground truth labels of corresponding categories are available, which means once we get the prototype of certain category c, we freeze it and do not update it in later steps.

Mask Matching Module. To avoid the weight shift of classifier, we propose the mask matching module to replace classifier. Specifically, given the extracted feature $\mathcal{F} \in \mathbb{R}^{H \times W \times C}$, we first predict n masks through the mask head that consists of MLP layers, which can be written as: $g_{mask}(\mathcal{F}) \in \mathbb{R}^{n \times H \times W}$. Each mask is regarded as the assembly of semantic information. Then, we can get mask features $\hat{m}$ according to each mask and extracted features with cross-product. Here, it is formulated as

$$\hat{m} = \mathcal{F} \otimes g_{mask}(\mathcal{F}). \tag{2}$$

We construct n mask features $\hat{m} \in \mathbb{R}^{n \times C}$ where each mask feature $i = 1, 2, ..., n$ contains a feature vector $\hat{m}_i$. Then we compute the similarity $S \in \mathbb{R}^{n \times C}$ between the prototype p_c of each class and mask feature $\hat{m}_i$ where $\mathcal{C}$ denotes classes. Each entry $s_{i,c} \in S$ is the cosine similarity between $\hat{m}_i$ and p_c, and the similarity is normalized to reflect the probability that the mask feature is similar to class c. The above can be defined as:

$$s_{i,c} = \frac{\hat{m}_i \cdot p_c}{||\hat{m}_i|| \cdot ||p_c||}, \tag{3}$$

$$\tilde{s}_{i,c} = \frac{\exp s_{i,c}}{\sum_{j=0}^{\mathcal{C}} \exp s_{i,j}}. \tag{4}$$

Then, our method applies similarity to predicted n masks for segmentation prediction $\hat{O}$, which is formulated as

$$\hat{O} = Softmax(\tilde{S} \otimes g_{mask}(\mathcal{F})). \tag{5}$$

Via mask matching module, the model captures the class similarity $\tilde{S}$ of mask features and embeds it into outputs $\hat{O}$.

3.3   Training Loss

Pseudo Labels. As mentioned above, the pixels labeled as background at current step may be labeled as foreground classes at previous steps or future steps, which leads to semantic background shift. Thus, we utilize pseudo-labeling strategy [11] to alleviate this problem. In our case, we use old model to generate pseudo-labels. Formally, at step t, given the old model f_{t-1}, we can get old classes outputs $\hat{O}^{t-1}$ and combine pseudo-labels and true labels as $\tilde{Y}^t$, which is formulated as

$$\tilde{Y}^t_{i,c} = \begin{cases} 1, & \text{if } Y^t_{i,c_b} = 0 \text{ and } c = \underset{c' \in \mathcal{C}_t}{\operatorname{argmax}} Y^t_{i,c'}, \\ 1, & \text{if } Y^t_{i,c_b} = 1 \text{ and } c = \underset{c' \in \mathcal{C}_{0:t-1}}{\operatorname{argmax}} \hat{O}^t_{i,c'}, \\ 0, & \text{otherwise.} \end{cases} \tag{6}$$

Then, we use the cross-entropy loss from combined labels $\tilde{Y}^t_{i,c}$ to train current step model f_t:

$$\mathcal{L}_{CE} = -\frac{\lambda}{HW} \sum_{i=1}^{HW} \sum_{c \in \mathcal{C}_{0:t}} \tilde{Y}^t_{i,c} \log O^t_{i,c} \tag{7}$$

Knowledge Distillation. As has been used in recent works [11,27], we use knowledge distillation to alleviate catastrophic forgetting. The feature distillation loss is formulated as

$$\mathcal{L}_{kd} = \sum_{l=1}^{L} ||\Theta(f^t_l(X^t)) - \Theta(f^{t-1}_l(X^t))||^2, \tag{8}$$

where l is the l-th layer of the network, $|| \cdot ||$ is the L2 distance the two sets of features and $\Theta(\cdot)$ is a function of summarizing spatial statistics of features. We select Local POD [11] as summarizing strategy $\Theta(\cdot)$.

The final loss function is the summation of cross-entropy loss and distillation loss:

$$\mathcal{L} = \mathcal{L}_{kd} + \mathcal{L}_{CE} \tag{9}$$

4   Experiments

4.1   Experimental Setting

Dataset. We follow the experimental setting of [27] and evaluate the performance of our proposed framework by standard image semantic segmentation dataset: Pascal-VOC 2012 [14] and ADE20k [43]. Pascal-VOC 2012 contains 20 foreground classes and background class. Its training and testing sets contain 10,582 and 1,449 images, respectively. ADE20k has 150 foreground classes, 20,210 training images, and 2,000 testing images.

CSS Protocols. In continual semantic segmentation, the training process consists of multiple steps. There are two different experimental settings: disjoint and overlapped. At step t, in the former, training samples only contain old classes $\mathcal{C}_{0:t-1}$ and current classes $\mathcal{C}_t$, while, in the latter, samples can contain future classes $\mathcal{C}_{t+1:T}$. Thus, we believe that the latter is more realistic and challenging and evaluate on the overlapped setup only.

For the Pascal-VOC 2012 dataset, we perform settings as follows: 15-5 setting with 2 steps (train 15 classes at step 0 and train 5 classes in the next step), 15-1 setting with 6 steps (train 15 classes at step 0 and train 1 class sequentially in later steps), 10-1 setting with 11 steps (train 10 classes at step 0 and train 1 class sequentially in later steps). For ADE20k, we perform settings 100-50, 100-10 and 100-5, similarly. Generally, longer setup with more steps is more challenging.

Evaluation Metrics. We use four mean Intersection-over-Union (mIoU) evaluation metrics. We compute mIoU at step 0 and other steps, which measures model's rigidity and plasticity respectively. Besides, we also compute the last step model's mIoU for all seen classes. Lastly, we compute the average of mIoU [11] at each step to reflect the performance over the entire continual learning process.

Baselines. We benchmark our method against the latest state-of-the-arts CSS methods including LwF [23], ILT [25], MiB [2], SDR [26], RCIL [41], PLOP [11], REMINDER [27]. For a fair comparison, all models do not allow the rehearsal memory where a limited quantity of previous classes samples can be rehearsed. Finally, we train a model with mixed training dataset including all classes, called *Joint*, which may constitute an upper bound for CSS methods.

Implementation Details. We select PLOP and REMINDER as *base method* respectively to compute prototypes in our experiments. L2MNet uses a Deeplab-V3 [5] architecture with a ResNet-101 [17] backbone pretrained on ImageNet [9] for all experiments. The output stride of Deeplab-v3 is set to 16. We apply the in-place activated batch normalization [1] as normalization layer. And we apply the same augmentation strategy as PLOP, e.g., horizontal flip and random crop. For all settings, the batch size is set to 24. And we optimize the network using SGD with an initial learning rate of 0.01 and a momentum value of 0.9. Also, we set the learning rate schedule following [11,27]. The number of mask head n is set to 100 for Pascal-VOC 2012 and 300 for ADE20k. Details of pseudo-label strategy and knowledge distillation follow the setting of PLOP.

4.2   Quantitative Evaluation

Table 1 shows quantitative experiments on VOC 15-5, 15-1 and 10-1. Our method's results are written as $\{\cdot\} + Ours$, where $\{\cdot\}$ denotes we select which method as *base method*. On 15-5, our method based on PLOP is on par with PLOP, as the performance of mask matching module is slightly worse than traditional classifier in non-continual semantic segmentation (i.e. step 0). Our method based on REMINDER is improved (+1.15%). On other settings, our method shows significant improvements on both old and new classes, whether based on PLOP or REMINDER. Especially, on the most challenging 10-1 setting, our method outperforms base method and recent SOTA approach (RCIL) a lot. The average mIoU shows that our method is more robust at each step. Table 2 shows

Table 1. Results on Pascal-VOC 2012 in mIoU (%). †: excerpted from [27]. Other results comes from re-implementation.

method	15-5(2 steps)				15-1(6 steps)				10-1(11 steps)			
	1–15	16–20	all	avg	1–15	16–20	all	avg	1–15	16–20	all	avg
LwF† [23]	60.8	36.6	55.0	–	6.0	3.9	5.5	–	8.0	2.0	4.8	–
ILT† [25]	67.8	40.6	61.3	–	9.6	7.8	9.2	–	7.2	3.7	5.5	–
MiB† [2]	76.4	49.4	70.0	–	38.0	13.5	32.2	–	20.0	20.1	20.1	–
SDR† [26]	76.3	50.2	70.1	–	47.3	14.7	39.5	–	32.4	17.1	25.1	–
RCIL† [41]	**78.8**	**52.0**	**72.4**	–	70.6	23.7	59.4	–	55.4	15.1	34.3	–
PLOP [11]	76.03	51.19	70.12	75.18	66.38	22.94	56.04	67.18	38.60	15.81	27.75	50.55
PLOP+Ours	76.57	49.81	70.20	75.20	70.64	18.41	58.20	67.30	63.35	21.18	43.27	56.21
REMINDER [27]	75.63	48.90	69.27	74.96	67.22	27.53	57.77	68.32	38.93	9.68	25.00	50.63
REMINDER+Ours	76.50	50.97	70.42	**75.29**	**72.97**	**29.67**	**62.66**	**69.86**	**64.40**	**23.34**	**44.85**	**57.96**
Joint†	78.2	78.0	78.2	–	79.8	72.6	78.2	–	79.8	72.6	78.2	–

Table 2. Results on ADE20k in mIoU (%). †: excerpted from [27]. Other results comes from re-implementation.

method	100-50(2 steps)				100-10(6 steps)				100-5(11 steps)			
	1–100	101–150	all	avg	1–100	101–150	all	avg	1–100	101–150	all	avg
ILT† [25]	18.3	14.8	17.0	–	0.1	2.9	1.1	–	0.1	1.3	0.5	–
MiB† [2]	40.7	17.7	32.8	–	38.3	11.3	29.2	–	36.0	5.6	25.9	–
RCIL† [41]	42.3	18.8	34.5	–	39.3	17.6	32.1	–	38.5	11.5	29.6	–
PLOP [11]	41.65	13.29	32.26	37.46	39.41	11.88	30.29	36.48	36.15	7.14	26.55	34.35
PLOP+Ours	42.65	19.27	34.91	38.78	41.18	16.74	33.09	37.54	40.27	13.61	31.44	36.24
REMINDER [27]	41.82	**21.92**	**35.23**	**38.94**	38.43	**20.55**	32.51	37.26	36.35	**14.77**	29.21	35.43
REMINDER+Ours	**42.80**	19.63	35.13	38.89	**41.43**	16.66	**33.23**	**37.58**	**40.50**	13.78	**31.65**	**36.35**
Joint†	44.3	28.2	38.9	–	44.3	28.2	38.9	–	44.3	28.2	38.9	–

Table 3. Performance on the Pascal-VOC 15-1 setting when using different predicting and distillation strategy.

mask head	distillation	1–15	16–20	all	avg
✓		70.22	18.7	57.95	67.19
	✓	50.59	6.08	39.99	54.31
✓	✓	70.64	18.41	58.20	67.30

results on the ADE20k 100-50, 100-10 and 100-5 settings. On 100-50 setting, our method outperforms REMINDER by 0.98% on old classes. On the long settings 100-10 (6 steps) and 100-5 (11 steps), our method outperforms on *1-100*, *all* and *avg* metrics.

Figure 2 shows the predictions for PLOP, REMINDER and our method with them on VOC 15-1 setting across time. All of them perform well at step 0. As the steps proceed, PLOP and REMINDER forget old knowledge gradually. When new categories train (green) and tv monitor (blue) are added at step 4 and step 5, we can observe severe weight shift on PLOP and REMINDER, some pixels are

Table 4. Performance on the Pascal-VOC 15-1 (top) and the ADE20k 100-5 (bottom) settings with different number of mask head n.

Dataset	n	1–15	16–20	1–100	101–150	all	avg
Pascal-VOC	50	69.22	22.84	–	–	58.18	67.33
	100	70.64	18.41	–	–	58.2	67.3
	200	69.35	21.94	–	—	58.06	67.46
ADE20k	100	–	–	38.58	9.67	29.01	34.14
	300	–	–	40.5	13.78	31.65	36.35

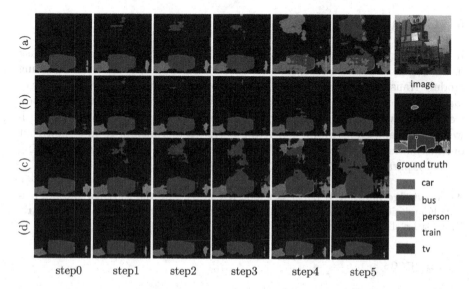

Fig. 2. Visualization of (a) PLOP, (b) PLOP+Ours, (c) REMINDER, (d) REMINDER+Ours predictions across steps in VOC 15-1. The classes after step 0 are potted plant, sheep, sofa, train and tv monitor respectively.

wrongly predicted as new classes. On the other hand, our method keeps much more stable and avoid that the model generates strong bias toward new classes.

4.3 Ablation Study

Effectiveness of Mask Matching Module. Here, we analyze the effect of mask matching module. Table 3 compares the mIoU of each ablation case on VOC 15-1 scenario based on PLOP. Firstly, based on the distillation of the feature extractor, we discuss whether the output logits of the mask head need to be distilled too. Although the mask head without distillation performs well on new classes, the one through distillation has better results on overall classes. Besides, when the mask head is removed, we compute the similarity between prototypes and per-pixel feature embedding from feature extractor. We obverse that the performance decreases a lot without mask head, which means mask head can greatly utilize feature information and mask-level similarity can lead to more stable learning.

Number of Mask Head. Table 4 shows the influence of the number n of mask head. We observe that the number of mask head has little effect on VOC 15-1 setting. 100 is enough for Pascal-VOC (20 classes), however, the accuracy of model on ADE20k (150 classes) decreases a lot when $n = 100$. We need to increase the number of mask head for datasets including more categories.

5   Conclusion

In this paper, we propose a new paradigm that leverages the power of mask matching in continual semantic segmentation. Our proposed mask matching module transfers a pixel-level prediction task into a mask-level feature matching task by computing the similarity between mask features and prototypes. Comparing with pixel-level prediction methods, our method can handle with weight shift and catastrophic forgetting better. Experimental results show that our method can achieve better balance of learning new knowledge and keeping old ones, particularly on long settings.

Acknowledgement. The paper is supported in part by the National Natural Science Foundation of China (62006036), and Fundamental Research Funds for Central Universities (DUT22LAB124, DUT22QN228).

References

1. Bulo, S.R., Porzi, L., Kontschieder, P.: In-place activated batchnorm for memory-optimized training of DNNs. In: Proceedings of the IEEE Conference on Computer Vision and Pattern Recognition, pp. 5639–5647 (2018)
2. Cermelli, F., Mancini, M., Bulo, S.R., Ricci, E., Caputo, B.: Modeling the background for incremental learning in semantic segmentation. In: Proceedings of the IEEE/CVF Conference on Computer Vision and Pattern Recognition, pp. 9233–9242 (2020)
3. Cha, S., Yoo, Y., Moon, T., et al.: SSUL: semantic segmentation with unknown label for exemplar-based class-incremental learning. Adv. Neural. Inf. Process. Syst. **34**, 10919–10930 (2021)
4. Chaudhry, A., et al.: Continual learning with tiny episodic memories (2019)
5. Chen, L.C., Papandreou, G., Schroff, F., Adam, H.: Rethinking atrous convolution for semantic image segmentation. arXiv preprint arXiv:1706.05587 (2017)
6. Cheng, B., Schwing, A., Kirillov, A.: Per-pixel classification is not all you need for semantic segmentation. Adv. Neural. Inf. Process. Syst. **34**, 17864–17875 (2021)
7. De Lange, M., et al.: A continual learning survey: defying forgetting in classification tasks. IEEE Trans. Pattern Anal. Mach. Intell. **44**(7), 3366–3385 (2021)
8. De Lange, M., Tuytelaars, T.: Continual prototype evolution: learning online from non-stationary data streams. In: Proceedings of the IEEE/CVF International Conference on Computer Vision, pp. 8250–8259 (2021)

9. Deng, J., Dong, W., Socher, R., Li, L.J., Li, K., Fei-Fei, L.: ImageNet: a large-scale hierarchical image database. In: 2009 IEEE Conference on Computer Vision and Pattern Recognition, pp. 248–255. IEEE (2009)

10. Ding, H., Jiang, X., Shuai, B., Liu, A.Q., Wang, G.: Semantic segmentation with context encoding and multi-path decoding. IEEE Trans. Image Process. **29**, 3520–3533 (2020)

11. Douillard, A., Chen, Y., Dapogny, A., Cord, M.: PLOP: learning without forgetting for continual semantic segmentation. In: Proceedings of the IEEE/CVF Conference on Computer Vision and Pattern Recognition, pp. 4040–4050 (2021)

12. Douillard, A., Chen, Y., Dapogny, A., Cord, M.: Tackling catastrophic forgetting and background shift in continual semantic segmentation. arXiv preprint arXiv:2106.15287 (2021)

13. Douillard, A., Cord, M., Ollion, C., Robert, T., Valle, E.: PODNet: pooled outputs distillation for small-tasks incremental learning. In: Vedaldi, A., Bischof, H., Brox, T., Frahm, J.-M. (eds.) ECCV 2020, Part XX. LNCS, vol. 12365, pp. 86–102. Springer, Cham (2020). https://doi.org/10.1007/978-3-030-58565-5_6

14. Everingham, M., Van Gool, L., Williams, C.K., Winn, J., Zisserman, A.: The pascal visual object classes (VOC) challenge. Int. J. Comput. Vision **88**, 303–338 (2010)

15. French, R.M.: Catastrophic forgetting in connectionist networks. Trends Cogn. Sci. **3**(4), 128–135 (1999)

16. Fu, J., et al.: Dual attention network for scene segmentation. In: Proceedings of the IEEE/CVF Conference on Computer Vision and Pattern Recognition, pp. 3146–3154 (2019)

17. He, K., Zhang, X., Ren, S., Sun, J.: Deep residual learning for image recognition. In: Proceedings of the IEEE Conference on Computer Vision and Pattern Recognition, pp. 770–778 (2016)

18. Hsu, Y.C., Liu, Y.C., Ramasamy, A., Kira, Z.: Re-evaluating continual learning scenarios: a categorization and case for strong baselines. arXiv preprint arXiv:1810.12488 (2018)

19. Huang, Z., et al.: Half-real half-fake distillation for class-incremental semantic segmentation. arXiv preprint arXiv:2104.00875 (2021)

20. Isele, D., Cosgun, A.: Selective experience replay for lifelong learning. In: Proceedings of the AAAI Conference on Artificial Intelligence, vol. 32 (2018)

21. Jung, H., Ju, J., Jung, M., Kim, J.: Less-forgetting learning in deep neural networks. arXiv preprint arXiv:1607.00122 (2016)

22. Kirkpatrick, J., et al.: Overcoming catastrophic forgetting in neural networks. Proc. Nat. Acad. Sci. **114**(13), 3521–3526 (2017)

23. Li, Z., Hoiem, D.: Learning without forgetting. IEEE Trans. Pattern Anal. Mach. Intell. **40**(12), 2935–2947 (2017)

24. Maracani, A., Michieli, U., Toldo, M., Zanuttigh, P.: Recall: replay-based continual learning in semantic segmentation. In: Proceedings of the IEEE/CVF International Conference on Computer Vision, pp. 7026–7035 (2021)

25. Michieli, U., Zanuttigh, P.: Incremental learning techniques for semantic segmentation. In: Proceedings of the IEEE/CVF International Conference on Computer Vision Workshops, pp. 0–0 (2019)

26. Michieli, U., Zanuttigh, P.: Continual semantic segmentation via repulsion-attraction of sparse and disentangled latent representations. In: Proceedings of the IEEE/CVF Conference on Computer Vision and Pattern Recognition, pp. 1114–1124 (2021)

27. Phan, M.H., Phung, S.L., Tran-Thanh, L., Bouzerdoum, A., et al.: Class similarity weighted knowledge distillation for continual semantic segmentation. In: Proceedings of the IEEE/CVF Conference on Computer Vision and Pattern Recognition, pp. 16866–16875 (2022)
28. Rebuffi, S.A., Kolesnikov, A., Sperl, G., Lampert, C.H.: ICARL: incremental classifier and representation learning. In: Proceedings of the IEEE Conference on Computer Vision and Pattern Recognition, pp. 2001–2010 (2017)
29. Rolnick, D., Ahuja, A., Schwarz, J., Lillicrap, T., Wayne, G.: Experience replay for continual learning. Adv. Neural Inf. Process. Syst. **32** (2019)
30. Singh, P., Mazumder, P., Rai, P., Namboodiri, V.P.: Rectification-based knowledge retention for continual learning. In: Proceedings of the IEEE/CVF Conference on Computer Vision and Pattern Recognition, pp. 15282–15291 (2021)
31. Sun, K., et al.: High-resolution representations for labeling pixels and regions. arXiv preprint arXiv:1904.04514 (2019)
32. Tao, X., Hong, X., Chang, X., Dong, S., Wei, X., Gong, Y.: Few-shot class-incremental learning. In: Proceedings of the IEEE/CVF Conference on Computer Vision and Pattern Recognition, pp. 12183–12192 (2020)
33. Thrun, S.: Lifelong learning algorithms. Learn. Learn **8**, 181–209 (1998)
34. Vaswani, A., et al.: Attention is all you need. Adv. Neural Inf. Process. Syst. **30** (2017)
35. Van de Ven, G.M., Tolias, A.S.: Three scenarios for continual learning. arXiv preprint arXiv:1904.07734 (2019)
36. Wu, Y., et al.: Large scale incremental learning. In: Proceedings of the IEEE/CVF Conference on Computer Vision and Pattern Recognition, pp. 374–382 (2019)
37. Yu, C., Wang, J., Peng, C., Gao, C., Yu, G., Sang, N.: Learning a discriminative feature network for semantic segmentation. In: Proceedings of the IEEE Conference on Computer Vision and Pattern Recognition, pp. 1857–1866 (2018)
38. Yu, L., Liu, X., Van de Weijer, J.: Self-training for class-incremental semantic segmentation. IEEE Trans. Neural Netw. Learning Syst. **34**, 9116–9127 (2022)
39. Yuan, Y., Chen, X., Wang, J.: Object-contextual representations for semantic segmentation. In: Vedaldi, A., Bischof, H., Brox, T., Frahm, J.-M. (eds.) ECCV 2020, Part VI. LNCS, vol. 12351, pp. 173–190. Springer, Cham (2020). https://doi.org/10.1007/978-3-030-58539-6_11
40. Zenke, F., Poole, B., Ganguli, S.: Continual learning through synaptic intelligence. In: International Conference on Machine Learning, pp. 3987–3995. PMLR (2017)
41. Zhang, C.B., Xiao, J.W., Liu, X., Chen, Y.C., Cheng, M.M.: Representation compensation networks for continual semantic segmentation. In: Proceedings of the IEEE/CVF Conference on Computer Vision and Pattern Recognition, pp. 7053–7064 (2022)
42. Zhang, C., Song, N., Lin, G., Zheng, Y., Pan, P., Xu, Y.: Few-shot incremental learning with continually evolved classifiers. In: Proceedings of the IEEE/CVF Conference on Computer Vision and Pattern Recognition, pp. 12455–12464 (2021)
43. Zhou, B., Zhao, H., Puig, X., Fidler, S., Barriuso, A., Torralba, A.: Scene parsing through ade20k dataset. In: Proceedings of the IEEE Conference on Computer Vision and Pattern Recognition, pp. 633–641 (2017)

Adaptive Channel Pruning
for Trainability Protection

Jiaxin Liu[1,2] , Dazong Zhang[4], Wei Liu[1,2(✉)], Yongming Li[3], Jun Hu[2], Shuai Cheng[2], and Wenxing Yang[1,2]

[1] School of Computer Science and Engineering, Northeastern University,
Liaoning 110167, China
[2] Neusoft Reach Automotive Technology Company, Liaoning 110179, China
lwei@reachauto.com
[3] College of Science, Liaoning University of Technology, Liaoing 121001, China
[4] BYD Auto Industry Company Limited, Shenzhen 518118, China

Abstract. Pruning is a widely used method for compressing neural networks, reducing their computational requirements by removing unimportant connections. However, many existing pruning methods prune pretrained models by using the same pruning rate for each layer, neglecting the protection of model trainability and damaging accuracy. Additionally, the number of redundant parameters per layer in complex models varies, necessitating adjustment of the pruning rate according to model structure and training data. To overcome these issues, we propose a trainability-preserving adaptive channel pruning method that prunes during training. Our approach utilizes a model weight-based similarity calculation module to eliminate unnecessary channels while protecting model trainability and correcting output feature maps. An adaptive sparsity control module assigns pruning rates for each layer according to a preset target and aids network training. We performed experiments on CIFAR-10 and Imagenet classification datasets using networks of various structures. Our technique outperformed comparison methods at different pruning rates. Additionally, we confirmed the effectiveness of our technique on the object detection datasets VOC and COCO.

Keywords: Convolutional neural networks · Trainability
preservation · Model compression · Pruning

1 Introduction

Deep convolutional neural networks (CNNs) are widely used in the field of computer vision, such as classification [6], object detection [4] and segmentation [16]. However, it is challenging to deploy many models on mobile devices due to the

---

This work is supported by the National Natural Science Foundation of China under Grant U22A2043, and the Unveiling the list of hanging (science and technology research special) of Liaoning province under Grant 2022JH1/10400030.

Q. Liu et al. (Eds.): PRCV 2023, LNCS 14434, pp. 137–148, 2024.
https://doi.org/10.1007/978-981-99-8549-4_12

models' high computational complexity and huge number of parameters, which place a heavy burden on the device's processing speed and memory. To address this issue, several model compression methods have been proposed. Pruning is one of the crucial directions.

Pruning compresses the number of parameters and computational cost of a model by removing unimportant connections in the model, and it can be divided into two categories according to the granularity of pruning: unstructured and structured pruning. Unstructured pruning methods prune the convolution kernel irregularly sparse [5] and need to be applied on hardware with special acceleration libraries to be accelerated. The pruned model of structured pruning method can be directly applied to mainstream hardware devices [9], where the channel pruning method is a mainstream structured pruning method.

Channel pruning methods can be divided into two types: static pruning and dynamic pruning, according to the method of filter importance calculation. The traditional static pruning methods have two main issues. The first is that the importance evaluation only considers the inherent characteristics of the trained model without taking into account the trainability of the pruned model, making it difficult to recover accuracy even after fine-tuning. Second, most methods use a global uniform pruning rate, but there is no fixed pruning rate applicable to each layer [1]. Dynamic pruning methods improve performance by introducing auxiliary modules that calculate the relevant channel salience score for each batch of input data and dynamically prune the unimportant parts of the network; however, they must save the entire network during inference and rely on activating different parts of the network to reduce computation without reducing memory consumption. Moreover, these methods introduce the auxiliary module only to calculate the salience score, ignoring its function of assisting network training. To address the problem presented above, we propose an adaptive static pruning method based on the similarity of convolutional kernels, motivated by the model trainability preservation methods [17,19]. Our method removes the redundant channels during training while maintaining the trainability of the original network and adaptively allocating the pruning rate of each layer. The main contribution of this paper consists of three aspects:

1. We design a similarity calculation module based on convolutional kernel weights to determine the pruned channel during training and maintain the trainability of the network.
2. We design an adaptive sparse control module that makes the discretization process of the saliency scores during training differentiable and assists the training of the network. The sparse control module interacts with the pruning rate regularization term in the loss function to adaptively assign the pruning rate of each layer.
3. The extensive experiments demonstrate that our method performs well on different tasks and different structured network models. Compared with existing pruning methods, the accuracy degradation of our method is lower for similar or larger pruning rates, demonstrating the effectiveness of our method.

2   Related Work

Pruning is a common model compression method, which can be divided into unstructured and structured pruning according to the granularity of pruning. Weight pruning is a common unstructured pruning technique that removes unimportant individual weights from the convolution kernel [5]. This causes the convolution kernel of the model to have irregular sparsity, making it difficult to apply the pruned model to general-purpose hardware or BLAS (Basic Linear Algebra Subprograms) libraries. The structure of the model after structured pruning is still regular and can avoid the application limitation of weight pruning. Typical structured pruning methods include kernel pruning [11], channel pruning [2,8,9,12,15], and layer pruning. Besides, in order to balance the computational efficiency and the performance of the pruned model, [21,22] proposed the N:M pruning method to retain up to N of the consecutive M weight parameters, and this sparse structure can achieve the acceleration effect on NVIDIA A100 GPUs.

Channel pruning can be divided into two types: static pruning and dynamic pruning. Static pruning methods generate a fixed compressed model for all input data. Among them, some static pruning methods evaluate the importance of the channel based on the inherent properties of the trained model. Li et al. [9] used the ℓ_1-norm of the convolutional kernel weight parameter as the importance evaluation index and prunes the convolutional kernels with small ℓ_1-norm values. FPGM [8] proposed a similarity calculation of convolutional kernels based on Euclidean distance to remove convolutional kernels with high similarity to other convolutional kernels. Hrank [12] concluded that the higher the rank of the feature map, the more information it contains, and confirmed that the rank of the feature map produced by the same convolution kernel is constant across all inputs. There are also some static pruning methods that identify unimportant convolutional kernels during training and output the pruned model after training. AutoPruner [14] used fully connected layers to predict the saliency score of the feature maps output from each convolutional layer and then used the sigmoid function to discretize the saliency score step by step. Rachwan et al. [15] proposed an efficient method to find the optimal submodel at the early stage of training based on gradient flow preservation.

Dynamic pruning differs from static pruning in that the complete network structure is kept, and the retained fraction of the network is dynamically modified during inference based on each input. FBS [3] proposed the squeeze-excitation module to predict the significance scores of channels, skipping the lower-scoring parts that contribute less to the results. FTWT [2] proposed a method to decouple pruning loss and target loss, utilizing an additional network to predict a fixed percentage of pruned channels for each layer. Since the dynamic pruning preserves the complete model structure, the compressed model after pruning will have higher representational power compared to the static pruning but also have a larger memory footprint.

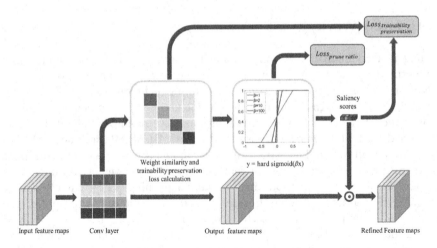

Fig. 1. Framework diagram of the method. For forward propagation, the saliency score of each channel in the output dimension is determined based on the similarity of the convolutional layer weights and then activated by the hard sigmoid function. The saliency score of the output after activation is utilized for the refining of the output feature map and the calculation of the pruning rate and weight orthogonalization loss.

3   Method

3.1   Method Framework and Motivation

Our method consists of two main components: the channel similarity calculation and trainability preservation module, and the sparsity control module. The overall structure and flow of the model are shown in Fig. 1. The design of the channel similarity calculation and trainability preservation modules is inspired by the methods [17,19]. The propagation of error signals and the network's convergence are both aided by the network's Jacobian's singular values being near to 1 [17]. For the fully-connected layer, the way it is implemented is that the gram matrix of weights is the identity matrix, as shown in Eq. (1):

$$
\begin{aligned}
\mathbf{y} &= \boldsymbol{W}\mathbf{x}, \\
||\mathbf{y}|| &= \sqrt{\mathbf{y}^{\mathrm{T}}\mathbf{y}} = \sqrt{\mathbf{x}^{\mathrm{T}}\boldsymbol{W}^{\mathrm{T}}\boldsymbol{W}\mathbf{x}} = ||\mathbf{x}||, \; iff. \boldsymbol{W}^{\mathrm{T}}\boldsymbol{W} = \boldsymbol{I},
\end{aligned}
\tag{1}
$$

I is the identity matrix. TPP [19] applies its extension to the convolutional layer. TPP computes the gram matrix of the convolutional layer weights W. The elements of the positions in the gram matrix associated with the pruned channels are added to the loss function to converge to zero, as shown in Eq. (2):

$$
\mathcal{L} = \sum_{l=1}^{L} ||\boldsymbol{W}_l\,\boldsymbol{W}_l^{\mathrm{T}} \odot (1 - \mathbf{m}\mathbf{m}^{\mathrm{T}})\,||_F^2, \; m_j = 0 \; if \; j \in S_l, \; else \; 1,
\tag{2}
$$

S_l stands for pruned channel.

Our paper designs a gram matrix-based channel similarity calculation module that is able to adaptively allocate the pruning rate of each layer according to the pruning target. To maintain the network's trainability, regularization constraints are applied on the associated parts of the pruned and retained channels in the loss function based on the salience score. The sparse control module activates and discretizes the channel similarity scores. During forward propagation, the salience scores are reweighted on the output feature maps to gradually reduce the transmission of information from channels with lower salience scores, and the gradient returned during backward propagation aids with model training. After the training is finished, the channels corresponding to the 0-valued channels are removed, and the pruned model is generated.

3.2 Channel Similarity Calculation and Trainability Preservation

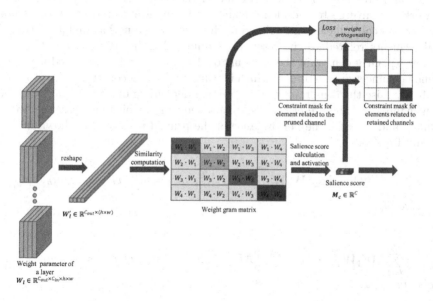

Fig. 2. Structure diagram of channel similarity calculation and trainability protection module.

The structure of the channel similarity calculation and trainability preservation module is shown in Fig. 2. This module is attached to the original network during training and removed when training is completed. When calculating the similarity between channels in the L_{th} convolutional layer of the model, the weights W_l are first resized into vectors in the output dimension to obtain W'_l, where C_{out}, C_{in}, and k correspond to the number of output channels, the number of input channels, and the size of the convolutional kernel, respectively. Then, the gram weight matrix is calculated, with each position of the matrix corresponding to the cosine similarity of the associated two channel weights. The gram matrix is normalized by row using softmax in order to compare the similarity difference between the parameters of each channel. The similarity of the channel to other channels is thus represented by the sum of the elements in each row, excluding the primary diagonal element. The larger the major diagonal element represents, the less similar the related channel is to other channels, and therefore the greater the possibility that it should be kept. Therefore, we take the major diagonal element of the gram matrix as the channel's salience score, M_c, and the whole process is shown in Eq. (3):

$$M_c = diag(softmax(W_l' W_l'^T)). \tag{3}$$

To preserve the trainability of the model during pruning, we also introduce regularization constraints on some elements of the gram matrix in the loss function. However, unlike TPP, we constrain the elements associated with both the pruned and preserved channels to better enable the gram matrix of the pruned network to converge to the identity matrix. In the gram matrix, we constrain the row and column elements related to the pruned channels as well as the diagonal elements related to the reserved channels. Another difference is that we do not explicitly constrain the elements of the gram matrix associated with the pruned channels to zero or one, which is still a too strong constraint. Instead, we gradually raise the constraint during training depending on the saliency score of each channel and the preset threshold parameter m_{th}, which may assist network training and improve the performance of the pruned network. As shown in Eq. (4) and Eq. (5):

$$\mathcal{L}_1 = \sum_{l=1}^{L} \|(1 - diag(W_1' W_1'^T) \odot M_c\|_F^2, \ m_{ci} = m_{ci}, \ if \ m_{ci} > m_{th}, \\ else \ m_{ci} = 0, \tag{4}$$

$$\mathcal{L}_2 = \sum_{l=1}^{L} \|W_1' W_1'^T \odot (1 - M_c^T M_c)\|_F^2, \ m_{ci} = m_{ci}, \ if \ m_{ci} < 1 - m_{th}, \\ else \ m_{ci} = 1. \tag{5}$$

3.3 Sparse Control and Optimization

To effectively utilize the filtering and trainability protection capabilities of the channel similarity calculation module, it is essential to gradually discretize the salience score in a differentiable manner during the training process. In this paper, we employ the hard sigmoid function, as illustrated in Eq. (6):

$$\sigma(\beta M_c) = min(max((\beta M_c) + 0.5, 0), 1), \tag{6}$$

where β is a scale parameter specific to each channel, which is adaptively adjusted based on the activation output value. When $\beta \to \infty$, the hard sigmoid function converges to the binarized sign function. To avoid the unstable data distribution during the early stages of training, which results in some of the salience scores being activated to 0 or 1, we set the scale parameter β to 1 during initialization for all channels. Meanwhile, we normalize the salience score M_c to a normal distribution with a mean of 0 and a standard deviation of 0.25 to limit the distribution range of M_c. As the training progresses, the difference in salience scores of different channels gradually expands and stabilizes, and the activation output of M_c gradually discretes to 0 or 1 after the modification of the scale parameter β. Where β is updated as shown in Eq. (7):

$$\beta = min(1 + \alpha\delta,\ 100), \qquad (7)$$

α and δ are the parameters that control the growth rate and the growth amount of β, respectively. where α is a constant value and δ is updated adaptively every T iterations according to the saliency score of each channel, as specified by Eq. (8):

$$\delta = \delta + 1 - |\ \sigma(\beta M_c) - 0.5|. \qquad (8)$$

To be able to prune the network based on the compression target and meet the computational complexity requirements of the model for various practical applications while ensuring the model's trainability. We add a simple but effective regularization term to the loss function, as shown in Eq. (9):

$$L = L_{\mathbf{Ori}} + \lambda(||\tfrac{F_p}{F_o} - p||_2 + \tau(L_1 + L_2)) \qquad (9)$$

The first term is the loss function of the original task, and the second term is the difference between the compression rate of the current model and the target compression rate and the parameter constraint loss of trainability protection. F_p, F_o, and p denote the computational complexity of the pruned model, the total computational complexity of the model, and the preset compression target, respectively. FLOPs are used as the metric to measure the computation of the model. τ and λ are balance parameters, where τ is a constant and λ is calculated as shown in Eq. (10):

$$\lambda = 100 \times \left| \tfrac{F_p}{F_o} - p \right| \qquad (10)$$

4 Experiments

4.1 Experiments Settings and Evaluation Metrics

We conducted experiments on the classification task and the object detection task, respectively, to evaluate the effectiveness of our proposed pruning method for model compression. The CIFAR-10 and Imagenet datasets are used with the VGG and ResNet series networks for the classification task model. On the VOC and COCO datasets, the target detection task was tested utilizing YOLOX-s networks. The threshold m_{th} used in the regularization loss calculation is set to

0.9. The scale parameters β used in the Hard sigmoid activation function are all set to 1 at initialization, its growth rate α is initialized to 0.0005, and the growth amount δ is initialized to 0. δ is updated every T iterations.

We evaluated the pruning rate using the percentage drop in FLOPs after pruning and the accuracy rate using the top-1 accuracy. Under similar pruning rates, we compare the accuracy declines of various methods.

4.2   Results on Imagenet

Table 1 shows the pruning results for ResN-18, ResN-34, and ResN-50 achieved by our method and some comparison methods on the ImageNet dataset. Our method achieves a lesser accuracy decline at higher pruning rates for the ResN-18 and ResN-34 pruning results. In particular, when the pruning rate reaches 51.68% on ResN-18 and 52.50% on ResN-34, the top-1 accuracy only drops by 1.15% and 1.13%, respectively. Our method demonstrates superior accuracy retention even at a pruning rate of 57.5% on ResN-50. When the pruning rate reaches 76.3%, its performance is comparable to BCNet. The comparison results show that our method preserves more rich features in the preserved channels than typical pruning methods that just consider the parameters of the trained model and partially adaptively search for the best structure. Figure 3 depicts the output features of the first convolutional layer of the ResNet-18 network's first residual module before and after pruning, and it can be observed that the features recovered from the maintained channels are more unique.

Table 1. The results of pruning ResNet on Imagenet

Model	Method	Dynamic	Baseline Top-1 Acc.(%)	Pruned Top-1 Acc.(%)	Top-1 Gap(%)	FLOPs↓(%)
ResNet-18	SFP [7]	✗	70.28	67.10	3.18	41.80
	FPGM [8]	✗	70.28	68.41	1.87	41.80
	PFP [10]	✗	69.74	65.65	4.09	43.10
	DSA [20]	✗	69.72	68.61	1.11	40.00
	ABCPruner [13]	✗	69.66	67.28	2.38	43.55
	Ours	✗	**69.75**	**69.11**	**0.64**	**43.80**
	FBS [3]	✔	70.71	68.71	2.54	49.50
	FTWT [2]	✔	69.76	67.49	2.27	51.56
	Ours	✗	**69.75**	**68.60**	**1.15**	**51.68**
ResNet-34	SFP [7]	✗	73.92	71.83	2.09	41.80
	FPGM [8]	✗	73.92	72.54	1.38	41.10
	FTWT [2]	✔	73.30	72.17	1.13	47.42
	Ours	✗	**73.38**	**72.62**	**0.76**	**49.20**
	FTWT [2]	✔	73.30	71.71	1.59	52.24
	Ours	✗	**73.38**	**72.25**	**1.13**	**52.50**
ResNet-50	FPGM [8]	✗	76.15	74.83	1.32	53.50
	ABCPruner [13]	✗	76.01	73.52	2.49	56.01
	Ours	✗	**76.01**	**74.82**	**1.19**	**57.50**
	Hrank [12]	✗	76.15	71.98	4.17	62.10
	BCNet [18]	✗	77.5	75.2	2.3	76.00
	Ours	✗	**76.01**	**73.70**	**2.31**	**76.30**

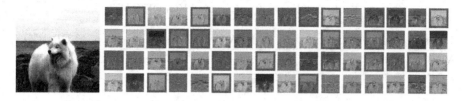

Fig. 3. The feature maps of the first convolutional layer of ResNet-18's first residual block The red box represents the channel that was preserved after pruning.

4.3 Results on Cifar-10

Table 2 presents pruning results for our proposed method and comparison methods on CIFAR-10 using VGG-16 and ResNet-56. Applying our method to VGG-16 at a 34.56% pruning rate resulted in a 0.2% increase in top-1 accuracy, which consistently outperformed other methods at higher pruning rates. Our approach achieves fewer top-1 accuracy drops for ResNet-56 network at similar pruning rates. For ResNet-56 network pruning by 67.80%, the accuracy drop is only 0.41%.

Table 2. The results of pruning ResNet on CIFAR-10

Model	Method	Dynamic	Baseline Top-1 Acc.(%)	Pruned Top-1 Acc.(%)	Top-1 Gap(%)	FLOPs↓(%)
VGG-16	PEFC [8]	✗	93.25	93.40	−0.15	34.20
	Ours	✗	**93.25**	**93.45**	**−0.20**	**34.56**
	FPGM [8]	✗	93.58	93.23	0.35	35.90
	Hrank [12]	✗	93.96	93.43	0.53	53.50
	FTWT [2]	✔	93.82	93.73	0.09	56.00
	Ours	✗	**93.25**	**93.18**	**0.07**	**54.00**
	Hrank [12]	✗	93.96	91.23	2.73	76.50
	FTWT [2]	✔	93.82	93.19	0.63	73.00
	Ours	✗	**93.25**	**92.74**	**0.51**	**77.87**
ResNet-56	SFP [7]	✗	93.59	92.26	1.33	52.60
	FPGM [8]	✗	93.59	93.49	0.10	52.60
	Hrank [12]	✗	93.26	93.17	0.09	50.00
	DSA [20]	✗	93.12	92.91	0.21	52.20
	Ours	✗	**93.05**	**93.00**	**0.05**	**54.20**
	FTWT [2]	✔	93.66	92.63	1.03	66.00
	Ours	✗	**93.05**	**92.64**	**0.41**	**67.80**

4.4 Results on YOLOX-s

To validate our proposed method on other vision tasks, we evaluated its efficacy on the lightweight object detection model YOLOX-s using the VOC and COCO datasets. Our experiments, as detailed in Table 3, showed that even after pruning, our method achieved high mAP values, underscoring its effectiveness.

Table 3. The results of pruning YOLOX-S [4]

Dataset	Baseline mAP$_{0.5:0.95}$	Pruned mAP$_{0.5:0.95}$	mAP$_{0.5:0.95}$ Gap	FLOPs↓(%)
VOC	59.07	58.28	0.79	41.20
COCO	39.60	37.96	1.64	39.80

4.5   Ablation

To validate our channel saliency score calculation method, we employed several comparison approaches and evaluated them on the ImageNet dataset using the ResNet-18 neural network. In the first variant, the inverse of the saliency score is taken by subtracting the original score from 1. In the second variant, the saliency score is set as a trainable parameter. Initially, it is randomly generated and then trained parallel to the network. Results in Table 4 showed that our proposed method had the least reduction in top-1 accuracy, while the Reverse method showed the most significant reduction. These results verified the effectiveness of our designed method in calculating the channel saliency score.

Table 4. Results of ablation study for the channel salience score calculation methods.

Method	Baseline Top-1 Acc.(%)	Pruned Top-1 Acc.(%)	Top-1 Gap(%)	FLOPs↓(%)
Ours	**69.75**	**69.11**	**0.64**	**43.80**
Reverse	69.75	67.86	1.89	43.60
Random	69.75	68.31	1.44	43.70

To evaluate the efficiency of our strategy, which constrains elements in the gram matrix associated with pruned and reserved channels using salience score guidance, we devised three comparison methods and ran experiments on ImageNet using the ResNet-18 network. The first approach constrained only the elements associated with pruned channels in the gram matrix using saliency score guidance without restricting diagonal elements corresponding to the retained channels. The second method removed saliency score guidance and directly constrained related elements to 0 or 1, according to the original settings. The third approach disregarded saliency score guidance while only constraining pruned channel elements in the gram matrix. Experimental results are presented in Table 5, which demonstrates the effectiveness of our proposed salience score bootstrap method and the joint constraint of pruned and reserved channel elements in protecting network trainability.

Table 5. Results of ablation study for the corresponding parameter constraint methods for the retained and pruned channels in the gram matrix. "Prune" refers to constraints on elements related to pruned channels; "Retain" refers to constraints on elements related to retained channels; and the "Salience Score" metric uses salience scores to bootstrap element constraints.

Method	Baseline Top-1 Acc.(%)	Pruned Top-1 Acc.(%)	Top-1 Gap(%)	FLOPs↓(%)
Prune + Retain + Salience Score (Ours)	**69.75**	**69.11**	**0.64**	**43.80**
Prune + Salience score	69.75	68.80	0.95	43.60
Prune + Retain	69.75	68.72	1.03	43.50
Prune	69.75	68.50	1.25	43.50

5   Conclusion

In this paper, we propose an adaptive channel pruning method that preserves trainability. Our approach utilizes a similarity calculation module that identifies redundant output channels during training while maintaining model trainability. Additionally, the sparsity control module assigns pruning rates adaptively based on the overall compression target. Our experimental results demonstrate that our method is effective for various visual tasks and various structural networks, achieving higher accuracy than existing pruning methods at similar pruning rates.

References

1. Chen, J., Zhu, Z., Li, C., Zhao, Y.: Self-adaptive network pruning. In: International Conference on Neural Information Processing, pp. 175–186 (2019)
2. Elkerdawy, S., Elhoushi, M., Zhang, H., Ray, N.: Fire together wire together: a dynamic pruning approach with self-supervised mask prediction. In: Proceedings of the IEEE/CVF Conference on Computer Vision and Pattern Recognition, pp. 12454–12463 (2022)
3. Gao, X., Zhao, Y., Dudziak, Ł., Mullins, R., Xu, C.Z.: Dynamic channel pruning: feature boosting and suppression. arXiv preprint: arXiv:1810.05331 (2018)
4. Ge, Z., Liu, S., Wang, F., Li, Z., Sun, J.: YOLOX: exceeding yolo series in 2021. arXiv preprint: arXiv:2107.08430 (2021)
5. Han, S., Pool, J., Tran, J., Dally, W.: Learning both weights and connections for efficient neural network. In: Advances in Neural Information Processing Systems, vol. 28 (2015)
6. He, K., Zhang, X., Ren, S., Sun, J.: Deep residual learning for image recognition. In: 2016 IEEE Conference on Computer Vision and Pattern Recognition (CVPR), Las Vegas, NV, USA, pp. 770–778 (2016)
7. He, Y., Kang, G., Dong, X., Fu, Y., Yang, Y.: Soft filter pruning for accelerating deep convolutional neural networks. IJCAI, 2234–2240 (2018)

8. He, Y., Liu, P., Wang, Z., Hu, Z., Yang, Y.: Filter pruning via geometric median for deep convolutional neural networks acceleration. In: IEEE/CVF Conference on Computer Vision and Pattern Recognition (CVPR), pp. 4335–4344 (2019)

9. Li, H., Kadav, A., Durdanovic, I., Samet, H., Graf, P.H.: Pruning filters for efficient ConvNets. In: International Conference on Learning Representations (2017)

10. Liebenwein, L., Baykal, C., Lang, H., Feldman, D., Rus, D.: Provable filter pruning for efficient neural networks. ICLR (2020)

11. Lin, M., Cao, L., Zhang, Y., Shao, L., Lin, C.W., Ji, R.: Pruning networks with cross-layer ranking & k-reciprocal nearest filters. IEEE Trans. Neural Netw. Learn. Syst. (2022)

12. Lin, M., et al.: HRank: filter pruning using high-rank feature map. In: 2020 IEEE/CVF Conference on Computer Vision and Pattern Recognition (CVPR), pp. 1526–1535 (2020)

13. Lin, M., Ji, R., Zhang, Y., Zhang, B., Wu, Y., Tian, Y.: Channel pruning via automatic structure search. In: Bessiere, C. (ed.) Proceedings of the Twenty-Ninth International Joint Conference on Artificial Intelligence, IJCAI-20, pp. 673–679. International Joint Conferences on Artificial Intelligence Organization (2020). main track

14. Luo, J.H., Wu, J.: AutoPruner: an end-to-end trainable filter pruning method for efficient deep model inference. Pattern Recogn. **107**, 107461 (2020)

15. Rachwan, J., Zügner, D., Charpentier, B., Geisler, S., Ayle, M., Günnemann, S.: Winning the lottery ahead of time: efficient early network pruning. In: International Conference on Machine Learning, pp. 18293–18309 (2022)

16. Romera, E., Alvarez, J.M., Bergasa, L.M., Arroyo, R.: ERFNet: efficient residual factorized ConvNet for real-time semantic segmentation. IEEE Trans. Intell. Transp. Syst. **19**(1), 263–272 (2018)

17. Saxe, A., McClelland, J., Ganguli, S.: Exact solutions to the nonlinear dynamics of learning in deep linear neural networks. In: International Conference on Learning Represenatations (2014)

18. Su, X., et al.: Searching for network width with bilaterally coupled network. IEEE Trans. Pattern Anal. Mach. Intell. **45**(7), 8936–8953 (2023)

19. Wang, H., Fu, Y.: Trainability preserving neural pruning. In: The Eleventh International Conference on Learning Representations (2023)

20. Ning, X., Zhao, T., Li, W., Lei, P., Wang, Yu., Yang, H.: DSA: more efficient budgeted pruning via differentiable sparsity allocation. In: Vedaldi, A., Bischof, H., Brox, T., Frahm, J.-M. (eds.) ECCV 2020. LNCS, vol. 12348, pp. 592–607. Springer, Cham (2020). https://doi.org/10.1007/978-3-030-58580-8_35

21. Zhang, Y., et al.: Learning best combination for efficient n:M sparsity. In: Oh, A.H., Agarwal, A., Belgrave, D., Cho, K. (eds.) Advances in Neural Information Processing Systems (2022)

22. Zhou, A., et al.: Learning n: M fine-grained structured sparse neural networks from scratch. ICLR (2021)

Exploiting Adaptive Crop and Deformable Convolution for Road Damage Detection

Yingduo Bai, Chenhao Fu, Zhaojia Li, Liyang Wang, Li Su, and Na Jiang[✉]

Capital Normal University, Beijing 100048, China
jiangna@cnu.edu.cn

Abstract. Road damage detection (RDD) based on computer vision plays an important role in road maintenance. Unlike conventional object detection, it is very challenging due to the irregular shape distribution and high similarity with the background. To address this issue, we propose a novel road damage detection algorithm from the perspective of optimizing data and enhancing feature learning. It consists of adaptive cropping, feature learning with deformable convolution, and a diagonal intersection over union loss function (XIOU). Adaptive cropping uses vanishing point estimation (VPE) to obtain the pavement reference position, and then effectively removes the redundant information of interference detection by cutting the raw image above the reference position. The feature learning module introduces deformable convolution to adjust the receptive field of road damage with irregular shape distribution, which will help enhance feature differentiation. The designed diagonal IOU loss function (XIOU) optimizes the road damage location by weighted calculation of the intersection and comparison between the predicted proposal and the groundtruth. Compared with existing methods, the proposed algorithm is more suitable for road damage detection task and has achieved excellent performance on authoritative RDD and CNRDD datasets.

Keywords: Road Damage Detection · Adaptive Image Cropping · Diagonal IOU Loss Function

1 Introduction

Road damage detection (RDD) is crucial for road maintenance, which can effectively ensure pavement safety. Due to the high similarity between road damages and the road background, as well as the irregular distribution of pixels that can represent damages, this task is very challenging. Different from early manual methods which is time-consuming and laborious, RDD based on deep learning can quickly and automatically analyze the collected images. Many algorithms that consider sensors or computer vision have been proposed and significantly improve the accuracy and efficiency of detection. Although such methods often

© The Author(s), under exclusive license to Springer Nature Singapore Pte Ltd. 2024
Q. Liu et al. (Eds.): PRCV 2023, LNCS 14434, pp. 149–160, 2024.
https://doi.org/10.1007/978-981-99-8549-4_13

exhibit high precision, the equipment used for data collection is expensive, limiting their widespread application. In contrast, methods that utilize RGB images captured by onboard cameras for detection are more efficient and cost-effective for the RDD task. These methods employ deep convolutional neural networks to extract damage features from the data collected by vehicle-mounted cameras.

Damage detection methods based on deep convolutional neural networks mainly include semantic segmentation methods and object detection methods. Semantic segmentation methods typically use an encoder-decoder structure to classify each pixel to obtain accurate damage areas, such as FCN [1], Unet [2], etc. However, the success of deep learning models in semantic segmentation comes at the cost of a heavy computational burden. Object detection methods are not only cost-effective but also simple and efficient because they do not require pixel-wise predictions. Instead, they only need to predict the position of a bounding box that encompasses the damage. Additionally, the cost of dataset annotation for semantic segmentation is much higher than that for object detection. Therefore, most research on road damage detection algorithms is based on object detection algorithms. Although conventional object detection methods have achieved high accuracy and performance in their fields, they cannot be directly applied to RDD tasks due to the differences between road damages and conventional target.

(a)   (b)

Fig. 1. Comparison between object detection and road damage detection. (a) comes from the COCO dataset, and (b) comes from the RDD2020 dataset.

As shown in Fig. 1. Conventional detection targets typically provide clear and highly informative visual data within the annotated box, with a significant proportion of pixels dedicated to the target. However, detecting road damages presents considerable challenges due to their slender and elongated shapes, which occupy only a fraction of the annotated box pixels. As a result of these intricacies, conventional object detection methods often yield unsatisfactory outcomes when employed for the RDD task.

Currently, most research on RDD task [3–8] uses conventional object detection algorithms [9,10] as the baseline for model adjustment or ensemble learning. Although ensemble learning can indeed improve accuracy, it also brings high computational costs. Additionally, existing methods have not taken into account the difficulties caused by the differences between damages and conventional targets. Firstly, due to the irregular shape distribution of damages, the network cannot accurately extract the features of damages. Secondly, the data is collected from onboard cameras, resulting in varying perspectives and a large rate of background. This information seriously interferes with the prediction performance of the network. Another crucial point is that the proportion of pixels occupied by damages within the annotated bounding boxes is very sparse. If a generic localization loss is used for supervised learning, it cannot achieve satisfactory supervision results.

To address these problems, our study proposes targeted algorithmic solutions that focus on road damage detection. We mainly make the following contributions:

1. Propose an adaptive image cropping data preprocessing method. It extracts the vanishing point in the images and then adaptively crops the raw image, which can effectively alleviate the interference of redundant information (sky, buildings, etc.) on RDD.
2. Improve feature learning module with deformable convolution(DCN). It exploits DCN to enhance feature differentiation, which can adapt to the irregular pixel distribution in road damages by adjusting receptive fields.
3. Design a diagonal intersection over union (XIOU) loss function. It calculates the Gaussian weighted distance between pixels in the prediction box and the groundtruth diagonal to achieve the intersection and union ratio, which can optimize the location and size of the prediction box.

2   Related Work

Deep learning methods can automatically extract various features of damage. In practice, object detection methods are divided into two categories: one-stage algorithms and two-stage algorithms. Two-stage algorithms first generate a series of candidate bounding boxes as samples, and then classify the samples through convolutional neural networks. Typical representatives of these algorithms include Faster R-CNN [9], Cascade R-CNN [11], etc. Hascoet et al. [3] used Faster-RCNN for damage detection, improved detection performance using label smoothing techniques, and demonstrated efforts to deploy models on local road networks. Vishwakarma et al. [4] introduced a tuning strategy of Faster-RCNN with ResNet and Feature Pyramid Network (FPN) backbone networks of different depths. In addition, they compared it with a single-stage YOLOv5 [10] model with a Cross Stage Partial network (CSPNet) backbone network. Pei et al. [5] applied Cascaded R-CNN to damage detection and proposed a consistency filtering mechanism (CFM) for self-supervised methods to fully utilize available unlabeled data.

However, one-stage algorithms treat target detection as a regression task, directly regress bounding boxes and predict multiple positions of categories throughout the entire image to obtain more comprehensive information, such as SSD [12], YOLO series [13,14], EfficientDet [15], etc. Zhang et al. [6] used YOLOv4 as the underlying network for damage detection and proposed the use of Generative Adversarial Networks (GAN) for data augmentation. The impact of data augmentation, transfer learning, and optimized anchors and combinations were evaluated. Mandal et al. [7] used YOLOv4 for damage detection and studied the effects of various backbone networks. Hu et al. [16] used YOLOv5 to detect cracks and compared the model sizes, detection speeds, and accuracy of four YOLOv5 versions. Guo et al. [8] proposed to replace the backbone network of the original YOLOv5s model with a lightweight MobileNetv3 network to reduce model parameters and GFLOPs, and optimize the model using coordinate attention. Zhang et al. [17] used edge detection combined with attention mechanism for damage detection and studied the differences in effect on different datasets.

Although the above damage detection algorithms have achieved good results, there are still some areas for improvement. The above methods only adopt traditional data augmentation strategies such as scaling, rotation, and random erasing. In fact, one of the simplest and most effective data preprocessing methods is to crop out the parts of the image that are not related to road damage detection. Zhang et al. [17] used fixed threshold clipping of images on their newly proposed CNRDD dataset. If image clipping is performed using the above method on datasets with multiple different perspectives, such as RDD, a large number of training samples will be lost. In addition, due to the sparse pixels taken up by damage in the annotated boxes, loss functions such as IOU, DIOU, CIOU, etc. cannot provide effective supervision. Therefore, this paper proposes a location loss function for damage detection tasks that can better reflect the learning effect of the network and improve detection performance.

3   Methods

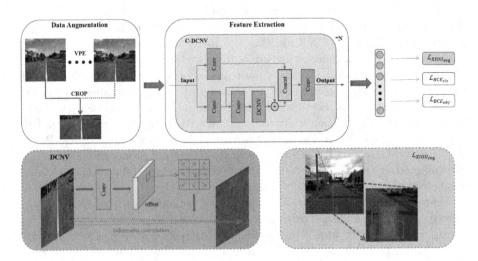

Fig. 2. Outline of our proposed method.

As shown in Fig. 2, the road damage detection method designed in this paper consists of three parts. In the first part, a vanishing point detection method is used to detect the vanishing point of the road in the image during the data preprocessing stage, and the image is cropped based on the vanishing point. In the second part, a C-DCNV module is introduced on the basis of the YOLOv5 model for feature extraction, which improves the effectiveness of extracting damage features. In the third part, a diagonal loss function that better fits the characteristics of road damage is used for network supervision during the supervised learning stage. Each part will be explained in detail below.

3.1   Adaptive Image Cropping Based on Vanishing Point Estimation

As mentioned in the previous text, the dataset collected using non-professional vehicle-mounted cameras contains images with various perspectives. If the original images are directly input into the neural network, a large amount of redundant information is simultaneously input, which increases the difficulty of training. Cropping the images can greatly avoid this problem. However, determining at what height to start cropping is a key issue. Therefore, this paper proposes an adaptive threshold cropping algorithm. Specifically, the Canny edge detection algorithm is used to generate the edge map of the image. The Hough transform is then used to screen out lines that meet the angle requirements from all edges in the edge map, and the intersection of the lines is calculated to obtain the vanishing point of the road in the image. The most suitable image cropping threshold can be obtained based on the height of the vanishing point. The cropping effect is shown in the Fig. 3.

Fig. 3. Visual result of adaptive image cropping based on VPE. (a) shows four raw images from the RDD2020 dataset, and (b) demonstrates their cropped images.

3.2   Feature Learning with Deformable Convolution

Convolution is widely used in the processing of data types such as images and speech, and is a core part of convolutional neural networks. The convolution operation uses a kernel to multiply its corresponding weights with each local region of the input image and then add them up to obtain a new output feature map. Here, "local region" refers to a rectangular range that is the same size as the convolution kernel, and the range slides from left to right and from top to bottom on the input image.

Although convolutional operation is a very important and efficient image processing method in convolutional neural networks. Due to the fixed shape and size of the convolution kernel, it is unable to adapt to the changes of object features in different scenarios. In the road damage detection task, The damages have elongated shape characteristics. Feature extraction for each fixed-size matrix range often fails to extract enough effective information, which resulting in poor detection results. This problem actually has a ready-made solution - deformable convolution.

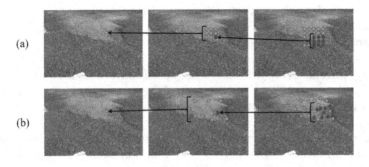

Fig. 4. Comparison of receptive fields between regular convolution and deformable convolution in extracting damage features. (a) refers to the receptive field of regular convolution, and (b) refers to the receptive field of deformable convolution.

As shown in Fig. 4, deformable convolution improves the flexibility of the convolution kernel by introducing an additional deformation branch to obtain learnable deformation parameters. The shape and position of the convolution kernel can be adaptively adjusted according to the input image. It can reach better recognize and segment objects in complex scenes. We have introduced deformable convolution into the feature extraction module of the YOLOv5 base model, which named C-DCNV. Through experiments, we have proved that adding the C-DCNV module to the last two stages of the feature extraction module can greatly improve the network detection accuracy.

3.3   Diagonal Intersection over Union Loss Function

The commonly used localization loss functions in object detection tasks are based on IOU calculation, such as IOU Loss, Giou Loss, DIoU Loss, and CIoU Loss. These loss functions consider factors such as overlap or distance between the predicted box and the ground truth box in different ways to balance the precision and stability of predictions. However, these loss functions are not effective in road damage detection tasks. Road damages occupy a very small proportion of pixels within the detection box, unlike conventional objects. Therefore, this paper proposes a diagonal loss function XIOU for road damage detection. XIOU is formulated as follows:

$$XIOU = \frac{\sum\limits_{(x,y)\in P\cap G} H(F(x,y))}{\sum\limits_{(x,y)\in G} H(F(x,y))} \tag{1}$$

$$F(x,y) = min\{d[(x,y), L_1], d[(x,y), L_2]\} \tag{2}$$

where G represents the set of all pixels contained in the ground truth box, which mathematical form is $\{(x,y)|x_{min} \leq x \leq x_{max}, y_{min} \leq y \leq y_{max}\}$. P represents all pixels contained in the prediction box. And then (x,y) represent the coordinates of each pixel in G or P. F(x,y) is the response function used to calculate the basic distance from any pixel to the diagonals of G. Where $L1$, $L2$ are the diagonals of G, and the function $d(.)$ calculates the vertical distance from all points in G to the diagonals of G. Due to the varying sizes of bounding boxes, the basic distance is normalized using the standard Gaussian function $H(.)$. The normalized and weighted result is then used to calculate the diagonal intersection over union loss according to the definition in Eq. 1. The visualization of the formula is shown in Fig. 5.

Fig. 5. Visualization of XIOU. The red portion in the image represents the weighted distribution used by XIOU. (Best viewed in color) (Color figure online)

As shown in Fig. 5. The green box represents the annotated bounding box, while the blue and purple boxes are two deliberately constructed different prediction boxes. The two prediction boxes have equal IOU and the same distance from the center point of the annotated box. When using the traditional IOU loss calculation, their loss values are approximately equal or slightly higher for the blue box due to differences in aspect ratio. However, in fact, the blue box detects more damage areas. When using the XIOU proposed in this paper for loss calculation, the weight covered by the blue box is much greater than that covered by the purple box. Therefore, by combining XIOU with other IOU loss for road defect detection, we have improved the detection performance.

4 Experiment

The experiments in this paper were conducted using one Tesla A100. The models were trained for 200 epochs on the RDD2020 dataset and CNRDD dataset, respectively. The experimental hyperparameter details are as follows.

Table 1. Implementation details of training.

Param	Value	Param	Value
lr_i	0.01	lr_f	0.1
momentum	0.937	weight_decay	0.0005
box_loss	0.05	cls_loss	0.3
obj_loss	0.7	scale	0.9
mosaic	1	mixup	0.1
batch_size	64	epoch	200

In Table 1, lr_i and lr_f represent the learning rates for different training stages. Box_loss, cls_loss, and obj_loss represent the weights for localization loss, classification loss, and confidence loss, respectively. The training process includes enabling mosaic augmentation and mixup augmentation.

4.1 Comparative Analysis of Different Datasets

The RDD2020 dataset is currently the most important public dataset for road damage detection tasks. It provides two test sets without ground truth and one training set containing four types of road damage labels (longitudinal cracks, transverse cracks, alligator cracks, and potholes). The CNRDD dataset is a Chinese road damage dataset proposed in 2022. The sample density of this dataset is higher and the classification is more detailed. We conducted experimental analyses on various state-of-the-art algorithms on both the RDD2020 and CNRDD datasets. The experimental results are shown in Table 2 and Table 3.

In Table 2, Test1 Score and Test2 Score represent the F1-Scores on two test sets of RDD2020. All the comparative data come from the evaluation system of the Global Road Damage Detection Challenge (GRDDC) based on RDD2020. As shown in Table 2, Our proposed method achieves higher Test1 Score and Test2

Table 2. Performance comparison on RDD2020 dataset.

Methods	Test1 Score	Test2 Score
titan_mu [7]	0.581	0.575
RICS [18]	0.565	0.547
AIRS-CSR [6]	0.554	0.541
CS17 [3]	0.541	0.543
AFSLF [17]	0.614	0.608
MN-YOLOv5 [8]	0.618	0.609
Ours	**0.626**	**0.621**

Table 3. Performance comparison on CNRDD dataset.

Methods	mAP50	F1-Score
YOLO v5 [10]	24.8	32.25
Faster R-CNN [9]	23.8	24.1
fcos [1]	18.4	21.41
Dongjuns [19]	25.2	22.67
AFSLF [17]	26.6	34.73
Ours	**32.7**	**37.1**

Score by 0.8% and 1.2% respectively compared to MN-YOLOv5. This demonstrates that the XIOU loss introduced in this paper effectively improves the network's learning performance in detecting damages under the similar strategy.

In Table 3, the experimental results of YOLO v5 [10], Faster R-CNN [9], fcos [1], and Dongjuns [19] were retrained using open source code on the CNRDD dataset. The data is the average of 10 repeated experiments. Comparing the experimental data in Table 3, it can be seen that the proposed method achieved significant performance improvement on the CNRDD dataset. After introducing the Adaptive Cropping Module, C-DCNV Module, and XIOU loss, our method achieves a 6.1% improvement in mAP50 compared to the previous state-of-the-art level. Furthermore, compared to the original YOLO v5, there is a substantial increase of 7.9% in mAP and 4.8% in F1-Score.

4.2   Ablation Analysis

In order to analyze the effects of adaptive threshold cropping, C-DCNV module, and using XIOU loss for supervised learning on road damage detection, we also conducted ablation studies on the CNRDD test set and the simulated RDD2020 test set. The simulated test set consists of 20% randomly selected data from the RDD2020 training set. YOLO v5 [10] serves as the baseline for the proposed method. Improved components are sequentially added to the baseline, and the experimental results are shown in Table 4.

Table 4. Evaluation of the effectiveness of each component.

	CNRDD		RDD2020	
	mAP50	F1-Score	mAP50	F1-Score
Baseline	24.8	32.2	54.8	57.3
+Adaptive Crop	27.2	34.5	55.7	59.6
+Adaptive Crop+C-DCNV	30.4	35.6	57.4	62.2
+Adaptive Crop+C-DCNV+XIOU	**32.7**	**37.1**	**58.1**	**63.1**

Compared to the baseline in Table 4, Our proposed method exhibits significant improvements on the RDD2020 and CNRDD datasets. Particularly, on the CNRDD dataset, we observe an increase of 7.9% in mAP50 and 4.9% in F1-Score. The table reveals that all three components of our proposed method contribute to enhancing the results to a certain extent. Specifically, the introduction of the C-DCNV module yields the greatest improvement. This highlights the effectiveness of deformable convolutions in accurately extracting features from irregularly shaped damages.

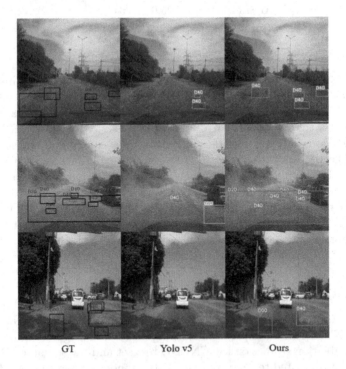

Fig. 6. Comparison between our proposed method and other methods.

In Fig. 6, from left to right, shows the annotated image of the training samples, the prediction results obtained by directly training on the original YOLOv5, and the prediction results obtained by our proposed method in this paper. It can be observed that our method significantly improves the detection performance of road damages.

5 Conclusion

This paper proposes an novel road damage detection algorithm from the perspective of data preprocessing, feature learning, and loss function. It first introduces VPE to adaptively crop the raw image, so as to reduce the interference of background redundant information on road damage detection. Secondly, this paper introduces the DCN to dynamically adjust the receptive field, which enables more accurate extraction of damage features through learnable offset parameters. In addition, a new loss function named XIOU is designed to optimize the position and size of the prediction box. On challenging CNRDD and RDD datasets, the proposed algorithm achieved significant performance improvements. On this basis, we will further focus on the research on the generalization ability and efficiency of road damage detection, in order to propose algorithms that can be applied to practical road maintenance as soon as possible.

Acknowledgments. This work was supported by the National Natural Science Foundation of China under Grant No.62002247 and the general project numbered KM202110028009 of Beijing Municipal Education Commission.

References

1. Yang, X., Li, H., Yu, Y., Luo, X., Huang, T., Yang, X.: Automatic pixel-level crack detection and measurement using fully convolutional network. Comput. Civ. Infrastruct. Eng. **33**, 1090–1109 (2018)
2. Yu, G., Dong, J., Wang, Y., Zhou, X.: RUC-Net: a residual-Unet-based convolutional neural network for pixel-level pavement crack segmentation. Sensors **23**, 53 (2023)
3. Hascoet, T., Zhang, Y., Persch, A., Takashima, R., Takiguchi, T., Ariki, Y.: Faster-RCNN monitoring of road damages: competition and deployment. In: Proceedings of the 2020 IEEE International Conference on Big Data (Big Data), Atlanta, GA, USA, 10–13 December 2020, pp. 5545–5552 (2020)
4. Vishwakarma, R., Vennelakanti, R.: CNN model tuning for global road damage detection. In: Proceedings of the 2020 IEEE International Conference on Big Data (Big Data), Atlanta, GA, USA, 10–13 December 2020, pp. 5609–5615 (2020)
5. Pei, Z., Lin, R., Zhang, X., Shen, H., Tang, J., Yang, Y.: CFM: a consistency filtering mechanism for road damage detection. In: Proceedings of the 2020 IEEE International Conference on Big Data (Big Data), Atlanta, GA, USA, 10–13 December 2020, pp. 5584–5591 (2020)
6. Zhang, X., Xia, X., Li, N., Lin, M., Song, J., Ding, N.: Exploring the tricks for road damage detection with a one-stage detector. In: Proceedings of the 2020 IEEE International Conference on Big Data (Big Data), Atlanta, GA, USA, 10–13 December 2020, pp. 5616–5621 (2020)

7. Mandal, V., Mussah, A.R., Adu-Gyamfifi, Y.: Deep learning frameworks for pavement distress classifification: a comparative analysis. In: Proceedings of the 2020 IEEE International Conference on Big Data (Big Data), Atlanta, GA, USA, 10–13 December 2020, pp. 5577–5583 (2020)
8. Guo, G., Zhang, Z.: Road damage detection algorithm for improved YOLOv5. Sci. Rep. **12**, 15523 (2022)
9. Ren, S., He, K., Girshick, R., Sun, J.: Faster R-CNN: towards real-time object detection with region proposal networks. IEEE Trans. Pattern Anal. Mach. Intell. **39**, 1137–1149 (2017)
10. Available online: https://github.com/ultralytics/yolov5. Accessed 5 Mar 2023
11. Cai, Z., Vasconcelos, N.: Cascade R-CNN: delving into high quality object detection. In: Proceedings of the IEEE Conference on Computer Vision and Pattern Recognition, Salt Lake City, UT, USA, 18–23 June 2018, pp. 6154–6162 (2018)
12. Liu, W., et al.: SSD: single shot MultiBox detector. In: Leibe, B., Matas, J., Sebe, N., Welling, M. (eds.) ECCV 2016. LNCS, vol. 9905, pp. 21–37. Springer, Cham (2016). https://doi.org/10.1007/978-3-319-46448-0_2
13. Bochkovskiy, A., Wang, C.-Y., Liao, H.-Y.M.: YOLOv4: optimal speed and accuracy of object detection. arXiv: arXiv:2004.10934 (2020)
14. Wang, C.-Y., Bochkovskiy, A., Liao, H.-Y.M.: YOLOv7: trainable bag-of-freebies sets new state-of-the-art for real-time object detectors. arXiv: arXiv:2207.02696 (2022)
15. Tan, M., Pang, R., Le, Q.V.: EffificientDet: scalable and effificient object detection. In: Proceedings of the IEEE/CVF Conference on Computer Vision and Pattern Recognition, Seattle, WA, USA, 13–19 June 2020, pp. 10778–10787 (2020)
16. Hu, G.X., Hu, B.L., Yang, Z., Huang, L., Li, P.: Pavement crack detection method based on deep learning models. Wirel. Commun. Mob. Comput. **2021**, 5573590 (2021)
17. Zhang, H., et al.: A new road damage detection baseline with attention learning. Appl. Sci. **12**, 7594 (2022)
18. Naddaf-Sh, S., Naddaf-Sh, M.M., Zargarzadeh, H., Kashanipour, A.R.: An efficient and scalable deep learning approach for road damage detection. arXiv: arXiv:2011.09577 (2020)
19. Jeong, D.: Road damage detection using YOLO with smartphone images. In: Proceedings of the 2020 IEEE International Conference on Big Data (Big Data), Atlanta, GA, USA, 10–13 December 2020, pp. 5559–5562 (2020)

Cascaded-Scoring Tracklet Matching for Multi-object Tracking

Yixian Xie, Hanzi Wang, and Yang Lu[✉]

Fujian Key Laboratory of Sensing and Computing for Smart City,
School of Informatics, Xiamen University, Xiamen, China
yixianxie@stu.xmu.edu.cn, {hanzi.wang,luyang}@xmu.edu.cn

Abstract. Multi-object tracking (MOT) aims at locating the object of interest in a successive video sequence and associating the same moving object frame by frame. Most existing approaches to MOT lack the integration of both motion and appearance information, which limits the effectiveness of tracklet association. The conventional approaches for tracklet association often struggle when dealing with scenarios involving multiple objects with indistinguishable appearances and irregular motions, leading to suboptimal performance. In this paper, we introduce an appearance-assisted feature warper (AFW) module and a motion-guided based target aware (MTA) module to efficiently utilize the appearance and motion information. Additionally, we introduce a cascaded-scoring tracklet matching (CSTM) strategy that seamlessly integrates the two modules, combining appearance features with motion information. Our proposed online MOT tracker is called CSTMTrack. Through extensive quantitative and qualitative results, we demonstrate that our tracker achieves efficient and favorable performance compared to several other state-of-the-art trackers on the MOTChallenge benchmark.

Keywords: Multiple Object Tracking · Tracklet Association · Detection · Cascaded Matching

1 Introduction

Multi-object tracking (MOT) tries to predict the tracklet of multiple specific objects frame by frame in a video sequence. It plays an essential role in numerous computer vision applications, such as autonomous vehicles [34], visual surveillance [22], video analysis [10], and more. As MOT involves perceiving the surrounding environment, applying it to autonomous driving can provide greater safety guarantees for the vehicle in terms of safe operation. In the domain of autonomous driving, the goal of MOT, whether in 3D or 2D, is to empower autonomous vehicles with the ability to precisely track objects in their surroundings, providing vital assistance in vehicle perception and decision-making processes. Although 3D MOT methods [19,31] offer superior capabilities for environment perception in autonomous driving scenarios, they also necessitate

© The Author(s), under exclusive license to Springer Nature Singapore Pte Ltd. 2024
Q. Liu et al. (Eds.): PRCV 2023, LNCS 14434, pp. 161–173, 2024.
https://doi.org/10.1007/978-981-99-8549-4_14

higher computing resources and sophisticated sensor data fusion technologies. Hence, our focus is on exploring 2D MOT [3,4], which encounters numerous challenges, including occlusion in densely populated scenes, objects with similar appearances, geometric variations, irregular motion patterns, and more.

Currently, most state-of-the-art MOT approaches are largely categorized into two paradigms that are tracking by detection and joint detection and tracking. The tracking by detection paradigm [2,8,9,24] breaks down the MOT problem into two distinct subtasks: object detection and tracklet association. Benefiting from the enhancement of object detection techniques, this paradigm is favored by most researchers due to its ability to achieve good tracking performance. Nevertheless, the tracking by detection paradigm treats object detection and tracklet association as separate tasks, which leads to a long inference time. As multi-task learning has made significant strides, the joint detection and tracking paradigm [7,11,14,30] has garnered more attention.

Joint detection and tracking trackers have been proposed to integrate object detection and tracklet association tasks within a unified tracking network by sharing most of the common parameters and computations. Several approaches applying the joint detection and tracking paradigm achieve an outstanding balance between speed and accuracy. However, we argue that there are two primary limitations when employing the joint detection and tracking paradigm for MOT: (1) excessive competition among different modules and (2) variations in the scopes of concern for different subtasks, whether they are focused on local details or global relationships.

Even though several refinements have been made in recent approaches to address the aforementioned issues, the practical tracking accuracy induced by this paradigm is still limited. To avoid this issue, we reconsider the tracking by detection paradigm, which was once popular and underappreciated. For the recent tracking by detection research, the Observation-Centric Simple Online and Realtime Tracking (OC-SORT) [3] is a motion-based SORT [2] tracker, which only exploits the motion feature to dispose of tracklet association. However, this approach has limitations when dealing with low frame rate video tracking tasks, especially when handling indistinguishable appearances and irregular motions. In this work, we present a simple yet powerful online tracking-by-detection tracker, named CSTMTrack, built upon OC-SORT tracker.

Our proposed CSTMTrack tackles this issue by implementing a cascaded-scoring tracklet matching (CSTM) strategy, incorporating an appearance-assisted feature warper (AFW) module and a motion-guided target-aware (MTA) module, to effectively track multiple objects in challenging scenarios such as background clutter, nonlinear motions, and geometric transformations. The AFW module establishes storage to save the cost vector of similarity matching between ID embeddings of detection bounding boxes and tracking targets, which can be used to aid inference on tracking offsets and better suppress background to enhance object discrimination in complex scenarios. In the MTA module, a novel IoU strategy is proposed, named EM-IoU, which enlarges the receptive field of detection bounding boxes when computing the intersection over union between two detection bounding boxes. Moreover, the MTA module adds a mask branch to the EM-IoU when the score of the candidate bounding box is low. This module is capable of deeply alleviating the challenges encountered when tracking in irregular motion scenarios.

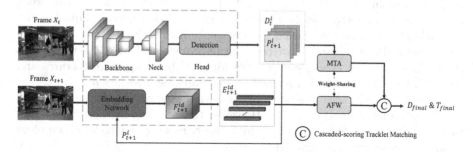

Fig. 1. The framework of our CSTMTrack. It consists of a detection and an embedding network. The input video frame is fed into the detection network to generate multiple bounding boxes, which are predicted to obtain the corresponding bounding box for the next frame. Meanwhile, the embedding network produces candidate embeddings. Ultimately, through the cascaded-scoring tracklet matching strategy, the MTA and AFW modules facilitate the association between detected bounding boxes and tracklets.

Our contributions are summarized as follows:

- We introduce a novel CSTM strategy that combines ID embedding and motion information using a cascaded matching approach, enabling more robust associations between tracklets.
- We propose two lightweight and plug-and-play modules to obtain more accurate associations in tracking multiple objects with indistinguishable appearances and irregular motions: the AFW and MTA modules.
- Experiments show that our CSTMTrack achieves efficient and favorable performance against the most state-of-the-art trackers on MOT17 and MOT20 benchmarks.

2 Related Work

2.1 Tracking by Detection

The tracking by detection paradigm [2,9,24,32] disentangles the MOT problem into two separate subtasks: The detection task first estimates the location of candidate bounding boxes in each frame by a detector, and then the tracklet association task uses feature information of motion or appearance to associate detection bounding boxes and tracklets. Numerous state-of-the-art multiple object trackers (such as [2,8,24]) have primarily focused on improving tracking accuracy rather than inference speed. For example, SORT [2] and DeepSORT [24] are the typical work of tracking by detection approach, and DeepSORT is an improvement on the SORT. These two approaches both use the Kalman filter to predict and update the movement of objects but also use the Hungarian algorithm to solve the assignment problem between objects. STRN [26] introduced a similarity learning network that establishes connections between tracklets and objects by encoding various spatio-temporal relationships. Despite its tracking results demonstrating top-tier performance, the model's complexity and computational demands leave room for improvement. UMA [28] tracker uses the Siamese network in single object tracking to obtain apparent features and motion information, and then realizes online matching association.

The advantage of the two-step trackers is that they can design the most appropriate model for each phase. However, the two-step method can significantly affect the inference speed of multi-object tracking, due to which employs separate computationally expensive cues.

2.2 Joint Detection and Tracking

The tracking by detection method results in slow inference speed due to separating object detection and feature extraction. In addition, the development of multi-task learning has been improved in deep learning. Therefore, the method of joint detection and tracking has attracted more and more attention, which has been proposed to integrate the detection and Re-ID task into a unified network by sharing most of the common parameters and computation. For example, FairMOT [33] not only uses a one-stage method to balance tracking speed and accuracy but also uses a multi-layer feature aggregation to alleviate a problem, which is detection and ID embedding requiring different feature dimensions. CenterTrack [35] localizes objects by tracking-conditioned detection and uses the detection results of the previous frame in the object motion estimation of the current frame. RMOT [25] uses language expressions as semantic cues to guide prediction of multi-object tracking. CSTrack [9] based on the one-stage method proposes a cross-correlation network and scale-aware attention network to improve the cooperative learning ability of detection and Re-ID.

Although the inference speed of the joint detection and tracking method is substantially higher than the tracking by detection method, its tracking accuracy is correspondingly lower. This is because the learned ID embedding is not optimal, which leads to a large number of ID switches. The key to solving this problem is to balance the contradiction between detection and Re-ID.

3 Proposed Method

In this section, we present a concrete introduction to our CSTMTrack, whose architecture is illustrated in Fig. 1. We first describe in detail the cascaded-scoring tracklet matching (CSTM) strategy that uses a cascading approach to combine appearance and motion information to tracklet association. Then, the motion-guided based target aware (MTA) and appearance-assisted feature warper (AFW) modules are detailed afterward, which mainly address tracking scenarios with irregular motions and indistinguishable appearances.

3.1   Cascaded-Scoring Tracklet Matching

The cascaded matching strategy [5,9,24,32] is a widely employed approach in the field of multi-object tracking (MOT), wherein samples that are easier to associate correctly are matched initially, followed by the matching of more challenging samples. In order to tackle the challenge of indistinguishable appearances, we propose a cascaded-scoring tracklet matching (CSTM) strategy that incorporates ID embedding and motion prediction information in two steps to optimize tracklet matching, instead of relying solely on cross-frame consistency measurement for tracklet association. Illustrated in Fig. 1, the input video frame X_t and $X_t + 1$ are processed through the detection and embedding networks, respectively. The frame X_t, once passed through the detection network, yields several bounding boxes denoted by D_t^i. These bounding boxes are then projected to predict their corresponding bounding boxes P_{t+1}^i for the subsequent frame. Concurrently, the embedding network generates candidate embeddings represented by E_{t+1}^{id}. Finally, our CSTM strategy further refines the outputs from the motion-guided target aware (MTA) and appearance-assisted feature warper (AFW) modules to derive the ultimate detection bounding boxes and tracklets.

In our tracker, we integrate appearance information using the CSTM strategy. Additionally, we utilize the MTA module to calculate a motion cost matrix. To effectively fuse appearance and motion information, we combine two suitable metric matrices, namely the IoU (Intersection over Union) matrix and the cosine matrix. The total cost matrix, denoted as C_t, is computed as a weighted sum of the appearance cost matrix A_t and the motion cost matrix M_t. The formulation of the total cost matrix is as follows:

$$C_t = \alpha A_t + (1 - \alpha) M_t, \tag{1}$$

where α denote the weight of appearance cost matrice. In particular, considering that the appearance features may be susceptible to crowding or blurry objects, we decrease the weight factor α when the detection confidence is low to maintain the correct feature vector.

Consequently, in our CSTM strategy, we fuse the ID embedding and motion prediction information in two stages. Initially, we prioritize the matching of detection bounding boxes and tracklets with the higher weight of the appearance cost matrix A_t when the IoU scores fall below a certain threshold. This helps mitigate the association error caused by the irregular motion of the target. Subsequently, we use the ID embedding similarity matrix with lower weight to fuse with the motion cost matrix M_t.

3.2 Motion-Guided Based Target Aware

When the tracked object exhibits irregular motion, the prediction error of the Kalman filter tends to increase. To mitigate this issue, we introduce the Motion-Guided Target Aware (MTA) module, which incorporates a novel IoU strategy and a straightforward mask branch. The novel IoU strategy, referred to as the Expanding Motion-Guided IoU (EM-IoU) strategy, is employed to expand the detection bounding boxes and enhance their overlap with the target's actual position.

For the EM-IoU strategy, we expand the receptive field of detection bounding boxes by setting a factor σ with the same position center and scale ratio for IoU matching. The strategy enlarges the matching space between detection and tracklet to better measure the geometric consistency [3,24,32,33]. Let $o = (x_1, y_1, x_2, y_2)$ denote the coordinate of an original detection bounding box and (x_1, y_1, x_2, y_2) be the top-left and down-right coordinate of the detection. Then, we use the newly defined coordinates $\tilde{o} = (x_1 - \sigma(x_2 - x_1)/2, y_1 - \sigma(y_2 - y_1)/2, x_2 + \sigma(x_2 - x_1)/2, y_2 + \sigma(y_2 - y_1/2)$ of the target box to compute the IoU of two detection bounding boxes.

Furthermore, we incorporate a mask branch to determine the admissibility of detections for tracklet association. Whenever a detection bounding box score falls below a specific threshold, it is considered an unreliable detection. The mask branch utilizes a binary variable matrix to downweight the final association score for low-scoring detections. In conclusion, the MTA module effectively addresses the challenges encountered during tracking in scenarios involving irregular motion patterns.

3.3 Appearance-Assisted Feature Warper

The joint detection and tracking paradigm has two major limitations. One of the limitations is that different components require different parameters: In the joint detection and tracking trackers, the information required by different components is obtained through shared feature vector parameters, which overlooks the inherent differences between detection and association tasks. In this way, it leads to excessive competition among different components. The other limitation is that different tasks require different scopes of concern: The joint detection and tracking trackers typically add a parallel branch after the backbone network of the detector for obtaining identity (ID) embedding information [13,15,23]. While these frameworks effectively mitigate the computational complexity, it disregards an essential difference between tracking and re-identification tasks. The difference is that the tracking task requires matching samples in the local domain, whereas the re-identification task requires searching and matching in the global domain. To avoid these issues, we adopt the tracking paradigm of tracking by detection, which divides the detection and extraction of the ID embedding into two separate models for higher performance.

When the tracked objects with irregular motion, the Kalman filter predictions tend to have higher uncertainty regarding the object's location. This uncertainty leads to a wider spread of probability mass in the state space and a smaller peak in the observed likelihood. To enhance object tracking in such scenarios, we introduce the Appearance-Assisted Feature Warper (AFW) module, which incorporates an embedding network to compute appearance features for tracking multiple objects. To strike a balance between speed and accuracy while reducing the computational complexity of the embedding network, we choose to train it offline to achieve simplicity and good discrimination. The structure of this embedding network is illustrated in Table 1.

The objective of the AFW module is to utilize the ID embedding features E_{t+1}^{id} as appearance clues to warp and propagate the predicted tracking offset O^{id}. We compute a cost vector matrix to measure the similarity matching between different ID embeddings. For each tracklet k, we maintain a gallery $E_k = e_k^{(i)}{}_k^N$ consisting of N ID embedding descriptors. The similarity matching values in the appearance space between the i-th tracklet and the j-th detection are computed as follows:

$$A_t(i,j) = \min\{1 - e_j^T e_k^{(i)} | e_k^{(i)} \in E_i\}, \tag{2}$$

We use the AFW module to facilitate the inference of tracking offsets by combining appearance information with motion features. This integration enables us to effectively suppress background interference and enhance the ability to distinguish objects in complex scenarios.

Table 1. Overview of the ID embedding network architecture.

Name	Input/Stride	Output
Conv 1	$3 \times 3/1$	$32 \times 128 \times 64$
Conv 2	$3 \times 3/1$	$32 \times 128 \times 64$
Max Pool	$3 \times 3/2$	$32 \times 64 \times 32$
Residual 1	$3 \times 3/1$	$32 \times 64 \times 32$
Residual 2	$3 \times 3/1$	$32 \times 64 \times 32$
Residual 3	$3 \times 3/2$	$64 \times 32 \times 16$
Residual 4	$3 \times 3/1$	$64 \times 32 \times 16$
Residual 5	$3 \times 3/2$	$128 \times 16 \times 8$
Residual 6	$3 \times 3/1$	$128 \times 16 \times 8$
Dense		128
L_2 normalization		128

4 Experiments

4.1 Experimental Setup

Datasets. We choose the MOT17 and MOT20 datasets from the MOTChallenge benchmark to evaluate our experiments. They are widely used for evaluating multi-object tracking algorithms. The MOT17 dataset is focuses on pedestrian tracking. It involves various challenging scenarios such as such as occlusion caused by crowds, blur caused by camera motion, and difficulty in motion prediction caused by nonlinear motion. The MOT20 dataset is oriented towards extremely dense crowd scenarios and realized an average of 246 pedestrians per frame, which is an extension of the MOT17 dataset.

Evaluation Metrics. To evaluate and compare MOT approaches, we exploit the commonly used metrics CLEAR. Including Multiple-Object Tracking Accuracy (MOTA), False Negative (FN), False Positive (FP), Mostly Tracked (MT), Mostly Lost (ML), and IDF1. Among them, MOTA is the most widely used evaluation matrix as it considers three evaluation metrics: FP, IDSW, and FN.

Implementation Details. To conduct a fair comparison experiment, we use experimental setup of OC-SORT tracker. In terms of the dataset, we train our tracker with the CrowdHuman dataset and half of the MOT17 training set, and we use the other half of the MOT17 training set as the validation set. When testing on the MOT17 test set, we add the Cityperson and ETHZ datasets for a new round of training to obtain a preferable tracking performance. To achieve our cascaded-scoring tracklet matching, we conduct grid research [1] to find the optimal combination of weight settings for motion and appearance features in the two stages on the training set. And following the OC-SORT tracker, we use linear interpolation to boost the online output.

4.2 Ablation Studies

To perform ablation studies optimally, we train our tracker using the CrowdHuman dataset and half of the MOT17 training set, and the other half of the MOT17 training set is used to evaluate the effectiveness of the AFW and MTA modules.

Influence of the AFW Module. Under the assumption that other parameters remain unchanged, we introduce the AFW module to the baseline tracker in order to assess its effectiveness. The results, as depicted in Table 2, demonstrate notable improvements in terms of MOTA and FN metrics with the inclusion of the AFW module in CSTMTrack. Particularly, MOTA shows a significant increase from 74.7% to 75.1%. Additionally, FN experiences a slight reduction from 3.14 to 3.06. Overall, the AFW module delivers decent performance, although there is a slight decrease in IDF1 and a minor increase in IDs. From the results presented above, it is evident that the AFW module effectively mitigates background interference and reduces instances of missed detection. Furthermore, it enhances the capability to discern objects within complex scenarios.

Influence of the MTA Module. As shown in Table 2, it is evident that the MTA module plays a crucial role in enhancing the performance of our CSTM-Track. The inclusion of the MTA module in our tracker results in higher MOTA and IDF1 scores compared to the baseline tracker. In particular, MOTA increases from 74.7% to 75.1%, and IDF1 improves from 76.5% to 77.6%. Additionally, both the number of IDs and FN demonstrate a decrease to some extent. The above results clearly demonstrate that the MTA module effectively addresses concerns such as trajectory loss and ID switching, particularly in complex scenarios.

Table 2. Ablation study on the MOT17 val set to evaluate the proposed AFW and MTA modules. The best results are shown in **bold**.

AFW	MTA	MOTA↑	IDF1↑	IDs↓	FN(10^4)↓
		74.7	76.5	846	3.14
✓		75.1	75.8	864	**3.06**
	✓	75.0	**77.6**	795	3.09
✓	✓	**75.1**	77.5	**780**	3.09

With the AFW and MTA module, our CSTMTrack improves MOTA by 0.4% and IDF1 by 1.0% and decreases IDs and FN compared with the baseline tracker in the MOT17 val test, respectively. The results affirm that both the AFW and MTA modules markedly enhance the performance of tracking.

Table 3. Comparison with state-of-the-art methods on the MOT17 test set with private detections. The best results are shown in **bold**.

Tracker	MOTA↑	IDF1↑	AssR↑	FN(10^4)↓
FairMOT [33]	73.7	72.3	63.6	11.7
GRTU [20]	74.9	75.0	65.8	10.8
TransTrk [17]	75.2	63.5	57.1	8.64
CSTrack [9]	74.9	72.6	63.2	11.4
QDTrack [12]	68.7	66.3	57.2	14.66
ReMOT [27]	77.0	72.0	61.7	9.36
PermaTr [18]	73.8	68.9	59.8	11.5
OC-SORT [3]	78.0	**77.5**	67.5	10.8
CSTMTrack (Ours)	**79.7**	77.3	**67.7**	**8.49**

Table 4. Comparison with state-of-the-art methods on the MOT20 test set with private detections. The best results are shown in **bold**.

Tracker	MOTA↑	IDF1↑	AssR↑	FN(10^4)↓
FairMOT [33]	61.8	67.3	60.7	8.89
GSDT [21]	67.1	67.5	52.7	13.5
TransTrk [17]	65.0	59.4	51.9	15.0
CSTrack [9]	66.6	68.6	57.6	14.4
RelationT [29]	67.2	70.5	66.1	10.5
MAA [16]	73.9	71.2	61.1	10.9
OC-SORT [3]	75.5	**75.9**	**67.5**	10.8
CSTMTrack (Ours)	**75.9**	73.9	65.9	**10.3**

4.3   Comparison with State-of-the-Art Methods

We conducted a comprehensive comparison of our proposed CSTMTrack with several state-of-the-art trackers on the MOTChallenge benchmark. It is important to note that all experimental results for these trackers utilize private detection. In order to ensure a fair comparison, we adopted equivalent experimental settings and training strategies as the baseline tracker OC-SORT. These include using the same detection network (e.g., YOLOX [6]), applying preprocessing techniques on input images, and setting the score threshold for the detection component, among other factors.

In order to achieve improved tracking performance, we incorporated the Cityperson and ETHZ datasets into our training process when evaluating the MOT17 and MOT20 test sets. The results, as shown in Table 3, highlight the superior performance of our CSTMTrack compared to several state-of-the-art trackers in terms of MOTA, AssR, and FN. Notably, our tracker exhibits a significant improvement in MOTA, surpassing the OC-SORT tracker by a relative gain of 1.7%. While our CSTMTrack may not achieve the best results in terms of IDF1 and AssR (as shown in Table 4), it consistently outperforms other trackers in terms of MOTA and FN scores. This also implies the effectiveness of our cascaded-scoring tracklet matching (CSTM) strategy for tracklet association. In addition to the tracking performance discussed above, we are also focused on the tracking speed of the algorithm. As different methods might have varying runtime environments, it is challenging to directly compare them. Therefore, we only conduct a rough comparison with the speed of OC-SORT. Our CSTM-Tracke achieves speeds of 355.1 Hz on MOT17 and 89.6 Hz on MOT20, while the OC-SORT tracker achieves only 29.0 Hz and 18.7 Hz, respectively. In general, experimental results tested on MOT17 and MOT20 show that our CSTM strategy improves both MOTA and FN metrics, illustrating the effectiveness of our tracker in tracking complex scenes. Moreover, from the comparison results of speed, our method successfully strikes a balance between speed and precision.

5   Conclusion

In this paper, we first analyze the pros and cons of two tracking paradigms, tracking by detection and joint detection and tracking. Then, we propose a simple yet robust online tracking by detection tracker, called CSTMTrack, which addresses the challenges of MOT with irregular motions and indistinguishable appearances through the cascaded-scoring tracklet matching strategy. Our proposed motion-guided based target aware and appearance-assisted feature warper modules are lightweight and can be simply plugged into other trackers to advance tracking performance. Moreover, this approach can be effectively applied to the scenario of autonomous driving. Experimental results demonstrate the effectiveness and efficiency of our CSTMTrack tracker, which achieves state-of-the-art performance on the MOT17 and MOT20 benchmarks.

Acknowledgements. This work was supported by the National Natural Science Foundation of China under Grant U21A20514, 62002302, by the FuXiaQuan National Independent Innovation Demonstration Zone Collaborative Innovation Platform Project under Grant 3502ZCQXT2022008, and by the China Fundamental Research Funds for the Central Universities under Grants 20720230038.

References

1. Bergstra, J., Bengio, Y.: Random search for hyper-parameter optimization. J. Mach. Learn. Res. **13**(2) (2012)
2. Bewley, A., Ge, Z., Ott, L., Ramos, F., Upcroft, B.: Simple online and realtime tracking. In: ICIP, pp. 3464–3468 (2016)
3. Cao, J., Pang, J., Weng, X., Khirodkar, R., Kitani, K.: Observation-centric sort: rethinking sort for robust multi-object tracking. In: CVPR, pp. 9686–9696 (2023)
4. Cheng, Y., Li, L., Xu, Y., Li, X., Yang, Z., Wang, W., Yang, Y.: Segment and track anything. arXiv preprint arXiv:2305.06558 (2023)
5. Gao, P., Ma, Y., Yuan, R., Xiao, L., Wang, F.: Learning cascaded siamese networks for high performance visual tracking. In: ICIP, pp. 3078–3082 (2019)
6. Ge, Z., Liu, S., Wang, F., Li, Z., Sun, J.: Yolox: exceeding yolo series in 2021. arXiv preprint arXiv:2107.08430 (2021)
7. Guo, S., Wang, J., Wang, X., Tao, D.: Online multiple object tracking with cross-task synergy. In: CVPR, pp. 8136–8145 (2021)
8. Huang, K., Chu, J., Qin, P.: Two-stage object tracking based on similarity measurement for fused features of positive and negative samples. In: PRCV, pp. 621–632 (2022)
9. Liang, C., Zhang, Z., Zhou, X., Li, B., Zhu, S., Hu, W.: Rethinking the competition between detection and reid in multiobject tracking. IEEE Trans. Image Process. **31**, 3182–3196 (2022)
10. Liu, Y., Yang, W., Yu, H., Feng, L., Kong, Y., Liu, S.: Background suppressed and motion enhanced network for weakly supervised video anomaly detection. In: PRCV, pp. 678–690 (2022)
11. Pang, B., Li, Y., Zhang, Y., Li, M., Lu, C.: Tubetk: adopting tubes to track multi-object in a one-step training model. In: CVPR, pp. 6308–6318 (2020)

12. Pang, J., et al.: Quasi-dense similarity learning for multiple object tracking. In: CVPR, pp. 164–173 (2021)
13. Peng, J., et al.: Chained-tracker: chaining paired attentive regression results for end-to-end joint multiple-object detection and tracking. In: Vedaldi, A., Bischof, H., Brox, T., Frahm, J.-M. (eds.) ECCV 2020. LNCS, vol. 12349, pp. 145–161. Springer, Cham (2020). https://doi.org/10.1007/978-3-030-58548-8_9
14. Saleh, F., Aliakbarian, S., Rezatofighi, H., Salzmann, M., Gould, S.: Probabilistic tracklet scoring and inpainting for multiple object tracking. In: CVPR, pp. 14329–14339 (2021)
15. Sommer, L., Krüger, W.: Usage of vehicle re-identification models for improved persistent multiple object tracking in wide area motion imagery. In: ICIP, pp. 331–335 (2022)
16. Stadler, D., Beyerer, J.: Modelling ambiguous assignments for multi-person tracking in crowds. In: WACV, pp. 133–142 (2022)
17. Sun, P., et al.: Transtrack: multiple object tracking with transformer. arXiv preprint arXiv:2012.15460 (2020)
18. Tokmakov, P., Li, J., Burgard, W., Gaidon, A.: Learning to track with object permanence. In: ICCV, pp. 10860–10869 (2021)
19. Wang, L., Hui, L., Xie, J.: Facilitating 3D object tracking in point clouds with image semantics and geometry. In: PRCV, pp. 589–601 (2021)
20. Wang, S., Sheng, H., Zhang, Y., Wu, Y., Xiong, Z.: A general recurrent tracking framework without real data. In: ICCV, pp. 13219–13228 (2021)
21. Wang, Y., Kitani, K., Weng, X.: Joint object detection and multi-object tracking with graph neural networks. In: ICRA, pp. 13708–13715 (2021)
22. Wang, Y., Li, C., Tang, J.: Learning soft-consistent correlation filters for RGB-T object tracking. In: PRCV, pp. 295–306 (2018)
23. Wang, Z., Zheng, L., Liu, Y., Li, Y., Wang, S.: Towards real-time multi-object tracking. In: Vedaldi, A., Bischof, H., Brox, T., Frahm, J.-M. (eds.) ECCV 2020. LNCS, vol. 12356, pp. 107–122. Springer, Cham (2020). https://doi.org/10.1007/978-3-030-58621-8_7
24. Wojke, N., Bewley, A., Paulus, D.: Simple online and realtime tracking with a deep association metric. In: ICIP, pp. 3645–3649 (2017)
25. Wu, D., Han, W., Wang, T., Dong, X., Zhang, X., Shen, J.: Referring multi-object tracking. In: CVPR, pp. 14633–14642 (2023)
26. Xu, J., Cao, Y., Zhang, Z., Hu, H.: Spatial-temporal relation networks for multi-object tracking. In: ICCV, pp. 3988–3998 (2019)
27. Yang, F., Chang, X., Sakti, S., Wu, Y., Nakamura, S.: Remot: a model-agnostic refinement for multiple object tracking. Image Vis. Comput. **106**, 104091 (2021)
28. Yin, J., Wang, W., Meng, Q., Yang, R., Shen, J.: A unified object motion and affinity model for online multi-object tracking. In: CVPR, pp. 6768–6777 (2020)
29. Yu, E., Li, Z., Han, S., Wang, H.: Relationtrack: relation-aware multiple object tracking with decoupled representation. IEEE Trans. Multimedia (2022)
30. Zeng, F., Dong, B., Zhang, Y., Wang, T., Zhang, X., Wei, Y.: MOTR: end-to-end multiple-object tracking with transformer. In: Avidan, S., Brostow, G., Cissé, M., Farinella, G.M., Hassner, T. (eds.) ECCV 2022. LNCS, vol. 13687, pp. 659–675. Springer, Cham (2022). https://doi.org/10.1007/978-3-031-19812-0_38
31. Zhang, M., Pan, Z., Feng, J., Zhou, J.: 3D multi-object detection and tracking with sparse stationary lidar. In: PRCV, pp. 16–28 (2021)

32. Zhang, Y., et al.: Bytetrack: multi-object tracking by associating every detection box. In: Avidan, S., Brostow, G., Cissé, M., Farinella, G.M., Hassner, T. (eds.) ECCV 2022. LNCS, vol. 13682, pp. 1–21. Springer, Cham (2022). https://doi.org/10.1007/978-3-031-20047-2_1

33. Zhang, Y., Wang, C., Wang, X., Zeng, W., Liu, W.: FairMOT: on the fairness of detection and re-identification in multiple object tracking. Int. J. Comput. Vision **129**, 3069–3087 (2021)

34. Zhao, K., et al.: Driver behavior decision making based on multi-action deep Q network in dynamic traffic scenes. In: PRCV, pp. 174–186 (2022)

35. Zhou, X., Koltun, V., Krähenbühl, P.: Tracking objects as points. In: Vedaldi, A., Bischof, H., Brox, T., Frahm, J.-M. (eds.) ECCV 2020. LNCS, vol. 12349, pp. 474–490. Springer, Cham (2020). https://doi.org/10.1007/978-3-030-58548-8_28

Boosting Generalization Performance in Person Re-identification

Lidong Cheng[1,2], Zhenyu Kuang[1], Hongyang Zhang[1], Xinghao Ding[1(✉)], and Yue Huang[1]

[1] Lab of Smart Data and Signal Processing, Xiamen University, Xiamen 361001, China
dxh@stu.xmu.edu.cn
[2] Institute of Artificial Intelligence, Xiamen University, Xiamen 361001, China

Abstract. Generalizable person re-identification (ReID) has gained significant attention in recent years as it poses greater challenges in recognizing individuals across different domains and unseen scenarios. Existing methods are typically limited to a single visual modality, making it challenging to capture rich semantic information across different domains. Recently, pre-trained vision-language models like CLIP have shown promising performances in various tasks by linking visual representations with their corresponding text descriptions. This enables them to capture diverse high-level semantics from the accompanying text and obtain transferable features. However, the adoption of CLIP has been hindered in person ReID due to the labels being typically index-based rather than descriptive texts. To address this limitation, we propose a novel Cross-modal framework wIth Conditional Prompt (CICP) framework based on CLIP involving the Description Prompt Module (DPM) that pre-trains a set of prompts to tackle the lack of textual information in person ReID. In addition, we further propose the Prompt Generalization Module (PGM) incorporates a lightweight network that generates a conditional token for each image. This module shifts the focus from being limited to a class set to being specific to each input instance, thereby enhancing domain generalization capability for the entire task. Through extensive experiments, we show that our proposed method outperforms state-of-the-art (SOTA) approaches on popular benchmark datasets.

Keywords: Person re-identification · Domain generalization · Vision-language model

1 Introduction

Person ReID is a popular research area in computer vision. It focuses on identifying the same individual from a vast collection of gallery images given a query

---

The work was supported in part by the National Natural Science Foundation of China under Grant 82172033, U19B2031, 61971369, 52105126, 82272071, 62271430, and the Fundamental Research Funds for the Central Universities 20720230104.

Q. Liu et al. (Eds.): PRCV 2023, LNCS 14434, pp. 174–185, 2024.
https://doi.org/10.1007/978-981-99-8549-4_15

image. Recently, there has been a growing interest in generalizable person ReID, driven by its significance of research and practical implications. This task studies the ability of a trained person ReID model to generalize to unseen scenarios and utilizes direct cross-dataset evaluation [5,23] for performance assessment. Most ReID methods employ ResNet [4] or ViT-based [3] models to extract visual features. However, they only rely on visual modality and ignore the rich semantic information. It has been proven by rigorous theory [6] that multi-modal learning is better than single modality.

Recently, there has been an increasing interest in vision-language learning approaches such as CLIP [18] and ALIGN [7], which establish connections between visual representations and their corresponding high-level language descriptions. These methods not only leverage larger datasets for training but also modify the pre-training task by aligning visual features with their textual counterparts. Therefore, the image encoder is able to capture diverse high-level semantics from the accompanying text and acquire transferable features, which possess a universal and generalized representation. This results in better performance for domain generalization tasks.

Specifically, in an image classification task, utilizing specific text labels paired with a prompt (e.g. "A photo of a...") can produce text descriptions. By employing distance metrics between image features and text features, the model determines the similarity or dissimilarity between the image and each category's textual representations. Subsequently, it classifies the image by associating it with the most relevant or similar text description. Building upon this idea, CoCoOp [28] introduces a lightweight visual network that is designed to generate an input-conditional token for each image. The additional network achieves further performance improvements. Text labels are necessary for CLIP and CoCoOp to create text descriptions for downstream tasks. Nevertheless, in many person ReID tasks, the labels consist of indexes rather than descriptive words for the images [12]. This has led to a limited adoption of the vision-language model in person ReID.

To tackle this issue, We propose a novel CICP framework based on CLIP, involving the DPM, which leverages a set of pre-trained prompts to generate text descriptions for person identities. Instead of relying solely on visual features, the model can now leverage the generated text descriptions to enhance its understanding of person identities. In addition, we further propose the PGM inspired by the ideas of [28]. This module incorporates an additional lightweight network that generates an input-conditional token for each image, making the prompt to be conditioned on each input instance. The proposed DPM can achieve better generalization by shifting its focus from specific class sets to individual input instance. Our approach exploits joint visual and language representation learning rather than relying solely on visual information. This results in the model tapping into high-level semantic information and acquiring domain-agnostic feature representations. Notably, our method achieves superior performance in domain generalization tasks.

In summary, this paper makes the following contributions:

- To the best of our knowledge, we are the first to apply CLIP to the generalizable person ReID task.
- We propose the CICP framework, which fully utilizes the cross-modal capabilities of CLIP. By pre-training a set of prompts in DPM to generate text descriptions and imposing constraints on the image encoder during the training phase.
- We propose the PGM, which conditions the prompt for each image and significantly enhances generalization capabilities.
- Through direct cross-dataset evaluations on three popular pedestrian datasets, we demonstrate that CICP outperforms the SOTA methods.

2   Related Work

2.1   Generalizable Person ReID

Generalizable person re-identification was first conducted in [5,23], where a direct cross-dataset evaluation was proposed as the benchmark for testing algorithms. In recent years, with the progress of deep learning, this task has attracted increasing attention. Most methods mainly adopted either a meta-learning pipeline or exploit domain-specific heuristics. Song et al. [20] proposed a domain-invariant mapping network following the meta-learning [2] pipeline. Jia et al. [8] utilized a combination of instance and feature normalization to mitigate style and content variances across datasets.

Nevertheless, these methods were focused on a single visual modality, which limits their potential for breakthroughs. In a novel way, we have innovatively applied CLIP to generalizable person ReID. Benefiting from a set of pre-trained prompts in DPM and proposed PGM, our method has surpassed SOTA on various datasets.

2.2   Vision-Language Learning

Compared to supervised pre-training on ImageNet, vision-language pre-training (VLP) has demonstrated substantial performance improvements in various downstream tasks by emphasizing the alignment between image and language. Notable practices in this domain include CLIP [18] and ALIGN [7], which employed image and text encoders in conjunction with two directional InfoNCE losses computed between their outputs during training. Building upon the foundations of CLIP, several studies [10,11] were proposed incorporating a wider range of learning tasks, such as image-to-text matching and masked image/text modeling.

Prompt or adapter-based tuning, inspired by recent advancements in NLP, has gained popularity in the field of vision. For instance, CoOp [29] introduced the concept of incorporating a learnable prompt for image classification. CoCoOp [28] explored a lightweight visual network that generates conditional tokens and utilizes learnable context vectors for each image.

Moreover, researchers have explored diverse downstream tasks for the application of CLIP. DenseCLIP [19] and MaskCLIP [27] utilized CLIP for per-pixel prediction in segmentation. EI-CLIP [16] and CLIP4CirDemo [1] employed CLIP for tackling retrieval problems. These CLIP-based methods made significant breakthroughs in their respective tasks. While CLIP-ReID [12] initially applies CLIP to single-domain image ReID and achieves promising results, it falls short in terms of generalization to unseen domains. We take it as our baseline and aim to make improvements upon it. Note that this paper focuses on the more practical and meaningful task of generalizable person ReID. To the best of our knowledge, no existing studies have specifically addressed this task based on CLIP.

3   Method

In this section, we first provide a brief introduction to the CLIP method in Sect. 3.1. Then, we present the overall framework of our proposed CICP in Sect. 3.2. Next, we discuss the design process of the prompt, which includes the Description Prompt Module (DPM), as well as the Prompt Generalization Module (PGM) in Sect. 3.3. Finally, we introduce the loss functions used in the pre-training and training stages in Sect. 3.4.

3.1   Review of CLIP

We first briefly introduce the CLIP architecture, which consists of two main parts: the image encoder and the text encoder. The image encoder extracts feature from the image and can be designed using a CNN such as ResNet-50 [4] or a ViT like ViT-B/16 [3]. Both of them have the ability to effectively summarize an image into a feature vector within the cross-modal embedding space. In addition, the text encoder is used to extract features from textual descriptions.

Let V_i and T_i represent the embeddings of an image and its corresponding text description in a cross-modal embedding space, where $i \in 1, 2, ..., N$ represents the index within a batch. The contrastive loss for image-to-text can be written as follow:

$$\mathcal{L}_{it}(i) = -\log \frac{\exp\left(\text{sim}\left(V_i, T_i\right)/\tau\right)}{\sum_{j=1}^{N} \exp\left(\text{sim}\left(V_i, T_j\right)\right)} \tag{1}$$

Then, the contrastive loss for text-to-image is defined as:

$$\mathcal{L}_{ti}(i) = -\log \frac{\exp\left(\text{sim}\left(V_i, T_i\right)/\tau\right)}{\sum_{j=1}^{N} \exp\left(\text{sim}\left(V_j, T_i\right)\right)} \tag{2}$$

where $\text{sim}(\cdot, \cdot)$ represents the cosine similarity and τ is a learned temperature parameter in Eq. 1 and Eq. 2.

Generally, CLIP transforms labels into text descriptions and generates embedding features V_i and T_i. However, for generalizable person ReID tasks, there are no specific text labels available, which makes it challenging to directly apply the CLIP method.

3.2 A Novel Cross-Modal Framework

Figure 1 illustrates the proposed CICP framework. The DPM tackles the lack of textual information in person ReID by pre-training a set of prompts. The PGM enhances the generalization capability of the prompts. It achieves this by incorporating an additional lightweight network that generates an input-conditional token for each image. This allows the prompt to be conditioned on each input instance, making it more adaptable and suitable for different scenarios.

The proposed CICP framework comprises a text encoder and an image encoder, using the widely adopted Transformer and ViT architectures as their backbone, respectively. For the image encoder, we utilize ViT-B/16, which consists of 12 transformer layers and a hidden size of 768 dimensions. To align with the output of the text encoder, we apply the linear projection to reduce the dimension of the image feature vector from 768 to 512.

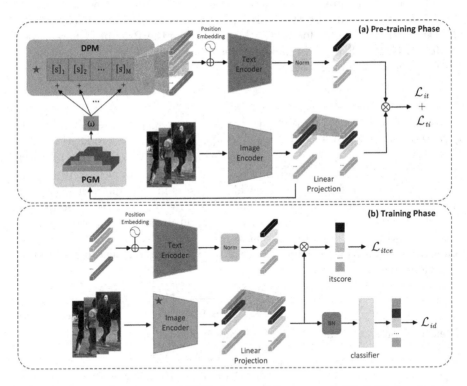

Fig. 1. Overview of our proposed CICP framework for person ReID. (a) In pre-training phase, we first input the person images to the image encoder and obtain corresponding image feature. Then, we pass it through the PGM to get a condition token ω which is combined with DPM and fed into the text encoder. (b) In training phase, we exploit the optimized prompt and only optimize the image encoder. The red pentagram represents the optimized parts during the pre-training or training phase. (Color figure online)

In the pre-training phase, as shown in Fig. 1(a), we keep the parameters of the text encoder and image encoder fixed and focus on optimizing the prompt in DPM and the lightweight network in PGM. Firstly, we input the pedestrian image to the image encoder and obtain the feature vector. Then, we pass it through the lightweight network to generate an input-conditional token. The token is combined with the prompt and fed into the text encoder, producing the text features. The model parameters are updated by calculating supervised contrastive losses $\mathcal{L}_{it}$ and $\mathcal{L}_{ti}$.

In the training phase, as illustrated in Fig. 1(b), we fix the parameters of the prompt and text encoder and only optimize the parameters in the image encoder. We feed the optimized prompt in DPM into the text encoder to obtain the corresponding text features and pass the image into the image encoder to obtain image features. Then, the similarity between the text features and image features is employed to compute the image-to-text cross-entropy loss $\mathcal{L}_{itce}$. Finally, the image encoder is optimized through back-propagation.

3.3   Prompt Design Process

In this section, we provide a detailed explanation of the prompt design process, which includes the proposed DPM and PGM.

Description Prompt Module. Firstly, we introduce DPM with class-specific prompts to learn text descriptions that are independent for each class. Specifically, the prompt is designed in the following form:

$$\text{Prompt} = \text{Template}([s]_1[s]_2...[s]_M, \text{person}) \tag{3}$$

where each $[s]_m (m \in 1, ..., M)$ represents a learnable context token with dimensions equal to the word embedding size of CLIP (i.e., 512). The hyperparameter M denotes the number of context tokens. By passing the prompt through the text encoder, we obtain class-specific context vectors for each class.

Prompt Generalization Module. Furthermore, to enhance the generalization performance of the prompt, we propose PGM. Specifically, we input the image feature into the lightweight network to get the corresponding input-conditional token. The token is then combined with the class-specific prompt to obtain an instance-conditional prompt.

Let $L_\theta(\cdot)$ denotes the lightweight network parameterized by θ, each instance-conditional context token is now obtained as $[s(x)]_m = [s]_m + \omega$, where $[s]_m$ represents the learnable context token mentioned earlier, and $\omega = L_\theta(x)$ represents the instance-specific representation generated by the $L_\theta(\cdot)$ for the input instance x. The instance-conditional prompt form can be expressed as:

$$\text{Prompt} = \text{Template}([s(x)]_1[s(x)]_2...[s(x)]_M, \text{person}) \tag{4}$$

During the pre-training process, both the context tokens $[s(x)]_m$ and the parameters of $L_\theta(\cdot)$ are updated.

3.4   Loss Function

We employ different loss functions during the pre-training and training phases.

In the pre-training phase, similar to CLIP, we utilize the supervised contrastive loss to optimize the instance-conditional prompt. Due to each ID sharing the same text description, we replace T_i with T_{y_i} in Eq. 1 and Eq. 2. The loss function is defined as:

$$\mathcal{L}_{it}(i) = -\log \frac{\exp\left(\operatorname{sim}\left(V_i, T_{y_i}(x)\right)/\tau\right)}{\sum_{j=1}^{N} \exp\left(\operatorname{sim}\left(V_i, T_{y_j}(x)\right)\right)} \tag{5}$$

where $T_i(x)$ represents the embedding of the instance-conditional prompt consisting of $[s(x)]_m$ in the cross-modal embedding space.

Furthermore, in the case of $\mathcal{L}_{ti}$, it is possible for different images in a batch to belong to the same person, resulting in T_{y_i} having more than one positive example. The specific form of the loss function is defined as follows:

$$\mathcal{L}_{ti}(y_i) = -\frac{1}{|H(y_i)|} \sum_{h \in H(y_i)} \log \frac{\exp\left(\operatorname{sim}\left(V_h, T_{y_i}(x)\right)/\tau\right)}{\sum_{j=1}^{N} \exp\left(\operatorname{sim}\left(V_j, T_{y_i}(x)\right)\right)} \tag{6}$$

where $H(y_i) = \{h \in 1...N : y_h = y_i\}$ represents the set of indices of all positive examples for T_{y_i} in the batch, and $|\cdot|$ denotes the cardinality of the set. By minimizing the loss, the gradients are back-propagated through the text encoder to optimize $[s(x)]_1 [s(x)]_2 \cdots [s(x)]_M$. Afterward, we obtain an optimized set of prompts, which are then stored in the model as part of its parameters.

In the training phase, following the mainstream person ReID methods, we apply the ID loss $\mathcal{L}_{id}$ and cross-entropy loss $\mathcal{L}_{ce}$ with label smoothing to optimize the image encoder. We represent p_k as the ID prediction of class k, and K as the number of identities. ID loss is formulated as:

$$\mathcal{L}_{id} = \sum_{k=1}^{K} -q_k \log(p_k) \tag{7}$$

For the optimized prompts obtained during the pre-training stage, we incorporate them into the cross-entropy loss. The specific formulation is as follows:

$$\mathcal{L}_{itce}(i) = \sum_{k=1}^{K} -q_k \log \frac{\exp\left(\operatorname{sim}\left(V_i, T_{y_k}\right)\right)}{\sum_{y_j=1}^{K} \exp\left(\operatorname{sim}\left(V_i, T_{y_j}\right)\right)} \tag{8}$$

The losses used in the pre-training and training phases can be summarized as follows:

$$\mathcal{L} = \begin{cases} \mathcal{L}_{it} + \mathcal{L}_{ti}, & \text{pre-training} \\ \mathcal{L}_{id} + \mathcal{L}_{itce}, & \text{training} \end{cases} \tag{9}$$

4   Experiments

4.1   Datasets and Evaluation Protocols

Our experiments are conducted on three large person ReID datasets: Market-1501 [25], DukeMTMC-reID [26], and MSMT17 [21]. Table 1 provides a detailed summary of the dataset information. The cross-dataset evaluation is conducted using these datasets, where the models are trained on the training subset of one dataset (except for MSMT17, where all images are used for training following [22,24]), and evaluated on the test subset of another dataset. We employ the cumulative matching characteristics (CMC) at Rank-1 (R-1) and the mean Average Precision (mAP) as performance evaluation metrics. All evaluations strictly follow the single-query evaluation protocol.

Table 1. Statistics of person datasets

Datasets	IDs	Images	Cameras
Market1501 [25]	1,501	32,668	6
DukeMTMC-reID [26]	1,812	36,441	8
MSMT17 [21]	4,101	12,6411	15

4.2   Implementation Details

We utilize the image encoder (VIT-B/16) and text encoder from CLIP as the backbone in our approach. The Adam method and warmup learning strategy are employed to optimize the model. In the pre-training phase, we initialize the learning rate to 3.5×10^{-4} with cosine learning rate decay and resize all images to 256×128. The batch size is set to 64, and no data augmentation methods are used. In the training phase, we employ common data augmentation techniques, including random cropping, horizontal flipping, and padding. The maximum number of training epochs is set to 60. Following [12], we gradually increase the learning rate from 5×10^{-7} to 5×10^{-6} over 10 epochs using a linear growth strategy. Subsequently, at the 30th and 50th epochs, the learning rate is reduced by a factor of 0.1.

4.3   Ablation Study

In order to demonstrate the effectiveness of our proposed DPM and PGM, we conduct ablation studies in the context of direct cross-dataset evaluation between the Market-1501 and DukeMTMC-reID datasets. The results are presented in Table 2. We take CLIP-ReID [12] on the generalizable person ReID task and use it as the baseline. By comparing the baseline with the DPM, which utilizes a set of pre-trained prompts, we find that DPM can effectively complement the lack of textual labels in the Person ReID task. The performance improvement is observed as 21.9% and 11.7% in R-1 under the Market→Duke and Duke→Market settings, respectively.

Furthermore, the results show that the addition of PGM leads to significant improvements in generalization. Specifically, we observe an increase of 10.6% and 10.3% in R-1 accuracy, as well as an improvement of 11.5% and 4.9% in mAP, under the Market→Duke and Duke→Market settings, respectively. This demonstrates that PGM enables better generalization by optimizing the prompt for each input instance.

Table 2. Ablation study

Method	Market→Duke				Duke→Market			
	R-1	R-5	R-10	mAP	R-1	R-5	R-10	mAP
baseline	34.6	49.6	57.0	18.9	51.9	68.7	74.9	24.0
+DPM	56.5	69.2	74.5	34.8	63.6	79.1	84.4	38.1
+PGM	**67.1**	**78.5**	**82.5**	**46.3**	**73.9**	**86.3**	**89.8**	**43.0**

Table 3. Comparsions against state-of-the-art methods

Method	Venue	Market→Duke				Duke→Market				MSMT17→Duke			
		R-1	R-5	R-10	mAP	R-1	R-5	R-10	mAP	R-1	R-5	R-10	mAP
OSNet-IBN [30]	ICCV'19	48.5	62.3	67.4	26.7	57.7	73.7	80.0	26.1	67.4	80.0	83.3	45.6
QAConv [13]	ECCV'20	48.8	-	-	28.7	58.6	-	-	27.6	69.4	-	-	52.6
SNR [9]	CVPR'20	55.1	-	-	33.6	66.7	-	-	33.9	69.2	-	-	49.9
MixStyle [32]	ICLR'21	47.5	62.0	67.1	27.3	58.2	74.9	80.9	29.0	68.0	80.6	85.1	50.2
OSNet-AIN [31]	TPAMI'21	52.4	66.1	71.2	30.5	61.0	77.0	82.5	30.6	71.1	83.3	86.4	52.7
TransMatcher [14]	NIPS'21	55.4	69.0	73.5	33.4	67.3	81.5	86.3	34.5	-	-	-	-
MetaBIN [2]	CVPR'21	55.2	69.0	74.4	33.1	69.2	83.1	87.8	35.9	-	-	-	-
MDA [17]	CVPR'22	56.7	-	-	34.4	70.3	85.2	89.6	38.0	71.7	-	-	52.4
QAConv-GS [15]	CVPR'22	59.8	72.0	76.8	37.0	73.0	85.9	89.5	39.3	71.5	84.2	**88.1**	54.5
CICP	Ours	**67.1**	**78.5**	**82.5**	**46.3**	**73.9**	**86.3**	**89.8**	**43.0**	**74.4**	**84.2**	87.7	**56.1**

4.4   Comparison with State-of-the-Art Methods

We conducted a comprehensive comparison between our proposed CICP and SOTA methods in generalizable person re-identification. The results of direct cross-dataset evaluation on three benchmark datasets are presented in Table 3. The comparison clearly demonstrates that CICP achieves significant improvements over the previous SOTA methods. Specifically, when training on the Market dataset and using the Duke dataset as the target dataset, we observe a remarkable improvement of 7.3% in R-1 accuracy and 9.3% in mAP. With

Duke→Market, the improvements are 0.9% in R-1 and 3.7% in mAP. Similarly, when training on the MSMT17 dataset and evaluating on the Duke dataset, we achieve 2.7% improvement in R-1 accuracy and 1.6% improvement in mAP. It is important to note that all the results presented here are obtained without employing any post-processing methods such as re-ranking.

4.5 Other Analysis

We evaluate the influence of the hyperparameter M on the performance of our method on the Duke and Market datasets. As illustrated in Fig. 2, it is evident that the optimal performance is achieved when M is set to 4. We observe that setting M to 1 does not sufficiently capture the textual descriptions, while a higher value of M (M = 10) results in excessive and unnecessary descriptions, leading to a significant drop in performance.

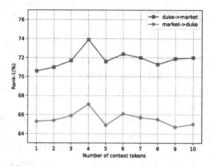

Fig. 2. Effect of hyperparameter M

5 Conclusion

In this paper, we propose a novel CICP framework to leverage textual information and design effective prompts to enhance model performance across different domains. The main contributions include introducing the Description Prompt Module (DPM) and the Prompt Generalization Module (PGM). Extensive experiments have shown that our approach outperforms previous SOTA methods. Future work may explore further improvements in network design and explore additional strategies for handling unseen domains.

References

1. Baldrati, A., Bertini, M., Uricchio, T., Del Bimbo, A.: Effective conditioned and composed image retrieval combining clip-based features. In: Proceedings of the IEEE/CVF Conference on Computer Vision and Pattern Recognition, pp. 21466–21474 (2022)

2. Choi, S., Kim, T., Jeong, M., Park, H., Kim, C.: Meta batch-instance normalization for generalizable person re-identification. In: Proceedings of the IEEE/CVF Conference on Computer Vision and Pattern Recognition, pp. 3425–3435 (2021)

3. Dosovitskiy, A., et al.: An image is worth 16x16 words: transformers for image recognition at scale. arXiv preprint arXiv:2010.11929 (2020)

4. He, K., Zhang, X., Ren, S., Sun, J.: Deep residual learning for image recognition. In: Proceedings of the IEEE Conference on Computer Vision and Pattern Recognition, pp. 770–778 (2016)

5. Hu, Y., Yi, D., Liao, S., Lei, Z., Li, S.Z.: Cross dataset person re-identification. In: Jawahar, C.V., Shan, S. (eds.) ACCV 2014. LNCS, vol. 9010, pp. 650–664. Springer, Cham (2015). https://doi.org/10.1007/978-3-319-16634-6_47

6. Huang, Y., Du, C., Xue, Z., Chen, X., Zhao, H., Huang, L.: What makes multimodal learning better than single (provably). Adv. Neural. Inf. Process. Syst. **34**, 10944–10956 (2021)

7. Jia, C., et al.: Scaling up visual and vision-language representation learning with noisy text supervision. In: International Conference on Machine Learning (2021)

8. Jia, J., Ruan, Q., Hospedales, T.M.: Frustratingly easy person re-identification: generalizing person re-id in practice. arXiv preprint arXiv:1905.03422 (2019)

9. Jin, X., Lan, C., Zeng, W., Chen, Z., Zhang, L.: Style normalization and restitution for generalizable person re-identification. In: Proceedings of the IEEE/CVF Conference on Computer Vision and Pattern Recognition, pp. 3143–3152 (2020)

10. Kim, W., Son, B., Kim, I.: ViLT: vision-and-language transformer without convolution or region supervision. In: International Conference on Machine Learning, pp. 5583–5594. PMLR (2021)

11. Li, J., Li, D., Xiong, C., Hoi, S.: BLIP: bootstrapping language-image pre-training for unified vision-language understanding and generation. In: International Conference on Machine Learning, pp. 12888–12900. PMLR (2022)

12. Li, S., Sun, L., Li, Q.: CLIP-ReID: exploiting vision-language model for image re-identification without concrete text labels. arXiv preprint arXiv:2211.13977 (2022)

13. Liao, S., Shao, L.: Interpretable and generalizable person re-identification with query-adaptive convolution and temporal lifting. In: Vedaldi, A., Bischof, H., Brox, T., Frahm, J.-M. (eds.) ECCV 2020. LNCS, vol. 12356, pp. 456–474. Springer, Cham (2020). https://doi.org/10.1007/978-3-030-58621-8_27

14. Liao, S., Shao, L.: Transmatcher: deep image matching through transformers for generalizable person re-identification. Adv. Neural. Inf. Process. Syst. **34**, 1992–2003 (2021)

15. Liao, S., Shao, L.: Graph sampling based deep metric learning for generalizable person re-identification. In: Proceedings of the IEEE/CVF Conference on Computer Vision and Pattern Recognition, pp. 7359–7368 (2022)

16. Ma, H., et al.: EI-CLIP: entity-aware interventional contrastive learning for e-commerce cross-modal retrieval. In: Proceedings of the IEEE/CVF Conference on Computer Vision and Pattern Recognition, pp. 18051–18061 (2022)

17. Ni, H., Song, J., Luo, X., Zheng, F., Li, W., Shen, H.T.: Meta distribution alignment for generalizable person re-identification. In: Proceedings of the IEEE/CVF Conference on Computer Vision and Pattern Recognition, pp. 2487–2496 (2022)

18. Radford, A., et al.: Learning transferable visual models from natural language supervision. Cornell University - arXiv (2021)

19. Rao, Y., et al.: Denseclip: language-guided dense prediction with context-aware prompting. In: Proceedings of the IEEE/CVF Conference on Computer Vision and Pattern Recognition, pp. 18082–18091 (2022)

20. Song, J., Yang, Y., Song, Y.Z., Xiang, T., Hospedales, T.M.: Generalizable person re-identification by domain-invariant mapping network. In: Proceedings of the IEEE/CVF Conference on Computer Vision and Pattern Recognition, pp. 719–728 (2019)

21. Wei, L., Zhang, S., Gao, W., Tian, Q.: Person transfer GAN to bridge domain gap for person re-identification. In: Proceedings of the IEEE Conference on Computer Vision and Pattern Recognition, pp. 79–88 (2018)

22. Yang, Q., Yu, H.X., Wu, A., Zheng, W.S.: Patch-based discriminative feature learning for unsupervised person re-identification. In: Proceedings of the IEEE/CVF Conference on Computer Vision and Pattern Recognition, pp. 3633–3642 (2019)

23. Yi, D., Lei, Z., Liao, S., Li, S.Z.: Deep metric learning for person re-identification. In: 2014 22nd International Conference on Pattern Recognition, pp. 34–39. IEEE (2014)

24. Yu, H.X., Zheng, W.S., Wu, A., Guo, X., Gong, S., Lai, J.H.: Unsupervised person re-identification by soft multilabel learning. In: Proceedings of the IEEE/CVF Conference on Computer Vision and Pattern Recognition, pp. 2148–2157 (2019)

25. Zheng, L., Shen, L., Tian, L., Wang, S., Wang, J., Tian, Q.: Scalable person re-identification: a benchmark. In: Proceedings of the IEEE International Conference on Computer Vision, pp. 1116–1124 (2015)

26. Zheng, Z., Zheng, L., Yang, Y.: Unlabeled samples generated by GAN improve the person re-identification baseline in vitro. In: Proceedings of the IEEE International Conference on Computer Vision, pp. 3754–3762 (2017)

27. Zhou, C., Loy, C.C., Dai, B.: Denseclip: extract free dense labels from clip. arXiv preprint arXiv:2112.01071 (2021)

28. Zhou, K., Yang, J., Loy, C.C., Liu, Z.: Conditional prompt learning for vision-language models. In: Proceedings of the IEEE/CVF Conference on Computer Vision and Pattern Recognition, pp. 16816–16825 (2022)

29. Zhou, K., Yang, J., Loy, C.C., Liu, Z.: Learning to prompt for vision-language models. Int. J. Comput. Vision **130**(9), 2337–2348 (2022)

30. Zhou, K., Yang, Y., Cavallaro, A., Xiang, T.: Omni-scale feature learning for person re-identification. In: Proceedings of the IEEE/CVF International Conference on Computer Vision, pp. 3702–3712 (2019)

31. Zhou, K., Yang, Y., Cavallaro, A., Xiang, T.: Learning generalisable omni-scale representations for person re-identification. IEEE Trans. Pattern Anal. Mach. Intell. **44**(9), 5056–5069 (2021)

32. Zhou, K., Yang, Y., Qiao, Y., Xiang, T.: Domain generalization with mixstyle. arXiv preprint arXiv:2104.02008 (2021)

Self-guided Transformer for Video Super-Resolution

Tong Xue, Qianrui Wang, Xinyi Huang, and Dengshi Li[✉] 🆔

School of Artificial Intelligence, Jianghan University, Wuhan 430056, China
reallds@jhun.edu.cn

Abstract. The challenge of video super-resolution (VSR) is to leverage the long-range spatial-temporal correlation between low-resolution (LR) frames to generate high-resolution (LR) video frames. However, CNN-based video super-resolution approaches show limitations in modeling using long-range dependencies and non-local self-similarity. In this paper. For further spatio-temporal learning, we propose a novel self-guided transformer for video super-resolution (SGTVSR). In this framework, we customize a multi-headed self-attention based on offset-guided window (OGW-MSA). For each query element on a low-resolution reference frame, the OGW-MSA enjoys offset guidance to globally sample highly relevant key elements throughout the video. In addition, we propose a feature aggregation module that aggregates the favorable spatial information of adjacent frame features at different scales as a way to improve the video reconstruction quality. Comprehensive experiments show that our proposed self-guided transformer for video super-resolution outperforms the state-of-the-art (SOTA) method on several public datasets and produces good results visually.

Keywords: Video super-resolution · Self-guided transformer · Multi-headed self-attention based on offset-guided window · Feature aggregation

1 Introduction

Video super-resolution (VSR) aims to recover a high-resolution (HR) video frames from a low-resolution (LR) video frames counter-part. Due to the increasing demand for high-quality video content in various application areas, such as surveillance, autonomous driving, and telemedicine, this technology is of great value in many practical applications. VSR differs dramatically in its approach from super resolution of a single image (SISR), which relies only on intra-frame spatial correlation to recover the spatial resolution of a single image. In contrast, VSR reconstructs HR from several consecutive adjacent LR frames. The challenge of VSR is to exploit the spatio-temporal coherence between LR frames for video frame reconstruction.

Q. Liu et al. (Eds.): PRCV 2023, LNCS 14434, pp. 186–198, 2024.
https://doi.org/10.1007/978-981-99-8549-4_16

Current CNN based VSR methods are mainly divided into two types: the first one [5, 17] aligns LR frames through motion generation and motion compensation. Their performance heavily depends on motion estimation and compensation accuracy. The other method [18] is based on 3D convolution, but it only explores local correlations during feature fusion and ignores non-local similarities between and within frames. Recently, recursive networks (RNNs) are also commonly used for VSR tasks to extract long-range temporal information [1, 14]. However, their performance depends heavily on the number of input frames, which is not friendly for videos lacking long-range temporal information.

In recent years, the advent of transformer has provided an alternative solution to alleviate the limitations of the aforementioned approaches. It can effectively captures the long-range spatio-temporal coherence of low-resolution frames by performing multi-headed self-attention computation on the input low-resolution frames. However, there are several major problems with using existing transformers directly for video super-resolution. On the one hand, the computational cost grows quadratically in the spatio-temporal dimension and is computationally prohibitive when using standard global transformer [2]. Second, the standard global transformer uses the entire image features as key elements, leading to self-attention focus on redundant key elements, which may lead to non-convergence of the model. On the other hand, the local window-based transformer [11] computes the self-attention within the window, which leads to a restricted perceptual field. When fast motion is present in low-resolution videos, the local window-based transformer [11] ignores key elements in all spatio-temporal neighborhoods with high similarity and sharper scenes. To summarize the main reasons for the above problems, previous transformers lacked guidance of motion information when computing self-attention.

To alleviate the above problems, in this paper, we propose a Self-Guided transformer for Video Super-Resolution (SGTVSR) framework. This framework constructs a robust pyramidal skeleton to achieve video super-resolution. In SGTVSR we formulate the multi-headed self attention module for offset-guided windows (OGW-MSA). For each query element on the reference frame, the offset-guided OGW-MSA shifts the unaligned and spatially sparse key elements in the video frame to regions with high correlation to the query element for sampling. These sampled key elements provide extremely relevant image prior information, which is critical for video super-resolution. The SGTVSR differs from previous visual transformers in that it neither samples redundant key elements nor suffers from a limited field of perception. While ensuring linear spatial complexity in the transformer self-attention module, and sampling out extremely relevant key elements to achieve efficient attention patterns. We also propose the feature aggregation module (FA), which adaptively aggregates the different scale spatial features of adjacent frames, to reduce the loss of spatial information due to downsampling, and thus to improve the video reconstruction quality. In summary our contributions are as follows:

1. We propose a new video super-resolution method, SGTVSR. To our knowledge, this is one of the works that successfully introduced transformer into the task of video super-resolution.
2. We propose a novel multi-headed self-attention mechanism, OGW-MSA.
3. We design a feature aggregation (FA) module that enhances the spatial details of intermediate frame features by aggregating the relevant features between adjacent frames. The goal of this module is to reduce the loss of spatial information introduced by downsampling operations and to improve the quality of video reconstruction.

2   Related Work

2.1   Video Super-Resolution

The spatio-temporal correlation between LR frames plays an important role in the VSR task. Previous VSR methods can be divided into explicit and implicit motion compensation methods. Most early methods [5,17] utilize explicit motion-based methods, which use optical flow to explicitly align frames. However, the estimated optical flow may be inaccurate. Meanwhile, the implicit motion compensation based approach [13,15] introduces deformable convolution to implicitly align temporal features and achieves impressive performance. Recurrent neural network-based VSRs use hidden states to convey relevant information from previous frames. The most representative ones, BasicVSR [1] and IconVSR [1], fuse past and future bidirectional hidden states for reconstruction and have been significantly improved. They try to make full use of the information of the whole sequence and update the hidden states simultaneously by reconstructing the weights of the network. However, due to the disappearance of gradients, this mechanism loses its long-term modeling capability to some extent.

2.2   Vision Transformers

Transformer is a dominant architecture in natural language processing (NLP) and has shown excellent performance in a variety of tasks. In recent years, the Vision Transformer model has also received increasing attention in the field of computer vision. Among them, Vision Transformer (ViT) [2] solves image classification problems by computing the attention between global image blocks, and it has good long-term modeling capability compared with convolutional neural networks (CNNs). Meanwhile, Liu et al. [11] proposed a new transformer-based visual task backbone, the Swin Transformer. This approach reduces the computational complexity by limiting the computation of attention to a local scale. However, the direct use of previous global or local transformers for video super-resolution may lead to high computational cost as well as the problem of restricted perceptual domain. To address the above issues, we propose self-guided transformer for video super-resolution.

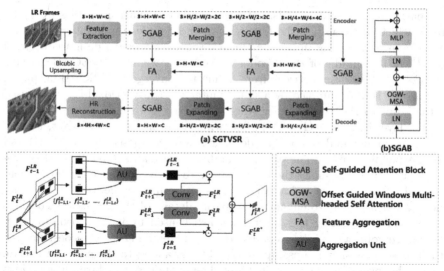

Fig. 1. Architecture of SGTVSR. (a) SGTVSR consists of Encoder, Feature Aggregation module (FA) and Decoder. SGTVSR is mainly built from several SGABs. (b) SGAB consists of normalization layer, OGW-MSA, and MLP layer. (c) FA enhances the spatial information of intermediate frames by aggregating locally relevant feature blocks from two adjacent frames of the encoder.

3 Our Method

In this section, we first outline the approach in Sect. 3.1. We introduce the multi-headed self-attention based on offset-guided window (OGW-MSA) module in Sect. 3.2. Our feature aggregation module is explained in Sect. 3.3.

3.1 Network Overview

Our proposed SGTVSR is a U-net type network structure. Figure 1(a) depicts the basic structure of our SGTVSR, which mainly consists of an encoder, a feature aggregation module and a decoder. Here the input is given as three consecutive LR frames $\{I_t^{LR}\}_{t=1}^3$ of size H×W×3, and the output of our VSR model is three consecutive HR frames $\{I_t^{HR}\}_{t=1}^3$ of size 4H×4W×3, where t denotes the timestamp of the frame, H and W denote the width and height of the frame, and 3 denotes the number of channels. Before entering the encoder we extract the features of the LR frames by the feature extraction module. The feature extraction module extracts features $\{F_t^{LR}\}_{t=1}^3$ through a residual network consisting of a convolutional layer with a kernel size of 3×3 and 2 residual blocks.

In the encoder stage, the input $\{F_t^{LR}\}_{t=1}^3$ generates hierarchical features through two self-guided attention modules (SGAB) and patch merging layers.

The patch merging layer consists of 4×4 stride convolutional layers with step size 2, which downsamples the feature map and doubles the number of feature map channels, resulting in a hierarchical design. This model design is inspired by [11]. In the middle of the decoder and encoder we add two SGABs as bottlenecks to prevent the network from failing to converge. The SGAB output in the encoder is downsampled by a step=2 interstep convolutional layer and used as the input to the next encoder stage. Thus, the feature map size of the i-th encoder layer is: $2^i C \times H/2^i \times W/2^i$. Following the design rules of U-Net, we customize a symmetric decoder. The decoder is also composed of two SGAB and patch expanding layers. The patch expanding layer consists of a 2×2 deconvolution layer that upsamples the feature map. To further enhance the final reconstructed HR frames, we add a feature aggregation module between the encoder and decoder as a way to enhance the spatial details of the reconstructed video frames while mitigating the information loss caused by model downsampling. After undergoing the decoder, we use the pixel- shuffle layer to reconstruct the HR frames. The bicubic upsampled $\{I_t^{LR}\}_{t=1}^3$ is added to the enlarged feature map in the form of a global residual join. The final output high-resolution video frames $\{I_t^{HR}\}_{t=1}^3$.

3.2 Multi-headed Self-attention Module Based on Offset-Guided Window (OGW-MSA)

Figure 1(b) depicts the basic unit of SGTVSR, i.e., the Self-Guided Attention Module (SGAB). The SGAB is composed of the normalization layer (LN), the multi-headed self-attention based on offset-guided window (OGW-MSA), and the MLP (Multi-Layer Perceptron). In this section, we mainly introduce OGW-MSA. Using the offsets generated by convolution as a guide when computing multihead self-attention, sampling key elements with high similarity from all spatio-temporal neighborhood frames.

The structure of OGW-MSA is shown in Fig. 2. For illustration, the input LR video frames feature F_t^{LR} is given here as the reference frame feature and F_{t+1}^{LR} as the spatio-temporal neighborhood frame feature. The patch block embedding in the middle of the reference frame F_t^{LR} becomes the query element $q_{i,j}^t$. OGW-MSA aims to model long-range spatio-temporal correlations and capture non-local self-similarity. For this purpose, OGW-MSA needs to sample the scene patches with high and clearer similarity to $q_{i,j}^t$ from the F_t^{LR} spatio-temporal neighborhood as key elements. Inspired by [13], the offset can adaptively change the perceptual field size to find the local features with high similarity. This idea can be transferred to multi-headed self-attention to find key elements $k_{i+\Delta x, j+\Delta y}^l$ with high similarity and clarity to $q_{i,j}^t$ by the guidance of offsets. We denote this as:

$$\Phi_{i,j}^t = \left\{ k_{i+\Delta x, j+\Delta y}^l \| l - t \mid \le r \right\} \tag{1}$$

$$(\Delta x, \Delta y) = \left[\Theta \left(F_t^{LR}, F_{t+1}^{LR} \right) (i, j) \right] \tag{2}$$

where $\boldsymbol{\Phi}_{i,j}^t$ is denoted as the set of sampled key elements. r denotes the time radius of the adjacent frame, and $(\Delta x, \Delta y)$ is denoted as the value of motion offset occurring from the reference frame F_t^{LR} to the adjacent frame F_{t+1}^{LR}. $|l{-}t| \leq r$ means that the maximum offset of the patch cannot exceed the time radius of the adjacent frame. $\boldsymbol{\Theta}$ denotes the offset generation network. (i,j) is the initial position code of the query element and the key element. OGW-MSA can be formulated as:

$$\text{OGW-MSA}\left(q_{i,j}^t, \boldsymbol{\Phi}_{i,j}^t\right) = \sum_{m=1}^{M} \mathbf{W}_m \sum_{k \in \boldsymbol{\Phi}_{i,j}^t} \mathbf{A}_{mq_{i,j}^t k} \mathbf{W}_m' k, \qquad (3)$$

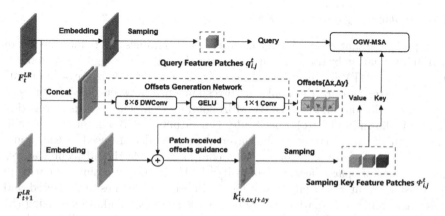

Fig. 2. Illustration of our multi-headed self-attention module for offset bootstrap windows (OGW-MSA). The offset generation network captures local features by 5×5 deep convolution. And GELU activation and 1×1 convolution are used to obtain 2D offset. In adjacent frames, OGW-MSA globally samples the key elements that are highly correlated with the query element (the number of query patches in the figure is 1, there are more query elements and key elements in practice).

For the convenience of simplifying the formula, $\boldsymbol{k}$ and $k_{i+\Delta x, j+\Delta y}^l$ are equivalent here. where M is the number of attentional heads. $\mathbf{W}_m$ and $\mathbf{W}_m'$ denote the learnable parameters in the network, $\mathbf{A}_{mq_{i,j}^t k}$ denotes the self-attention of the mth head, and the formula is expressed as:

$$\mathbf{A}_{mq_{i,j}^t k} = \text{softmax}_{k \in \boldsymbol{\Phi}_{i,j}^t} \left(\frac{q_{i,j}^t k^T}{\sqrt{d}} \right) v, \qquad (4)$$

Where v denotes the value in self-attention, where d=C/M denotes the dimensionality of each head. the computational cost of OGW-MSA is similar to the corresponding part in Swin Transformer [11]. The only additional overhead comes from the subnetwork used to generate the offsets. The Global-MSA and OGW-MSA computational costs as:

$$O(\text{Global} - \text{MSA}) = 4(\text{HW})\text{C}^2 + 2(\text{HW})^2\text{C}, \tag{5}$$

$$O(\text{OGW} - \text{MSA}) = 2(\text{HW})\text{C}^2 + 2\text{N}^2(\text{HW})\text{C} + \text{NC}, \tag{6}$$

Where N is the number of sampled patches. NC is the additional computation used to generate the offset sub-network. Our OGW-MSA is linearly related to the size of the input image compared to the global self-attention. Compared with [11], we do not need to compute the self-attention for each local window and have a larger and more flexible perceptual field. OGW-MSA can change the size of the perceptual field depending on the number of offsets, which provides more flexibility and robustness.

3.3 Feature Aggregation (FA)

To improve the spatial detail of reconstructed video frames and reduce the loss of spatial information by downsampling, we introduce a feature aggregation module between the SGTVSR encoder and decoder. This multiscale aggregation strategy can better capture the nonlocal correspondence between video features of different resolutions and help OGW-MSA computation. The feature aggregation module is shown in Fig. 1(c). Specifically, the output of SGAB in the encoder and the output of the patch expanding layer in the decoder are used as the inputs for each scale feature aggregation. The core is to divide the adjacent features into two sets of spatial feature blocks $[\mathbf{f}_{t-1}^{LR}]$, $[\mathbf{f}_{t+1}^{LR}]$, and by cosine similarity in the most relevant S spatial feature blocks in adjacent frames are selected and these most relevant S spatial feature blocks are aggregated adaptively using the Aggregation Unit (AU) proposed by [8]. Here is an example of $[\mathbf{f}_{t+1}^{LR}]$ and the intermediate frame local feature $\mathbf{f}_t^{LR}$:

$$\cos\left([\mathbf{f}_{t+1}^{LR}], \mathbf{f}_t^{LR}\right) = \frac{[\mathbf{f}_{t+1}^{LR}]}{\|[\mathbf{f}_{t+1}^{LR}]\|} \cdot \frac{\mathbf{f}_t^{LR}}{\|\mathbf{f}_t^{LR}\|} \tag{7}$$

$$\mathbf{f}_{t+1}^{LR} = \text{AU}\left([\mathbf{f}_{t+1,1}^{LR}, \mathbf{f}_{t+1,2}^{LR}, \ldots, \mathbf{f}_{t+1,S}^{LR}]\right) \tag{8}$$

Where $\left([\mathbf{f}_{t+1,1}^{LR}, \mathbf{f}_{t+1,2}^{LR}, \ldots, \mathbf{f}_{t+1,S}^{LR}]\right)$ denotes the S most relevant spatial feature blocks divided by adjacent frames $F_{(t+1)}^{LR}$. $\cos\left([\mathbf{f}_{t+1}^{LR}], \mathbf{f}_t^{LR}\right)$ represents the cosine similarity. $\mathbf{f}_{t+1}^{LR}$ is the aggregated spatially enhanced feature blocks, and similarly we can obtain $\mathbf{f}_{t-1}^{LR}$ in this way. Subsequently, in order to implement the adaptive aggregation strategy, we will assign adaptive weights to these spatially enhanced feature blocks. The final spatially enhanced intermediate frame local features $\mathbf{f}_t^{LR*}$ are obtained by multiplying and summing the weights:

$$W_1 = \text{Conv}\left[F_t^{LR}, F_{t-1}^{LR}\right], W_2 = \text{Conv}\left[F_t^{LR}, F_{t+1}^{LR}\right] \tag{9}$$

$$\mathbf{f}_t^{LR*} = f_{t-1,s}^{LR} \odot W_1 + f_{t+1,s}^{LR} \odot W_2 \tag{10}$$

where W_1, W_2 are the spatially adaptive weights of $f_{t-1,s}^{LR}$, $f_{t+1,s}^{LR}$, respectively, which are obtained from F_t^{LR} with F_{t-1}^{LR}, F_{t+1}^{LR} through convolution layers, respectively. The spatially enhanced intermediate frames F_t^{LR*} are obtained by performing this local feature aggregation several times.

4   Experiments

4.1   Datasets and Experimental Settings

We trained on the most commonly used VSR datasets, **Vimeo-90K** [17], and **REDS** [12]. For REDS [12], it was published in the NTIRE19 challenge [12]. It contains a total of 300 video sequences, of which 240 are used for training, 30 for validation and 30 for testing. Each sequence contains 100 frames with a resolution of 720 × 1280. To create testsets, we select four sequences in the original dataset as the testset, called REDS4 [12]. We selected the remaining 266 sequences from the training and validation sets as the training set. For Vimeo-90K [17], it contains 64,612 sequences for training and 7,824 sequences for testing. Each sequence contains 7 frames with a resolution of 448 × 256. Peak signal-to-noise ratio (PSNR) and structural similarity (SSIM) [16] were used as evaluation metrics. The model is trained using 2 V100 GPUs with 200 training rounds. our SGTVSR divides the video features into 64 × 64 patches as input with a batch size of 4. we train the model with Gaussian blurred LR images and downsample them in the reconstruction part using a scale factor of × 4. Our model is trained by the Adam optimizer [6], set to $\beta_1 = 0.9$ and $\beta_2 = 0.99$. The initial value of the learning rate is set to 1×10^{-4}. The 2D filters are all set to a size of 3 × 3. The basic channel number C is 64. We use the Charbonnier penalty function [7] as the loss function, defined as: $L = \sqrt{\|I_{GT} - I_{HR}\|^2 + \varepsilon^2}$ where ε is set to 1×10^{-3}.

Table 1. Quantitative comparison (PSNR/SSIM) on the REDS4 [12] dataset for 4× video super-resolution. The results are tested on RGB channels. Red indicates the best and blue indicates the second best performance. Frame indicates the number of input frames required to perform an inference, and "r" indicates to adopt the recurrent structure.

Method	Frame	Clip_000	Clip_011	Clip_015	Clip_020	Average
Bicubic	1	24.55/0.6489	26.06/0.7261	28.52/0.8034	25.41/0.7386	26.14/0.7292
TOFlow [17]	7	26.52/0.7540	27.80/0.7858	30.67/0.8609	26.92/0.7953	27.98/0.7990
EDVR [15]	5	27.78/0.8156	31.60/0.8779	33.71/0.9161	29.74/0.8809	30.71/0.8726
MuCAN [8]	5	27.99/0.8219	27.99/0.8219	33.90/0.9170	29.78/0.8811	30.88/0.8750
BasicVSR [1]	r	28.39/0.8429	32.46/0.8975	34.22/0.9237	30.60/0.8996	31.42/0.8909
IconVSR [1]	r	28.55/0.8478	32.89/0.9024	34.54/0.9270	30.80/0.9033	31.67/0.8948
TTVSR [10]	r	28.82/0.8566	33.47/0.9100	35.01/0.9325	31.17/0.9094	32.12/0.9021
SGTVSR	r	29.06/0.8604	33.72/0.9163	35.23/0.9359	31.42/0.9131	32.36/0.9064

4.2   Comparisons with State-of-the-Art Methods

Quantitative Evaluation. Quantitative comparison To evaluate the performance, we compared the proposed SGTVSR with 13 VSR methods. We evaluated our model on four testsets: Vid4 [9], UDM10 [18], Vimeo-90K-T [17], REDS4 [12]. For a fair comparison, we used the same downsampling operation for all HR frame reconstruction parts in the test. As shown in Table 1, our SGTVSR is compared with other SOTA methods in the very challenging VSR dataset REDS4 [12]. The table shows that our SGTVSR achieves a PSNR result of 32.36 dB and outperforms the best performing TTVSR [10] by 0.24 dB. Both SGTVSR and TTVSR [10] are based on transformer video super-resolution model, which indicates that our SGTVSR is more capable of reconstructing HR frames using video long-term temporal correlation compared to [10].

To further validate the generalization ability of SGTVSR, we trained SGTVSR on the Vimeo-90K dataset [17] and evaluated the results on the Vid4 [9], UDM10 [18], and Vimeo-90K-T [17] datasets, respectively. As shown in Table 2, the PSNR reached 28.63 dB, 40.63 dB, and 38.26 dB on the Vid4 [9], UDM10 [18], and Vimeo-90K-T [17] testsets, respectively, outperforming all other SOTA methods. When evaluated on the Vimeo-90K-T [17] dataset with 7 frames in each test sequence, SGTVSR shows quite good visual results, indicating that SGTVSR is able to model the information in long program columns well.

Table 2. Quantitative comparison (PSNR/SSIM) on Vid4 [9], UDM10 [18] and Vimeo-90K-T [17] dataset for 4× video super-resolution.Red indicates the best and blue indicates the second best performance

Method	Vid4	UDM10	Vimeo-90K-T	Param(M)
Bicubic	21.80/0.5246	28.47/0.8253	31.30/0.8687	-
TOFlow [17]	25.85/0.7659	36.26/0.9438	34.62/0.9212	-
FRVSR [14]	26.69/0.8103	37.03/0.9537	35.64/0.9319	5.1
DUF [5]	27.38/0.8329	38.48/0.9605	36.87/0.9447	5.8
RBPN [4]	27.17/0.8205	38.66/0.9596	37.20/0.9458	12.2
RLSP [3]	27.48/0.8388	38.48/0.9606	36.49/0.9403	-
EDVR [15]	27.85/0.8503	39.40/0.9663	37.33/0.9484	20.6
TDAN [13]	26.86/0.8140	38.19/0.9586	36.31/0.9376	17.2
PFNL [18]	27.16/0.8355	38.74/0.9627	36.14/0.9363	-
MuCAN [8]	-	-	37.32/0.9465	13.6
BasicVSR [1]	27.96/0.8553	39.96/0.9694	37.53/0.9498	6.3
IconVSR [1]	28.04/0.8570	40.03/0.9694	37.84/0.9524	8.7
TTVSR [10]	28.40/0.8643	40.41/0.9712	37.92/0.9526	6.8
SGTVSR(Ours)	28.63/0.8691	40.63/0.9735	38.26/0.9541	9.8

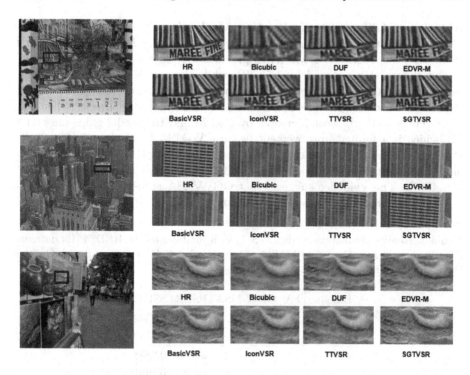

Fig. 3. Visual results on the testset for 4× scaling factor.

Qualitative Comparison. To further compare the visual quality of the different methods, we show in Fig. 3 the visual results produced by SGTVSR and other SOTA methods on different test sets. It can be observed that SGTVSR shows a significant improvement in visual quality, especially for areas with detailed textures. For example, in the second row of Fig. 3, SGTVSR can recover more texture details from the tall buildings in the city.

4.3 Ablation Study

In this section, we performed an ablation study in the REDS4 dataset. The "Baseline" model was derived by directly removing the proposed FA and OGW-MSA modules from our SGTVSR.

Break-Down Ablation. We first perform a decomposition ablation to investigate the impact of each module on performance. We directly use convolutional layers instead of OGW-MSA as our "Baseline" model. The results are shown in Table 3. The Baseline model yields 29.76 dB. 1.22 dB and 1.48 dB improvement is achieved after applying FA and OGW-MSA, respectively. The model gain is 2.60 dB when both OGW-MSA and FA are used, and the results show the effectiveness of FA and OGW-MSA.

Ablation of Different Self-attention Mechanism. We compare our OGW-MSA with other self-attention mechanisms in Table 4. The Baseline model yields 29.76 dB while costing 6.1 M. When global MSA [2] is used, the video super-resolution model is reduced by 1.52 dB, while the cost is 72.34 M. The main reason is that the global MSA samples too many redundant key elements and requires a large amount of computational and memory resources. When W-MSA [11] is used, the cost increases by 2.02 M and the model gains only 0.66 dB, which is a limited improvement because W-MSA [11] only calculates self-attention within a specific location window. When using OGW-MSA, the model only increased the computational cost by 3.17 M while yielding an improvement of 1.48 dB. This evidence suggests that OGW-MSA is more effective than W-MSA [11] in terms of video super-resolution.

Table 3. Results of ablation studies of SGTVSR modules on the REDS4 [12] dataset. FA: Feature Aggregation, OGW-MSA: Multi-Headed Self-Attention Based On the Offset-guided Window Module.

Baseline	FA	OGW-MSA	PSNR	SSIM
✓			29.76	0.8342
✓	✓		30.98	0.8749
✓		✓	31.24	0.8838
✓	✓	✓	**32.36**	**0.9064**

Table 4. Ablation study of using different self-attention mechanisms.

Method	Baseline	Global MSA	W-MSA	OGW-MSA
PSNR	29.76	28.24	30.42	**31.24**
SSIM	0.8342	0.8124	0.8654	**0.8838**
Params(M)	6.1	72.34	8.12	9.27

5   Conclusion

In this paper, we propose SGTVSR, a novel transformer-based video super-resolution method. In SGTVSR, we customize a novel multi-headed self-attention mechanism, OGW-MSA. Guided by the motion offsets, OGW-MSA samples spatially sparse but highly correlated key elements corresponding to video frames in similar and well-defined scene blocks, thus enlarging the spatio-temporal perceptual domain of the self-attention mechanism in transformer. In addition, we propose a feature aggregation module (FA) to aggregate the favorable spatial information of the features of adjacent frames at different scales, thereby improving the spatial details of the reconstructed video frames and reducing the loss of spatial information caused by downsampling. Comprehensive experiments show that our SGTVSR outperforms the SOTA method and produces good visual effects in video super-resolution.

References

1. Chan, K.C., Wang, X., Yu, K., Dong, C., Loy, C.C.: BasicVSR: the search for essential components in video super-resolution and beyond. In: Proceedings of the IEEE/CVF Conference on Computer Vision and Pattern Recognition, pp. 4947–4956 (2021)
2. Dosovitskiy, A., et al.: An image is worth 16x16 words: transformers for image recognition at scale. arXiv preprint arXiv:2010.11929 (2020)
3. Fuoli, D., Gu, S., Timofte, R.: Efficient video super-resolution through recurrent latent space propagation. In: 2019 IEEE/CVF International Conference on Computer Vision Workshop (ICCVW), pp. 3476–3485. IEEE (2019)
4. Haris, M., Shakhnarovich, G., Ukita, N.: Recurrent back-projection network for video super-resolution. In: Proceedings of the IEEE/CVF Conference on Computer Vision and Pattern Recognition, pp. 3897–3906 (2019)
5. Jo, Y., Oh, S.W., Kang, J., Kim, S.J.: Deep video super-resolution network using dynamic upsampling filters without explicit motion compensation. In: Proceedings of the IEEE Conference on Computer Vision and Pattern Recognition, pp. 3224–3232 (2018)
6. Kingma, D.P., Ba, J.: Adam: a method for stochastic optimization. arXiv preprint arXiv:1412.6980 (2014)
7. Lai, W.S., Huang, J.B., Ahuja, N., Yang, M.H.: Deep laplacian pyramid networks for fast and accurate super-resolution. In: Proceedings of the IEEE Conference on Computer Vision and Pattern Recognition, pp. 624–632 (2017)
8. Li, W., Tao, X., Guo, T., Qi, L., Lu, J., Jia, J.: MuCAN: multi-correspondence aggregation network for video super-resolution. In: Vedaldi, A., Bischof, H., Brox, T., Frahm, J.-M. (eds.) ECCV 2020. LNCS, vol. 12355, pp. 335–351. Springer, Cham (2020). https://doi.org/10.1007/978-3-030-58607-2_20
9. Liu, C., Sun, D.: On Bayesian adaptive video super resolution. IEEE Trans. Pattern Anal. Mach. Intell. **36**(2), 346–360 (2013)
10. Liu, C., Yang, H., Fu, J., Qian, X.: Learning trajectory-aware transformer for video super-resolution. In: Proceedings of the IEEE/CVF Conference on Computer Vision and Pattern Recognition, pp. 5687–5696 (2022)
11. Liu, Z., et al.: Swin transformer: hierarchical vision transformer using shifted windows. In: Proceedings of the IEEE/CVF International Conference on Computer Vision, pp. 10012–10022 (2021)
12. Nah, S., et al.: NTIRE 2019 challenge on video deblurring and super-resolution: dataset and study. In: Proceedings of the IEEE/CVF Conference on Computer Vision and Pattern Recognition Workshops (2019)
13. Tian, Y., Zhang, Y., Fu, Y., Xu, C.: TDAN: temporally-deformable alignment network for video super-resolution. In: Proceedings of the IEEE/CVF Conference on Computer Vision and Pattern Recognition, pp. 3360–3369 (2020)
14. Vemulapalli, R., Brown, M., Sajjadi, S.M.M.: Frame-recurrent video super-resolution, US Patent 10,783,611 (2020)
15. Wang, X., Chan, K.C., Yu, K., Dong, C., Change Loy, C.: EDVR: video restoration with enhanced deformable convolutional networks. In: Proceedings of the IEEE/CVF Conference on Computer Vision and Pattern Recognition Workshops (2019)
16. Wang, Z., Bovik, A.C., Sheikh, H.R., Simoncelli, E.P.: Image quality assessment: from error visibility to structural similarity. IEEE Trans. Image Process. **13**(4), 600–612 (2004)

17. Xue, T., Chen, B., Wu, J., Wei, D., Freeman, W.T.: Video enhancement with task-oriented flow. Int. J. Comput. Vision **127**, 1106–1125 (2019)
18. Yi, P., Wang, Z., Jiang, K., Jiang, J., Ma, J.: Progressive fusion video super-resolution network via exploiting non-local spatio-temporal correlations. In: Proceedings of the IEEE/CVF International Conference on Computer Vision, pp. 3106–3115 (2019)

SAMP: Sub-task Aware Model Pruning with Layer-Wise Channel Balancing for Person Search

Zimeng Wu[1,2], Jiaxin Chen[1,2(✉)], and Yunhong Wang[1,2]

[1] State Key Laboratory of Virtual Reality Technology and Systems, Beihang University, Beijing, China
{zimengwu,jiaxinchen,yhwang}@buaa.edu.cn
[2] School of Computer Science and Engineering, Beihang University, Beijing, China

Abstract. The deep convolutional neural network (CNN) has recently become the prevailing framework for person search. Nevertheless, these approaches suffer from the high computational cost, raising the necessity of compressing deep models for applicability on resource-restrained platforms. Despite of the promising performance achieved in boosting efficiency for general vision tasks, current model compression methods are not specifically designed for person search, thus leaving much room for improvement. In this paper, we make the first attempt in investigating model pruning for person search, and propose a novel loss-based channel pruning approach, namely Sub-task Aware Model Pruning with Layer-wise Channel Balancing (SAMP). It firstly develops a Sub-task aware Channel Importance (SaCI) estimation to deal with the inconsistent sub-tasks, *i.e.* person detection and re-identification, of person search. Subsequently, a Layer-wise Channel Balancing (LCB) mechanism is employed to progressively assign a minimal number of channels to be preserved for each layer, thus avoiding over-pruning. Finally, an Adaptive OIM (AdaOIM) loss is presented for pruning and post-training via dynamically refining the degraded class-wise prototype features by leveraging the ones from the full model. Experiments on CUHK-SYSU and PRW demonstrate the effectiveness of our method, by comparing with the state-of-the-art channel pruning approaches.

Keywords: Person search · Model compression · Channel pruning · Channel importance estimation

1 Introduction

Given a query person, person search aims to locate and identify the specific target from a gallery of uncropped scene images [16,26], which has a wide range of applications in practice. In the past decade, deep neural networks have become the most prevailing framework for person search, and substantially promote the accuracy [4,6,7]. Nevertheless, the high computational cost of deep models significantly degrades the inference efficiency, severely impeding their application

Q. Liu et al. (Eds.): PRCV 2023, LNCS 14434, pp. 199–211, 2024.
https://doi.org/10.1007/978-981-99-8549-4_17

in various real scenarios, especially when being deployed on resource-restrained platforms such as unmanned vehicles and mobile devices.

A few efforts have been devoted to improving the inference efficiency of deep models for person search, most of which concentrate on employing light-weight backbone networks such as MobilieNet [13]. However, they heavily rely on hand-crafted network architecture, which are not flexible, especially when encountering with distinct types of hardware.

Recently, structured model pruning [5,19] has emerged as a promising way to accelerate the inference speed of deep neural networks by estimating the channel importance based on the learning loss and trimming the less important ones. As the scale or the floating point operations (FLOPs) of the pruned model can be controlled by a predefined pruning ratio, it is flexible and friendly for deployment, which therefore has been extensively studied in the computer vision community [5,18]. Nevertheless, current structured pruning approaches are mostly designed for the general fundamental computer vision tasks such as classification and detection. There are several limitations when straightforwardly applying them to the person search task without specific refinement. Firstly, representative deep models for person search consists of multiple sub-task heads *i.e.* pedestrian detection and person re-identification, with a shared backbone (*e.g.* ResNet-50). As the detection focuses on exploring person commonness while the re-identification relies on person uniqueness, the two sub-tasks are intrinsically inconsistent, making the importance of a particular channel divergent for different sub-tasks. Existing works coarsely estimate the importance based on the overall loss, and ignore the distinctness between sub-tasks, thus incurring estimation bias. Secondly, current pruning methods are empirically prone to overly prune a few layers while leaving redundancy for the rest ones, seriously deteriorating the representation capacity. Thirdly, the online instance matching (OIM) loss is widely used in person search, which dynamically extracts class-wise prototype feature for each person identity to construct negative/positive pairs for representation learning. However, when being used in model pruning, the discriminativeness of the prototype features is degraded as the model is progressively trimmed, leading to a drop in performance. The above limitations of existing approaches leave much room for improvement.

To address these issues, we propose a novel channel pruning approach, namely Sub-task Aware Model Pruning with Layer-wise Channel Balancing (SAMP). Specifically, a Sub-task aware Channel Importance (SaCI) estimation method is firstly developed to deal with the inconsistent sub-tasks, via separately computing the channel importance for detection and re-identification followed by a hybrid integration. The Layer-wise Channel Balancing (LCB) mechanism is successively employed to progressively assign a minimal number of channels to be preserved for each layer, thus avoiding over-pruning. Finally, an Adaptive OIM (AdaOIM) loss is presented for pruning and finetuning (post training) via dynamically refining the degraded class-wise prototype features by leveraging the ones from the full model.

The main contribution of this paper is summarized in three-fold:

1) To the best of our knowledge, we make the first attempt in investigating model pruning for person search, and specifically design a novel channel pruning approach for person search, dubbed as Sub-task Aware Model Pruning with Layer-wise Channel Balancing (SAMP).
2) We develop the Sub-task aware Channel Importance (SaCI), the Layer-wise Channel Balancing (LCB) mechanism and an Adaptive OIM (AdaOIM) loss to deal with the sub-task inconsistency, over-pruning and prototype feature degradation of the OIM loss, respectively.
3) We extensively evaluate the performance of our method on two widely used benchmarks for person search, and achieve the state-of-the-art performance.

2   Related Work

End-to-End Person Search. Person search encompasses two sub-tasks: person detection and re-identification (ReID). Xiao et al. [24] deal with these subtasks into a unified network and employ the OIM loss. Li et al. [15] propose SeqNet, which refines the bounding boxes for pedestrian detection by leveraging Faster-RCNN detection head before ReID. Yan et al. [27] deal with person search in the one-step two-stage framework, and promote the efficiency by removing RPN and the anchor box generation process. Li et al. [14] investigate the unsupervised domain adaptation for person search. Jaffe et al. [12] propose to filter areas that pedestrians are unlikely to appear. Despite of the improved accuracy, most of these approaches depend on a complex backbone with high computational cost, thus lacking efficiency for real applications.

Model Compression. Model compression aims to decrease the computation cost and accelerate the inference speed of a given deep model, while maintaining the accuracy [3]. Current approaches for model compression can be roughly divided into the following: efficient network design [10], model pruning [18], quantization [8], knowledge distillation [20], and low-rank decomposition [30]. Due to its promising performance and flexibility, we concentrate on model pruning for promoting the efficiency of person search in this paper.

Basically, model pruning involves three stages, including pre-training, pruning, and finetuning [1]. A lot of efforts have been devoted to pruning for general classification and object detection tasks, by proposing distinct pruning principles in different pruning granularity. In regards of pruning granularity, current approaches can be categorized as unstructured pruning [9,11] and structured pruning [5,19], depending on whether the set of weights for pruning conforms to a regular structure. Concretely, structured pruning, especially channel pruning, is extensively studied as it is hardware-friendly for practical deployment [5,18]. As for pruning principles, prevailing techniques are primarily divided into importance criteria or parameter optimization. The approaches based on importance criteria focus on estimating the contribution of prunable units to the performance, and trim those with relatively low importance. These criteria span from

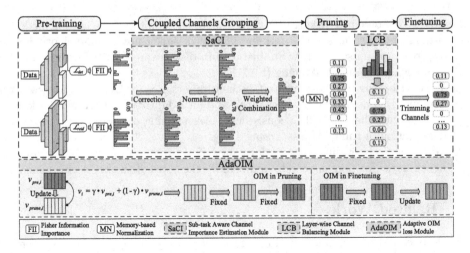

Fig. 1. Illustration on the framework of the proposed SAMP method.

metrics like absolute value [5] and loss function [18] to customized measurements such as discrimination [17] and layer correlation [28]. The methods based on parameter optimization mainly leverage network optimization processes to autonomously learn the unit importance. GReg [22] facilitates weight divergence via evolving regularization coefficients. Slim [29] employs switchable BN layers for acquiring varied parameter quantities during training. DNCP [33] presents a learning method to predict the channel retention probabilities.

However, most existing approaches are designed for general purpose, and neglect to specifically develop effective model compression method for person search. Recently, a few works make some attempts. Li et al. [13] adopt MobileNet as the backbone for person detection to accelerate inference. Xie et al. [25] present a model compression pipeline for person ReID, by combining pruning and knowledge distillation based fine-tuning. Nevertheless, they either directly apply the off-the-shelf techniques or focus on one of the sub-tasks (*e.g.* detection or ReID), which is not fully optimized for the complete person search task, thus leaving much room for improvement.

3   The Proposed Method

3.1   Framework Overview

Given a pre-trained full model, as Fig. 1 shows, our approach follows the same framework as depicted in [18], which consists of three successive steps, including coupled channels grouping, pruning channels based on channel importance estimation and fine-tuning the pruned model.

Coupled Channels Grouping. Typical convolutional neural networks for person search contain coupled channels especially in the residual blocks, the feature pyramid networks, and the ROI alignment block. To deal with the coupled structures, we adopt the layer grouping method [18] by dividing channels into different groups, where channels within the same group share the pruning mask.

Pruning Based on Channel Importance Estimation. We employ the Fisher information [18] to estimate the importance s_i of the i-th channel by applying Taylor expansion to the change of loss $\mathcal{L}$:

$$s_i = \mathcal{L}(m - e_i) - \mathcal{L}(m) \approx -e_i^T \nabla_m \mathcal{L} + \frac{1}{2} e_i^T (\nabla_m^2 \mathcal{L}) e_i = -g_i + \frac{1}{2} H_{ii}. \quad (1)$$

Here, m is the channel mask vector and e_i is the one-hot vector with the i-th element being 1. g is the gradient of $\mathcal{L}$ w.r.t m and H is the Hessian matrix. Channel pruning is conducted by dot production between the feature map A and m, i.e. $\tilde{A} = A \odot m$. Assuming that the model is convergent, i.e. $g = \nabla_m \mathcal{L} \approx 0$, given a batch of size N, we can deduce the following formula:

$$g_i = \frac{\partial \mathcal{L}}{\partial m_i} = \frac{1}{N} \sum_{n=1}^{N} \frac{\partial \mathcal{L}_n}{\partial m_i} \approx 0, \quad (2)$$

where $\mathcal{L}_n$ is the loss w.r.t. the n-th sample.

According to Fisher information [18], the second-order derivative can be approximately formulated as the square of the first-order derivative as below

$$H_{ii} = \frac{\partial^2 \mathcal{L}}{\partial m_i^2} \approx -\frac{\partial^2}{\partial m_i^2} E[\log p(y|x)] = E[-\frac{\partial \log p(y|x)}{\partial m_i}]^2 \approx \frac{1}{N} \sum_{n=1}^{N} (\frac{\partial \mathcal{L}_n}{\partial m_i})^2. \quad (3)$$

By extending $m \in \mathbb{R}^c$ to $\widetilde{m} \in \mathbb{R}^{c \times h \times w}$, where c, h, w represent the channel size, the height and width of the feature map respectively, the extended channel pruning of the n-th sample can be written as $\tilde{A}_n = A_n \odot \widetilde{m}$. The corresponding gradient becomes $\nabla_{\widetilde{m}} \mathcal{L}_n = A_n \odot \nabla_{\tilde{A}_n} \mathcal{L}_n$. Based on Eq. (1)–Eq. (3), and applying the grouping operation of coupled channels, s_i is estimated as below

$$s_i \propto \sum_{n=1}^{N} \left(\sum_{x \in X} \frac{\partial \mathcal{L}_n}{\partial m_i^x} \right)^2, \quad (4)$$

where X is the index set of group that the i-th channel belongs to.

Besides the importance, we also consider the computation cost by memory-based normalization as depicted in [18], i.e. $s_i := s_i / \Delta M$ where $\Delta M = n \times h \times w$ approximates the related FLOPs. The channel with the least s_i is trimmed.

Finetuning. After pruning, the trimmed model is finetuned based on the training data, generating the final compressed model for practical deployment.

3.2 Sub-task Aware Channel Importance Estimation

As depicted in Sect. 3.1, the channel importance s_i is determined by the loss function $\mathcal{L}$, which is usually the sum of the pedestrian detection loss $\mathcal{L}_{det}$ and

ReID loss $\mathcal{L}_{reid}$ in person search. However, they usually have inconsistent targets, as the former focuses on person commonness and the later relies on person uniqueness, raising challenges for importance estimation. For instance, gradients from the two sub-tasks may be significantly high but numerically opposite. In this case, the channel importance is high for either of the two sub-tasks, but becomes low for the overall task. To address this issue, we propose a novel sub-task aware channel importance estimation (SaCI) method, by ensembling the relative significance of sub-tasks.

Concretely, we first compute the channel importance s_{det} and s_{reid} for the detection and ReID sub-tasks by performing Eq. (4) on $\mathcal{L}_{det}$ and $\mathcal{L}_{reid}$, respectively. SaCI then calculates the overall importance as below:

$$s_{new} = (1 + \beta^2) \cdot \frac{s_{det} \cdot s_{reid}}{\beta^2 \cdot s_{det} + s_{reid}}, \tag{5}$$

where β is the balancing weight.

Since some layers (*i.e.* the sub-task head) are uniquely used for a sub-task, making their importance to another task 0, which leads to the ineffectiveness of Eq. (5). In this case, SaCI utilizes a rectification mechanism. Without loss of generality, assuming that channel C is only used for ReID, *i.e.* when $s_{det}(C) = 0$, SaCI modifies it as $s_{det}(C) := \max_{j}\{s_{det}(j)/F_j\} \times F_C$, where $F_j = h_k \times w_k \times h_o \times w_o \times c_o$ is the FLOPs of the j-th channel. Such rectification avoids 0, and is not sensitive to the feature map size as the maximal sub-task channel importance of all active channels is normalized by FLOPs of the channel.

Moreover, the magnitudes of channel importance for distinct sub-tasks are often in different scales. Therefore, we additionally adopt a scale normalization strategy. As observing the long-tail distribution of channel importance, we select the top $\alpha\%$ largest channel importance for normalization. Specifically, the channel importance for the ReID sub-task is normalized as $s'_{reid} = s_{reid} \cdot \frac{s_{det,top-\alpha\%}}{s_{reid,top-\alpha\%}}$, and the final channel importance becomes $s'_{new} = (1 + \beta^2) \cdot \frac{s_{det} \cdot s'_{reid}}{\beta^2 \cdot s_{det} + s'_{reid}}$.

3.3   Layer-Wise Channel Balancing

To address the over-pruning problem, the layer-wise channel balancing (LCB) is presented by stopping pruning layers that are with FLOPs below given thresholds, which is defined as the ratio of the minimal number of untrimmed channels w.r.t the one of channels in the full model before pruning.

Specifically, we adopt different thresholds for the backbone networks and the non-backbone ones, denoted by T_B and T_{NB}, respectively. Since the backbone networks is usually more important, we initialize T_B to a large value, and consistently decrease T_{NB} as per the size S of the pruning model. Concretely, the thresholds change per iteration step as below, which are also shown in Fig. 2(a):

$$\begin{cases} T_B := T_B, & \text{if some non-backbone layers are above } T_{NB}, \\ T_B := T_B - 0.05, & \text{if no non-backbone layers are above } T_{NB}, \\ T_{NB} := \sqrt{S}/2. \end{cases} \tag{6}$$

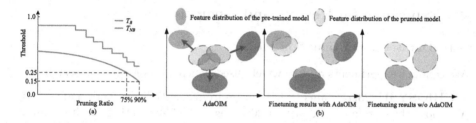

Fig. 2. (a) The change of thresholds used in LCB w.r.t. the pruning ratio. (b) Illustration on the proposed AdaOIM loss.

As shown in Fig. 2(a), the thresholds are progressively decreased without sharp drops, thus avoiding pruning layers that with extremely few channels.

3.4   Adaptive OIM Loss for Model Pruning and Finetuning

The OIM loss [24] is a widely used component of the ReID loss, which builds prototype feature for each identity and formulates the probability of an input feature x belong to the i-th identity as below:

$$p_i = \frac{\exp(v_i^T \cdot x/\tau)}{\Sigma_{j \in \text{LUT}} \exp(v_j^T \cdot x/\tau) + \Sigma_{k \in \text{CQ}} \exp(u_k^T \cdot x/\tau)}, \qquad (7)$$

where v_i is prototype feature for the i-th identity, and τ is the hyperparameter.

Usually, $\{v_i\}$ guides training discriminative features and is updated online based on x during pruning, which however is degraded, since the presentation capability of the pruning model decreases as more channels are trimmed.

To deal with this issue, we propose an adaptive OIM (AdaOIM) loss. As illustrated in Fig. 2(b), AdaOIM leverages the prototypes generated by the pre-trained model $\{v_{\text{pre},i}\}$ to update $\{v_i\}$, formulated as below:

$$v_i = \gamma \cdot v_{\text{pre},i} + (1 - \gamma) \cdot v_{\text{prune},i}, \qquad (8)$$

where $v_{\text{prune},i}$ is the prototype by using the pruning model, and γ is the weight.

According to the principle of the loss-based methods, the estimation of channel importance depends on the convergence assumption. If the pre-trained prototype is employed in the very early stage, the pruning model is accelerated to be deviated from the convergence assumption. To avoid this issue, in the pruning stage, we only use $\{v_{\text{prune},i}\}$ to update $\{v_i\}$ in the early stage, and progressively introduce $\{v_{\text{pre},i}\}$. Specifically, we initialize γ as 0, and gradually increase it to 1. In the finetuning stage, the situation becomes opposite. Therefore, we completely use the pre-training prototypes $\{v_{\text{pre},i}\}$ in the early stage of finetuning, and release this constraint to speed up the convergence of training in the late stage. Concretely, we initialize γ in Eq. (8) as 1, and set it to 0 in the late stage.

4 Experimental Results and Analysis

4.1 Dataset and Evaluation Metric

We conduct experiments on two widely used benchmarks for person search.

CUHK-SYSU [23] contains 18,184 images from 8,432 distinct person identities. We adopt the standard split with 11,206 images of 5,532 identities in the training set, and 2,900 query persons as well as 6,978 gallery images in the test set. **PRW** [32] includes 11,816 images with 43,110 annotated bounding boxes from 932 person identities. 5,704 images with 482 identities are used for training, and 2,057 query persons as well as 6,112 gallery images are utilized for testing.

Evaluation Metric. We evaluate the performance of our method by comparing the widely used mean Average Precision (mAP) [31] and the top 1 (Top-1), top 5 (Top-5) accuracies [21] of the deep person search model compressed to the same FLOPs budgets (*e.g.* 25% and 10% FLOPs of the full model).

4.2 Implementation Details

We select SeqNet [15] and AlignPS [27] with ResNet-50 backbone as the representatives for the two-stage and one-stage frameworks for person search, respectively. Regarding the hyperparameters, we set α and β in Eq. (5) to fixed values of 1 and 0.5 with SeqNet, respectively. T_B in Eq. (6) is initialized as 0.9 and 0.8 for SeqNet and AlignPS, respectively. γ in Eq. (8) is initialized as 0, is dynamically increased by 0.1 (from 0.2) for every 5% decrease of FLOPs when it is lower than 80%. In the pre-training stage, we follow the original training settings. Specifically, for SeqNet, the batch size is set to 4. The learning rate is initialized as 0.0024 and is linearly warmed up at epoch 1 and decreased by 10% at epoch 16 within totally 20 epochs. For AlignPS, the batch size is set to 2. The learning rate is initialized as 0.0005 and is linearly warmed up at epoch 1 and decreased by 10% at epoch 16 and 22 respectively within 24 epochs in all. For both of the two models, we use stochastic gradient descent with a momentum of 0.9 and weight decay as the optimizer. In the pruning stage, the original training settings are most reused, except for setting the batch size to 1. Moreover, the initial learning rate is 0.0001 and the warm-up strategy is cancelled, and pruning is performed every 10 training steps. In the finetuning stage, the initial learning rate is set to 10% of the initial learning rate of full model, and the rest of the training settings follow the original ones. All the experiments are implemented based on MMDetection [2] on an NVIDIA GeForce RTX 3090 Ti GPU.

4.3 Comparison with the State-of-the-Art Approaches

We compare our method with the representative state-of-the-art channel pruning approaches, including Slim [29] based on parameter optimization, Group-Fisher [18] based on loss criterion, and DepGraph [5] based on the absolute value criterion and parameter optimization.

Table 1. Comparison of mAP (%), Top-1/5 accuracies (%) and FLOPs by using various model pruning approaches based on *SeqNet*. '-' indicates that DepGraph doesn't converge in pruning, thus failing to generate the final results. Best in bold.

Method	CUHK-SYSU			PRW			FLOPs
	mAP	Top-1	Top-5	mAP	Top-1	Top-5	
Full model	93.39	94.07	90.00	46.10	82.45	91.30	100%
Slim [29]	82.00	83.14	92.21	32.94	75.26	87.26	25%
GroupFisher [18]	91.55	92.52	97.21	39.98	78.95	89.26	25%
DepGraph [5]	43.58	46.48	60.32	7.17	35.66	51.92	25%
SAMP (Ours)	**91.90**	**92.66**	**97.45**	**40.48**	**79.10**	**89.35**	25%
Slim [29]	74.61	76.14	87.83	27.37	72.29	84.25	10%
GroupFisher [18]	90.05	90.69	96.66	35.87	77.25	88.14	10%
DepGraph [5]	6.99	7.06	12.71	-	-	-	10%
SAMP (Ours)	**90.84**	**91.24**	**96.90**	**37.43**	**77.35**	**88.72**	10%

Table 2. Comparison of mAP (%), Top-1/5 accuracies (%) and FLOPs by using various model pruning approaches based on *AlignPS*. Best in bold.

Method	CUHK-SYSU			PRW			FLOPs
	mAP	Top-1	Top-5	mAP	Top-1	Top-5	
Full model	92.56	93.07	97.76	45.03	81.28	90.86	100%
Slim [29]	58.03	58.17	77.31	5.47	39.82	55.23	25%
GroupFisher [18]	92.23	**92.72**	97.28	41.68	81.14	**90.57**	25%
SAMP (Ours)	**92.36**	92.45	**97.55**	**42.43**	81.14	90.28	25%
Slim [29]	50.94	50.86	71.07	3.64	32.62	46.86	10%
GroupFisher [18]	89.12	88.90	96.45	36.87	78.17	88.77	10%
SAMP (Ours)	**91.18**	**91.52**	**97.28**	**37.60**	**78.61**	**88.92**	10%

Since our work is the first one that investigates model pruning for efficient person search, we re-implement the compared methods, and evaluate their performance. We mainly evaluate on the two-stage person search framework, *i.e.* SeqNet. As displayed in Table 1, GroupFisher reaches the highest accuracy with the same FLOPs among the compared methods. Our method is established upon the framework of GroupFisher and further consistently promotes its performance. Particularly, when trimming the model to extremely low FLOPs (*i.e.* 10%), the proposed SAMP method obtains 0.79% and 1.56% improvements in mAP on the CUHK-SYSU and PRW datasets, respectively, by dealing with the inconsistency of sub-tasks in estimating channel importance via SaCI, and mitigating the over-purning via the layer-wise channel balancing mechanism.

It is worth noting both Slim and DepGraph fail to generalize well to person search. As for Slim, it is prone to generate redundant features when updating the memory bank of the OIM loss, thus significantly degrading the overall performance. In regards of DepGraph, it imposes strong sparsity restrictions on all convolution layers, incurring severe over-pruning within the backbone, thus resulting in sharp drop in accuracy or even worse misconvergence when adopting a low pruning ratio of (*e.g.* 10%).

Table 3. Ablation study of mAP, Top-1/5 accuracies and FLOPs on the main components of the proposed method based on SeqNet on the CUHK-SYSU dataset.

Method	mAP	Top-1	Top-5	FLOPs
Baseline	91.55	92.52	97.21	25%
+SaCI	91.66	92.03	97.45	25%
+LCB	91.76	92.55	97.31	25%
+AdaOIM	91.83	92.52	**97.48**	25%
Full method	**91.90**	**92.66**	97.45	25%
Baseline	90.05	90.69	96.66	10%
+SaCI	90.28	90.69	96.72	10%
+LCB	90.80	91.52	96.83	10%
+AdaOIM	90.80	**91.59**	96.59	10%
Full method	**90.84**	91.24	**96.90**	10%

On Generalizability to Distinct Frameworks for Person Search. Besides the two-stage framework for person search such as SeqNet, we further evaluate the scalability of the proposed SAMP method on AlignPS, the representative of the alternative one-stage framework. Compared to the two-stage framework, AlignPS utilizes a shared backbone and a unified task head for detection and re-identification, thus being more challenging in determining which channels are critical for the two sub-tasks to maintain the overall accuracy. As a consequence, as shown in Table 2, the performance of Slim is remarkably deteriorated, and DepGraph fails to converge, thus not being reported. In contrast, our method performs steadily, achieving 2.06% and 0.73% improvements in mAP with 10% FLOPs on the CUHK-SYSU and PRW datasets, respectively, compared to the other model pruning approaches.

4.4 Ablation Study

We further evaluate the effectiveness of the main components of our method, including the SaCI estimation, the LCB mechanism, and the AdaOIM loss. As our method is based on GroupFisher, it is selected as the 'Baseline' for comparison.

As summarized in Table 3, all proposed components boost the overall performance, with 25% and 10% FLOPs. And a combination of them generally reaches the best performance in most cases. It is worth noting that despite only being evaluated on person search, the proposed SaCI is generally applicable for the multi-modal and multi-task problems. LCB can be also extended to a broad range of tasks, as the over-pruning issue prevails in classification and detection.

5 Conclusion

In this paper, we investigate model pruning for efficient person search, and design a novel approach dubbed as Sub-task Aware Model Pruning with layer-wise channel balancing (SAMP). A sub-task aware channel importance estimation is devised to address the inconsistency between the detection and re-identification sub-tasks. A layer-wise channel balancing mechanism is developed to mitigate over-pruning. Finally, an adaptive OIM loss is presented to address the prototype degradation issue. Experiments on two widely used benchmarks clearly validate the effectiveness of the proposed approach, reaching 0.79% and 1.56% improvements in mAP when using SeqNet on CUHK-SYSU and PRW, respectively.

Acknowledgements. This work is supported by the National Key R&D Program of China (2021ZD0110503), the National Natural Science Foundation of China (62202034), the Research Program of State Key Laboratory of Virtual Reality Technology and Systems, and the grant No. KZ46009501.

References

1. Chang, X., Li, Y., Oymak, S., et al.: Provable benefits of overparameterization in model compression: from double descent to pruning neural networks. In: AAAI Conference on Artificial Intelligence, pp. 6974–6983 (2021)
2. Chen, K., Wang, J., Pang, J., et al.: MMDetection: open MMLab detection toolbox and benchmark. arXiv preprint arXiv:1906.07155 (2019)
3. Deng, L., Li, G., Han, S., et al.: Model compression and hardware acceleration for neural networks: a comprehensive survey. Proc. IEEE **108**(4), 485–532 (2020)
4. Doering, A., Chen, D., Zhang, S., et al.: PoseTrack21: a dataset for person search, multi-object tracking and multi-person pose tracking. In: IEEE/CVF Conference on Computer Vision and Pattern Recognition, pp. 20963–20972 (2022)
5. Fang, G., Ma, X., Song, M., et al.: Depgraph: towards any structural pruning. In: IEEE/CVF Conference on Computer Vision and Pattern Recognition, pp. 16091–16101 (2023)
6. Feng, D., Yang, J., Wei, Y., et al.: An efficient person search method using spatio-temporal features for surveillance videos. Appl. Sci. **12**(15), 7670 (2022)
7. Gao, C., Cai, G., Jiang, X., et al.: Conditional feature learning based transformer for text-based person search. IEEE Trans. Image Process. **31**, 6097–6108 (2022)
8. Han, S., Mao, H., Dally, W.J.: Deep compression: compressing deep neural network with pruning, trained quantization and huffman coding. In: International Conference on Learning Representations (2016)
9. Han, S., Pool, J., Tran, J., et al.: Learning both weights and connections for efficient neural network. In: Advances in Neural Information Processing Systems, pp. 1135–1143 (2015)

10. Howard, A.G., Zhu, M., Chen, B., et al.: Mobilenets: efficient convolutional neural networks for mobile vision applications. arXiv preprint arXiv:1704.04861 (2017)

11. Hu, H., Peng, R., Tai, Y.W., et al.: Network trimming: a data-driven neuron pruning approach towards efficient deep architectures. arXiv preprint arXiv:1607.03250 (2016)

12. Jaffe, L., Zakhor, A.: Gallery filter network for person search. In: IEEE/CVF Winter Conference on Applications of Computer Vision, pp. 1684–1693 (2023)

13. Li, J., Liang, F., Li, Y., et al.: Fast person search pipeline. In: IEEE International Conference on Multimedia and Expo, pp. 1114–1119 (2019)

14. Li, J., Yan, Y., Wang, G., et al.: Domain adaptive person search. In: Avidan, S., Brostow, G., Cissé, M., Farinella, G.M., Hassner, T. (eds.) ECCV 2022. LNCS, vol. 13674, pp. 302–318. Springer, Cham (2022). https://doi.org/10.1007/978-3-031-19781-9_18

15. Li, Z., Miao, D.: Sequential end-to-end network for efficient person search. In: AAAI Conference on Artificial Intelligence, pp. 2011–2019 (2021)

16. Lin, X., Ren, P., Xiao, Y., et al.: Person search challenges and solutions: a survey. In: International Joint Conference on Artificial Intelligence, pp. 4500–4507 (2021)

17. Liu, J., Zhuang, B., Zhuang, Z., et al.: Discrimination-aware network pruning for deep model compression. IEEE Trans. Pattern Anal. Mach. Intell. **44**(8), 4035–4051 (2021)

18. Liu, L., Zhang, S., Kuang, Z., et al.: Group fisher pruning for practical network compression. In: International Conference on Machine Learning, pp. 7021–7032 (2021)

19. Meng, F., Cheng, H., Li, K., et al.: Pruning filter in filter. In: Advances in Neural Information Processing Systems, pp. 17629–17640 (2020)

20. Mirzadeh, S.I., Farajtabar, M., Li, A., et al.: Improved knowledge distillation via teacher assistant. In: AAAI Conference on Artificial Intelligence, pp. 5191–5198 (2020)

21. Russakovsky, O., Deng, J., Su, H., et al.: Imagenet large scale visual recognition challenge. Int. J. Comput. Vision **115**, 211–252 (2015)

22. Wang, H., Qin, C., Zhang, Y., et al.: Neural pruning via growing regularization. In: International Conference on Learning Representations (2021)

23. Xiao, T., Li, S., Wang, B., et al.: End-to-end deep learning for person search. arXiv preprint arXiv:1604.01850 (2016)

24. Xiao, T., Li, S., Wang, B., et al.: Joint detection and identification feature learning for person search. In: IEEE Conference on Computer Vision and Pattern Recognition, pp. 3415–3424 (2017)

25. Xie, H., Jiang, W., Luo, H., et al.: Model compression via pruning and knowledge distillation for person re-identification. J. Ambient. Intell. Humaniz. Comput. **12**, 2149–2161 (2021)

26. Xu, Y., Ma, B., Huang, R., et al.: Person search in a scene by jointly modeling people commonness and person uniqueness. In: ACM International Conference on Multimedia, pp. 937–940 (2014)

27. Yan, Y., Li, J., Qin, J., et al.: Anchor-free person search. In: IEEE/CVF Conference on Computer Vision and Pattern Recognition, pp. 7690–7699 (2021)

28. Yeom, S.K., Seegerer, P., Lapuschkin, S., et al.: Pruning by explaining: a novel criterion for deep neural network pruning. Pattern Recogn. **115**, 107899 (2021)

29. Yu, J., Yang, L., Xu, N., et al.: Slimmable neural networks. In: International Conference on Learning Representations (2018)

30. Yu, X., Liu, T., Wang, X., et al.: On compressing deep models by low rank and sparse decomposition. In: IEEE Conference on Computer Vision and Pattern Recognition, pp. 7370–7379 (2017)
31. Zheng, L., Shen, L., Tian, L., et al.: Scalable person re-identification: a benchmark. In: IEEE International Conference on Computer Vision, pp. 1116–1124 (2015)
32. Zheng, L., Zhang, H., Sun, S., et al.: Person re-identification in the wild. In: IEEE Conference on Computer Vision and Pattern Recognition, pp. 1367–1376 (2017)
33. Zheng, Y.J., Chen, S.B., Ding, C.H., et al.: Model compression based on differentiable network channel pruning. IEEE Trans. Neural Netw. Learn. Syst. (2022)

MKB: Multi-Kernel Bures Metric for Nighttime Aerial Tracking

Yingjie He, Peipei Kang, Qintai Hu$^{(\boxtimes)}$, and Xiaozhao Fang

Guangdong University of Technology, Guangzhou 510006, China
huqt8@gdut.edu.cn

Abstract. In recent years, many advanced visual object tracking algorithms have achieved significant performance improvements in daytime scenes. However, when these algorithms are applied at night on unmanned aerial vehicles, they often exhibit poor tracking accuracy due to the domain shift. Therefore, designing an effective and stable domain adaptation tracking framework is in great demand. Besides, as the complexity of the tracking task scene and a large number of samples, it is difficult to accurately characterize the feature differences. Moreover, the method should avoid excessive manual parameters. To solve above three challenges, this study proposes a metric-based domain alignment framework for nighttime object tracking of unmanned aerial vehicles. It achieves transfer learning from daytime to nighttime by introducing a multi-kernel Bures metric (MKB). MKB quantifies the distribution distance by calculating the difference between two domain covariance operators in the latent space. MKB also uses a linear combination of a series of kernel functions to accommodate complex samples of different scenes and categories, enhancing the ability of the metric to represent various sequences in the latent space. Besides, we introduce an alternating update mechanism to optimize the weights of kernels avoiding manual parameters. Experimental results demonstrate that the proposed method has significant advantages in improving tracking accuracy and robustness at nighttime.

Keywords: Domain Adaptation · Cross-domain · Kernel Bures Metric · Visual Tracking · Aerial Tracking

1 Introduction

Recently, the applications of unmanned aerial vehicles (UAV) have become increasingly widespread in many fields [1,15], such as autonomous flight, search

This work was supported in part by the National Natural Science Foundation of China under Grant No. 62237001, 62202107, and 62176065, in part by the Guangdong Provincial National Science Foundation under Grant No. 2021A1515012017, in part by Science and Technology Planning Project of Guangdong Province, China, under Grant No. 2019B110210002, in part by the Guangdong Basic and Applied Basic Research Foundation under Grant No. 2021B1515120010, in part by Huangpu International Sci&Tech Cooperation Foundation of Guangzhou under Grant No. 2021GH12.

Q. Liu et al. (Eds.): PRCV 2023, LNCS 14434, pp. 212–224, 2024.
https://doi.org/10.1007/978-981-99-8549-4_18

and rescue, automatic tracking, intelligent monitoring, and security surveillance. With the support of large-scale datasets, more sophisticated AI models and frameworks make UAV more intelligent, efficient, and accurate. Particularly in the task of object tracking, researchers have constructed many inspiring models [18].

Existing airborne robot object tracking methods are designed with novel and ingenious structures [2,5], but the sample datasets used by these models are mostly collected under good lighting conditions. If these models are used in nighttime scenes, it will be found that the existing tracking methods perform poorly because of the domain shift between the training set and the task, which will seriously hinder the application of these algorithms [19]. Generally speaking, applying these tracking models to different domains, there may be problems such as distribution shift or domain differences. So, we need to reduce the domain differences between the source domain and the target domain, enabling the model to better adapt to the target domain.

To solve the domain shift problem, a large-scale labeled dataset for night-time can be used to retrain the backbone. However, creating such a large labeled dataset requires a huge amount of manual labor, making it challenging to execute. A more feasible approach would be employing a large-scale labeled daytime dataset and a small-scale unlabeled nighttime dataset together, and integrating unsupervised domain adaptation (UDA) techniques into the training of existing tracking models [20]. Such a strategy would effectively reduce the domain shift between datasets, and bring in improved tracking performance in nighttime scenarios.

Ye et al. [20] proposed the unsupervised domain adaptation for nighttime aerial tracking (UDAT), which uses a domain adversarial network to align the feature distributions between the daytime domain and the nighttime domain. However, due to the instability tracking through adversarial mechanism, especially when there are drastic changes in tracking scenes, the target may be lost or the tracking may be incorrect when the target is blurry or occluded. The metric-based adaptation methods can reduce the difference between domains by defining a well-designed domain discrepancy function, which facilitates more accurate transfer. Wasserstein distance can accurately measure the difference between two distributions from a geometric perspective, as the latent geometric information is hidden in the cost function. Recently, Zhang et al. [22] extended the Wasserstein distance to Reproducing Kernel Hilbert Space (RKHS) and introduced the corresponding kernel Bures distance. As the tracking task involves complex and diverse scenes and samples, these metrics are limited in accurately characterizing these complex and irregular samples. Moreover, the hyperparameters during training could determine the performance of the network. The massive scale of daytime data involved in tracking network training makes tuning parameters time-consuming and laborious.

Overall, this task still faces the following three challenges: (1) Constructing a stable and accurate domain adaptation method to reduce the feature differences between night and day. (2) Accurately characterizing domain differences in complex scenes and large scale samples. (3) Learning hyperparameters adaptively. To address the above challenges, we propose a metric-based domain alignment nighttime object tracking method, which reduces domain differences by defining a well-designed domain discrepancy function to achieve stable and accurate transfer and enhance the stability of tracking at nighttime. We also introduce the multi-kernel mechanism of the Bures metric, which uses a linear combination of a series of kernel functions to accommodate samples of different scenes and categories, thereby enhancing the ability of the metric to represent various sequences in the latent space. Finally, we further propose an automatic parameter updating strategy to optimize the linear combination of kernels, allowing the algorithm to automatically obtain the combination of kernels and accurate feature representation.

The main contributions are as follows:

- We propose a metric-based domain alignment nighttime object tracking framework. This framework utilizes metric function to enhance the accuracy and robustness of UAV nighttime tracking.
- In this study, a novel multi-kernel Bures metric is proposed along with a parameter auto-update strategy. This distance measure effectively characterizes complex samples and accurately measures domain differences.
- We conduct a series of comparative and visualization experiments, which validates the effectiveness of the proposed method.

2   Methodology

This framework is based on UDA. The source domain is a large-scale dataset with good lighting conditions and annotations, while the target domain is a small-scale dataset with poor lighting conditions and no annotations. As shown in Fig. 1, data from both domains are put into the shared parameter backbone to obtain template and search features. The framework can be divided into two parallel processes. (1) The alignment process takes the search features as input to a domain alignment layer module and calculates the corresponding multi-kernel Bures loss to reduce the domain difference and improve tracking performance in nighttime drone scenarios. (2) The tracking process takes both template and search patch features as input into the tracking header module and calculates the classification and regression loss to capture the mutual information as much as possible, thus improving overall tracking performance.

The multi-kernel Bures metric (MKB) is based on the kernel Bures metric, so we first introduce the kernel Bures metric.

2.1   Kernel Bures Metric

The kernel Bures metric is represented in reproducing kernel Hilbert space (RKHS), we provide the concept of RKHS. Let $\mathcal{X}$ be a non-empty set and $\mathcal{H}$ be an RKHS defined on $\mathcal{X}$. Let k be a reproducing kernel defined on $\mathcal{H}$, satisfying the following conditions: $(1)\forall x \in \mathcal{X}, k(\cdot, x) \in \mathcal{H}$ $(2)\forall x \in \mathcal{X}, f \in \mathcal{H}$,

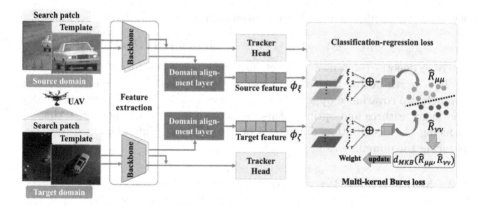

Fig. 1. The proposed framework. The left side shows the input data from the source and target domains. The source domain dataset consists of well-lit tracking data, while the target domain dataset consists of nighttime tracking data captured from a drone perspective. After feature extraction by the backbone network, the features are input to the domain feature alignment layer and the tracker. Then, both domain features are mapped into latent space by a set of kernel functions (ϕ_ξ, ϕ_ζ), and the corresponding estimated covariance operators $\hat{R}_{\mu\mu}$ and $\hat{R}_{\nu\nu}$ are calculated. Finally, the multi-kernel Bures loss and classification-regression loss are calculated, respectively.

$\langle f, k(\cdot, x) \rangle = f(x)$. Also, let ψ be corresponding feature map, i.e., $\psi(x) = k(\cdot, x)$. Then, $\forall x \in \mathcal{X}$ and $\forall y \in \mathcal{X}$, it holds that $\langle \psi(x), \psi(y) \rangle_{\mathcal{H}} = k(x, y)$.

The kernel Bures measure is derived from the kernel Wasserstein distance that measures the similarity between two probability distributions by computing the transport distance from one to other distribution in RKHS [22]. The kernel Wasserstein distance is defined as

$$d_{KW}^2(\psi_{\#}\mu, \psi_{\#}\nu) = \inf_{\gamma \in \Pi(\psi_{\#}\mu \times \psi_{\#}\nu)} \int_{\mathcal{H} \times \mathcal{H}} c^2(x, y) d\gamma(x, y), \qquad (1)$$

where μ and ν are probability measures on $\mathcal{X}$, $\psi_{\#}\mu$ and $\psi_{\#}\nu$ are pushforward measures of μ and ν, Π is the set of joint probability measures on $\mathcal{H} \times \mathcal{H}$, γ is a joint probability measure of Π, and $c(\cdot, \cdot)$ is 2-norm on RKHS, so $c^2(x, y) = \|\psi(x) - \psi(y)\|_{\mathcal{H}}^2 = \langle \psi(x), \psi(x) \rangle_{\mathcal{H}} + \langle \psi(y), \psi(y) \rangle_{\mathcal{H}} - 2\langle \psi(x), \psi(y) \rangle_{\mathcal{H}} = k(x, x) + k(y, y) - 2k(x, y)$.

The kernel Wasserstein distance computes the minimal transport distance between two distributions on $\mathcal{H}$, and it has wide applications in computer vision. Particularly, if we consider that the implicit feature maps are injective and the mean values of the distributions are equal, then the kernel Wasserstein distance has a corresponding form that is the kernel Bures distance as follows:

$$d_{KB}^2(R_{\mu\mu}, R_{\nu\nu}) = Tr(R_{\mu\mu} + R_{\nu\nu} + R_{\mu\nu}), \qquad (2)$$

where R represents the covariance operator on $\mathcal{H}$, which is defined as follows:

$$R_{\mu\nu} = E_{X \sim \mu, Y \sim \nu}((\psi(X) - m_\mu) \otimes (\psi(Y) - m_\nu)), \tag{3}$$

where $\otimes$ denotes the tensor product. m_μ and m_ν are the corresponding means.

It is worth noting that d_{KB} measures the dispersion between distributions through two covariance operators, so we can calculate the d_{KB} between the day and night features and optimize this distance through neural networks, thereby reducing the distribution differences of the day and night data features, i.e., reducing the domain gap.

2.2   Multi-Kernel Bures Metric

The Bures distance in single kernel mode expresses a single aspect and is suitable for small-scale datasets. However, due to the complexity and diversity of the tracking scenarios and the large size of the data, the performance of the Bures distance in single kernel mode is generally poor. Therefore, we extend the kernel Bures distance to a multi-kernel distance to more accurately characterize these complex data and make the distance suitable for large-scale datasets, thus improving the accuracy of the distance.

Let $X = \{x_1, x_2..., x_m\}$ and $Y = \{y_1, y_2, ..., y_n\}$ represent the samples of the source domain and target domain, respectively. m and n represent the number of samples in the source domain and target domain, respectively. Since it is a multi-kernel mechanism, we need to find a set of source mappings $\phi_\xi(x_i) = \{\xi_p \phi_p(x_i)\}_{p=1}^r$ and a set of target domain mappings $\phi_\zeta(y_i) = \{\zeta_p \phi_p(y_i)\}_{p=1}^r$, where r is the number of mappings, besides, ξ and ζ are weight parameters that need to be learned, defined as $\xi = [\xi_1, \xi_2, ..., \xi_r]^T$ and $\zeta = [\zeta_1, \zeta_2, ..., \zeta_r]^T$. Let K_{XY} be the kernel matrix, and the elements are defined as

$$(K_{XY})_{ij} = \sum_{p=1}^r k_p(x_i, y_j) = \sum_{p=1}^r \xi_p \zeta_p \phi_p(x_i) \phi_p(y_j). \tag{4}$$

Similarly, the elements in the kernel matrices K_{XX} and K_{YY} are defined as follows:

$$(K_{XX})_{ij} = \sum_{p=1}^r \xi_p^2 \phi_p(x_i) \phi_p(x_j), \tag{5}$$

$$(K_{YY})_{ij} = \sum_{p=1}^r \zeta_p^2 \phi_p(y_i) \phi_p(y_j). \tag{6}$$

Let matrix $\Phi = [\sum_{p=1}^r \phi_p(x_1), ..., \sum_{p=1}^r \phi_p(x_m)]$, and the matrix $\Psi = [\sum_{p=1}^r \phi_p(y_1), ..., \sum_{p=1}^r \phi_p(y_n)]$. Moreover, let the centering matrices $C_m = I_{m \times m} - \frac{1}{m} 1_m 1_m^T$ and $C_n = I_{n \times n} - \frac{1}{n} 1_n 1_n^T$, where 1 is a column vector of ones. Then, the empirical multi-kernel Bures metric is defined as follows:

$$d_{MKB}^2(\hat{R}_{\mu\mu}, \hat{R}_{\nu\nu}) = \frac{1}{m} Tr(K_{XX} C_m) + \frac{1}{n} Tr(K_{YY} C_n) - \frac{2}{\sqrt{mn}} ||C_m K_{XY} C_n||_*, \tag{7}$$

where $\hat{R}_{\mu\mu}$ and $\hat{R}_{\nu\nu}$ are estimated covariance operators, which are defined as $\hat{R}_{\mu\mu} = \frac{1}{m}\Phi C_m \Phi^T$ and $\hat{R}_{\nu\nu} = \frac{1}{n}\Psi C_n \Psi^T$, respectively. $|| \cdot ||_*$ is denoted as the nuclear norm, and specifically $||B||_* = Tr(\Sigma)$, where Σ is a semi-positive definite diagonal matrix whose diagonal elements are the singular values of matrix B.

With the extension of multiple kernels, manually setting the weight parameters ξ and ζ can seriously hinder the usability of this distance measure. So we introduce an alternating update mechanism to simultaneously identify the kernel combination and weights [16]. The above problem needs to solve:

$$\min_{\xi,\zeta} \sum_{p=1}^{r} \frac{1}{m}\xi_p^2 a_p + \frac{1}{n}\zeta_p^2 b_p - \frac{2}{\sqrt{mn}}\xi_p c_p \zeta_p,$$
$$s.t.\xi^T \mathbf{1} = 1, \zeta^T \mathbf{1} = 1, \tag{8}$$

where $a_p = Tr((K_{XX})_p C_m)$, $b_p = Tr((K_{YY})_p C_n)$, and $c_p = ||C_m (K_{XY})_p C_n||_*$, and the subscript p represents the p-th kernel matrix. This can be further organized into the following problem:

$$\min_{\xi,\zeta} \xi^T Z_1 \xi + \zeta^T Z_2 \zeta + \xi^T Z_3 \zeta,$$
$$s.t.\xi^T \mathbf{1} = 1, \zeta^T \mathbf{1} = 1, \tag{9}$$

where, $Z_1 = \frac{1}{m}diag(a_1, ..., a_r)$, $Z_2 = \frac{1}{n}diag(b_1, ..., b_r)$, and $Z_3 = \frac{-2}{\sqrt{mn}}diag(c_1, ..., c_r)$. The above problem can be solved by following step:

1) Update ξ with ζ fixed:

$$\min_{\xi} \xi^T Z_1 \xi + \zeta^T Z_3 \xi, s.t.\xi^T \mathbf{1} = 1. \tag{10}$$

2) Update ζ with ξ fixed:

$$\min_{\zeta} \zeta^T Z_1 \zeta + \xi^T Z_2 \zeta,$$
$$s.t.\zeta^T \mathbf{1} = 1. \tag{11}$$

The problem (10) and (11) are convex quadratic programming problems (QP) that can be solved directly by using a QP solver [4].

The multi-kernel Bures distance is suitable for tracking datasets, as the linear combination of multiple kernel functions allows various sample features to be represented by different kernel functions with different weights, enhancing the accuracy of large-scale datasets such as object tracking. The weights of the linear combination can be solved using the above optimal problem, greatly facilitating the use of multi-kernel Bures metrics.

2.3   Objective Loss

Classification and regression loss are only calculated in the source domain, as the model is based on unsupervised adaptation. In order to ensure tracking performance, we use an existing SiamCAR [6] as a tracker head, which calculates the classification loss, regression loss, and the distance between the center of the ground-truth bounding box and the predicted bounding box. The sum of these three losses is denoted as $\mathcal{L}_{Tracker}$.

Multi-kernel Bures loss aligns the search features from different domains to ensure consistency in the generated features. Specifically, we denote the search patches from the source data set as $T_s = [t_1^s, ..., t_m^s]$ and the search patches from the target domain data set as $T_t = [t_1^t, ..., t_n^t]$. Through the backbone network and domain align layer $\mathcal{G}$, we can obtain the source and target domain search features as $\mathcal{G}(T_s)$ and $\mathcal{G}(T_t)$, respectively. Next, we substitute $\mathcal{G}(T_s)$ and $\mathcal{G}(T_t)$ into Eq. (7) for X and Y, and calculate the d_{MKB} as the multi-kernel Bures loss $\mathcal{L}_{MKB}$. So, the total loss of our proposed framework is given by

$$\mathcal{L}_{total} = \mathcal{L}_{Tracker} + \lambda \mathcal{L}_{MKB}, \tag{12}$$

where λ is a weight parameter.

3   Experiments

3.1   Implementation Details

Our method is implemented on the Ubuntu system with Tesla V100 32 GB GPU for training. We use pre-trained ResNet-50 as the backbone network and pre-trained SiamCAR as the tracking head. And we use a ResNet-18 network as a domain alignment layer, with an output feature of 128 dimensions. We adopt the Adam optimizer with an initial learning rate of 1.5×10^{-3} and λ set as 10^{-5}. The entire training process lasts for 20 epochs and takes about 12 h. To ensure the fairness of the experiment, we use pre-training datasets, ImageNet VID [12] and GOT-10K [8] as the source domain, and evaluate models with one-pass evaluation (OPE) in official pre-trained parameters.

A total of 17 SOTA methods are compared, i.e., Dsiam [7], LUDT [15], SiameseFC [1], SiamDW-FC [23], SiamAPN [5], SE-SiamFC [14], Ocean [24], SiamFC++ [17], SiamAPN [5], UpdateNet [21], D3S [11], SiamRPN++ [9], DaSiamRPN [25], HiFT [2], SiamCAR [6], SiamBAN [3], and UDAT-CAR [20]. Besides, our method is marked as MKB-CAR.

3.2   Evaluation Datasets

NAT2021 [20] is a nighttime scene dataset based on the perspective of UAV. One of the test sets is the NAT2021-test, which contains 180 video segments. The other is NAT2021-test-L, which contains 23 video segments, where the suffix "L" indicates that these are long videos. **UAVDark70** [10] is a benchmark for visual object tracking under challenging low light and nighttime conditions, specifically designed for UAV scenes. The dataset consists of 70 video sequences.

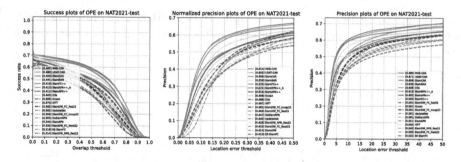

Fig. 2. Comparison of the success rate, normalized precision, and precision plots on the NAT2021-test dataset.

3.3 Comparison Results

We evaluated our proposed MKB-CAR method on the NAT2021-test dataset. The comparison results are shown in Fig. 2. Our method achieved an normalized precision of 61.4%, which is 1.49% higher than the previous best method. It is worth noting that MKB-CAR not only outperforms other methods in terms of average accuracy, but also achieves 1.04% and 1.34% higher success rate and precision rate than UDAT-CAR, respectively. Experimental results show that the multi-kernel Bures metric can improve the accuracy of nighttime object tracking by narrowing the feature distribution gap between day and night.

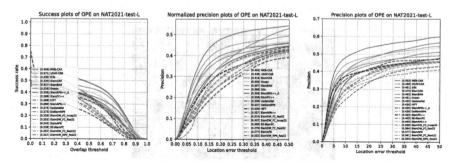

Fig. 3. Comparison of the success rate, normalized precision, and precision plots on the NAT2021-test dataset.

We conducted experiments on the NAT-test-L dataset as shown in Fig. 3. This dataset is mainly used to test the robustness and long-term performance of trackers in tracking nighttime objects. Due to factors such as complex scene changes, occlusion, and object motion pattern changes, tracking nighttime objects has become more challenging. Our method outperforms the second-place UDAT-CAR (0.445) by an absolute advantage of 4.8% in normalized precision. Meanwhile, MKB-CAR ranks first with significant advantages in all three evaluation. The experimental results indicate that MKB-CAR improves night-time recognition by narrowing the gap, and also maintains stable and robust tracking performance.

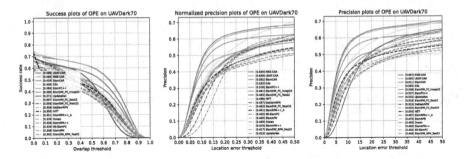

Fig. 4. Comparison of the success rate, normalized precision, and precision plots on the UAVDark70 dataset.

We tested our method on the UAVDark70 dataset as shown in Fig. 4. It can be used to test the algorithm's generalization performance, which refers to its ability to perform on unseen data. Our method is 1.2% higher than UDAT-CAR in success rate. This indicates that our domain alignment technique is better than GAN, because the distance we proposed has a more accurate description of domain distribution differences.

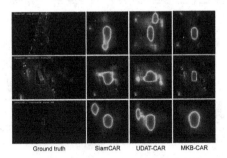

Fig. 5. Heatmap visualization. The first column represents the ground truth. The following columns correspond to SiamCAR, UDAT-CAR, and MKB-CAR, respectively.

3.4   Visualization

We utilized Grad-CAM [13] to generate heatmaps for Siam-CAR, UDAT-CAR, and MKB-CAR, as illustrated in Fig. 5. Comparing the heatmaps, it is evident that MKB-CAR demonstrates superior performance in accurately predicting and discerning the positioning and contours of objects.

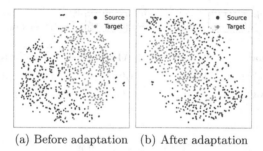

(a) Before adaptation   (b) After adaptation

Fig. 6. The T-SNE visualization. Subfigure (a) shows the features before domain alignment, while subfigure (b) shows the features after alignment learning.

We also used T-SNE algorithm to visualize the search features. As shown in Fig. 6, the figure shows the feature visualization of SiamCAR (left) and MKB-CAR (right). Blue points represent source domain samples, and orange points represent target domain samples. In SiamCAR, it can be clearly seen that there is a significant domain gap between the orange and blue points, and the source domain and target domain are respectively located on both sides of the domain gap. So there is domain shift in the night-time tracking task. In MKB-CAR, it can be clearly seen that the multi-kernel Bures distance narrows the feature distribution of source domain samples and target domain samples.

3.5   Ablation Study

Table 1. The ablation results on NAT2021-test. Source represents the training on the source domain dataset, KB represents kernel Bures distance, and MKB represents multi-kernel Bures distance.

Source	KB	MKB	Prec.	Norm.Prec	Succ.
✓			0.467	0.588	0.650
✓	✓		0.482	0.609	0.676
✓		✓	0.485	0.614	0.680

Table 1 shows the performance of the tracker improved with the addition of a domain adaptation module (KB or MKB) compared with using only the source domain data, indicating that domain adaptation can enhance night-time tracking performance. With the introduction of the multi-kernel mechanism and parameter update mechanism in MKB, the normalized precision and success rate were improved by 0.8% and 0.6%, respectively. This indicates that MKB is better suited for processing large-scale datasets and accurately characterizing the distribution of these datasets in latend space, thereby more accurately aligning the distributions.

Table 2. Tracker speed comparison. Comparison table of tracker speed test on three datasets NAT-test, NAT-test-L and UAVDark70. "fps" is frames per second.

Method	NAT2021-test	NAT2021-test-L	UAVDark70
SiamCAR	30.93 fps	31.01 fps	29.79 fps
MKB-CAR	31.17 fps	30.96 fps	30.98 fps

To verify that our proposed framework does not increase the inference time of the tracker when the domain alignment layer is not loaded, we conducted speed tests. Table 2 shows that SiamCAR and proposed MKB-CAR on the NAT2021-test, NAT2021-test-L, and UAVDark70. The experimental results demonstrate that the proposed framework maintains the tracking speed of the tracker.

4   Conclusion

This study proposes multi-kernel Bures metric for nighttime aerial tracking. The framework achieves transfer learning from daytime to nighttime scenarios by introducing multi-kernel Bures metric, which quantifies the distribution distance by computing the difference between covariance operators of the two domains in the latent space. MKB also utilizes a linear combination of a series of kernel functions to adapt to complex samples from different scenes and categories, enhancing the metric's ability to represent various sequences in the latent space. Additionally, an alternate updating mechanism is introduced to optimize the weights of the kernels, avoiding excessive manual parameter tuning. Experimental results validate the effectiveness and advancement of this approach. The study also provides new insights for future research on UAV nighttime tracking.

References

1. Bertinetto, L., Valmadre, J., Henriques, J.F., Vedaldi, A., Torr, P.H.S.: Fully-convolutional siamese networks for object tracking. In: Hua, G., Jégou, H. (eds.) ECCV 2016. LNCS, vol. 9914, pp. 850–865. Springer, Cham (2016). https://doi.org/10.1007/978-3-319-48881-3_56
2. Cao, Z., Fu, C., Ye, J., Li, B., Li, Y.: HiFT: hierarchical feature transformer for aerial tracking. In: IEEE International Conference on Computer Vision (ICCV) (2021)
3. Chen, Z., Zhong, B., Li, G., Zhang, S., Ji, R.: Siamese box adaptive network for visual tracking. In: IEEE Conference on Computer Vision and Pattern Recognition (CVPR) (2020)
4. Diamond, S., Boyd, S.: CVXPY: a python-embedded modeling language for convex optimization. J. Mach. Learn. Res. (JMLR) **17**(1), 2909–2913 (2016)
5. Fu, C., Cao, Z., Li, Y., Ye, J., Feng, C.: Siamese anchor proposal network for high-speed aerial tracking. In: IEEE International Conference on Robotics and Automation (ICRA). IEEE (2021)

6. Guo, D., Wang, J., Cui, Y., Wang, Z., Chen, S.: Siamcar: siamese fully convolutional classification and regression for visual tracking. In: IEEE Conference on Computer Vision and Pattern Recognition (CVPR) (2020)
7. Guo, Q., Feng, W., Zhou, C., Huang, R., Wan, L., Wang, S.: Learning dynamic siamese network for visual object tracking. In: IEEE International Conference on Computer Vision (ICCV) (2017)
8. Huang, L., Zhao, X., Huang, K.: GOT-10k: a large high-diversity benchmark for generic object tracking in the wild. IEEE Trans. Pattern Anal. Mach. Intell. (TPAMI) **43**(5), 1562–1577 (2021)
9. Li, B., Wu, W., Wang, Q., Zhang, F., Xing, J., Yan, J.: SiamRPN++: evolution of siamese visual tracking with very deep networks. In: IEEE Conference on Computer Vision and Pattern Recognition (CVPR) (2019)
10. Li, B., Fu, C., Ding, F., Ye, J., Lin, F.: Adtrack: target-aware dual filter learning for real-time anti-dark UAV tracking. In: IEEE International Conference on Robotics and Automation (ICRA) (2021)
11. Lukezic, A., Matas, J., Kristan, M.: D3S-a discriminative single shot segmentation tracker. In: IEEE Conference on Computer Vision and Pattern Recognition (CVPR) (2020)
12. Russakovsky, O., et al.: Imagenet large scale visual recognition challenge. Int. J. Comput. Vis. (IJCV) **115**, 211–252 (2015)
13. Selvaraju, R.R., Cogswell, M., Das, A., Vedantam, R., Parikh, D., Batra, D.: Grad-CAM: visual explanations from deep networks via gradient-based localization. In: IEEE International Conference on Computer Vision (ICCV) (2017)
14. Sosnovik, I., Moskalev, A., Smeulders, A.W.: Scale equivariance improves siamese tracking. In: IEEE/CVF Winter Conference on Applications of Computer Vision (WACV) (2021)
15. Wang, N., Zhou, W., Song, Y., Ma, C., Liu, W., Li, H.: Unsupervised deep representation learning for real-time tracking. Int. J. Comput. Vis. (IJCV) **129**, 400–418 (2021)
16. Wang, R., Lu, J., Lu, Y., Nie, F., Li, X.: Discrete and parameter-free multiple kernel k-means. IEEE Trans. Image Process. (TIP) **31**, 2796–2808 (2022)
17. Xu, Y., Wang, Z., Li, Z., Yuan, Y., Yu, G.: SiamFC++: towards robust and accurate visual tracking with target estimation guidelines. In: AAAI Conference on Artificial Intelligence (AAAI) (2020)
18. Yang, L., Liu, R., Zhang, D.D., Zhang, L.: Deep location-specific tracking. In: ACM International Conference on Multimedia (MM) (2017)
19. Ye, J., Fu, C., Cao, Z., An, S., Zheng, G.Z., Li, B.: Tracker meets night: a transformer enhancer for UAV tracking. IEEE Robot. Autom. Lett. (RA-L) **7**, 3866–3873 (2022)
20. Ye, J., Fu, C., Zheng, G., Paudel, D.P., Chen, G.: Unsupervised domain adaptation for nighttime aerial tracking. In: IEEE Conference on Computer Vision and Pattern Recognition (CVPR) (2022)
21. Zhang, L., Gonzalez-Garcia, A., Weijer, J.v.d., Danelljan, M., Khan, F.S.: Learning the model update for siamese trackers. In: IEEE International Conference on Computer Vision (ICCV) (2019)
22. Zhang, Z., Wang, M., Nehorai, A.: Optimal transport in reproducing kernel hilbert spaces: theory and applications. IEEE Trans. Pattern Anal. Mach. Intell. (TPAMI) **42**(7), 1741–1754 (2019)
23. Zhang, Z., Peng, H.: Deeper and wider siamese networks for real-time visual tracking. In: IEEE Conference on Computer Vision and Pattern Recognition (CVPR) (2019)

224   Y. He et al.

24. Zhang, Z., Peng, H., Fu, J., Li, B., Hu, W.: Ocean: object-aware anchor-free track-ing. In: Vedaldi, A., Bischof, H., Brox, T., Frahm, J.-M. (eds.) ECCV 2020. LNCS, vol. 12366, pp. 771–787. Springer, Cham (2020). https://doi.org/10.1007/978-3-030-58589-1_46
25. Zhu, Z., Wang, Q., Li, B., Wu, W., Yan, J., Hu, W.: Distractor-aware siamese networks for visual object tracking. In: European Conference on Computer Vision (ECCV) (2018)

Deep Arbitrary-Scale Unfolding Network for Color-Guided Depth Map Super-Resolution

Jialong Zhang[1] , Lijun Zhao[1(✉)] , Jinjing Zhang[2], Bintao Chen[1],
and Anhong Wang[1]

[1] Institute of Digital Media and Communication, Taiyuan University of Science and
Technology, Taiyuan 030024, China
`leejun@tyust.edu.cn`
[2] Data Science and Technology, North University of China, Jiancaoping District,
Taiyuan 030051, China

Abstract. Although color-guided Depth map Super-Resolution (DSR) task has made great progress with the help of deep learning, this task still suffers from some issues: 1) many DSR networks are short of good interpretability; 2) most of the popular DSR methods cannot achieve arbitrary-scale up-sampling for practical applications; 3) dual-modality gaps between color image and depth map may give rise to texture-copying problem. As for these problems, we build a new joint optimization model for two tasks of high-low frequency decomposition and arbitrary-scale DSR. According to alternatively-iterative update formulas of the solution for these two tasks, the proposed model is unfolded as Deep Arbitrary-Scale Unfolding Network (DASU-Net). In the DASU-Net, we propose a Continuous Up-Sampling Fusion (CUSF) module to address two problems of arbitrary-scale feature up-sampling and dual-modality inconsistency during color-depth feature fusion. A large number of experiments have demonstrated that the proposed DASU-Net achieves more significant reconstruction results as compared with several state-of-the-art methods.

Keywords: High-low frequency decomposition · Depth map super-resolution · Arbitrary-scale up-sampling · Deep explainable network

1 Introduction

At present, color-guided Depth map Super-Resolution (DSR) has attracted more and more people's attention, which comes from two reasons: 1) affordable color camera can easily capture high-quality and High-Resolution (HR) color images; 2) similar boundary information from color image can enhance discontinuously-geometric boundaries of depth map. However, their inconsistent structure and details may lead to texture-copying problem in the DSR task [7]. In order to solve these problems, traditional methods in the early stage have achieved excellent results. For example, image filtering-based DSR methods use adaptive weights from color-range domain and depth-range domain to perform local filtering on depth map [1,3,6]. In markov random field-based methods [2], color images are used to obtain the weight of the smoothness item to measure the relevance of the neighboring pixels, so as to regularize HR depth map reconstruction. Although traditional methods have good interpretability, the performances

© The Author(s), under exclusive license to Springer Nature Singapore Pte Ltd. 2024
Q. Liu et al. (Eds.): PRCV 2023, LNCS 14434, pp. 225–236, 2024.
https://doi.org/10.1007/978-981-99-8549-4_19

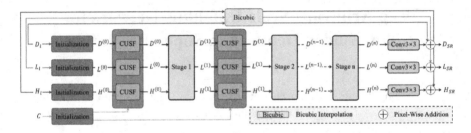

Fig. 1. The diagram of the proposed DASU-Net.

of these methods need to be improved due to the limitations of hand-made filters and unlearnable weighting rules.

Because deep learning-based methods have strong feature representation capabilities by supervised learning with the large-scale training dataset, their performances are often better than those of traditional methods. For instance, *Hui et al.* proposed a hierarchical DSR method to fuse multi-scale color features and depth features for depth map reconstruction [4]. *Kim et al.* designed an explicitly-sparse and spatially-variant convolutional kernels to deal with differential spatial change of details and structures of depth and color images, which effectively improved the performance of DSR tasks [5]. *Zhong et al.* proposed an attention-based hierarchical multi-modal fusion network, in which multi-modal attention model effectively selected consistent structure information between color image and depth map [12]. Unlike the above black-box-based networks, *Zhou et al.* embedded the structural prior of optimized model into the deep learning network and achieved better image Super-Resolution (SR) results [13]. Although the above methods have achieved good results in fixed factor up-sampling, they cannot meet the arbitrary-scale up-sampling requirement of practical applications. *Wang et al.* achieved continuous arbitrary-scale up-sampling by using coordinate query method, and embedded spatial geometric information into DSR model to aggregate single modal features, but they ignored the importance of feature fusion and model interpretability [9]. Therefore, to simultaneously alleviate three problems of network interpretability, arbitrary-scale DSR and texture-copying, we build a novel joint optimization model and unfold it as Deep Arbitrary-Scale Unfolding Network (DASU-Net).

In summary, our contributions are listed below: 1) we propose a novel joint optimization model based on High-Low Frequency (H-LF) decomposition and arbitrary-scale DSR, which is unfolded as DASU-Net to solve the problems of black-box network, continuous super-resolution and texture-copying in DSR task; 2) we reasonably transform each iterative operator of optimization model into three sub-networks, including Depth map reconstruction sub-Network (DNet), High-frequency map reconstruction sub-Network (HNet) and Low-frequency map reconstruction sub-Network (LNet); 3) to address dual-modality inconsistency during color-depth fusion and to achieve continuous arbitrary-scale up-sampling, we propose a Continuous Up-Sampling Fusion (CUSF) module based on feature enhancement and multiple strategies fusion.

The rest of this paper is organized as follows. In Section 2, the proposed method is described. Experimental results are presented in Section 3. In Section 4, conclusion is discussed.

2   The Proposed Method

2.1   Problem Formulation

The degradation mathematical models of HR Low Frequency (LF) map $L \in \mathbb{R}^{1 \times M \times N}$, HR High Frequency (HF) map $H \in \mathbb{R}^{1 \times M \times N}$ and HR depth map $D \in \mathbb{R}^{1 \times M \times N}$ can be linearly expressed as:

$$\begin{cases} D_l = A_d K_d D + n_d, \\ L_l = A_{lf} K_{lf} L + n_{lf}, \\ H_l = A_{hf} K_{hf} H + n_{hf}, \end{cases} \tag{1}$$

where the observed $D_l \in \mathbb{R}^{1 \times M/m \times N/m}$, $L_l \in \mathbb{R}^{1 \times M/m \times N/m}$ and $H_l \in \mathbb{R}^{1 \times M/m \times N/m}$ are obtained by down-sampling D, L and H m times. $n_{(\cdot)}$, $A_{(\cdot)}$ and $K_{(\cdot)}$ denote additive noise, down-sampling operators and blurring kernel.

In order to filter out the texture-copying artifacts caused by color-depth information inconsistency in DSR task, combining Eq.(1), we propose a new joint optimization model for two tasks of H-LF decomposition about depth map and arbitrary-scale DSR, which can be mathematically written as:

$$\begin{aligned} \underset{L,H,D}{argmin} \ &\frac{1}{2}\|D_l - A_d K_d D\|_2^2 + \alpha\Phi_1(D) + \eta\frac{1}{2}\|D - L - H\|_2^2 \\ &+ \frac{1}{2}\beta_1\|L_l - A_{lf}K_{lf}L\|_2^2 + \frac{1}{2}\beta_2\|L - D * g_l\|_2^2 + \beta_3\Phi_2(L) \\ &+ \frac{1}{2}\gamma_1\|H_l - A_{hf}K_{hf}H\|_2^2 + \frac{1}{2}\gamma_2\|H - D * g_h\|_2^2 + \gamma_3\Phi_3(H). \end{aligned} \tag{2}$$

Here $\| \cdot \|_2^2$ is the square of l_2-norms. $\Phi_1(\cdot)$, $\Phi_2(\cdot)$ and $\Phi_3(\cdot)$ are three implicit regularization terms. In addition, α, η, β_1, β_2, β_3, γ_1, γ_2 and γ_3 are trade-off parameters. g_h and g_l denote high-pass filter and low-pass filter, while $*$ is convolution operation.

The terms unrelated to LF features L can be ignored in Eq.(2), and then we use proximal gradient descent algorithm to obtain the iterative formula about L:

$$\begin{cases} L^{(n)} = prox_{\Phi_2}(L^{(n-1)} - \delta_1 \nabla f_1(L^{(n-1)})), \\ \nabla f_1(L^{(n-1)}) = \beta_1(A_{lf}K_{lf})^T(L_l - A_{lf}K_{lf}L^{(n-1)}) \\ + \beta_2(L^{(n-1)} - D^{(n-1)} * g_l) + \eta(D^{(n-1)} - L^{(n-1)} - H^{(n-1)}), \end{cases} \tag{3}$$

where $prox_{\Phi_2}(\cdot)$ is the proximal operator with regard to non-differentiable term Φ_2, ∇ represents the gradient operator, and δ_1 is the updating step size. $(\cdot)^T$ represents transpose operation. In the same way, the HF features of H can be iteratively updated as follows:

$$\begin{cases} H^{(n)} = prox_{\Phi_3}(H^{(n-1)} - \delta_2 \nabla f_2(H^{(n-1)})), \\ \nabla f_2(H^{(n-1)}) = \gamma_1(A_{hf}K_{hf})^T(H_l - A_{hf}K_{hf}H^{(n-1)}) \\ + \gamma_2(H^{(n-1)} - D^{(n-1)} * g_h) + \eta(D^{(n-1)} - L^{(n)} - H^{(n-1)}), \end{cases} \tag{4}$$

where $prox_{\Phi_3}(\cdot)$ represents the proximal operator with regard to non-differentiable term Φ_3, and δ_2 is the updating step size. The updated HF and LF features are used as prior conditions to assist in deducing depth features D:

$$\begin{cases} D^{(n)} = prox_{\Phi_1}(D^{(n-1)} - \delta_3 \nabla f_3(D^{(n-1)})), \\ \nabla f_3(D^{(n-1)}) = (A_d K_d)^T (D_l - A_d K_d D^{(n-1)}) \\ \quad + \beta_2(g_l)^T * (L^{(n)} - D^{(n-1)} * g_l) \\ \quad + \gamma_2(g_h)^T * (H^{(n)} - D^{(n-1)} * g_h) \\ \quad + \eta(D^{(n-1)} - L^{(n)} - H^{(n)}), \end{cases} \tag{5}$$

where $prox_{\Phi_1}(\cdot)$ represents the proximal operator with regard to non-differentiable term Φ_1, and δ_3 is the updating step size.

2.2 Algorithm Unfolding

We unfold the optimization model Eq.(3), Eq.(4) and Eq.(5) into Low-frequency map reconstruction sub-Network (LNet), High-frequency map reconstruction sub-Network (HNet) and Depth map reconstruction sub-Network (DNet) to build an explicable DASU-Net for solving the problems of DSR. The overall architecture of our proposed DASU-Net is shown in Fig. 1, which follows the iterative update of the optimization model. DASU-Net inputs D_l, L_l, H_l and color map $C \in \mathbb{R}^{3 \times M \times N}$, while it outputs the SR depth map $D_{SR} \in \mathbb{R}^{1 \times M \times N}$, the SR LF map $L_{SR} \in \mathbb{R}^{1 \times M \times N}$ and the SR HF map $H_{SR} \in \mathbb{R}^{1 \times M \times N}$. Following the literature of [9], DASU-Net up-samples to the full-resolution in two steps. Specifically, the feature map is up-sampled through CUSF module to intermediate resolution and target resolution before and after Stage 1. In addition, CUSF module also shoulders the responsibility of multi-modal fusion. In DASU-Net, the initialization module plays a role in expanding the number of channels and preliminary feature extraction, including one 1×1 convolutional layer and two Residual Blocks (ResBlock). In the n-th stage, the proposed network receives $D^{(n-1)}$, $L^{(n-1)}$, $H^{(n-1)}$, while it outputs $D^{(n)}$, $L^{(n)}$ and $H^{(n)}$. Each stage includes three sub-networks: LNet, HNet and DNet, as shown in Fig. 2, and the updating order is LNet $\Rightarrow$ HNet $\Rightarrow$ DNet. To prevent information loss, three bypasses with Bicubic interpolation are used to assist LR image reconstruction. In order to better understand the principle of DASU-Net, the next step is to analyze the n-th stage of the optimization model in Eq.(3), Eq.(4) and Eq.(5), and unfold them into sub-networks.

Low-Frequency Map Reconstruction Sub-Network (LNet). In order to obtain LF feature $L^{(n)}$ in the n-the stage through LNet, we use convolutional modules to approximate some operators in Eq.(3). $A_{lf}K_{lf}$ is learned by two ResBlocks and one down-sampling operation. It should be noted that the up-sampling and down-sampling of the proposed method are achieved by grid sampler [9], which is the key to achieve arbitrary-scale sampling. $(A_{lf}K_{lf})^T$ is the inverse operation of $A_{lf}K_{lf}$, so it is approximated by two ResBlocks and one up-sampling operation. We use convolutional kernels of 3×3 and 5×5 sizes to predict g_l. We propose a Multi-Scale Regular Module (MSRM) to approximate the proximal operator $prox_{\Phi_2}(\cdot)$, which includes one 7×7, one 5×5,

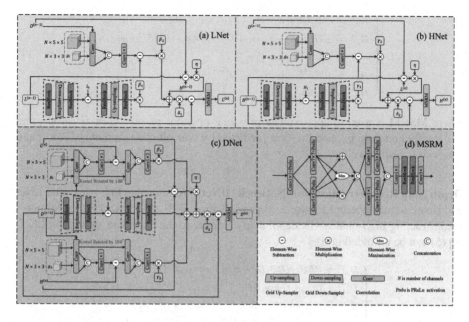

Fig. 2. The structures of LNet (a), HNet (b), DNet (c), and MSRM (d).

three 3×3, four 1×1 convolution kernels and two ResBlocks to extract multi-scale features. In addition, MSRM has three fusion strategies to mix different scales features, as shown in Fig. 2 (d). Based on the above analysis, as shown in Fig. 2 (a), Eq.(3) can be unfolded as LNet:

$$
\begin{cases}
L_a^{(n)} = \beta_1 RB(Up(RB(L_l - RB(Down(RB(L^{(n-1)})))))), \\
L_b^{(n)} = \beta_2(L^{(n-1)} - Conv_1([D^{(n-1)} * g_{l3}, D^{(n-1)} * g_{l5}])), \\
L_c^{(n)} = \eta(D^{(n-1)} - L^{(n-1)} - H^{(n-1)}), \\
L_d^{(n)} = L^{(n-1)} - \delta_1(L_a^{(n)} + L_b^{(n)} + L_c^{(n)}), \\
L^{(n)} = MSRM(L_d^{(n)}),
\end{cases}
\tag{6}
$$

where RB, Up, $Down$, $[\cdot]$, $Conv_1$, g_{l3} and g_{l5} represent ResBlock, grid up-sampler, grid down-sampler, the concatenation along channel dimension, 1×1 convolution, convolutional kernels of 3×3 and 5×5 sizes respectively. g_{l3} and g_{l5} jointly compose g_l. $Conv_1$ plays a role in changing the number of channels. Additionally, convolutional operation $*$ is represented by Conv in Fig. 2.

How-Frequency Map Reconstruction Sub-Network (HNet). The optimized model in Eq.(4) is unfolded to obtain HNet for predicting HF features, as shown in Fig. 2 (b), and the unfolding process of Eq.(4) is similar to LNet. HNet can be written as:

$$\begin{cases} H_a^{(n)} = \gamma_1 RB(Up(RB(H_l - RB(Down(RB(H^{(n-1)})))))), \\ H_b^{(n)} = \gamma_2(H^{(n-1)} - Conv_1([D^{(n-1)} * g_{h3}, D^{(n-1)} * g_{h5}])), \\ H_c^{(n)} = \eta(D^{(n-1)} - L^{(n)} - H^{(n-1)}), \\ H_d^{(n)} = H^{(n-1)} - \delta_2(H_a^{(n)} + H_b^{(n)} + H_c^{(n)}), \\ H^{(n)} = MSRM(H_d^{(n)}), \end{cases} \qquad (7)$$

where MSRM is used to approximate proximal operator $prox_{\Phi_3}(\cdot)$ in $H^{(n)}$. g_{h3} and g_{h5} jointly compose g_h, and they are predicted by convolutional kernels of 3×3 and 5×5 sizes respectively.

Depth Map Reconstruction Sub-Network (DNet). As shown in Fig. 2 (c), after LNet and HNet, we can obtain LF prior $L^{(n)}$ and HF prior $H^{(n)}$, which is a key condition for reconstructing high-quality HR depth maps. Similar to the unfolding strategy in LNet, Eq.(5) can be unfolded into:

$$\begin{cases} D_a^{(n)} = RB(Up(RB(D_l - RB(Down(RB(D^{(n-1)})))))), \\ D_{b1}^{(n)} = L^{(n)} - Conv_1([D^{(n-1)} * g_{l3}, D^{(n-1)} * g_{l5}]), \\ D_{b2}^{(n)} = \beta_2 Conv_1([(g_{l3})^T * D_{b1}^{(n)}, (g_{l5})^T * D_{b1}^{(n)}]), \\ D_{c1}^{(n)} = H^{(n)} - Conv_1([D^{(n-1)} * g_{h3}, D^{(n-1)} * g_{h5}]), \\ D_{c2}^{(n)} = \gamma_2 Conv_1([(g_{h3})^T * D_{c1}^{(n)}, (g_{h5})^T * D_{c1}^{(n)}]), \\ D_d^{(n)} = \eta(D^{(n-1)} - L^{(n)} - H^{(n)}), \\ D_e^{(n)} = H^{(n-1)} - \delta_3(D_a^{(n)} + D_{b2}^{(n)} + D_{c2}^{(n)} + D_d^{(n)}), \\ D^{(n)} = MSRM(D_e^{(n)}), \end{cases} \qquad (8)$$

where MSRM is used to approximate proximal operator $prox_{\Phi_1}(\cdot)$. $(g_l)^T$ and $(g_h)^T$ are the $180°$ rotation of g_l and g_h.

2.3   Continuous Up-Sampling Fusion (CUSF)

As shown in Fig. 3 (a), CUSF module is designed to achieve two functionalities including continuous arbitrary-scale up-sampling and multi-modal fusion. Note that we use the same up-sampler as done in [9]. For simplicity, in CUSF, the specific feature related to color map is described as guidance feature, while the target feature is extracted from LF feature, HF feature and depth feature. Specifically, the proposed Feature Enhancement and Multiple Strategies Fusion (FEMSF) component is used to perform multi-modal fusion, as shown in Fig. 3 (b). Finally, three convolution layers and two Res-Blocks are serialized to merge different features.

As shown in Fig. 3 (b), we first introduce Dual Affinity Matrix (DAM) [10] to filter impurities in guidance features G to obtain the enhanced guidance feature E. In multiple strategies fusion, the spatial features of E are first condensed to 1×1 tensor through global max-pooling and global average- pooling to get global information, and then the global information are concatenated in the channel dimension and processed by two convolutional layers. Afterwards, the softmax and split function utilizes

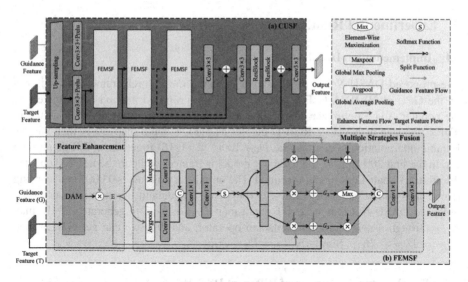

Fig. 3. The structures of CUSF (a) and FEMSF (b).

global information to assign weights to each fusion strategy. The assigned weight in an element-wise way multiply E and then its output in an element-wise way is added with G to obtain the modulated guidance features G_i, $i = 1, 2, 3$. Then three different fusion strategies for T and G_i are performed including element-wise addition, element-wise multiplication and element-wise maximization operations. Element-wise maximization highlights the most prominent information between different modalities, which is beneficial for boundary information enhancement. Element-wise multiplication represents "AND" operation, that is, when both modal signals are strong, a strong signal will be outputted. A weak signal from another side will weaken the output information, which has a strong effect on removing textures. Element-wise addition is used to supplement the information loss problem caused by the above two fusion strategies. Three fusion strategies are concatenated along the channel dimension, and then the output is adaptively weighted with the convolutional layer.

2.4 Loss Function

The $L1$ norm is utilized to regularize the D_{SR}, L_{SR} and H_{SR} prediction learning of the proposed DASU-Net, which can be written as $Loss = \frac{1}{N} \sum_{i=1}^{N} \|D_{SR,i} - D_{GT,i}\|_1 + \xi_1 \frac{1}{N} \sum_{i=1}^{N} \|L_{SR,i} - L_{GT,i}\|_1 + \xi_2 \frac{1}{N} \sum_{i=1}^{N} \|H_{SR,i} - H_{GT,i}\|_1$. Here, i denote the i-th pixel of the image, the total pixel number of the image is N. D_{GT}, L_{GT} and H_{GT} are Ground Truth (GT) depth map, GT HF map and GT LF map. ξ_1 and ξ_2 are used to make trade-off between the reconstruction of various sub-networks.

3   Experimental Results

3.1   Implementation Details

The proposed DASU-Net is implemented by using the Pytorch framework. It is run on the GPU of NVIDIA GeForce RTX 3090. The proposed method is optimized by the popular Adam optimizer. During the training phase, we randomly crop image to 256×256 size of NYU-v2 RGB-D dataset as the GT depth map D_{GT}, then we use Bicubic interpolation to degrade the D_{GT} by $m \subseteq (1, 16)$ times as the LR depth map D_l. In the first 200 epochs, the fixed $8\times$ model is trained, while in the last 200 epochs, the arbitrary-scale DSR model is trained. The initial learning rate is 0.0001 and learning rate decays with a factor of 0.2 every 60 epochs. The default iteration number of the proposed DASU-Net is $n = 2$. GT depth map D_{GT} is used to generate GT HF map H_{GT} through 4-neighborhood Laplacian operator, and then we can get GT LF map $L_{GT} = D_{GT} - H_{GT}$. Similarly, $L_l = D_l - H_l$.

3.2   The Quality Comparison of Different DSR Methods

We train the proposed method by using the first 1000 pairs RGB-D images in the NYU-v2 dataset, and test it by using the last 499 pairs RGB-D images from NYU-v2 dataset, 30 pairs RGB-D images from Middlebury dataset, and 6 pairs RGB-D images from Lu dataset to evaluate the performance of the proposed DASU-Net. Structural Similarity (SSIM), Peak Signal-to-Noise Ratio (PSNR) and Root Mean Square Error (RMSE) are used to assess the performance of the proposed DASU-Net at arbitrary-scale upsampling. A lot of excellent methods are used to compare with our methods, including GF [3], DMSG [4], FDKN [5], DKN [5], DCTNet [11], AHMF [12], DJFR [7], MADUNet [13], JIIF [8] and GeoDSR [9]. In some tables, black and underlined fonts indicate the best performance and the second best performance respectively.

As shown in Table 1, our DASU-Net has a significant advantage at both $8\times$ and $16\times$ up-sampling, except for ranking fourth at $4\times$ up-sampling on the NYU-v2 dataset. The Middlebury and Lu RGB-D datasets are used to validate generalization ability. Testing on the Middlebury and Lu RGB-D datasets, proposed DASU-Net ranks first in terms of RMSE values, which further indicating that our proposed method is state-of-the-art. We provide the average performance of the $4\times$, $8\times$ and $16\times$ up-sampling in the last three columns of Table 1. At three scale factors, the average RMSE values of DASU-Net are decreased by 0.05, 0.10 and 0.29 compared to the second ranked GeoDSR [9] respectively. In order to further demonstrate the progressiveness of our method in arbitrary-scale DSR, we show in Table 2 the RMSE values up-sampling on non-integer factors for three datasets. The proposed DASU-Net achieves state-of-the-art performance in all up-sampling factors, which indicates that our method has extremely strong generalization ability. And the parameter number of GeoDSR [9] is three times higher than ours. It is worth noting that $17.05\times$ is the up-sampling factor outside of our training setting $(1 - 16\times)$.

In order to intuitively demonstrate the advantages of our method, we show the 1012-nd visualization result and error map of the NYU-v2 dataset in $8\times$ up-sampling, as shown in Fig. 4. Due to the limited feature mapping capabilities of traditional methods

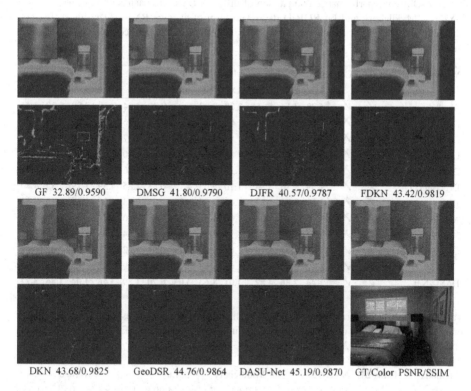

Fig. 4. Comparison of 8× up-sampling visualization, error map results and color image of 1012-nd image of the NYU-v2 RGB-D dataset. The brighter the area in the error map, the greater the pixel error.

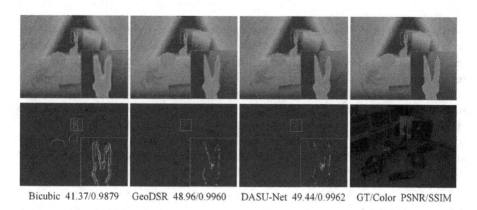

Fig. 5. Comparison of 3.75× up-sampling visualization, error map results and color image of 1-st image of the Lu RGB-D dataset. The brighter the area in the error map, the greater the pixel error.

Table 1. Objective performance comparison of different DSR approaches on integer factors on NYU-v2, Middlebury and Lu RGB-D dataset in term of average RMSE (The lower, the better). **Intp** and **Cont** indicate whether networks have interpretability and continuous up-sampling function.

Methods	Intp	Cont	NYU-v2			Middlebury			Lu			Average		
			4×	8×	16×	4×	8×	16×	4×	8×	16×	4×	8×	16×
Bicubic	✓	✓	4.28	7.14	11.58	2.28	3.98	6.37	2.42	4.54	7.38	2.99	5.22	8.44
DJFR [7]	✗	✗	2.80	5.33	9.46	1.68	3.24	5.62	1.65	3.96	6.75	2.04	4.18	7.28
DMSG [4]	✗	✗	3.02	5.38	9.17	1.88	3.45	6.28	2.30	4.17	7.22	2.40	4.33	7.17
FDKN [5]	✗	✗	1.86	3.58	6.96	1.08	2.17	4.50	0.82	2.10	5.05	1.25	2.62	5.50
DKN [5]	✗	✗	1.62	3.26	6.51	1.23	2.12	4.24	0.96	2.16	5.11	1.27	2.51	5.29
DCTNet [11]	✗	✗	1.59	3.16	5.84	1.10	2.05	4.19	0.88	1.85	4.39	1.19	2.35	4.80
JIIF [8]	✗	✗	**1.37**	2.76	5.27	1.09	1.82	3.31	0.85	1.73	4.16	1.10	2.10	4.24
AHMF [12]	✗	✗	<u>1.40</u>	2.89	5.64	1.07	**1.63**	3.14	0.88	1.66	<u>3.71</u>	1.11	2.06	4.16
MADUNet [13]	✓	✗	1.51	3.02	6.23	1.15	1.69	3.23	0.90	1.74	3.86	1.18	2.15	4.44
GeoDSR-small [9]	✗	✓	1.48	2.73	5.10	<u>1.04</u>	1.73	3.19	0.82	1.62	4.11	1.11	2.02	4.13
GeoDSR [9]	✗	✓	1.42	<u>2.62</u>	<u>4.86</u>	<u>1.04</u>	<u>1.68</u>	<u>3.10</u>	<u>0.81</u>	<u>1.59</u>	3.92	<u>1.09</u>	<u>1.96</u>	<u>3.96</u>
DASU-Net	✓	✓	1.44	**2.56**	**4.82**	**1.01**	**1.63**	**3.05**	**0.70**	**1.39**	**3.46**	**1.04**	**1.86**	**3.77**

Table 2. Objective performance comparison of different DSR approaches on non-integer factors on NYU-v2, Middlebury and Lu RGB-D dataset in term of average RMSE (The lower, the better) and Parameters (Paras, $1M = 10^6$).

Methods	Paras(M)	NYU-v2		Middlebury		Lu		Average	
		14.60×	17.05×	14.60×	17.05×	14.60×	17.05×	14.60×	17.05×
Bicubic	-	10.90	11.93	6.00	6.52	7.20	7.96	8.03	8.80
GeoDSR [9]	5.52	<u>4.56</u>	<u>5.18</u>	<u>2.77</u>	<u>3.17</u>	<u>3.54</u>	<u>4.24</u>	<u>3.62</u>	<u>4.19</u>
DASU-Net	1.70	**4.51**	**5.12**	**2.75**	**3.16**	**3.21**	**3.88**	**3.49**	**4.05**

such as GF [3], the reconstructed depth map exhibits excessively surface smooth, and its error map are brightest. Although DMSG [4] and DJFR [7] rely on deep learning, simple structures of their network result in structure distortion of the reconstructed results. GeoDSR [9] uses a similar to the 'AND' operation to fuse color-depth features, which has a good effect on texture removal, but it will inevitably weaken the reconstruction of boundary information, resulting in significant pixel errors around the reconstructed boundaries. We also show the results of 3.75× up-sampling from Lu RGB-D dataset in Fig. 5. From this figure, it can be found that the reconstruction results of ours are closer to GT image, since the proposed DASU-Net inherit the natural structure of optimized model.

3.3　Ablation Study

In this section, we train various variants on the NYU-v2 RGB-D dataset and use the Lu RGB-D dataset as the validation dataset to verify the effectiveness of the proposed method. To demonstrate the effectiveness of CUSF module, we sequentially perform grid up-sampling, concatenation of channel dimensions, and several ResBlocks to replace CUSF, and the results are shown in Table 3. In addition, MSRM is replaced by several ResBlocks to learn proximal operators, and its performance is shown in Table 3.

Table 3. The ablation study of DASU-Net in the Lu RGB-D dataset (The lower the RMSE, the better, and the higher the PSNR and SSIM, the better).

CUSF	MSRM	RMSE		PSNR		SSIM	
		$14.60\times$	$17.05\times$	$14.60\times$	$17.05\times$	$14.60\times$	$17.05\times$
✗	✓	3.50	4.08	36.49	35.31	0.9819	0.9784
✓	✗	3.24	3.89	36.69	35.59	0.9829	0.9792
✓	✓	**3.21**	**3.88**	**37.98**	**36.51**	**0.9837**	**0.9800**

We test two up-sampling factors in terms of PSNR and SSIM metrics, one within the training distribution ($14.60\times$) and one outside the training distribution ($17.05\times$). When both CUSF and MSRM exist in the proposed network, the network performance reaches its better level. If any module is replaced by ResBlock, the performances significantly decrease, especially for CUSF module. From the above results, it can conclude that our proposed CUSF and MSRM play different roles in performance improvement.

4   Conclusion

To alleviate network interpretability, texture-copying and arbitrary-scale super-resolution problems in DSR task, we propose an optimization model, and unfold it into DASU-Net for arbitrary-scale DSR. In addition, with the help of CUSF module, DASU-Net can effectively aggregate dual-modality features and adapt to arbitrary-scale up-sampling for DSR task. Specifically, CUSF module first effectively narrows the gap between the dual-modality, and a multiple strategy fusion component is introduced to effectively highlight and complement the advantages and disadvantages of different fusion strategies. We have demonstrated the rationality of our method, and numerous experiments have shown that the depth map reconstruction performance of our method is significantly improved for arbitrary-scale DSR task.

Acknowledgements. This work was supported by National Natural Science Foundation of China Youth Science Foundation Project (No.62202323), Fundamental Research Program of Shanxi Province (No.202103021223284), Taiyuan University of Science and Technology Scientific Research Initial Funding (No.20192023, No.20192055), Graduate Education Innovation Project of Taiyuan University of Science and Technology in 2022 (SY2022027), National Natural Science Foundation of China (No.62072325).

References

1. Barron, J.T., Poole, B.: The fast bilateral solver. In: Leibe, B., Matas, J., Sebe, N., Welling, M. (eds.) Computer Vision ?C ECCV 2016. pp. 617C632. Springer International Publishing, Cham (2016)

2. Diebel, J., Thrun, S.: An application of markov random fields to range sensing. In: Proceedings of the 18th International Conference on Neural Information Processing Systems. p. 291298. NIPS05, MIT Press, Cambridge, MA, USA (2005)

3. He, K., Sun, J., Tang, X.: Guided image filtering. IEEE Transactions on Pattern Analysis and Machine Intelligence 35(6), 1397 C1409 (2013). https://doi.org/10.1109/TPAMI.2012.213

4. Hui, T.W., Loy, C.C., Tang, X.: Depth map super-resolution by deep multi-scale guidance. In: Leibe, B., Matas, J., Sebe, N., Welling, M. (eds.) Computer Vision C ECCV 2016. pp. 353C369. Springer International Publishing, Cham (2016)

5. Kim, B., Ponce, J., Ham, B.: Deformable kernel networks for joint image filtering. International Journal of Computer Vision 129(2), 579 C 600 (2021), https://doi.org/10.1007/s11263-020-01386-z

6. Kopf, J., Cohen, M.F., Lischinski, D., Uyttendaele, M.: Joint bilateral upsampling. In: ACM SIGGRAPH 2007 Papers. SIGGRAPH 07, Association for Computing Machinery, New York, NY, USA (2007)

7. Li, Y., Huang, J.B., Ahuja, N., Yang, M.H.: Joint image filtering with deep convolutional networks. IEEE transactions on pattern analysis and machine intelligence 41(8), 1909C1923 (2019)

8. Tang, J., Chen, X., Zeng, G.: Joint implicit image function for guided depth super-resolution. In: Proceedings of the 29th ACM International Conference on Multi-media. ACM (oct 2021). DOI: https://doi.org/10.1145/3474085.3475584

9. Wang, X., Chen, X., Ni, B., Tong, Z., Wang, H.: Learning continuous depth representation via geometric spatial aggregator (2022), https://doi.org/10.48550/arXiv.2212.03499

10. Zhang, Z., Zheng, H., Hong, R., Xu, M., Yan, S., Wang, M.: Deep color consistent network for low-light image enhancement. In: 2022 IEEE/CVF Conference on Computer Vision and Pattern Recognition (CVPR). pp. 1889C1898 (2022). https://doi.org/10.1109/CVPR52688.2022.00194

11. Zhao, Z., Zhang, J., Xu, S., Lin, Z., Pfister, H.: Discrete cosine transform network for guided depth map super-resolution. In: 2022 IEEE/CVF Conference on Computer Vision and Pattern Recognition (CVPR). pp. 5687C5697 (2022). DOI: https://doi.org/10.1109/CVPR52688.2022.00561

12. Zhong, Z., Liu, X., Jiang, J., Zhao, D., Chen, Z., Ji, X.: High-resolution depth maps imaging via attention-based hierarchical multi-modal fusion. IEEE Transactions on Image Processing 31, 648C663 (2022). DOI: https://doi.org/10.1109/TIP.2021.3131041

13. Zhou, M., Yan, K., Pan, J., Ren, W., Xie, Q., Cao, X.: Memory-augmented Deep Unfolding Network for Guided Image Super-resolution. arXiv e-prints arXiv:2203.04960 (Feb 2022)

SSDD-Net: A Lightweight and Efficient Deep Learning Model for Steel Surface Defect Detection

Zhaoguo Li[1,2,3] , Xiumei Wei[1,2,3], and Xuesong Jiang[1,2,3,4]([✉])

[1] Key Laboratory of Computing Power Network and Information Security, Ministry of Education, Shandong Computer Science Center, Qilu University of Technology (Shandong Academy of Sciences), Jinan, China
`jxs@qlu.edu.cn`
[2] Shandong Engineering Research Center of Big Data Applied Technology, Faculty of Computer Science and Technology, Qilu University of Technology (Shandong Academy of Sciences), Jinan, China
[3] Shandong Provincial Key Laboratory of Computer Networks, Shandong Fundamental Research Center for Computer Science, Jinan, China
[4] State Key Laboratory of High-end Server & Storage Technology, Jinan, China

Abstract. Industrial defect detection is a hot topic in the computer vision field. At the same time, it is hard work because of the complex features and various categories of industrial defects. To solve the above problem, this paper introduces a lightweight and efficient deep learning model (SSDD-Net) for steel surface defect detection. At the same time, in order to improve the efficiency of model training and inference in the XPU distributed computing environment, parallel computing is introduced in this paper. First, a light multiscale feature extraction module (LMFE) is designed to enhance the model's ability to extract features. The LMFE module employs three branches with different receptive fields to extract multiscale features. Second, a simple effective feature fusion network (SEFF) is introduced to be the neck network of the SSDD-Net to achieve efficient feature fusion. Extensive experiments are conducted on a steel surface defect detection dataset, NEU-DET, to verify the effectiveness of the designed modules and proposed model. And the experimental results demonstrate that the designed modules are effective. Compared with other SOTA object detection models, the proposed model obtains optimal performance (73.73% in mAP@0.5) while keeping a small number of parameters (3.79M).

Keywords: deep learning · steel surface defect detection · feature extraction · feature fusion

1 Introduction

Steel has wide use in industry, so the surface quality inspection of steel is important work. However, the surface defect of steel has complex features and various categories. Using manual quality inspection, it is difficult to find out the steel surface defects and distinguish the defect categories. Furthermore, manual quality inspection would waste a lot of manpower and be very expensive. Studying a method to replace the manual quality inspection is a crucial and important problem.

In recent years, deep learning has been developed rapidly [1]. The object detection models based on deep learning appear endless [2]. They are divided into two kinds: two-stage detectors and one-stage detectors. The well-known two-stage detectors include R-CNN [3], Fast R-CNN [4], and Faster R-CNN [5]. And the classical one-stage detectors include SSD [6], RetinaNet [7], and the YOLO series [8–13]. Because of their high accuracy, deep learning methods are widely used in the detection of steel surface defects. Wang et al. [14] proposed a few-shot defect detection framework for Steel Surface Defect Detection. Hatab et al. [15] used YOLO network to achieve steel surface defect detection. In [16], Wang et al. proposed a real-time steel surface defect detection technology based on the YOLO-v5 detection network. In [17], a YOLOv4 defect detection algorithm based on weighted fusion was proposed. It uses a GAN network to generate a mask map in real time, and performs weighted fusion of the feature maps in YOLOv4. Metal surface defect detection has always been an important branch of target detection. Deng et al. [18] used the Cascade-YOLOv4 (C-YOLOv4) network model to achieve iron surface crack detection. Kou et al. [19] developed an end-to-end detection model based on YOLO-V3 to detect steel surface defect. However, due to the complex multi-scale characteristics of steel surface defects, it is difficult for the existing steel surface defect detection models to achieve good detection results. Moreover, the above models often do not handle the multi-scale feature fusion problem of steel surface defects well.

Aiming at the detection of steel surface defects, this paper proposes a lightweight and efficient deep learning model (SSDD-Net). The steel surface defects are multiscale. To enhance the multiscale feature extraction ability of the model, this paper designs a light multiscale feature extraction module (LMFE). The LMFE uses three different branches with multiscale convolution kernels to extract multiscale features. To achieve effective feature fusion, this paper designs a simple effective feature fusion network (SEFF). The SEFF network causes the middle feature map to interact with the top and bottom feature maps, which achieves the feature fusion of stronger location information and stronger semantic information and benefits the detection performance of the model.

The main contributions of this paper are as follows:

– A lightweight and efficient deep learning model (SSDD-Net) is proposed to detect steel surface defects. Extensive experiments demonstrate that the proposed model is superior to other SOTA object detectors in the detection of steel surface defects.
– A light multiscale feature extraction module (LMFE) with three different branches is designed. The LMFE module employs multiscale receptive fields to enhance the model's ability to extract multiscale features.
– A simple effective feature fusion network (SEFF) is introduced to be the neck network of the proposed model and achieve efficient feature fusion operation.

Several experiments prove that the SEFF network has better performance than other well-known neck networks in the task of steel surface defect detection.

The rest of this paper is organized as follows. The proposed method and modules are presented in Sect. 2. Datasets used in the experiments, performance evaluation metrics, and extensive experimental results are provided in Sect. 3. Finally, Sect. 4 concludes the paper.

2 Methods

In this section, the proposed SSDD-Net is introduced in detail. The overview diagram of the SSDD-Net is shown in Fig. 1. The SSDD-Net is composed of three parts: the backbone network, the neck network, and the head network. The backbone network is used to extract features and pass the obtained multiple feature maps (P3, P4, and P5) to the neck network. The neck network is responsible for fusing features. And the head network outputs the predicted values, which include locations, confidence, and classes.

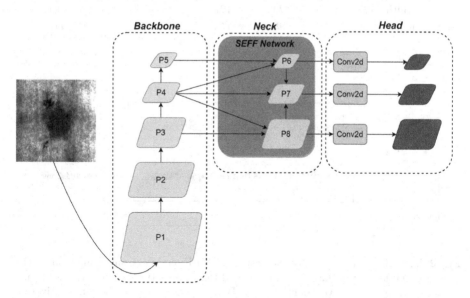

Fig. 1. The overview diagram of the proposed SSDD-Net. Conv2d is the convolutional layer. P represents the feature map.

2.1 LMFE: Light Multiscale Feature Extraction Module

Steel surface defects have multiscale sizes, the average sizes of each defect category in the steel surface defect detection dataset (NEU-DET [20]) are shown in Table 1. To enhance the model's ability to extract multiscale features, this paper designs a light multiscale feature extraction module (LMFE) as shown in Fig. 2. The LMFE has three different branches, and the receptive fields of the three branches are multiscale. The first branch and second branch first use a 1×1 Conv module to reduce channel's amount. Then, the first branch uses three cascade 3×3 depthwise convolutional layer [21] to extract features. The purpose of the residual connection operation [22] is to enhance the memory capability of the LMFE module for the features of the front layers. Unlike the first branch, the second branch uses three cascade 5×5 depthwise convolutional layer to extract features. The third branch only has a 1×1 Conv module. So, the outputs of the three branches are multiscale features. Then, the outputs of the three branches are spliced together. The designed LMFE module employs three multiscale branches to extract features, which can effectively enhance the model's ability to extract multiscale features.

Table 1. The average size of each category in the NEU-DET dataset.

	crazing	patches	inclusion	pitted_surface	rolled-in_scale	scratches
Average width	125	56	29	127	73	61
Average height	75	83	85	173	78	115

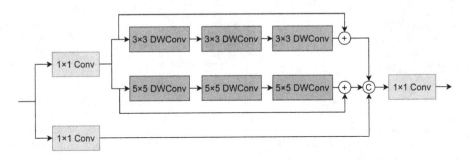

Fig. 2. The structure diagram of the designed LMFE module. Conv is made up of the convolutional layer, the BatchNorm layer, and the SiLU activation function. DWConv includes a depthwise convolutional layer, a BatchNorm layer, and the SiLU activation function.

2.2   SEFF: Simple Effective Feature Fusion Network

The stronger semantic information is beneficial for predicting categories of defects, and the powerful location information is proper for predicting where the defects are. Effectively fusing the stronger semantic information and location information is helpful to improve the detection performance of the model. To achieve effective feature fusion in steel defect detection, this paper builds a novel feature fusion method, as shown in Fig. 3, namely the SEFF network. Among the feature maps generated in the backbone network, the larger feature map has stronger location information, and the smaller feature map has more semantic information. So, the middle feature map is their compromise, it has stronger location information than the smaller feature map and more semantic information than the larger feature map. The SEFF network uses the middle feature maps as a medium for large feature maps and small feature maps to transmit information. The middle feature map P4 is sent to the upper branch and the lower branch for feature fusion, respectively. The SEFF uses P4 to transmit the stronger location information to the powerful semantic information of the upper branch and the stronger semantic information to the powerful location information of the lower branch. The SEFF achieves the feature fusion of stronger location information and stronger semantic information in both the upper and lower branches. In addition, information is transferred top-to-middle and bottom-to-middle in the vertical direction. The top-to-middle path passes the stronger semantic information to the middle feature map, and the bottom-to-middle path passes the stronger location information to the middle feature map. So, the vertical direction also achieves the feature fusion of stronger location information and stronger semantic information. At the same time, P4 is sent to the vertical direction of the SEFF for feature fusion. The overview of the SEFF network is presented as follows:

$$P6 = Concat(P5, Downsample(P4)) \tag{1}$$

$$P8 = Concat(Upsample(P4), P3) \tag{2}$$

$$P7 = Concat(Upsample(P6), P4, Downsample(P8)) \tag{3}$$

where the *Upsample* operation represents a 1×1 Conv and a nearest neighbor interpolation operation, and the *Downsample* operation uses a 3×3 Conv with a stride of 2.

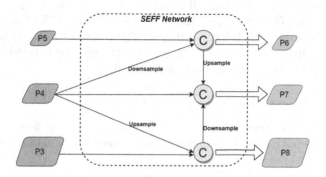

Fig. 3. The overview diagram of the designed SEFF network.

2.3 SSDD-Net

This paper employs the designed LMFE module and the SEFF network to build a lightweight and efficient deep learning model (SSDD-Net) for steel surface defect detection. The structure diagram of the SSDD-Net is shown in Fig. 4. First, the backbone network is composed of the Conv module with a stride of 2, the designed LMFE module, and the SPPF module of the YOLOv5 model [12]. The Conv module with a stride of 2 is responsible for the downsampling operation. The LMFE module is used to extract features. And the function of the SPPF module is to fuse multiscale information. Second, the neck network is the designed SEFF, which can achieve effective feature fusion. Finally, the YOLO head is used as the head network of the SSDD-Net.

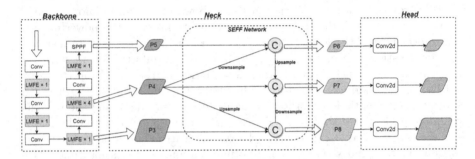

Fig. 4. The structure of the proposed SSDD-Net. Conv includes convolutional layer, BatchNorm layer, and SiLU activation function. Conv2d represents the convolutional layer.

3 Experiments and Analysis

3.1 Implementation Details

All experiments are conducted on an NVIDIA A100 GPU, the Python version is 3.8.15, and the PyTorch version is 1.12.1. Before being fed into the model, the images are resized to 640 × 640. All models are trained for 500 epochs in total. The optimizer is SGD, and the learning rate is set to 0.01. The momentum is set to 0.937, and the weight decay is 0.0005. The training batch size is set to 16, and the test batch size is set to 32. Moreover, the proposed model accelerates computational efficiency through parallel computing.

3.2 Evaluation Metrics

To verify the effectiveness of the designed modules and the proposed SSDD-Net, this paper uses mAP and the number of parameters as evaluation metrics. The specific equations are shown as follows:

$$P = \frac{TP}{TP + FP} \tag{4}$$

$$R = \frac{TP}{TP + FN} \tag{5}$$

$$AP = \int_0^1 P(R)dR \times 100\% \tag{6}$$

$$mAP = \frac{\sum_{i=1}^{N} AP}{N} \tag{7}$$

where TP is true positive, which means the model predicts a defect and the prediction is correct. FP represents false positive, which means that the model predicts a defect but the prediction is incorrect. FN stands for false negative, which means that the defects are not detected. AP denotes the detection performance of the model for one category, and mAP represents the detection performance of the model for all categories.

3.3 Dataset

To verify the effectiveness of the proposed modules and the model, this study uses the steel surface defect detection dataset, NEU-DET [20], to conduct experiments. Some examples of steel surface defects are shown in Fig. 5. There are 6 categories in the dataset, namely 'crazing', 'patches', 'inclusion', 'pitted_surface', 'rolled-in_scale', 'scratches'. The dataset has 1800 images in total. We divided the dataset, and 1448 images are in the training dataset and 352 images are in the test dataset.

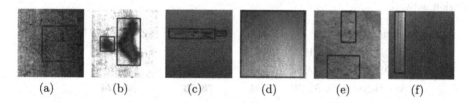

(a)   (b)   (c)   (d)   (e)   (f)

Fig. 5. Some examples of the NEU-DET dataset. (a)–(f) is 'crazing', 'patches', 'inclusion', 'pitted_surface', 'rolled-in_scale', 'scratches', respectively.

3.4 Ablation Studies

To verify the effectiveness of the designed LMFE module and SEFF network, several ablation experiments are conducted on the NEU-DET dataset. The experiments are set up in two groups. First, the STDD-Net with no neck is trained on the dataset, which is called Model A. Second, the designed SEFF network is integrated into Model A to be the neck network of Model A. The model is called Model B. The experimental results are shown in Table 2. From Table 2, it can be seen that the Model A obtains 71.06% in mAP@0.5. This is enough to prove the effectiveness of the designed LMFE module because the Model A is composed of LMFE modules. And the effectiveness of the LMFE module is due to its multiscale receptive fields. Furthermore, we can see from Table 2, the recall increases 2.87%, the mAP@0.5 increases 2.67%, and the mAP@0.5:0.95 increases 0.87% after integrating the designed SEFF network into Model A. Therefore, the designed SEFF network is effective as the neck network of the STDD-Net to conduct feature fusion.

Table 2. The ablation experiments on the NEU-DET dataset.

Model	Model Structure	Recall (%)	mAP@0.5 (%)	mAP@0.5:0.95 (%)
A	SSDD-Net with no neck	68.31	71.06	37.37
B (SSDD-Net)	A + SEFF	71.18(+2.87)	73.73(+2.67)	38.24(+0.87)

To further verify the effectiveness of the SEFF network, we set up three groups of experiments on the NEU-DET dataset. The three models are SSDD-Net with FPN [23], SSDD-Net with PANet [24], and SSDD-Net with the proposed SEFF, respectively. The experimental results are shown in Table 3, and it can be seen that the SEFF network obtains the best performance. Therefore, the proposed SEFF network is more superior than other neck networks.

Table 3. The comparison experiments between our SEFF network and other feature fusion networks.

Model	Model Structure	Recall (%)	mAP@0.5 (%)	mAP@0.5:0.95 (%)
A	SSDD-Net with FPN	68.62	73.05	37.34
B	SSDD-Net with PANet	67.56	72.39	37.88
C	SSDD-Net with SEFF	**71.18**	**73.73**	**38.24**

3.5 Comparison with Other SOTA Methods

Extensive comparison experiments are conducted on the NEU-DET dataset [20] to verify the effectiveness of the proposed SSDD-Net. The chosen models to be compared include lightweight models, the YOLOv3-tiny [11], and the YOLOv5s [12], and heavy models, YOLOv3 [11], YOLOv3-spp, YOLOv5m [12], and YOLOv5l [12]. The experimental results are shown in Table 4 and Fig. 6. From Table 4, it can be seen that our SSDD-Net obtains optimal performance while keeping the smallest number of parameters. It achieves 73.73% mAP@0.5 and keeps 3.79M parameters. Compared with the other SOTA models, the proposed SSDD-Net is lighter and stronger. Although compared with the YOLOv5l, the SSDD-Net still obtains better performance. At the same time, the parameters of the SSDD-Net are much smaller than that of YOLOv5l. In conclusion, the proposed SSDD-Net is a lightweight and efficient deep learning model for steel surface defect detection.

Table 4. The comparison experiments with other models on the NEU-DET dataset.

Model	Neck	Param (M)	Recall (%)	mAP@0.5 (%)	mAP@0.5:0.95 (%)
YOLOv3	FPN	61.55	67.13	68.37	34.67
YOLOv3-tiny	FPN	8.68	62.82	53.3	22.39
YOLOv3-spp	FPN	62.6	65.32	69.94	36.1
YOLOv5s	PANet	7.04	69.24	70	36.41
YOLOv5m	PANet	20.89	67.35	70.93	36.99
YOLOv5l	PANet	46.17	66.79	71.91	37.65
SSDD-Net	**SEFF**	**3.79**	**71.18**	**73.73**	**38.24**

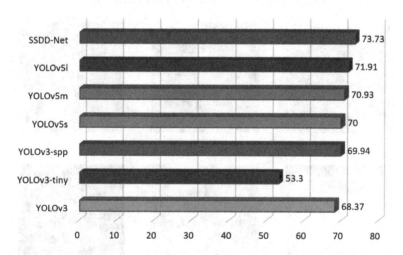

Fig. 6. The mAP@0.5 comparison of our SSDD-Net and other SOTA methods.

3.6 Comprehensive Performance of SSDD-Net

The mAP@0.5 curve of our SSDD-Net is shown in Fig. 7. It can be clearly seen that accuracy increases as the epoch increases. Some detection results of the SSDD-Net are shown in Fig. 8, and we can find that the SSDD-Net has good detection performance for the steel surface defects. The reason is that the SSDD-Net employs the designed LMFE module to enhance its multiscale feature extraction ability and uses the proposed SEFF as the neck network to achieve effective feature fusion.

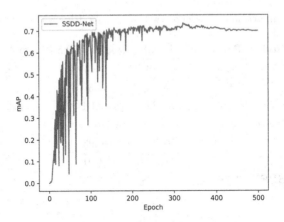

Fig. 7. The mAP@0.5 curve of our SSDD-Net.

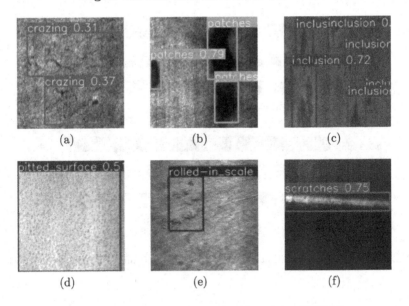

Fig. 8. Some detection results of the proposed SSDD-Net.

4 Conclusion

This paper proposes a lightweight and efficient deep learning model for steel surface defect detection. First, to enhance the model's ability to extract multiscale features, this paper designs a light multiscale feature extraction module (LMFE), which employs three branches with different receptive fields to extract multiscale features. Second, this paper introduces the simple effective feature fusion network (SEFF) to achieve effective feature fusion. Extensive experiments are conducted on the NEU-DET dataset to verify the effectiveness of the designed modules and the proposed SSDD-Net. And the experimental results show that the designed modules are effective, and the proposed SSDD-Net obtains optimal performance while keeping the smallest number of parameters. In the future, we will further optimize the model's structure to make it have better performance.

Acknowledgements. This work was supported by the project ZR2022LZH017 supported by Shandong Provincial Natural Science Foundation.

References

1. LeCun, Y., Bengio, Y., Hinton, G.: Deep learning. nature **521**(7553), 436–444 (2015)
2. Zou, Z., Shi, Z., Guo, Y., Ye, J.: Object detection in 20 years: a survey. arXiv preprint arXiv:1905.05055 (2019)
3. Girshick, R., Donahue, J., Darrell, T., Malik, J.: Rich feature hierarchies for accurate object detection and semantic segmentation. In: Proceedings of the IEEE Conference on Computer Vision and Pattern Recognition, pp. 580–587 (2014)
4. Girshick, R.: Fast R-CNN. In: Proceedings of the IEEE International Conference on Computer Vision, pp. 1440–1448 (2015)
5. Ren, S., He, K., Girshick, R., Sun, J.: Faster R-CNN: towards real-time object detection with region proposal networks. In: Advances in Neural Information Processing Systems, vol. 28 (2015)
6. Liu, W., et al.: SSD: single shot multibox detector. In: Leibe, B., Matas, J., Sebe, N., Welling, M. (eds.) ECCV 2016. LNCS, vol. 9905, pp. 21–37. Springer, Cham (2016). https://doi.org/10.1007/978-3-319-46448-0_2
7. Lin, T.Y., Goyal, P., Girshick, R., He, K., Dollár, P.: Focal loss for dense object detection. In: Proceedings of the IEEE International Conference on Computer Vision, pp. 2980–2988 (2017)
8. Redmon, J., Divvala, S., Girshick, R., Farhadi, A.: You only look once: unified, real-time object detection. In: Proceedings of the IEEE Conference on Computer Vision and Pattern Recognition, pp. 779–788 (2016)
9. Redmon, J., Farhadi, A.: YOLO9000: better, faster, stronger. In: Proceedings of the IEEE Conference on Computer Vision and Pattern Recognition, pp. 7263–7271 (2017)
10. Bochkovskiy, A., Wang, C.Y., Liao, H.Y.M.: YOLOV4: optimal speed and accuracy of object detection. arXiv preprint arXiv:2004.10934 (2020)
11. Redmon, J., Farhadi, A.: Yolov3: An incremental improvement. arXiv preprint arXiv:1804.02767 (2018)

12. Ultralytics: YOLOv5 v6.2. https://github.com/ultralytics/yolov5. Accessed 17 Aug 2022
13. Wang, C.Y., Bochkovskiy, A., Liao, H.Y.M.: YOLOv7: trainable bag-of-freebies sets new state-of-the-art for real-time object detectors. arXiv preprint arXiv:2207.02696 (2022)
14. Wang, H., Li, Z., Wang, H.: Few-shot steel surface defect detection. IEEE Trans. Instrum. Meas. **71**, 1–12 (2021)
15. Hatab, M., Malekmohamadi, H., Amira, A.: Surface defect detection using YOLO network. In: Arai, K., Kapoor, S., Bhatia, R. (eds.) IntelliSys 2020. AISC, vol. 1250, pp. 505–515. Springer, Cham (2021). https://doi.org/10.1007/978-3-030-55180-3_37
16. Wang, L., Liu, X., Ma, J., Su, W., Li, H.: Real-time steel surface defect detection with improved multi-scale YOLO-v5. Processes **11**(5), 1357 (2023)
17. Wang, C., Xu, J., Liang, X., Yin, D.: Metal surface defect detection based on weighted fusion. In: 2020 International Conference on Virtual Reality and Visualization (ICVRV), pp. 179–184. IEEE (2020)
18. Deng, H., Cheng, J., Liu, T., Cheng, B., Sun, Z.: Research on iron surface crack detection algorithm based on improved YOLOv4 network. J. Phys: Conf. Ser. **1631**, 012081 (2020)
19. Kou, X., Liu, S., Cheng, K., Qian, Y.: Development of a YOLO-v3-based model for detecting defects on steel strip surface. Measurement **182**, 109454 (2021)
20. He, Y., Song, K., Meng, Q., Yan, Y.: An end-to-end steel surface defect detection approach via fusing multiple hierarchical features. IEEE Trans. Instrum. Meas. **69**(4), 1493–1504 (2019)
21. Howard, A.G., et al.: MobileNets: efficient convolutional neural networks for mobile vision applications. arXiv preprint arXiv:1704.04861 (2017)
22. He, K., Zhang, X., Ren, S., Sun, J.: Deep residual learning for image recognition. In: Proceedings of the IEEE Conference on Computer Vision and Pattern Recognition, pp. 770–778 (2016)
23. Lin, T.Y., Dollár, P., Girshick, R., He, K., Hariharan, B., Belongie, S.: Feature pyramid networks for object detection. In: Proceedings of the IEEE Conference on Computer Vision and Pattern Recognition, pp. 2117–2125 (2017)
24. Liu, S., Qi, L., Qin, H., Shi, J., Jia, J.: Path aggregation network for instance segmentation. In: Proceedings of the IEEE Conference on Computer Vision and Pattern Recognition, pp. 8759–8768 (2018)

Effective Small Ship Detection with Enhanced-YOLOv7

Jun Li[1], Ning Ding[2], Chen Gong[1], Zhong Jin[1], and Guangyu Li[1]

[1] Key Laboratory of Intelligent Perception and Systems for High -Dimensional Information of Ministry of Education, School of Computer Science and Engineering, Nanjing University of Science and Technology, Nanjing, China
{jun_li,chen.gong,zhongjin,guangyu.li2017}@njust.edu.cn
[2] National Ocean Technology Center, Tianjin, China

Abstract. Small ship detection is widely used in marine environment monitoring, military applications and so on, and it has gained increasing attentions both in industry and academia. In this paper, we propose an effective small ship detection algorithm with enhanced-YOLOv7. Specifically, to reduce the feature loss of small ships and the impact of marine environment, we firstly design a small object-aware feature extraction module by considering both small-scale receptive fields and multi-branch residual structures. In addition, we propose a small object-friendly scale-insensitive regression scheme, to strengthen the contributions of both bounding box distance and difficult samples on regression loss as well as further increase learning efficiency of small ship detection. Moreover, based on the formulated penalty model, we design a geometric constraint-based Non-Maximum Suppression (NMS) method, to effectively decrease small ship detection omission rate. Finally, extensive experiments are implemented, and corresponding results confirm the effectiveness of the proposed algorithm.

Keywords: Small Ship · Improved YOLOv7 Network · Efficiency Detection

1 Introduction

The field of deep learning-based object detection algorithms has undergone rapid development [2,4,7,9,11,12,15,17], leading to significant advancements in ship detection. This progress has positioned deep learning-based ship detection as a critical pillar in the realm of intelligent maritime applications [14]. These applications encompass various domains, such as maritime rescue and autonomous sailing. The majority of deep learning-based ship detectors primarily focus on conventionally sized ships. However, there has been relatively limited research conducted on small ships. Due to their characteristic of exhibiting limited appearance information, the task of detecting small ships becomes more challenging.

Supported by the National Science Fund of China under Grant 62006119.

Consequently, small ship detection remains an unresolved and demanding problem in the field. To tackle this problem, a series of works have been proposed.

Bai [1] proposes MTGAN, a framework comprising a generator and a discriminator. The generator is responsible for up-sampling low-resolution images, the discriminator assesses super-resolution image patches. The two components engage in a competitive process to obtain details of small objects. Deng [3] proposes an extended feature pyramid network with an extra high-resolution pyramid level specialized for small object detection, which uses deconvolution to obtain features specifically for small object. Rabbi [10] applies EESRGAN to alleviate the lack of high-frequency edge information of small objects in the reconstruction of low-resolution images, improving the detection performance of small objects.

Feature pyramid network (FPN) [6] integrates the feature information of objects of different scales through the top-down horizontal connection structure, detects objects of corresponding scales on the feature map of different depths, thus improving the detection ability of small objects. PANet [8] adds a bottom-up structure to strengthen feature fusion, and enriches the feature hierarchy through a bidirectional path. The feature layer further improves the fusion of small objects. QueryDet [19] designs a cascade sparse query strategy to use the location information on the low-resolution feature map to guide the efficient detection of small objects on the high-resolution feature map. EfficientDet [15] proposes BiFPN, which uses bidirectional path and weight feature fusion to enhance the importance of small object features in fusion. However, the limited appearance information of small objects often leads to the loss of their semantic details during multi-scale feature processing. This, in turn, makes small ship detection a challenging task. Another contributing factor is the discrepancy between the receptive field scale of conventional anchor-based object detectors and the scale sensitivity of the Intersection over Union (IOU) metric.

To address the issues, we propose an effective small ship detection algorithm with enhanced-YOLOv7. Specifically, we firstly propose a small object-aware feature extraction module that mitigates feature loss during multi-scale processing and reduces the impact of marine background noise by incorporating small-scale receptive fields and multi-branch residual structures. Additionally, to address the challenges associated with scale sensitivity in sample allocation and learning efficiency for small ships, We design a small object-friendly scale-insensitive regression scheme. This scheme enhances the contributions of bounding box distance and difficult samples. Furthermore, we construct a geometric constraint-based NMS method. Based on the formulated penalty model, the method reduce the influence of area factors, leading to a decrease in the omission rate of small ships. Our main contributions are as follows:

- We design a small object-aware feature extraction module, to effectively reduce the feature loss of small ships during multi-scale processing and alleviate the impact of marine background noise by incorporating small-scale receptive fields and multi-branch residual structures.

- In order to overcome the difficulties of sample allocation and low learning efficiency of small ships, we propose a small object-friendly scale-insensitive regression scheme to strengthen the contributions of both bounding box distance and difficult samples.
- So as to further decrease the omission rate of small ships, we construct a geometric constraint-based NMS method by considering the formulated penalty model.

2   Method

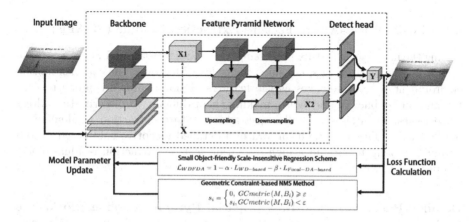

Fig. 1. The pipeline of an effective small ship detection algorithm with enhanced-YOLOv7. X denotes the small object-aware feature extraction module, including a detailed receptive field-based spatial pyramid pooling block represented by X1 and a high-resolution feature extraction residual block represented by X2. Y denotes the geometric constraint-based NMS method.

To address the challenges in detecting small ships, which arise due to their occupation of fewer pixels and the lack of appearance information, we propose an effective small ship detection algorithm with enhanced-YOLOv7. The structure is depicted in Fig. 1 and comprises several components: a backbone network for initial feature extraction, a small object-aware feature extraction residual module, a feature pyramid network for extracting enhanced features at multiple scales, a geometric constraint-based non-maximum suppression (NMS) method, and detection heads that generate the final predictions. Our proposed modules, namely the small object-friendly scale-insensitive regression scheme and the geometric constraint-based NMS method, enhance the accuracy of small ship detection through gradient calculations and backpropagation during training.

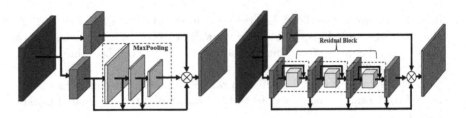

(a) The detailed receptive field-based spatial pyramid pooling module

(b) The high-resolution feature extraction residual module

Fig. 2. The overview of a small object-aware feature extraction residual module.

2.1   Small Object-Aware Feature Extraction Module (SOAFE)

During the multi-scale feature processing, the characteristics of fewer pixels make it easy for the features of small ships to get lost. Moreover, the complex marine environment poses challenges as the features of small ships are prone to contamination by background noise, making them difficult to discern. To address these issues, we design a small object-aware feature extraction residual module (SOAFE). This module incorporates a detailed receptive field-based spatial pyramid pooling block and a high-resolution feature extraction residual block, as illustrated in Fig. 2.

Detailed Receptive Field-Based Spatial Pyramid Pooling Block. The block (Fig. 2a) is positioned within the deep, low-resolution feature map. It operates on features through two branches: one conducts a convolution operation to adjust the channels, while the other deviates from the original parallel structure. The latter branch incorporates three consecutive maximum pooling layers with smaller $N \times N$ kernel sizes and leverages the detailed receptive field to capture local feature information specific to small ships. This approach mitigates the loss of feature information caused by inconsistent receptive field sizes. Simultaneously, the cascading structure overcomes the limitation of disconnected parallel branches by integrating the feature information acquired from all pooling layers. This integration facilitates the better preservation and acquisition of small ship feature information. Consequently, the block effectively addresses the issue of feature loss during multi-scale feature processing, thereby bolstering the model's capacity to learn distinctive features associated with small ships.

High-Resolution Feature Extraction Residual Block. The block shown in Fig. 2b is situated within the shallow, high-resolution feature map. This positioning is deliberate as the feature map offers more comprehensive and influential feature information pertaining to small ships. The block operates on features using two branches: one conducts a standard convolution operation to adjust the channels, while the other consists of a cascade of three residual blocks. This multi-branch structure effectively extracts the features specific to small ships. Furthermore, the integration of feature information from multiple residual blocks enhances the richness of semantic information obtained. Consequently, the block prioritizes the extraction of small ship features and mitigates the negative impact of complex marine backgrounds, thus exhibiting robustness.

2.2 Small Object-Friendly Scale-Insensitive Regression Scheme (SOFSIR)

In anchor-based object detectors, a commonly utilized regression scheme is based on the Intersection over Union (IOU) metric. Nevertheless, the IOU metric exhibits scale sensitivity and is highly susceptible to positional deviations in the case of small objects, which has two main draw-backs: (i) Firstly, the network relies on the IOU metric as the threshold for allocating positive and negative samples during feature learning. However, the characteristics of small ships, such as their limited pixel representation, often lead to conflicting labels. Consequently, the features of positive and negative samples for small ships become indistinguishable, impeding network convergence. (ii) Secondly, small ships pose challenges as difficult training samples due to their limited pixel coverage and insufficient appearance information. Nevertheless, the impact of geometric factors, including the IOU metric, is magnified during backpropagation. Consequently, there is an increased penalty imposed on small ships, constraining their gradient gain and impeding efficient learning for this category of objects.

To mitigate the issue (i), considering that the Wasserstein distance owns the scale invariance, we model the two bounding boxes b_i and b_{gt} into two-dimensional Gaussian distribution $\mathcal{N}_i$ and $\mathcal{N}_{gt}$ respectively, and use the Wasserstein distance between the two Gaussian distributions to represent the similarity between the two bounding boxes b_i and b_{gt} [18], which is given by:

$$WD\left(\mathcal{N}_i, \mathcal{N}_{gt}\right) = \left\|\left(\left[x_i, y_i, \frac{w_i}{2}, \frac{h_i}{2}\right]^{\mathrm{T}}, \left[x_{gt}, y_{gt}, \frac{w_{gt}}{2} \cdot \frac{h_{gt}}{2}\right]^{\mathrm{T}}\right)\right\|_2^2, \quad (1)$$

where x and y denote the horizontal and vertical coordinates of two center points, respectively, w and h denote the width and height of the two bounding boxes, respectively, and $\|\cdot\|_2$ is the Euclidean norm.

We employ exponential form normalization to conduct regression calculations, which is defined as follows:

$$\mathcal{L}_{WD-based} = 1 - \exp\left(-\frac{\sqrt{WD\left(\mathcal{N}_i, \mathcal{N}_{gt}\right)}}{C}\right), \quad (2)$$

where C is a constant and is set to the average size of the objects in the dataset.

By adopting this approach, we enhance the scale insensitivity and ensure accurate label allocation, thereby facilitating network convergence.

To mitigate the issue (ii), For reducing the contribution of geometric factors and simple samples to the loss function, we firstly construct the loss function of the two-layer attention mechanism $\mathcal{L}_{DA-based}$ based on distance attention $\mathcal{DA}_{IoU}$, which is defined as:

$$\mathcal{L}_{DA-based} = \mathcal{DA}_{IoU}\mathcal{L}_{IoU}, \tag{3}$$

where L_{IoU} denotes the loss function based on the IOU metric, and $\mathcal{DA}_{IoU}$ denotes the distance attention, which is defined as follows:

$$\mathcal{DA}_{IoU} = \exp\left(\frac{(x - x_{gt})^2 + (y - y_{gt})^2}{(W_g^2 + H_g^2)^*}\right), \tag{4}$$

where x and y denote the horizontal and vertical coordinates of two center points, respectively. W_g and H_g represent the width and height of the smallest enclosing box that covers both boxes respectively. the superscript * indicates that the metrics do not participate in the gradient calculation of backpropagation.

The distance attention $\mathcal{DA}_{IoU}$ focuses on the distance factors [16], enhances the gradient gain of small ships, and relieves the punishment of $\mathcal{L}_{IoU}$ on small ships. At the same time, $\mathcal{L}_{IoU}$ is used to limit the $\mathcal{DA}_{IoU}$ of high-quality bounding boxes, which reduces the contributions of simple samples. Moreover, the monotonic focusing mechanism [7] is used to focus more on difficult samples, especially small ships. The formula is as follows:

$$\mathcal{L}_{Focal-DA-based} = \left(\frac{\mathcal{L}_{IoU}^*}{\overline{\mathcal{L}}_{IoU}}\right)^\gamma \mathcal{L}_{DA-based}, \gamma > 0, \tag{5}$$

where $\overline{\mathcal{L}}_{IoU}$ denotes the normalizing factor,which is the mean value of L_{IoU}.

The dynamic update of $\overline{\mathcal{L}}_{IoU}$ can maintain the gradient gain at a high level, alleviating the issue of slow convergence during the latter stages of training. This approach focuses both distance factors and difficult samples, ultimately enhancing the learning efficiency for small ships. Finally, the formula for the small object-friendly scale-insensitive regression scheme (SOFSIR) is defined as:

$$\mathcal{L}_{WDFDA} = 1 - \alpha \cdot L_{WD-based} - \beta \cdot L_{Focal-DA-based}, \tag{6}$$

where α and β represent the proportion of the two functions $L_{WD-based}$ and $L_{Focal-DA-based}$ in SOFSIR respectively.

2.3   Geometric Constraint-Based Non-Maximum Suppression Method (GCNMS)

Prior to generating the results, the non-maximum suppression (NMS) method is employed to eliminate redundant objects and retain only the local maximum object by comparing the IOU metric with a predefined threshold value. However, due to the limited pixel representation of small ships, even a slight positional change can lead to significant errors in the IOU metric, resulting in inadvertent deletions. Consequently, the omission rate of small ships tends to increase.

To address these challenges, a geometric constraint-based NMS method (GCNMS) is constructed. This method redesigns the threshold by incorporating a formulated penalty model, thus reducing the reliance on the IOU metric and enhancing the recall rate of small ships. Firstly, we establish the Euclidean distance between the two bounding boxes, thereby diminishing the influence of area factors and rendering it more suitable for small ships with limited pixel coverage. Secondly, to mitigate the issue of aspect ratio disparity hindering the fitting of small ships, we introduce penalty terms for height h and width w. These terms enable the calculation of the differences in height and width between the two bounding boxes [20], ultimately minimizing the disparity and improving the model's ability to accurately fit small ships. The GCNMS formula is as follows:

$$s_i = \begin{cases} 0, & GCmetric\,(M, B_i) \geqslant \varepsilon \\ s_i, & GCmetric\,(M, B_i) < \varepsilon \end{cases}, \tag{7}$$

where si represents the detection score of the i-th bounding box, Bi represents the i-th bounding box, M represents the bounding box with the highest detection score, and ε represents the threshold value. The metric formula $GCmetric\,(M, B_i)$ is as follows:

$$GCmetric\,(M, B_i) = IoU\,(M, B_i) - \frac{\rho^2\,(m, b_i)}{c^2} - \frac{\rho^2\,(w, w_i)}{C_w^2} - \frac{\rho^2\,(h, h_i)}{C_h^2}, \tag{8}$$

where m and b_i represent the center point of the bounding box with the highest detection score and the i-th bounding box, respectively. $\rho\,(\cdot)$ represents the Euclidean distance. c represents the area of the smallest enclosing box that covers both boxes. C_w and C_h are the width and height of the smallest enclosing box.

We enhance the significance of distance factors while diminishing the influence of area factors, thereby creating a measurement approach that is better suited for evaluating the similarity between small ships with fewer pixels. This adjustment effectively reduces the omission rate associated with small ships.

3   Experiments

3.1   Experimental Settings

Datasets. We evaluate the performance of our proposed algorithm on two datasets: (i)SeaShips [14]: a widely recognized large-scale maritime surveillance dataset comprising precisely annotated visual images. This dataset consists of 6 classes, with 7,000 publicly available images. We divide the dataset into training, validation, and test sets in a ratio of 1:1:2. (ii)Pascal VOC2007: a renowned object detection dataset. We extract all ship objects from this dataset, resulting in a subset of 549 images containing only boat labels. The dataset is partitioned into training, validation, and test sets at a ratio of 9:1:1.

Table 1. Detection results on Seaships.

Model	AP	$AP_{0.5}$	$AP_{0.75}$	AP_s	AR_s
SSD300(VGG16) [9]	0.588	0.935	0.673	0.013	0.114
SSD300(MobileNetv2)	0.494	0.891	0.491	0.004	0.057
Faster-RCNN(ResNet50) [12]	0.591	0.949	0.658	0.031	0.050
Faster-RCNN(VGG16)	0.588	0.946	0.650	0.019	0.036
CenterNet [4]	0.697	0.961	0.824	0.138	0.150
YOLOv3 [11]	0.572	0.941	0.631	0.114	0.143
YOLOv4 [2]	0.506	0.921	0.506	0.072	0.114
YOLOv5	0.649	0.952	0.762	0.101	0.179
YOLOv7 [17]	0.675	0.960	0.794	0.085	0.150
Ours	**0.711**	**0.962**	**0.836**	**0.178**	**0.232**

Implementation Details. All experiments are conducted on an NVIDIA RTX 2080Ti GPU, CUDA version is 10.0, cuDNN version is 7.5.1, and PyTorch version is 1.2.0. All models are trained for 300 epochs with batch size of 4, an initial learning rate of 1e–2, which is then reduced to a minimum of 1e–4 using a cosine annealing algorithm. We utilize the sgd optimizer with momentum 0.937 and weight decay 5e–4. YOLOv7 is the original network of the proposed algorithm. We set $N = 5$, $C = 1.0$, $\gamma = 0.5$, $\alpha = 0.8$, $\beta = 0.2$ and $\varepsilon = 0.4$. To demonstrate the efficacy of the proposed algorithm, we conduct the experimental comparison with other classical object detectors on Seaships and Pascal VOC2007.

Evaluation Indicators. We adopt evaluation indicators of COCO dataset, including AP, $AP_{0.5}$, $AP_{0.75}$, AP_s and AR_s. AP is the average mAP across different IoU thresholds IoU $= \{0.5, 0.55, \cdots, 0.95\}$, $AP_{0.5}$ and $AP_{0.75}$ are APs at IoU threshold of 0.5 and 0.75, respectively. AP_s and AR_s are used to evaluate the precision and recall of small-scale object (less than 32×32 pixels), respectively.

3.2 Quantitative Analysis

Table 1 presents the performance of various algorithms on Seaships, our proposed algorithm significantly outperforms the baseline and achieves the best performance. Specifically, our algorithm improves 9.3% and 8.2% over the original network in AP_s and AR_s, respectively. In comparison to SSD, which utilizes a multi-scale approach for feature processing, our algorithm with SOAFE reduces the loss of small ship by considering detailed receptive fields and multi-branch structures, and achieves an impressive increase of 16.5% and 17.4% in AP_s. Furthermore, SOFSIR focuses both distance factors and difficult samples, when compared to the renowned two-stage object detector Faster-RCNN, our algorithm showcases notable enhancements of 14.7% and 15.9% in AP_s, along with 18.2% and 19.6% improvements in AR_s. Moreover, in comparison to the NMS-based YOLO series network, GCNMS effectively diminishes the influence of

Table 2. Detection results on Pascal VOC2007.

Model	AP	$AP_{0.5}$	$AP_{0.75}$	AP_s	AR_s
SSD300(VGG16) [9]	0.267	0.603	0.238	0.005	0.150
SSD300(MobileNetv2)	0.199	0.515	0.104	0.025	0.100
Faster-RCNN(ResNet50) [12]	0.284	0.567	0.221	0.202	0.200
Faster-RCNN(VGG16)	0.311	0.628	0.273	0.151	0.150
CenterNet [4]	0.280	0.192	0.602	0.151	0.150
RetinaNet [7]	0.373	0.656	0.398	0.151	0.150
EfficientDet [15]	0.318	0.609	0.304	0.177	0.200
YOLOv3 [11]	0.272	0.642	0.202	0.101	0.100
YOLOv4 [2]	0.258	0.636	0.231	0.101	0.100
YOLOv5	0.325	0.710	0.284	0.202	0.200
YOLOv7 [17]	0.361	0.753	0.330	0.119	0.250
Ours	**0.409**	**0.779**	**0.426**	**0.210**	**0.400**

Table 3. Ablation experimental results of module on Seaships.

Model	AP	$AP_{0.5}$	$AP_{0.75}$	AP_s	AR_s
Baseline(YOLOv7)	0.675	0.960	0.794	0.085	0.150
+SOAFE	0.671	0.960	0.795	0.116	0.200
+SOAFE+SOFSIR	**0.716**	0.961	**0.843**	0.145	0.200
+SOAFE+GCNMS	0.678	0.960	0.803	0.150	0.179
Ours	0.711	**0.962**	0.836	**0.178**	**0.232**

area factors, our algorithm outperforms the best-performing YOLOv5 by 5.3% in terms of AR_s. Furthermore, when compared to CenterNet, an anchor-free detector without NMS, our algorithm achieves an 8.2% improvement in AR_s.

To verify the universality of the proposed algorithm, the same comparative experiments are done on Pascal VOC2007. As shown in Table 2, our algorithm also attains the best performance. Specifically, our algorithm reaches 21.0% and 40.0% in AP_s and AR_s, which is 9.1% and 15.0% higher than that of the original YOLOv7. When compared to the EfficientDet with BiFPN integration, our algorithm, aided by SOAFE, addresses the issue of information loss for small ships, resulting in a 3.3% improvement in AP_s. Moreover, in comparison to RetinaNet, which utilizes focal loss to prioritize difficult samples, our algorithm demonstrates remarkable enhancements of 5.9% and 25.0% in AP_s and AR_s, respectively. This improvement stems from SOFSIR's consideration of not only difficult samples but also distance factors. Furthermore, when compared to the anchor-free detector CenterNet, GCNMS significantly enhances our algorithm's performance by 20% and 25% in terms of AR_s, respectively. The experimental results conclusively demonstrate the effectiveness of the proposed algorithm in the field of small ship detection.

Table 4. Ablation study on the penalty terms in NMS on Seaships.

Model	AP	$AP_{0.5}$	$AP_{0.75}$	AP_s	AR_s
NMS-based	0.675	0.960	0.794	0.085	0.150
GIOU-based NMS [13]	**0.678**	0.961	**0.803**	0.157	0.200
DIOU-based NMS [21]	0.676	**0.962**	0.796	0.101	0.150
CIOU-based NMS [21]	0.678	0.960	0.799	0.150	0.207
SIOU-based NMS [5]	0.673	0.961	0.792	0.122	0.150
Ours	0.677	0.960	0.797	**0.161**	**0.229**

3.3 Ablation Studies

Ablation experiments are conducted on Seaships to validate the effectiveness of each module in our proposed algorithm. As shown in Table 3, the results demonstrate the positive impact of each module. SOAFE successfully mitigates feature loss issues for small ships, resulting in a 3.1% improvement in AP_s and a 5.0% improvement in AR_s. Similarly, SOFSIR addresses the scale sensitivity problem, enhancing the learning efficiency for small ships and leading to a 2.9% increase in AP_s, thus improving the overall accuracy. Lastly, the addition of GCNMS on the aforementioned modules effectively reduces the omission rate of small ships, resulting in a 3.3% increase in AP_s and a 3.2% increase in AR_s. Experimental results validate the significant improvements achieved by the proposed modules in enhancing the detection performance of small ships to varying degrees.

We perform ablation experiments on the penalty terms of the NMS method. As shown in Table 4, GIOU-based NMS incorporates a penalty based on the area of the enclosing box, but geometry still dominates. DIOU-based NMS introduces the Euclidean distance between the two bounding boxes, but it lacks sufficient consideration of distance factors. CIOU-based NMS further includes the aspect ratio, but the difference in aspect ratio poses challenges in accurately fitting small ships. Similarly, SIOU-based NMS encounters the same issue. In contrast, GCNMS considers distance factors and separately calculates the difference of height and width, enabling a better fit for the shape of small ships. Our algorithm reaches 16.1% and 22.9% in AP_s and AR_s, which is 7.6% and 7.9% higher than that of the original algorithm. Experimental results show that the proposed penalty model in GCNMS is effective in the application of small ship detection.

3.4 Qualitative Analysis

Figure 3 demonstrations the performance of our proposed algorithm compared to other classical algorithms in detecting small ships. The observations indicate that CenterNet, YOLOv4, and YOLOv7 tend to miss small ships, while YOLOv5 exhibits a biased bounding box. In contrast, our algorithm enhances the feature extraction capability and learning efficiency for small ships, effectively addressing the issue of missed detections. Consequently, our proposed algorithm effortlessly resolves these challenges. It can be concluded that our algorithm outperforms the others in accurately detecting small ships.

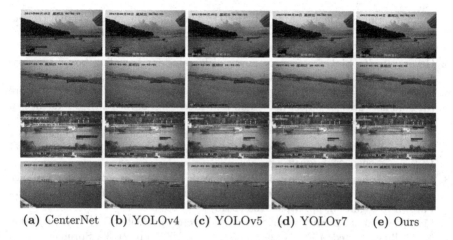

(a) CenterNet (b) YOLOv4 (c) YOLOv5 (d) YOLOv7 (e) Ours

Fig. 3. Qualitative comparison of different algorithms on Seaships.

4 Conclusion

In this paper, we have proposed an effective small ship detection algorithm with enhanced-YOLOv7. Specifically, a small object-aware feature extraction module has been firstly designed, which effectively addresses the challenge of feature loss during multi-scale processing and takes into account the impact of the marine environment by considering both small-scale receptive fields and multi-branch residual structures. Additionally, we have proposed a small object-friendly scale-insensitive regression scheme. This scheme tackles the difficulties of sample allocation and low learning efficiency by strengthening the contributions of both bounding box distance and difficult samples. Furthermore, we have constructed a geometric constraint-based Non-Maximum Suppression method, which significantly reduces the omission rate of small ships by incorporating the formulated penalty model. Experimental results demonstrate the effectiveness of our proposed algorithm. In the future, we would like to further improve the effectiveness of the small ship detection in the extreme marine environments.

References

1. Bai, Y., Zhang, Y., Ding, M., Ghanem, B.: SOD-MTGAN: small object detection via multi-task generative adversarial network. In: Ferrari, V., Hebert, M., Sminchisescu, C., Weiss, Y. (eds.) ECCV 2018. LNCS, vol. 11217, pp. 210–226. Springer, Cham (2018). https://doi.org/10.1007/978-3-030-01261-8_13
2. Bochkovskiy, A., Wang, C.Y., Liao, H.Y.M.: YOLOv4: optimal speed and accuracy of object detection. arXiv preprint arXiv:2004.10934 (2020)
3. Deng, C., Wang, M., Liu, L., Liu, Y., Jiang, Y.: Extended feature pyramid network for small object detection. IEEE Trans. Multimedia **24**, 1968–1979 (2021)

4. Duan, K., Bai, S., Xie, L., Qi, H., Huang, Q., Tian, Q.: CenterNet: keypoint triplets for object detection. In: Proceedings of the IEEE/CVF International Conference on Computer Vision, pp. 6569–6578 (2019)

5. Gevorgyan, Z.: SIoU Loss: more powerful learning for bounding box regression. arXiv preprint arXiv:2205.12740 (2022)

6. Lin, T.Y., Dollár, P., Girshick, R., He, K., Hariharan, B., Belongie, S.: Feature pyramid networks for object detection. In: Proceedings of the IEEE Conference on Computer Vision and Pattern Recognition, pp. 2117–2125 (2017)

7. Lin, T.Y., Goyal, P., Girshick, R., He, K., Dollár, P.: Focal loss for dense object detection. In: Proceedings of the IEEE International Conference on Computer Vision, pp. 2980–2988 (2017)

8. Liu, S., Qi, L., Qin, H., Shi, J., Jia, J.: Path aggregation network for instance segmentation. In: Proceedings of the IEEE Conference on Computer Vision and Pattern Recognition, pp. 8759–8768 (2018)

9. Liu, W., et al.: SSD: single shot multibox detector. In: Leibe, B., Matas, J., Sebe, N., Welling, M. (eds.) ECCV 2016. LNCS, vol. 9905, pp. 21–37. Springer, Cham (2016). https://doi.org/10.1007/978-3-319-46448-0_2

10. Rabbi, J., Ray, N., Schubert, M., Chowdhury, S., Chao, D.: Small-object detection in remote sensing images with end-to-end edge-enhanced GAN and object detector network. Remote Sens. **12**(9), 1432 (2020)

11. Redmon, J., Farhadi, A.: YOLOv3: an incremental improvement. arXiv preprint arXiv:1804.02767 (2018)

12. Ren, S., He, K., Girshick, R., Sun, J.: Faster r-CNN: towards real-time object detection with region proposal networks. Advances in neural information processing systems, vol. 28 (2015)

13. Rezatofighi, H., Tsoi, N., Gwak, J., Sadeghian, A., Reid, I., Savarese, S.: Generalized intersection over union: a metric and a loss for bounding box regression. In: Proceedings of the IEEE/CVF Conference on Computer Vision and Pattern Recognition, pp. 658–666 (2019)

14. Shao, Z., Wu, W., Wang, Z., Du, W., Li, C.: SeaShips: a large-scale precisely annotated dataset for ship detection. IEEE Trans. Multimedia **20**(10), 2593–2604 (2018)

15. Tan, M., Pang, R., Le, Q.V.: EfficientDet: scalable and efficient object detection. In: Proceedings of the IEEE/CVF Conference on Computer Vision and Pattern Recognition, pp. 10781–10790 (2020)

16. Tong, Z., Chen, Y., Xu, Z., Yu, R.: Wise-IoU: bounding box regression loss with dynamic focusing mechanism. arXiv preprint arXiv:2301.10051 (2023)

17. Wang, C.Y., Bochkovskiy, A., Liao, H.Y.M.: YOLOv7: trainable bag-of-freebies sets new state-of-the-art for real-time object detectors. arXiv preprint arXiv:2207.02696 (2022)

18. Wang, J., Xu, C., Yang, W., Yu, L.: A normalized gaussian Wasserstein distance for tiny object detection. arXiv preprint arXiv:2110.13389 (2021)

19. Yang, C., Huang, Z., Wang, N.: QueryDet: cascaded sparse query for accelerating high-resolution small object detection. In: Proceedings of the IEEE/CVF Conference on Computer Vision and Pattern Recognition, pp. 13668–13677 (2022)

20. Zhang, Y.F., Ren, W., Zhang, Z., Jia, Z., Wang, L., Tan, T.: Focal and efficient IOU loss for accurate bounding box regression. Neurocomputing **506**, 146–157 (2022)

21. Zheng, Z., Wang, P., Liu, W., Li, J., Ye, R., Ren, D.: Distance-IoU Loss: faster and better learning for bounding box regression. In: Proceedings of the AAAI Conference on Artificial Intelligence, vol. 34, pp. 12993–13000 (2020)

PiDiNeXt: An Efficient Edge Detector Based on Parallel Pixel Difference Networks

Yachuan Li[1], Xavier Soria Poma[2,3], Guanlin Li[1], Chaozhi Yang[1],

Qian Xiao[1], Yun Bai[1], and Zongmin Li[1(✉)]

[1] China University of Petroleum (East China), Qingdao, Shandong 266500, China
{liyachuan,liguanlin,xiaoqian,baiyun}@s.upc.edu.cn,
yang.chaozhi@foxmail.com, lizongmin@upc.edu.cn
[2] ESPOCH Polytechnic University, Data Science Research Group, Riobamba, Ecuador
[3] National University of Chimborazo, Riobamba 060110, Ecuador
xavier.soria@unach.edu.ec

Abstract. The Pixel Difference Network (PiDiNet) is well-known for its success in edge detection. Combining traditional operators with deep learning, PiDiNet achieves competitive results with fewer parameters. However, the complex and inefficient choice of traditional edge detection operators hinders PiDiNet's further development. Therefore, we propose a novel lightweight edge detector called PiDiNeXt, which combines traditional edge detection operators with deep learning-based model in parallel to solve the operators choice problem and further enrich features. The results of experiments on BSDS500 and BIPED datasets demonstrate that PiDiNeXt outperforms PiDiNet in terms of accuracy. Moreover, we employ the reparameterization technique to prevent the extra computational cost caused by the multi-branch construction. This enables PiDiNeXt to achieve an inference speed of 80 FPS, comparable to that of PiDiNet. Furthermore, the lightweight version of PiDiNeXt can achieve an inference speed of over 200 FPS, meeting the needs of most real-time applications. The source code is available at https://github.com/Li-yachuan/PiDiNeXt.

Keywords: Pixel difference · Edge detection · Reparameterization

1 Introduction

Edges that contain vital semantic information and eliminate texture noise are one of the most fundamental components of images. Therefore, edge detection plays a crucial role in many higher-level computer vision tasks such as salient detection [35], semantic segmentation [32], and depth map prediction [23], to name a few.

Edge detection has been concerned for a long time, and researchers have studied edge detection as early as the last century [4,22]. Early classical methods usually rely on the first or second order derivatives of images to detect edge, such as Sobel [24], Prewitt [19], and Canny [3]. Later, learning-based methods [9,11] further improve the accuracy of edge detection by utilizing various gradient information [16,17,30]. These methods make full use of the local cues of images to detect edges and achieve initial

Q. Liu et al. (Eds.): PRCV 2023, LNCS 14434, pp. 261–272, 2024.
https://doi.org/10.1007/978-981-99-8549-4_22

success. However, some edges and textures have very similar performance in local features, and can only be distinguished by global semantic information. The lack of global information limits the performance of the traditional edge detection operators.

Deep learning-based edge detectors [5,12,13,18,31] have achieved disruptive results in recent years, which obtain more sufficient global information through end-to-end learning. And traditional edge detection operators have become dated and are gradually being forgotten by researchers. However, deep learning-based edge detectors are far from perfect. One of the biggest problems is the lack of inductive bias. The features required for edge detection must be statistically induced from large-scale datasets. As a result, the edge detectors are forced to rely on the large pretrained backbone, which is memory and energy consuming.

In order to avoid the dependence on the large pretrained backbone, Su *et al.* [26] combine deep learning-based edge detectors with traditional edge detection operators and propose a pixel difference network (PiDiNet), which can surpass the recorded result of human perception with a more lightweight network by modeling the relationship of local features. PiDiNet has achieved great success, but the choice of edge detection operators become an issue. To ensure the inference speed, PiDiNet is designed to be VGG-style with single-branch. Therefore, each layer of the model can only choose one kind of operator, which makes the choice of operators a pain in PiDiNet.

To address the issue in PiDiNet, we propose a simple, lightweight yet effective edge detector named PiDiNeXt. We extend PiDiNet to a multi-branch structure, where each branch uses different traditional edge detection operators to induce the biases of the model. All operators are integrated into the model in parallel, which effectively avoids the complex operator choice problem and further enrich the features. And we introduce reparameterization technology [7], which can convert PiDiNeXt into a single-branch VGG-style model during inference, thereby ensuring the inference speed of PiDiNeXt.

2   Related Work

2.1   The Development of Deep Learning Based Edge Detection

Holistically-nested Edge Detection (HED) [31] introduces the deep supervision mechanism to edge detection and learns multi-scale predictions holistically. HED employs pretrained VGG16 as backbone and achieves the best performance at the time. Numerous works [12,13] follow this setup to get a fair comparison. In recent years, with the emergence of higher performance feature extraction models, the backbone of edge detection has better options. DexiNed [25] uses a densely connected module, UAED [36] chooses EfficientNet as backbone, and EDTER [20] introduces the vision Transformer model. Their accuracy are significantly better than those VGG-based methods. At the same time, researchers try to design lightweight architectures to achieve efficient edge detection [28,29]. While they have a large gap in detection accuracy compared to general methods.

In the field of edge detection, it seems to be a consensus that model accuracy is positively correlated with model size and pre-trained model is essential. This is mainly due to the lack of manual induction bias, and large models is needed to learn the distribution of data from large-scale datasets. It is the constant pursuit of edge detection researchers to learn how to achieve higher performance with a more lightweight network.

2.2 Review of Pixel Difference Convolution

To eliminate the dependence on large models and pretrained backbone, Su *et al.* [26] propose Pixel Difference Convolution, in which the deep learning-based model is combined with traditional edge detection operators to explicitly incorporate inductive bias into the model. Specifically, Pixel Difference Convolution contains three forms, namely Pixel Difference Convolution based on Central Differences (CPDC), Pixel Difference Convolution based on Angular Difference (APDC) and Pixel Difference Convolution based on Radial Difference (PRDC), which are used to capture the difference between the current feature and surrounding features, the difference between surrounding features, and the difference between surrounding features and more distant features.

The backbone of PiDiNet [26] consists of 16 convolutional layers. In order to ensure inference speed, the whole network is a single-branch model, so each convolutional layer only employs one kind of pixel difference convolution or vanilla convolution to extract features. As a result, PiDiNet has a total of 4^{16} alternative structures. Experimentally choosing the best of them seems to be an impossible task. In PiDiNet [32], only 14 structures are tried, and the experiments conducted are clearly not sufficient. Thus, it is important to explore the possibility of avoiding the need to select different operators in PiDiNet.

3   Method

PiDiNeXt has different architectures for training and inference; we first present the architecture for training and then show how PiDiNeXt can be converted to a single-branch VGG-style model by reparameterization during the inference.

3.1   PiDiNeXt Architecture

As shown in the Fig. 1, PiDiNeXt employs 16 pixel difference blocks as the backbone, which are evenly divided into four stages by three 2×2 pooling layers. The output features of each stage are collected and called multi-scale features. After enhanced by compact dilation convolution based modules (CDCM) and compact spatial attention modules (CSAM), respectively, the multi-scale features are used to obtain side edges by dimension reduction and resize. The side edges are transformed into the final edge by a 1×1 convolution. Both the side edges and the final edge are constrained by the supervision information.

CDCM and CSAM are two compact feature enhancement modules that enhance multi-scale features through parallel dilation convolution and attention, as shown in Fig. 2(A) and (B). They are presented in PiDiNet [26] and we stick with this setup.

Following PiDiNet, PiDiNeXt is divided into three versions: PiDiNeXt, PiDiNeXt-small, and PiDiNeXt-tiny, with the number of channels set to 20, 30, and 60 respectively. For further inference speed improvement, the CDCM and CSAM modules

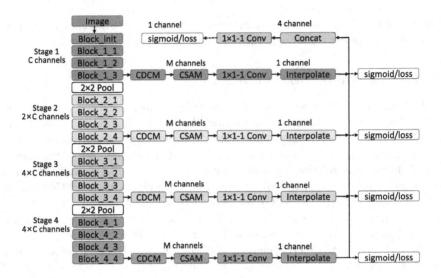

Fig. 1. PiDiNeXt architecture.

can be removed to obtain the corresponding lightweight versions named PiDiNeXt-L, PiDiNeXt-small-L, and PiDiNeXt-tiny-L. The six versions of PiDiNeXt are in one-to-one correspondence with PiDiNet.

The main innovation of PiDiNeXt is in the backbone. The backbone block of PiDiNet is a residual structure that incorporates a kind of feature operator, as shown in Fig. 2(C). To avoid the choice of feature operators and further enrich the features, we propose a Parallel Pixel Difference Module (PPDM), in which multiple feature operators are integrated into the deep learning-based module in parallel. And 1×1 convolution is utilized to fully select and fuse the features. The enhanced feature is fused with the original feature through the residual structure to ensure the stability of the back-propagation gradient. The structure of PPDM is shown in Fig. 2(D). To speed up the inference, the normalization layers are rejected in PiDiNeXt. Therefore, to ensure the stability of the gradient, multiple parallel features are fused by averaging. Experimental results show that this parallel structure can significantly improve the performance of the model.

4   PiDiNeXt Reparameterization

PPDM can improve the network performance, but the parallel structure will increase the computational cost and inference delay, which greatly reduces the efficiency of the model. To overcome this drawback, PPDM is converted to vanilla convolutional layers with the help of reparameterization technique [7, 26], resulting in PiDiNeXt almost as efficient as PiDiNet. The process of reparameterization can be expressed by the following equations:

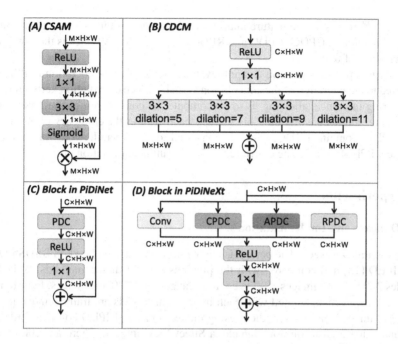

Fig. 2. Detail of PiDiNeXt. *PDC* in (C) means Pixel difference convolution, is one of four operators mentioned in Sect. 2.2. *Conv* in (D) means vanilla convolution.

$$Y = AVE(Conv(\boldsymbol{X}), CPDC(\boldsymbol{X}), APDC(\boldsymbol{X}), PRDC(\boldsymbol{X}))$$
$$= AVE(x_1 \cdot w_1^v + x_2 \cdot w_2^v + x_2 \cdot w_2^v...,$$
$$(x_1 - x_5) \cdot w_1^c + (x_2 - x_5) \cdot w_2^c + (x_3 - x_5) \cdot w_2^c...;$$
$$(x_1 - x_2) \cdot w_1^a + (x_2 - x_3) \cdot w_2^a + (x_3 - x_6) \cdot w_2^a...,$$
$$(x_1' - x_1) \cdot w_1^{r'} + (x_2' - x_2) \cdot w_2^{r'} + (x_3' - x_3) \cdot w_3^{r'}...)$$
$$= AVE(x_1 \cdot w_1^v + x_2 \cdot w_2^v + x_2 \cdot w_2^v...,$$
$$x_1 \cdot w_1^c + ... + x_4 \cdot w_4^c + x_5 \cdot (- \sum_{i=\{i=\{1,2,3,4,6,7,8,9\}} \cdot w_i)..., \quad (1)$$
$$x_1 \cdot (w_1^a - w_4^a) + x_2 \cdot (w_2^a - w_1^a) + x_3 \cdot (w_3^a - w_2^a)...,$$
$$x_1' \cdot w_1^{r'} + x_1 \cdot (-w_1^{r'}) + x_2' \cdot w_2^{r'} + x_2 \cdot (-w_2^{r'})...)$$
$$= x_1' \cdot AVE(w_1^{r'}) + x_2' \cdot AVE(w_2^{r'}) + x_3' \cdot AVE(w_3^{r'})$$
$$+ x_1 \cdot AVE(w_1^v + w_1^c + w_1^a - w_4^a - w_1^{r'}) + ...$$
$$+ x_5 \cdot AVE(w_5^v - \sum_{i=\{1,2,3,4,6,7,8,9\}} \cdot w_i^c) + ...$$

Where X, Y represent the feature matrices of the input and output. v, c, a and r mean vanilla convolution, CPDC, APDC and RPDC, respectively. AVE means the average of the four kinds of features.

Through reparameterization technology, the complex feature operation in PPDM is converted to the operation of the convolution kernel W. The convolution kernel remains fixed during inference, allowing us to precompute the convolution operation and only perform the vanilla convolution. This key approach ensures the efficiency of our model. Since PPDM contains PRDC, which is converted to 5×5 vanilla convolution during inference, PPDM will also be converted to 5×5 vanilla convolution during inference.

5   Experiments

5.1   Datasets and Implementation

Experimental Datasets. The performance of our method is validated on BSDS500 [1] and BIPED [18] and compared with the previous state-of-the-art methods. BSDS500 includes 200 training images, 100 validation images, and 200 test images. Four to nine annotated maps are associated with each image. Our models are trained and validated using 300 images, and we test them using the test images. BIPED contains 250 carefully annotated high-resolution Barcelona Street View images. There are 200 images for training and validation, and 50 images for testing. For a fair comparison, we use the same data augmentation method as RCF [14] for BSDS500, and the same as DexiNed [25] for BIPED.

Parameter Settings. Our method is implemented on the PyTorch library. All the experiments are conducted on a single 2080 Ti GPU with 11 GB memory. The threshold γ is set to 0.3 for BSDS500. And γ is useless for BIPED, due to the ground truth being binary annotations. λ is set to 1.1 for both BSDS500 and BIPED. Adam optimizer is adopted. A standard Non-Maximum Suppression (NMS) is performed to produce the final edge maps. Other parameters that are not specified are consistent with PiDiNet [26].

Performance Metrics. During evaluation, F-measures of both Optimal Dataset Scale (ODS) and Optimal Image Scale (OIS) are reported for all datasets. Since efficiency is one of the main focuses in this paper, all the models are compared based on the evaluations from single scale images. More detailed information on performance metrics can be referred to previous works [14,26].

5.2   Comparison with the State of the Arts

BSDS500 Dataset: We compare our method with prior edge detectors including both traditional ones and recently proposed CNN based ones, as summarized in Table 1 and Fig. 3. The ODS of PiDiNeXt is 0.2% higher than that of PiDiNet, reaching 0.809. Additionally, PiDiNeXt maintains an inference speed of 80 FPS, surpassing the great

Table 1. The performance of edge detectors with BSDS500 dataset. *Param.* means the number of parameters. ‡ indicates the speeds with our implementations based on a NVIDIA RTX 2080 Ti GPU. † indicates the cited GPU speeds.

Method		Pub. 'year'	ODS	OIS	Param. (M)	FPS
*	Human (reference)	-	0.803	0.803	-	-
Classic	Canny [3]	PAMI'86	0.611	0.676	-	28
	Pb [15]	PAMI'04	0.672	0.695	-	-
	SCG [21]	NeurIPS'12	0.739	0.758	-	-
	SE [8]	PAMI'14	**0.746**	0.767	-	12.5
	OEF [10]	CVPR'15	**0.746**	**0.770**	-	2/3
General deep learning	HFL[2]	ICCV'15	0.767	0.788	-	5/6†
	CEDN [34]	CVPR'16	0.788	0.804	-	10†
	HED [31]	IJCV'17	0.788	0.807	**14.6**	**78‡**
	CED [27]	CVPR'17	0.794	0.811	-	-
	LPCB [6]	ECCV'18	0.808	0.824	-	30†
	RCF [14]	PAMI'19	0.806	0.823	14.8	67‡
	BDCN [12]	PAMI'22	**0.820**	**0.838**	16.3	47‡
	FCL [33]	NN'22	0.815	0.834	16.5	-
Lightweight	TIN1 [28]	ICIP'20	0.749	0.772	0.08	-
	TIN2 [28]	ICIP'20	0.772	0.795	0.24	-
	FINED3-Inf [29]	ICME'21	0.788	0.804	1.08	124†
	FINED3-Train [29]	ICME'21	0.790	0.808	1.43	99†
	PiDiNet [26]	ICCV'21	0.807	0.823	0.71	92‡
	PiDiNeXt	(Ours)	**0.809** (0.002↑)	**0.824** (0.001↑)	0.78 (0.07↑)	80‡ (12↓)
	PiDiNet-L [26]	ICCV'21	0.800	0.815	0.61	128‡
	PiDiNeXt-L	(Ours)	0.801 (0.001↑)	0.819 (0.004↑)	0.68 (0.07↑)	116‡ (12↓)
	PiDiNet-small [26]	ICCV'21	0.798	0.814	0.18	148‡
	PiDiNeXt-small	(Ours)	0.801 (0.003↑)	0.816 (0.002↑)	0.22 (0.04↑)	139‡ (9↓)
	PiDiNet-small-L [26]	ICCV'21	0.793	0.809	0.16	**212‡**
	PiDiNeXt-small-L	(Ours)	0.796 (0.003↑)	0.813 (0.004↑)	0.20 (0.04↑)	202‡ (10↓)
	PiDiNet-Tiny [26]	ICCV'21	0.789	0.806	0.08	152‡
	PiDiNeXt-Tiny	(Ours)	0.795 (0.006↑)	0.811 (0.005↑)	0.11 (0.03↑)	144‡ (8↓)
	PiDiNet-Tiny-L [26]	ICCV'21	0.787	0.804	**0.07**	215‡
	PiDiNeXt-Tiny-L	(Ours)	0.790 (0.003↑)	0.806 (0.002↑)	0.10 (0.03↑)	207‡ (8↓)

majority of VGG-based methods (such as RCF and LPCB) in both accuracy and speed. PiDiNeXt achieves state-of-the-art results compared with the lightweight models. Furthermore, PiDiNeXt is based on the same setup as PiDiNet and no longer relies on the ImageNet pretraining. The whole model is trained from random initialization. Compared to PiDiNet, PiDiNeXt has a little more parameters and is slightly slower because PPDM is a converted 5×5 convolution while CPDC and APDC in PiDiNet can be converted into 3×3 convolutions. Even so, the fastest version PiDiNeXt-Tiny-L, can also achieve comparable prediction performance with more than 200 FPS, further demonstrating the effectiveness of our methods. Another interesting observation is that the accuracy improvement brought by PiDiNeXt becomes smaller as the model gets larger.

PiDiNeXt-L improves ODS by 0.1% over PiDiNet-L and PiDiNeXt by 0.2% over PiDiNet, which are obviously no match for the improvement of the lighter models. This is because PiDiNeXt indirectly increases the size of the model while introducing the multi-branch structure. When the model is large enough but trained with limited data, further increasing the model size may cause the over-fitting problem, which is not conducive to the improvement of the performance of the model. Su *et al.* [26] verify the existence of the over-fitting through ablation experiments, which confirms our conjecture. We also show some qualitative results in Fig. 4. Compared with the results of PiDiNet, our results achieve sharper edges while contain less irrelevant cue.

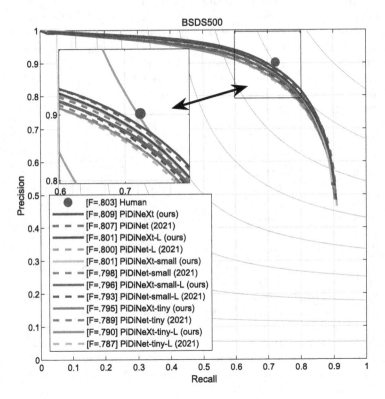

Fig. 3. Precision-Recall curves of our models compared with PiDiNet versions on BSDS500 dataset.

BIPED Dataset: The comparison results on the BIPED dataset are illustrated in Table 2. The situation is similar to that on BSDS, where the accuracy of all versions of PiDiNeXt is improved to varying degrees over the corresponding PiDiNet. The ODS of the brightest PiDiNeXt-Tiny-L is 0.5% higher than PiDiNet-Tiny-L. There are two characteristics worth noting in the experimental results of BIPED. Firstly, because the size of images on BIPED dataset is 1281×720, which is much larger than the

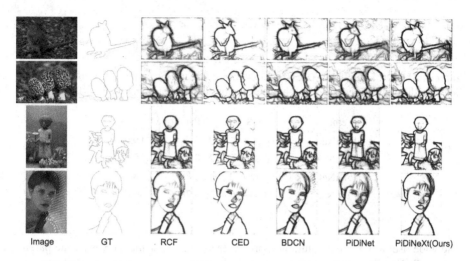

| Image | GT | RCF | CED | BDCN | PiDiNet | PiDiNeXt(Ours) |

Fig. 4. Example results of our method and other works on BSDS500 dataset, including RCF [14], CED [27], BDCN [12] and PiDiNet [26]

images of BSDS500, the inference speed of the model is significantly reduced. Even the most lightweight PiDiNeXt-Tiny-L can only achieve 52 FPS. Secondly, the accuracy improvement is larger for the lightweight PiDiNeXt, which is consistent with the performance on the BSDS500 dataset. It illustrates that the overfitting problem due to the lack of large pre-trained models and insufficient training data is widespread.

Table 2. Comparison with PiDiNet [26] on BIPED dataset. Param. means the number of parameters. ‡ indicates the speeds with our implementations based on a single 2080 Ti GPU. Red indicates improved performance and green indicates decreased performance.

Method	ODS	OIS	Param. (M)	FPS
RCF [14]	0.886	0.892	**14.8**	-
BDCN [12]	0.894	0.899	16.3	-
DexiNed-f [25]	**0.895**	**0.900**	35.3	-
PiDiNet [26]	0.891	0.896	0.71	24‡
PiDiNeXt (Ours)	**0.893** (0.002↑)	**0.900** (0.004↑)	0.78 (0.07↓)	20‡ (4↓)
PiDiNet-L [26]	0.889	0.896	0.61	32‡
PiDiNeXt-L (Ours)	0.890 (0.001↑)	0.899 (0.003↑)	0.68 (0.07↓)	29‡ (3↓)
PiDiNet-small [26]	0.886	0.893	0.18	37‡
PiDiNeXt-small (Ours)	0.887 (0.001↑)	0.896 (0.003↑)	0.22 (0.04↓)	35‡ (2↓)
PiDiNet-small-L [26]	0.880	0.889	0.16	53‡
PiDiNeXt-small-L (Ours)	0.884 (0.004↑)	0.895 (0.006↑)	0.20 (0.04↓)	50‡ (3↓)
PiDiNet-Tiny [26]	0.882	0.889	0.08	38‡
PiDiNeXt-Tiny (Ours)	0.884 (0.002↑)	0.893 (0.004↑)	0.11 (0.03↓)	36‡ (2↓)
PiDiNet-Tiny-L [26]	0.875	0.884	**0.07**	**54‡**
PiDiNeXt-Tiny-L (Ours)	0.880 (0.005↑)	0.890 (0.006↑)	0.10 (0.03↓)	52‡ (2↓)

5.3   Ablation Study

Table 3. Ablation Study on BSDS500 dataset. *Param.* means the number of parameters in millions. ‡ indicates the speeds with our implementations based on a single 2080 Ti GPU.

Method	ODS	OIS	Param. (M)	FPS
PiDiNet	0.804	0.820	0.71	92
PiDiNeXt-	0.806	0.822	0.71	92
PiDiNet [v × 16]	0.789	0.804	0.78	80
PiDiNeXt	**0.809**	**0.824**	0.78	80

The wide use of 5 × 5 convolution improves the accuracy while increasing the parameters of PiDiNeXt. To verify the effect of model size on the accuracy, we designed two sets of ablation experiments, and the results are shown in Table 3. *PiDiNet* represents the method proposed by Su*et al.* [26], and the results of our reproduction are slightly lower than the official results. *PiDiNeXt-* denotes PiDiNeXt using RPCD only in the 3rd, 7th, 11th, and 15th parallel modules. *PiDiNeXt-*'s model structure during inference is exactly consistent with PiDiNet. It can be observed that the multi-branch structure is still able to improve the accuracy of PiDiNet even the model size is the same.

In order to make the comparison fair, we increased the parameters of PiDiNet. *PiDiNet [v × 16]* means PiDiNet that only uses RPCD. Due to RPCD will be converted to 5 × 5 convolution during inference, the model sizes of *PiDiNeXt* and *PiDiNet [v × 16]* are consistent. While the accuracy of PiDiNet decreases instead of increasing, which indicates that only increasing parameters is useless, and the accuracy can only be improved by obtaining richer features through careful design, which is consistent with the conclusions of the previous work [26].

These sets of ablation experiments imply that the success of PiDiNeXt depends on the novel multi-branching structure rather than the increase of parameters.

6   Conclusion

In conclusion, the contribution in this paper is two-fold: Firstly, we propose a multi-branch structure Parallel Pixel Difference Module (PPDM), which combines traditional edge detection operators and deep learning-based model in parallel. It solves the issue of feature choice in PiDiNet and further enrich features. Secondly, We introduce reparameterization into PiDiNeXt to prevent the extra computational cost caused by the multi-branch construction of PPDM. As a result, PiDiNeXt keeps the inference efficiency similar to PiDiNet, while the accuracy of multiple versions is improved across the board.

Limitation and Outlook. Due to the limited data and training without pre-trained model, the overfitting problem still exists in PiDiNeXt, which limits the upper bound of PiDiNeXt. Since the parallel structure of multiple branches does not affect the speed during inference, it is reasonable to try to incorporate more traditional edge detection operators into PiDiNeXt.

Acknowledgements. This work is partly supported by National key r&d program (Grant no. 2019YFF0301800), National Natural Science Foundation of China (Grant no. 61379106), the Shandong Provincial Natural Science Foundation (Grant nos. ZR2013FM036, ZR2015FM011).

References

1. Arbelaez, P., Maire, M., Fowlkes, C., Malik, J.: Contour detection and hierarchical image segmentation. IEEE Trans. Pattern Anal. Mach. Intell. **33**(5), 898–916 (2010)
2. Bertasius, G., Shi, J., Torresani, L.: High-for-low and low-for-high: efficient boundary detection from deep object features and its applications to high-level vision. In: 2015 IEEE International Conference on Computer Vision, ICCV 2015, Santiago, Chile, 7–13 December 2015, pp. 504–512. IEEE Computer Society (2015)
3. Canny, J.: A computational approach to edge detection. IEEE Trans. Pattern Anal. Mach. Intell. **8**(6), 679–698 (1986)
4. Davis, L.S.: A survey of edge detection techniques. Comput. Graph. Image Process. **4**(3), 248–270 (1975)
5. Deng, R., Liu, S.: Deep structural contour detection. In: Proceedings of the 28th ACM International Conference on Multimedia, pp. 304–312 (2020)
6. Deng, R., Shen, C., Liu, S., Wang, H., Liu, X.: Learning to predict crisp boundaries. In: Ferrari, V., Hebert, M., Sminchisescu, C., Weiss, Y. (eds.) ECCV 2018. LNCS, vol. 11210, pp. 570–586. Springer, Cham (2018). https://doi.org/10.1007/978-3-030-01231-1_35
7. Ding, X., Zhang, X., Ma, N., Han, J., Ding, G., Sun, J.: RepVGG: making VGG-style ConvNets great again. In: Proceedings of the IEEE/CVF Conference on Computer Vision and Pattern Recognition, pp. 13733–13742 (2021)
8. Dollár, P., Zitnick, C.L.: Fast edge detection using structured forests. IEEE Trans. Pattern Anal. Mach. Intell. **37**(8), 1558–1570 (2014)
9. Dollár, P., Zitnick, C.: Fast edge detection using structured forests. arXiv Computer Vision and Pattern Recognition, June 2014
10. Hallman, S., Fowlkes, C.C.: Oriented edge forests for boundary detection. In: Proceedings of the IEEE Conference on Computer Vision and Pattern Recognition, pp. 1732–1740 (2015)
11. Hallman, S., Fowlkes, C.: Oriented edge forests for boundary detection. Cornell University - arXiv, December 2014
12. He, J., Zhang, S., Yang, M., Shan, Y., Huang, T.: BDCN: bi-directional cascade network for perceptual edge detection. IEEE Trans. Pattern Anal. Mach. Intell. **44**(1), 100–113 (2022)
13. Liu, L., Zhao, L., Long, Y., Kuang, G., Fieguth, P.: Extended local binary patterns for texture classification. Image Vis. Comput. **30**(2), 86–99 (2012)
14. Liu, Y., et al.: Richer convolutional features for edge detection. IEEE Trans. Pattern Anal. Mach. Intell. **41**(08), 1939–1946 (2019)
15. Martin, D.R., Fowlkes, C.C., Malik, J.: Learning to detect natural image boundaries using local brightness, color, and texture cues. IEEE Trans. Pattern Anal. Mach. Intell. **26**(5), 530–549 (2004)
16. Martin, D., Fowlkes, C., Malik, J.: Learning to detect natural image boundaries using brightness and texture. In: Neural Information Processing Systems, January 2002
17. Martin, D., Fowlkes, C., Malik, J.: Learning to detect natural image boundaries using local brightness, color, and texture cues. IEEE Trans. Pattern Anal. Mach. Intell. **26**, 530–549 (2004)

18. Poma, X.S., Riba, E., Sappa, A.: Dense extreme inception network: towards a robust CNN model for edge detection. In: Proceedings of the IEEE/CVF Winter Conference on Applications of Computer Vision, pp. 1923–1932 (2020)
19. Prewitt, J.M., et al.: Object enhancement and extraction. Picture Process. Psychopictorics **10**(1), 15–19 (1970)
20. Pu, M., Huang, Y., Liu, Y., Guan, Q., Ling, H.: EDTER: edge detection with transformer. In: Proceedings of the IEEE/CVF Conference on Computer Vision and Pattern Recognition, pp. 1402–1412 (2022)
21. Ren, X., Bo, L.: Discriminatively trained sparse code gradients for contour detection. In: Proceedings of the 25th International Conference on Neural Information Processing Systems, vol. 1, pp. 584–592 (2012)
22. Sharifi, M., Fathy, M., Mahmoudi, M.T.: A classified and comparative study of edge detection algorithms. In: Proceedings of the International Conference on Information Technology: Coding and Computing, pp. 117–120. IEEE (2002)
23. Shengjie, Z., Garrick, B., Xiaoming, L.: The edge of depth: explicit constraints between segmentation and depth. In: Proceedings of the IEEE/CVF Conference on Computer Vision and Pattern Recognition, pp. 13116–13125 (2020)
24. Sobel, I., Feldman, G., et al.: A 3×3 isotropic gradient operator for image processing. Presented at the Stanford Artificial Intelligence Project (SAIL), pp. 271–272 (1968)
25. Soria, X., Sappa, A., Humanante, P., Akbarinia, A.: Dense extreme inception network for edge detection. Pattern Recogn. **139**, 109461 (2023)
26. Su, Z., et al.: Pixel difference networks for efficient edge detection. In: Proceedings of the IEEE/CVF International Conference on Computer Vision, pp. 5117–5127 (2021)
27. Wang, Y., Zhao, X., Huang, K.: Deep crisp boundaries. In: 2017 IEEE Conference on Computer Vision and Pattern Recognition (CVPR), July 2017
28. Wibisono, J.K., Hang, H.M.: Traditional method inspired deep neural network for edge detection. In: 2020 IEEE International Conference on Image Processing (ICIP), pp. 678–682. IEEE (2020)
29. Wibisono, J.K., Hang, H.M.: FINED: fast inference network for edge detection. In: 2021 IEEE International Conference on Multimedia and Expo (ICME), pp. 1–6. IEEE Computer Society (2021)
30. XiaoFeng, R., Bo, L.: Discriminatively trained sparse code gradients for contour detection. In: Neural Information Processing Systems, December 2012
31. Xie, S., Tu, Z.: Holistically-nested edge detection. Int. J. Comput. Vis. **125**(1), 3–18 (2017)
32. Xu, J., Xiong, Z., Bhattacharyya, S.P.: PIDNet: a real-time semantic segmentation network inspired from PID controller. arXiv preprint arXiv:2206.02066 (2022)
33. Xuan, W., Huang, S., Liu, J., Du, B.: FCL-Net: towards accurate edge detection via fine-scale corrective learning. Neural Netw. **145**, 248–259 (2022)
34. Yang, J., Price, B., Cohen, S., Lee, H., Yang, M.H.: Object contour detection with a fully convolutional encoder-decoder network. In: Proceedings of the IEEE Conference on Computer Vision and Pattern Recognition, pp. 193–202 (2016)
35. Zhao, J.X., Liu, J.J., Fan, D.P., Cao, Y., Yang, J., Cheng, M.M.: EGNet: edge guidance network for salient object detection. In: Proceedings of the IEEE/CVF International Conference on Computer Vision, pp. 8779–8788 (2019)
36. Zhou, C., Huang, Y., Pu, M., Guan, Q., Huang, L., Ling, H.: The treasure beneath multiple annotations: an uncertainty-aware edge detector. In: Proceedings of the IEEE/CVF Conference on Computer Vision and Pattern Recognition, pp. 15507–15517 (2023)

Transpose and Mask: Simple and Effective Logit-Based Knowledge Distillation for Multi-attribute and Multi-label Classification

Yuwei Zhao, Annan Li$^{(\boxtimes)}$, Guozhen Peng, and Yunhong Wang

State Key Laboratory of Virtual Reality Technology and Systems,
School of Computer Science and Engineering, Beihang University,
Beijing 100191, China
{yuweizhao,liannan,guozhen_peng,yhwang}@buaa.edu.cn

Abstract. Knowledge distillation (KD) improves a student network by transferring knowledge from a teacher network. Although KD has been extensively studied in single-labeled image classification, it is not well explored under the scope of multi-attribute and multi-label classification. We observe that the logit-based KD method for the single-label scene utilizes information from multiple classes in a single sample, but we find such logits are less informative in the multi-label scene. To address this challenge in the multi-label scene, we design a *Transpose* method to extract information from multiple samples in a batch instead of a single sample. We further note that certain classes may lack positive samples in a batch, which can negatively impact the training process. To address this issue, we design another strategy, the *Mask*, to prevent the influence of negative samples. To conclude, we propose **T**ranspose and **M**ask **K**nowledge **D**istillation (TM-KD), a simple and effective logit-based KD framework for multi-attribute and multi-label classification. The effectiveness of TM-KD is confirmed by experiments on multiple tasks and datasets, including pedestrian attribute recognition (PETA, PETA-zs, PA100k), clothing attribute recognition (Clothing Attributes Dataset), and multi-label classification (MS COCO), showing impressive and consistent performance gains.

Keywords: Knowledge distillation · multi-attribute classification · multi-label classification

1 Introduction

Recent deep learning research has revealed that increasing the model capacity properly often leads to improved performance [6,7,10]. Nevertheless, the larger model usually comes with potential drawbacks, such as long training time, increased inference latency, and high GPU memory consumption.

To address this issue, knowledge distillation (KD) [11,21] is often employed, which utilizes a strong teacher network to transfer knowledge to a relatively

Q. Liu et al. (Eds.): PRCV 2023, LNCS 14434, pp. 273–284, 2024.
https://doi.org/10.1007/978-981-99-8549-4_23

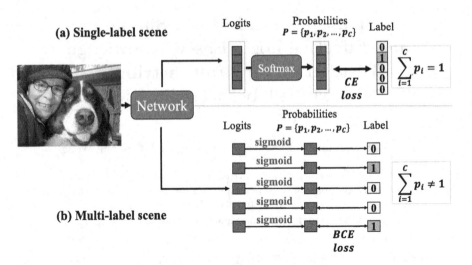

Fig. 1. The different training objectives in the single-label scene and the multi-label scene. Due to the property of the softmax function, logits in the single-label scene sum up to 1, while logits in the multi-label scene don't have such property.

weaker student network. KD has already been extensively studied in the single-labeled image classification task, where each sample has only one label. However, it has not been well investigated in the multi-attribute and multi-label classification (we use the multi-label scene to refer to them in this paper), where each sample may have multiple labels.

In this paper, we focus on exploring logit-based Knowledge Distillation (KD) in the context of the multi-label scene. The logit-based KD method stands out due to its simplicity in both idea and implementation, its independence from the backbone model structure, as well as its relatively low computational overhead compared to other KD methods [9]. In this method, the term *logits* refers to the outputs of the neural network's final layer, which will then be fed into a softmax function.

However, for the following reasons, we do not directly employ the logit-based KD method [11] (which we call vanilla KD below) commonly used in single-label image classification.

As shown in Fig. 1, in the single-label scene, a softmax function is applied to the logits to generate predictions in terms of probabilities. As the sum of prediction for all classes equal 1, the logits of different classes in a single sample become highly correlated, and therefore, contain interdependent information. This entanglement of logits has also been noted by Decoupled KD [28]. However, in the context of the multi-label scene, the logits of different classes are used to calculate loss with their own labels and do not explicitly interact with the logits of other classes. Such lack of interaction between logits of different classes weaken the information contained in the relation of logits of different classes. Since the vanilla KD distills exactly the relation of logits of different classes within a sample, directly employing the vanilla KD in the multi-label scene leads to a reduced amount of information.

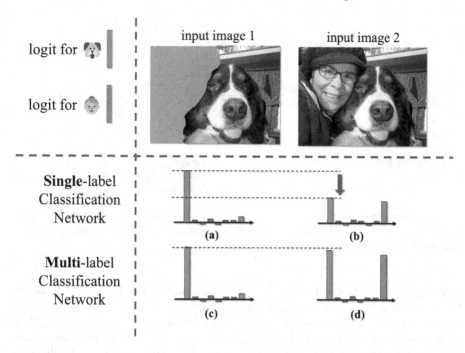

Fig. 2. In the single-label scene, the logit of a same dog can be affected by the presence of the person, as shown in **(a)** and **(b)**. Instead, in the multi-label scene, the logit of a same dog tends to be more stable and less influenced by the other classes, as shown in **(c)** and **(d)**.

On the contrary, when evaluating logits from multiple samples, logits of the same class across multiple samples in the multi-label scene are more comparable than those in the single-label scene. To demonstrate this idea, we present an example in Fig. 2, where two input images contain the exactly same dog, with the only difference being a person in the second image. In a single-label scene, the softmax function results in $\Sigma_{i=1}^{C} p_i = 1$, which makes the rise in person lead to the decrease in the dog. However, in the multi-label scene, such a decrease is not as significant because the logit of the person does not explicitly influence the logit of the dog. To conclude, if we look into the logits of the same class in different samples, their values are not comparable in the single-label scene but are comparable in the multi-label scene.

Inspired by the aforementioned two observations, we propose to distill knowledge in logits from the same class and different samples, rather than from the same sample and different classes. We refer to this strategy as the *Transpose*.

We also note that the multi-label scene typically exhibits a class imbalance, where most classes or attributes contain fewer (or significantly fewer) positive samples compared to negative samples. Therefore, under our transpose strategy, there are more negative samples than vanilla KD [11]. We assume that the relative relation in negative samples is less informative, since from two positive

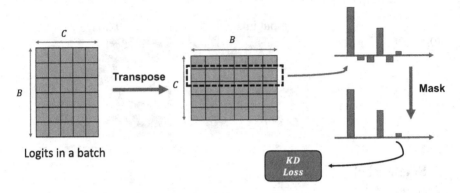

Fig. 3. Illustration of our proposed **T**ranspose and **M**ask **K**nowledge **D**istillation (TM-KD), where logits are first transposed and are then masked with 0 according to negative samples predicted by the teacher.

samples the network can learn from salient information (for example, the cat in the i-th image is more salient than the j-th image) but such information does not exist in two negative samples (since they simply contain no cat). We thus propose another strategy to fill all position whose logits in the teacher network is negative (negative samples predicted by the teacher) with zero before distillation, which we refer to the *Mask* strategy.

As illustrated in Fig. 3, based on the above analysis, we propose **T**ranspose and **M**ask **K**nowledge **D**istillation (TM-KD), which is a simple but effective logit-based knowledge distillation method. We further validate the effectiveness of our method on three tasks and five datasets, which shows TM-KD is better than both the vanilla student network and the student network with vanilla logit-based KD.

2   Related Work

Knowledge Distillation. Knowledge distillation (KD), proposed by Hinton et al. [11], aims to utilize a strong teacher network for a better student network. KD methods can be roughly divided into logit-based [11,13,15,26] and feature-based [2,21]. KD is originally proposed for single-label image classification but recent studies also show the effectiveness of KD in other tasks like object detection [4,29], semantic segmentation [19,24], graph neural network [8,25], anomaly detection [3] and some low-level tasks [23].

Multi-attribute and Multi-label Learning. Due to the limited space, it is hard to provide a detailed overview of each task. Here we provide some representative methods for these tasks. Label2Label [14] proposes a language modeling framework for clothing attribute recognition and pedestrian attribute recognition. JLAC [22] exploits graph neural network on top of convolution neural network for better results for pedestrian attribute recognition. Query2Label [17] proposes a simple transformer for better modelling multi-label classification. Note that our work doesn't focus on the state-of-the-art performance of each dataset and thus is orthogonal to these works.

KD for the Multi-label Scene. So far, KD for the multi-label scene is not well explored. Liu et al. [20] leverages the extra information from weakly-supervised detection for KD in the multi-label scene. Zhang et al. [27] proposes a feature-based method for KD in the multi-label scene exploiting class activation maps. On the contrary, our work focuses on better logit-based KD and uses no auxiliary model like the object detector.

3   Method

3.1   Preliminaries

A training batch with B samples and C classes for the multi-label scene can be described as $D = \{(x_i, y_i), i = 1, 2, ..., B\}$, where x_i is the i-th image in a batch and $y_i \in \{0, 1\}^C$ is a binary vector with length C, the lables for i-th sample. We used y_{ij} to represent the j-th attribute label for i-th sample and $y_{ij} = 1$ for a positive sample, $y_{ij} = 0$ for a negative sample.

Then, a classification network f is trained to predict a vector $z_i \in \mathbb{R}^C$ for the i-th sample and $z \in \mathbb{R}^{B \times C}$ is called **logits**. In the multi-label scene, each separate logit is then fed into *sigmoid* function, and then calculate the binary cross-entropy (BCE) loss. The above process can be formally defined as:

$$z_i = f(x_i) \tag{1}$$

$$p_{ij} = \frac{1}{(1 + e^{-z_{ij}})} \tag{2}$$

$$\mathcal{L}_{\text{BCE}} = -\frac{1}{BC} \sum_{i=1}^{B} \sum_{j=1}^{C} y_{ij} log(p_{ij}) + (1 - y_{ij}) log(1 - p_{ij}) \tag{3}$$

And if KD is applied during the training, the final loss can be represented as:

$$\mathcal{L} = \mathcal{L}_{\text{BCE}} + \lambda \mathcal{L}_{\text{KD}} \tag{4}$$

where λ is a hyperparameter to balance BCE loss and KD loss.

Below we will show the different designs of $\mathcal{L}_{\text{KD}}$ in vanilla KD and our TM-KD.

3.2   Vanilla KD

Apart from the student network f^s, KD uses a stronger teacher network f^t trained beforehand to help the student network. The logits of them can be represented as $z^s \in \mathbb{R}^{B \times C}$ and $z^t \in \mathbb{R}^{B \times C}$.

Directly using vanilla KD from the single-label classification, Kullback-Leibler (KL) divergence loss is used to minimize the discrepancy of probabilities from different classes in the same sample, which is:

$$p_i^t = softmax(z_i^t/\tau), \quad p_i^s = softmax(z_i^s/\tau), \quad i = 1, 2, ..., B \tag{5}$$

$$\mathcal{L}_{\mathrm{KD}} = \frac{1}{B}\sum_{i=1}^{B} KL(p_i^t, p_i^s) \tag{6}$$

where τ is a hyperparameter to adjust the smoothness of two probabilistic distributions.

Algorithm 1: PyTorch-style pseudocode for vanilla KD and TM-KD.

```
T = 1
KLdiv = nn.KLDivLoss(reduction="batchmean")
def KD_from_logit(logit_stu, logit_tea):
  log_prob = F.log_softmax(logits_stu/T, dim=1)
  prob = F.softmax(logits_tea/T, dim=1)
  loss = KLdiv(log_prob, prob)
  return loss

def vanillaKD(logit_stu, logit_tea):
  # logit_stu.shape : [B, C]
  return KD_from_logit(logit_stu, logit_tea)

def TM_KD(logit_stu, logit_tea):
  # logit_stu.shape : [B, C]
  mask = (logit_tea[0]<0).int()
  logits_stu = logits_stu.masked_fill(mask, 0)
  logits_tea = logits_tea.masked_fill(mask, 0)
  logits_stu = logits_stu.permute(1,0)
  logits_tea = logits_tea.permute(1,0)
  # logit_stu.shape : [C, B]
  return KD_from_logit(logit_stu, logit_tea)
```

3.3 TM-KD

As illustrated in Fig. 3, our TM-KD consist of two strategies, i.e. the Transpose and the Mask respectively.

For the Mask, to alleviate the influence of useless information in negative samples for a class, we set the position to zero if the corresponding logits in the teacher network are negative. By doing so, the teacher network only distills the knowledge in positive samples from its perspective. Formally:

$$\hat{z}_{ij}^* = \begin{cases} z_{ij}^*, & z_{ij}^t \geq 0 \\ 0, & z_{ij}^t \leq 0 \end{cases} \tag{7}$$

where $* \in \{s, t\}$. Note the student network and teacher network share the same mask from the teacher network.

For the Transpose, we no longer distill from the different classes in the same sample, but from the different samples in the same class, which is done by:

$$p_j^t = softmax(\hat{z}_{*j}^t/\tau), \quad p_j^s = softmax(\hat{z}_{*j}^s/\tau), \quad j = 1, 2, ..., C \tag{8}$$

$*$ can be any i where $1 \leq i \leq B$, and $\hat{z}_{*j}^t, \hat{z}_{*j}^s \in \mathbb{R}^B$ can be viewed as logits of the same class in different samples. We have analyzed why they contain more information in the multi-label scene in the introduction.

We also provide a pseudocode of vanilla KD and TM-KD in Algo. 1.

4 Experiments

In this section, we validate the performance of our TM-KD on three tasks and five datasets. Our TM-KD consistently demonstrates impressive performance across all the datasets. In addition, we conduct ablation studies to demonstrate the effectiveness of our Transpose and Mask strategy.

4.1 Experimental Setting

We conduct our experiment under two KD settings with ResNet [10], where ResNet-101 serves as the teacher model for teaching ResNet-50, and ResNet-50 serves as the teacher model for teaching ResNet-18. We train a teacher first and then utilize it to help train a student. The only exception is that we used ResNet-101 as the teacher and ResNet-34 as the student for the MS COCO dataset. Our code is on top of the codebase by Jia et al. [12] and the teacher network is retrained instead of loaded. Below, we'll present more experimental details.

Datasets. We evaluate on pedestrian attribute recognition using PETA [5], PETA-zs [5,12], and PA100k [18], clothing attribute recognition using the Clothing Attributes Dataset [1], and multi-label classification using MS COCO [16]. We list the statistics of used datasets in Table 1. We utilize default dataset split and more details can be found in their original paper and codebase by Jia et al. [12]. Note that for the clothing attributes attribute dataset, we use only 22 out of 26 attributes, where we exclude attributes of *sleeve length, neckline, category* and *gender*.

Implementation Details. For KD hyperparameters, we set $\lambda = 20$ in Eq. 4 and set $\tau = 1$ in Eq. 8 for all experiments. It turns out that the order of magnitude of λ and τ (in the power of 10) does affect the results, but when λ and τ are in the same order of magnitude, the exact values of them don't affect the result. Since we implemented our method on top of the codebase by Jia et al. [12], we used most of the default settings from it. It's guaranteed that hyperparameters for different methods on the same dataset are also the same.

Table 1. Statistics of 5 used datasets from 3 different tasks. **PAR**: pedestrian attribute recognition. **CAR**: clothing attribute recognition. **MLC**: multi-label classification. N_{train}: number of samples in train set. N_{test}: number of samples in test set. N_{attr}: number of attributes used in this dataset.

Task	Dataset	N_{train}	N_{test}	N_{attr}
PAR	PETA	11400	7600	35
	PETA-zs	15067	3933	35
	PA100k	90000	10000	26
CAR	Clothing Attributes	1500	356	22
MLC	MS COCO	82081	40137	80

Metrics. Following the routine in previous works, We report mean accuracy (mA) and $micro-F1$ for pedestrian attribute recognition and clothing attribute recognition datasets and report mA for the multi-label classification dataset. Since ReduceLROnPlateau learning rate scheduler is used following the codebase by Jia et al. [12], we report the metrics after the first epoch of learning rate reducing to 10^{-5} for pedestrian attribute recognition datasets, and we report those of clothing attribute recognition dataset for first reduction to 10^{-6}. For the multi-label classification dataset, we report the metrics at the last (30) epoch.

Table 2. The results for pedestrian attribute recognition, Δ_* represent the performance gains from our TM-KD compared with the baseline w/o KD and the baseline with vanilla KD. The rightmost column reports the average of 3 datasets.

KD Setting	KD Method	PETA		PETA-zs		PA100k		Avg.	
		mA	F1	mA	F1	mA	F1	mA	F1
ResNet50 ↓ ResNe18	Teacher	84.21	86.21	71.07	72.12	80.27	87.32	78.51	81.88
	w/o KD[12]	81.76	84.85	70.70	71.54	77.97	85.92	76.81	80.77
	Vanilla KD[11]	81.95	85.18	69.29	71.70	78.54	86.33	76.59	81.07
	TM-KD(Ours)	84.75	85.02	73.54	71.77	81.66	85.96	79.98	80.91
	$\Delta_{w/o}$	+2.99	+0.17	+2.84	+0.23	+3.69	+0.04	+3.17	+0.14
	$\Delta_{vanilla}$	+2.80	-0.16	+4.25	+0.07	+3.12	-0.37	+3.39	-0.16
ResNet101 ↓ ResNet50	Teacher	85.13	87.11	71.84	72.25	80.66	87.59	79.21	82.31
	w/o KD[12]	84.21	86.21	71.07	72.12	80.27	87.32	78.51	81.88
	Vanilla KD[11]	84.04	86.35	71.07	72.76	80.22	87.34	78.44	82.15
	TM-KD(Ours)	86.72	85.83	73.40	72.62	82.74	87.29	80.95	81.91
	$\Delta_{w/o}$	+2.51	-0.38	+2.33	+0.50	+2.47	-0.03	+2.44	+0.03
	$\Delta_{vanilla}$	+2.68	-0.52	+2.33	-0.14	+2.52	-0.05	+2.51	-0.24

4.2 Main Results

Pedestrian Attribute Recognition. We report our result in Table 2. It can be seen that vanilla KD has negligible influence on baseline. And we compared TM-KD with the baseline w/o KD and with vanilla KD in Δ_* rows. Our TM-KD has impressive and consistent gains on all datasets w.r.t mA.

When it comes to $F1$, our model isn't as outstanding as its performance w.r.t mA but still gets an overall positive delta on the average performance compared to the baseline w/o KD. We argue that mA is the main metric for pedestrian attribute recognition since it calculates the mean accuracy over classes, while $micro - F1$ treats all samples equally and thus can't well reflect the model's performance in class imbalance scene. Jia et al. [12] also note that the trade-off exists between mA and $F1$, and they show that by changing the weight function we can control the trade-off lean to mA or $F1$ to some extents.

Table 3. Our results on the Clothing Attributes Dataset. Note that our TM-KD has even helped student's performance surpass their corresponding teacher.

KD Setting	Teacher	w/o KD	Vanilla KD	TM-KD(Ours)
	(mA/F1)	(mA/F1)	(mA/F1)	(mA/F1)
R50→R18	65.2/48.7	62.3/41.0	61.1/39.5	**68.7/50.2**
R101→R50	66.4/50.0	65.2/48.7	64.5/47.4	**71.3/55.2**

Clothing Attribute Recognition. Our results on clothing attribute recognition are presented in Table 3, wherein our TM-KD demonstrates more remarkable performance. Surprisingly, our ResNet-18 student, trained by ResNet-50, outperforms even the ResNet-101 teacher. Additionally, the ResNet-50 teached by ResNet-101 with our TM-KD also achieves significantly better results compared to all other methods. On the contrary, the vanilla KD approach leads to performance degradation for both ResNet-18 and ResNet-50.

One possible reason for such remarkable performance may be the fact that the Clothing Attributes Dataset contains a very limited number of samples (recall Table 1). Intuitively, when the training data is extremely insufficient to train a network, the network has even more potential to progress. Consider two college students, and in their final exams one gets a D^- grade while another gets an A grade. If we teach them in the same way, apparently the former will progress more.

Table 4. Our results on the MS COCO dataset.

ResNet-101	ResNet-34		
	w/o KD [12]	vanilla KD [11]	TM-KD(ours)
83.04	78.81	77.11	**80.83**

Multi-label Classification. As shown in Table 4, our results also boost the performance of the student ResNet-34 (+2.02) on the MS COCO multi-label dataset, contrary to the negative impact caused by vanilla KD. Although the improvement may not be significant, considering the difficulty of this dataset and performance degradation of the vanilla KD, the result is still quite impressive.

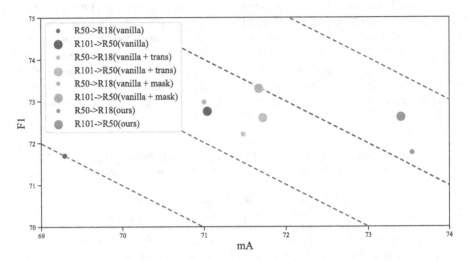

Fig. 4. Ablation of the proposed Transpose strategy and Mask strategy on the PETA-zs dataset. The dark blue dashed line assumes the 1:1 trade-off between mA and F1. (Color figure online)

4.3   Ablation Study

To evaluate the effectiveness of the two proposed strategies, we conduct an ablation study by incorporating one of them into the vanilla KD method. The corresponding results are presented in Fig. 4. As discussed previously in Sect. 4.2, there exists a trade-off between mA and F1 scores in pedestrian attribute recognition. To provide a better visual representation, we assume an equal trade-off ratio of 1:1 and plot a dark blue dashed line. Under this assumption, points on the same line are considered equally effective. It can be seen that applying only one of our strategies can also improve the performance compared with vanilla KD, validating the effectiveness of the proposed two strategies. And when used together in our TM-KD, the performance becomes even better.

5   Conclusion

In this paper, we analyze the logits in the single-label scene and the multi-label scene and then propose TM-KD (**T**ranspose and **M**ask **K**nowledge **D**istillation), a simple and effective logit-base KD method for multi-attribute and multi-label classification. The proposed method is evaluated on five datasets of three tasks. While vanilla KD usually brings nearly no improvement and sometimes even degradation, TM-KD gets impressive and consistent results on all datasets, validating the effectiveness of TM-KD.

Acknowledgment. This work was supported by National Natural Science Foundation of China under Grant U20B2069.

References

1. Chen, H., Gallagher, A., Girod, B.: Describing clothing by semantic attributes. In: Fitzgibbon, A., Lazebnik, S., Perona, P., Sato, Y., Schmid, C. (eds.) ECCV 2012. LNCS, vol. 7574, pp. 609–623. Springer, Heidelberg (2012). https://doi.org/10.1007/978-3-642-33712-3_44
2. Chen, P., Liu, S., Zhao, H., Jia, J.: Distilling knowledge via knowledge review. In: Proceedings of the IEEE/CVF Conference on Computer Vision and Pattern Recognition, pp. 5008–5017 (2021)
3. Cheng, H., Yang, L., Liu, Z.: Relation-based knowledge distillation for anomaly detection. In: Ma, H., et al. (eds.) PRCV 2021. LNCS, vol. 13019, pp. 105–116. Springer, Cham (2021). https://doi.org/10.1007/978-3-030-88004-0_9
4. Dai, X., et al.: General instance distillation for object detection. In: Proceedings of the IEEE/CVF Conference on Computer Vision and Pattern Recognition, pp. 7842–7851 (2021)
5. Deng, Y., Luo, P., Loy, C.C., Tang, X.: Pedestrian attribute recognition at far distance. In: Proceedings of the 22nd ACM International Conference on Multimedia, pp. 789–792 (2014)
6. Devlin, J., Chang, M.W., Lee, K., Toutanova, K.: Bert: pre-training of deep bidirectional transformers for language understanding. arXiv preprint arXiv:1810.04805 (2018)
7. Dosovitskiy, A., et al.: An image is worth 16×16 words: transformers for image recognition at scale. In: International Conference on Learning Representations (2021). https://openreview.net/forum?id=YicbFdNTTy
8. Feng, K., Li, C., Yuan, Y., Wang, G.: Freekd: free-direction knowledge distillation for graph neural networks. In: Proceedings of the 28th ACM SIGKDD Conference on Knowledge Discovery and Data Mining, pp. 357–366 (2022)
9. Gou, J., Yu, B., Maybank, S.J., Tao, D.: Knowledge distillation: a survey. Int. J. Comput. Vision **129**, 1789–1819 (2021)
10. He, K., Zhang, X., Ren, S., et al.: Deep residual learning for image recognition. In: CVPR (2016)
11. Hinton, G., Vinyals, O., Dean, J.: Distilling the knowledge in a neural network. arXiv preprint arXiv:1503.02531 (2015)
12. Jia, J., Huang, H., Chen, X., Huang, K.: Rethinking of pedestrian attribute recognition: a reliable evaluation under zero-shot pedestrian identity setting. arXiv preprint arXiv:2107.03576 (2021)
13. Jin, Y., Wang, J., Lin, D.: Multi-level logit distillation. In: Proceedings of the IEEE/CVF Conference on Computer Vision and Pattern Recognition, pp. 24276–24285 (2023)

14. Li, W., Cao, Z., Feng, J., Zhou, J., Lu, J.: Label2label: a language modeling framework for multi-attribute learning. In: Computer Vision-ECCV 2022: 17th European Conference, Tel Aviv, Israel, 23–27 October 2022, Proceedings, Part XII, pp. 562–579. Springer, Heidelberg (2022). https://doi.org/10.1007/978-3-031-19775-8_33

15. Li, Z., et al.: Curriculum temperature for knowledge distillation. In: Proceedings of the AAAI Conference on Artificial Intelligence, vol. 37, pp. 1504–1512 (2023)

16. Lin, T.-Y., et al.: Microsoft COCO: common objects in context. In: Fleet, D., Pajdla, T., Schiele, B., Tuytelaars, T. (eds.) ECCV 2014. LNCS, vol. 8693, pp. 740–755. Springer, Cham (2014). https://doi.org/10.1007/978-3-319-10602-1_48

17. Liu, S., Zhang, L., Yang, X., Su, H., Zhu, J.: Query2label: a simple transformer way to multi-label classification. arXiv preprint arXiv:2107.10834 (2021)

18. Liu, X., et al.: Hydraplus-net: attentive deep features for pedestrian analysis. In: Proceedings of the IEEE International Conference on Computer Vision, pp. 350–359 (2017)

19. Liu, Y., Shu, C., Wang, J., Shen, C.: Structured knowledge distillation for dense prediction. IEEE Trans. Pattern Anal. Mach. Intell. (2020)

20. Liu, Y., Sheng, L., Shao, J., Yan, J., Xiang, S., Pan, C.: Multi-label image classification via knowledge distillation from weakly-supervised detection. In: Proceedings of the 26th ACM International Conference on Multimedia, pp. 700–708 (2018)

21. Romero, A., Ballas, N., Kahou, S.E., Chassang, A., Gatta, C., Bengio, Y.: Fitnets: hints for thin deep nets. arXiv preprint arXiv:1412.6550 (2014)

22. Tan, Z., Yang, Y., Wan, J., Guo, G., Li, S.Z.: Relation-aware pedestrian attribute recognition with graph convolutional networks. In: Proceedings of the AAAI Conference on Artificial Intelligence, vol. 34, pp. 12055–12062 (2020)

23. Wang, N., Cui, Z., Li, A., Su, Y., Lan, Y.: Multi-priors guided dehazing network based on knowledge distillation. In: Pattern Recognition and Computer Vision: 5th Chinese Conference, PRCV 2022, Shenzhen, China, 4–7 November 2022, 2022, Proceedings, Part IV, pp. 15–26. Springer, Heidelberg (2022). https://doi.org/10.1007/978-3-031-18916-6_2

24. Wang, Y., Zhou, W., Jiang, T., Bai, X., Xu, Y.: Intra-class feature variation distillation for semantic segmentation. In: Vedaldi, A., Bischof, H., Brox, T., Frahm, J.-M. (eds.) ECCV 2020. LNCS, vol. 12352, pp. 346–362. Springer, Cham (2020). https://doi.org/10.1007/978-3-030-58571-6_21

25. Yang, Y., Qiu, J., Song, M., Tao, D., Wang, X.: Distilling knowledge from graph convolutional networks. In: Proceedings of the IEEE/CVF Conference on Computer Vision and Pattern Recognition, pp. 7074–7083 (2020)

26. Zhang, Y., Xiang, T., Hospedales, T.M., Lu, H.: Deep mutual learning. In: Proceedings of the IEEE Conference on Computer Vision and Pattern Recognition, pp. 4320–4328 (2018)

27. Zhang, Y., Qin, Y., Liu, H., Zhang, Y., Li, Y., Gu, X.: Knowledge distillation from single to multi labels: an empirical study. arXiv preprint arXiv:2303.08360 (2023)

28. Zhao, B., Cui, Q., Song, R., Qiu, Y., Liang, J.: Decoupled knowledge distillation. In: Proceedings of the IEEE/CVF Conference on Computer Vision and Pattern Recognition, pp. 11953–11962 (2022)

29. Zheng, Z., et al.: Localization distillation for dense object detection. In: Proceedings of the IEEE/CVF Conference on Computer Vision and Pattern Recognition, pp. 9407–9416 (2022)

CCSR-Net: Unfolding Coupled Convolutional Sparse Representation for Multi-focus Image Fusion

Kecheng Zheng, Juan Cheng, and Yu Liu(✉)

Department of Biomedical Engineering, Hefei University of Technology, Hefei 230009, China
yuliu@hfut.edu.cn

Abstract. Multi-focus image fusion aims to generate an all-in-focus image from multiple partially focused images of the same scene captured with different focal settings. In this paper, we present a coupled convolutional sparse representation (CCSR) model for multi-focus image fusion. Instead of being solved by an iterative thresholding algorithm, the proposed CCSR model is unfolded into a learnable neural network (termed as CCSR-Net) using the deep unfolding technique, taking the advantages of both traditional methods and deep-learning (DL)-based ones. Based on the CCSR-Net, a new multi-focus image fusion method with good interpretability is further proposed. Experimental results on two popular datasets show that the proposed method can obtain the state-of-the-art performance in terms of both visual quality and objective assessment.

Keywords: Multi-focus image fusion · Convolutional sparse representation · Deep unfolding · Convolutional neural network

1 Introduction

Due to limited depth-of-field (DOF) of optical lenses, in some cases, it is difficult to capture an image where all the objects are in focus. Consequently, objects within the DOF have sharp appearance while the others tend to be blurred. Multi-focus image fusion, which offers an effective way to address this problem, aims to generate an all-in-focus image from multiple partially focused images of the same scene captured with different focal settings [1]. This technique is of great significance in many fields such as digital photography and optical microscopy. In recent years, various multi-focus image fusion methods have been proposed, which can roughly be categorized into traditional methods and deep learning (DL)-based methods.

Traditional multi-focus image fusion methods can be further divided into transform domain methods (e.g., multi-scale transform-based ones [2–4] and spare representation-based ones [5–7]) and spatial domain methods (e.g., block-based ones [8–10] and pixel-based ones [11–13]). Among them, owing to the robust signal modeling ability, sparse representation (SR) has become a popular model in the study of multi-focus image fusion [14]. Particularly, SR can not only be used as a transform framework in transform domain methods [5], but also be adopted as a feature extraction approach to generate activity level measures in spatial domain methods [11]. Since sparsity is an inherent property of images, the SR-based methods have solid theoretical foundations with good interpretability and tend to produce promising fusion results. In addition to the conventional patch-wise SR-based fusion methods (i.e., modeling overlapping image patches) [5,6], Liu et al. introduced a convolutional sparse representation (CSR)-based image fusion method [7], in which sparse coding is performed over the entire image to achieve globally optimized representations, leading to better capacity in detail preservation and higher robustness to mis-registration. However, the SR-based methods usually suffer from low computational efficiency due to the adoption of iterative optimization algorithms.

In the last few years, DL-based fusion methods have become a mainstream in multi-focus image fusion, with a number of methods being proposed based on convolutional neural networks (CNNs) [15–20], generative adversarial networks (GANs) [21–23], Transformer models [24,25], etc. In comparison to traditional methods, DL-based methods can achieve more effective feature representation ability and avoid the difficulty of manual designs. Moreover, with the aid of GPU acceleration, the DL-based methods usually have quite high computational efficiency during inference processes. However, most existing DL-based fusion models lack of interpretability, which may affect their generalization ability. In fact, a recent comparative study on multi-focus image fusion has demonstrated that the performance of most existing DL-based methods is not always superior or even slightly inferior to the state-of-the-art traditional methods [26].

In this paper, we attempt to bridge the above gap between DL-based methods and traditional methods, so as to obtain their complementary advantages. In particular, a coupled CSR (CCSR) model for multi-focus image fusion is presented. Instead of being solved by an iterative thresholding algorithm, the proposed CCSR model is unfolded into a learnable neural network using the deep unfolding technique [27]. As a result, an interpretable DL-based fusion model is obtained. It can preserve the strong feature learning ability and high inference speed of DL models, while with good interpretability benefited from traditional SR-based approaches. In summary, the main contribution of this paper is two-fold:

1. An image fusion-oriented coupled CSR model is presented and unfolded into a neural network. This interpretable DL-based fusion model, termed as CCSR-Net, combines the advantages of DL-based and traditional methods.
2. A new multi-focus image fusion approach based on the CCSR-Net is proposed. Experimental results demonstrate that the proposed method can obtain the state-of-the-art performance in terms of both visual quality and objective assessment.

2   Background

Sparse coding is a technique to represent a signal by a few atoms in a dictionary. Specifically, we assume that $x = D\alpha$, where $x \in R^n$ denotes the signal, $D \in R^{n \times M}, n \leq M$, is the dictionary, and the sparse coding $\alpha \in R^M$ can be obtained by solving the minimization problem

$$\min_{\alpha} \frac{1}{2}\|x - D\alpha\|_2^2 + \lambda\|\alpha\|_1, \tag{1}$$

where λ is the regularization parameter, and the regularization term with ℓ_1-norm is to promote sparsity. In the image domain, sparse coding can be extended to convolutional sparse coding (CSC), defined as:

$$\min_{A_m} \frac{1}{2}\left\|X - \sum_{m=1}^{M} D_m * A_m\right\|_2^2 + \lambda \sum_{m=1}^{M} \|A_m\|_1, \tag{2}$$

where $X \in R^{n_1 \times n_2}$ represents the input image. $D_m \in R^{p_1 \times p_2}$, $m = 1, \ldots, M$, are the atoms of the convolution dictionary $\mathcal{D} \in R^{p_1 \times p_2 \times M}$. $A_m \in R^{n_1 \times n_2}$, $m = 1, \ldots, M$, are the sparse feature maps, and $*$ is the convolutional operator.

CSC can be improved if prior knowledge of side information on the target image is known. Given the sparse feature map of the side information $B_m \in R^{n_1 \times n_2}$, it is assumed that A_m should be similar to B_m. Taking such prior knowledge into account, the side information guided CSC model can be formulated by inserting an extra penalty into Eq. (2), shown as:

$$\min_{A_m} \frac{1}{2}\left\|X - \sum_{m=1}^{M} D_m * A_m\right\|_2^2 + \lambda \sum_{m=1}^{M} (\|A_m\|_1 + \|A_m - B_m\|_1). \tag{3}$$

3   The Proposed Method

3.1   Model Formulation

For a certain target image, the sparse feature map can be obtained by Eq. 3 when the sparse feature map of the guide image is known, so that the constraint in Eq. 3 is a unidirectional flow of information. Howevre, for multi-focus image fusion or even image fusion tasks, the information of source images is complementary, and during the process of integrating shch complementary information into one fused image, the corresponding flow should be bidirectional and symmetric. From this perspective, a coupled convolutional sparse representation of source images is constructed, so that the sparse feature maps of the source images (*i.e.* two source images) can be mutually constrained, or guided, to form a bidirectional flow. Specifically, for two source image X and Y, the sparse feature map of the source image X is given as the side information of the source image Y. In turn, the sparse feature map of the source image Y is given as the side information of the source image X. Therefore, the optimization problem is formulated as:

$$\min_{U_m} \frac{1}{2} \left\| X - \sum_{m=1}^{M} C_m * U_m \right\|_2^2 + \lambda \sum_{m=1}^{M} \left(\| U_m \|_1 + \| U_m - V_m \|_1 \right), \quad (4)$$

$$\min_{V_m} \frac{1}{2} \left\| Y - \sum_{m=1}^{M} D_m * V_m \right\|_2^2 + \lambda \sum_{m=1}^{M} \left(\| V_m \|_1 + \| V_m - U_m \|_1 \right), \quad (5)$$

where $X \in R^{n_1 \times n_2}$, $Y \in R^{n_1 \times n_2}$ represents the two source images. $C_m \in R^{p_1 \times p_2}$, $D_m \in R^{p_1 \times p_2}$, $m = 1, \ldots, M$, are the atoms of the convolutional dictionary $C \in R^{p_1 \times p_2 \times M}$ and the convolutional dictionary $D \in R^{p_1 \times p_2 \times M}$, respectively. $U_m \in R^{n_1 \times n_2}$, $V_m \in R^{n_1 \times n_2}$, $m = 1, \ldots, M$, are the corresponding sparse feature maps, and $*$ is the convolutional operator.

Equation 4 and Eq. 5 are standard CSC problems guided by side information, and can be solved by a learned multi-modal CSC (LMCSC) algorithm proposed in [28], shown as follows:

$$\mathcal{U}^t = P_\gamma \left(\mathcal{U}^{t-1} - \mathcal{Q}_u * \mathcal{R}_u * \mathcal{U}^{t-1} + \mathcal{S}_u * X \ ; \mathcal{V}^{t-1} \right), \quad (6)$$

$$\mathcal{V}^t = P_\gamma \left(\mathcal{V}^{t-1} - \mathcal{Q}_v * \mathcal{R}_v * \mathcal{V}^{t-1} + \mathcal{S}_v * Y ; \mathcal{U}^t \right), \quad (7)$$

where $\mathcal{U} \in R^{p_1 \times p_2 \times M}$, $\mathcal{V} \in R^{p_1 \times p_2 \times M}$, are the sparse feature maps of the source images X and Y, respectevely. $\mathcal{Q}_u \in R^{p_1 \times p_2 \times c \times M}$, $\mathcal{R}_u \in R^{p_1 \times p_2 \times c \times M}$, $\mathcal{S}_u \in R^{p_1 \times p_2 \times c \times M}$, $\mathcal{Q}_v \in R^{p_1 \times p_2 \times c \times M}$, $\mathcal{R}_v \in R^{p_1 \times p_2 \times c \times M}$, $\mathcal{S}_v \in R^{p_1 \times p_2 \times c \times M}$, are the learnable convolutional layers. γ is the learnable parameter. c is the channel number of each source image. t is the number of iterations. $P_\gamma(x; s)$ is the piecewise soft thresholding function, defined as:

If $s \geq 0$,

$$P_\gamma(x; s) = \begin{cases} x + 2\gamma, & x < -2\gamma \\ 0, & -2\gamma \leq x \leq 0 \\ x, & 0 < x < s \\ s, & s \leq x \leq s + 2\gamma \\ x - 2\gamma, & x \geq s + 2\gamma \end{cases} \quad (8)$$

If $s < 0$,

$$P_\gamma(x; s) = \begin{cases} x + 2\gamma, & x < s - 2\gamma \\ s, & s - 2\gamma \leq x \leq s \\ x, & s < x < 0 \\ 0, & 0 \leq x \leq 2\gamma \\ x - 2\gamma, & x \geq 2\gamma. \end{cases} \quad (9)$$

After iteratively solving to obtain the sparse feature map of each source image, the corresponding activity level measurement is follow-by conducted. Specifically, the content of $\mathcal{U}$ or $\mathcal{V}$ at position (x, y) in the spatial domain is denoted by $\mathcal{U}_{1:M}$ or $\mathcal{V}_{1:M}$. According to [7], the ℓ_1-norm of $\mathcal{U}_{1:M}(x, y)$ and

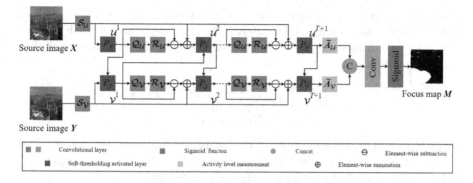

Fig. 1. The structure of the proposed CCSR-Net model.

$\mathcal{V}_{1:M}(x, y)$ are employed to measure the activity level of the two source images. Therefore, the initial activity level map $\boldsymbol{A_U}(x, y)$ and $\boldsymbol{A_V}(x, y)$ are given as:

$$\boldsymbol{A_U}(x, y) = \|\boldsymbol{\mathcal{U}}_{1:M}(x, y)\|_1, \tag{10}$$

$$\boldsymbol{A_V}(x, y) = \|\boldsymbol{\mathcal{V}}_{1:M}(x, y)\|_1. \tag{11}$$

To improve the robustness to misalignment and noises, a window-based averaging strategy is used to obtain the final activity level map:

$$\bar{\boldsymbol{A}}_{\boldsymbol{U}}(x, y) = \frac{\sum_{p=-r}^{r} \sum_{q=-r}^{r} \boldsymbol{A_U}(x+p, y+q)}{(2r+1)^2}, \tag{12}$$

$$\bar{\boldsymbol{A}}_{\boldsymbol{V}}(x, y) = \frac{\sum_{p=-r}^{r} \sum_{q=-r}^{r} \boldsymbol{A_V}(x+p, y+q)}{(2r+1)^2}, \tag{13}$$

where r is the window size. A larger r will be more robust to error information, but meanwhile leading to lose some detail information. In multi-focus image fusion, the position of the object edges in a certain source image is not always the same as that in other source images due to their different sharpness. In this paper, r is relatively large and set to 9.

The structure of our proposed CCSR-Net model is shown in Fig. 1. After obtaining the two active level maps that contain focus information of each source image, these two active level maps of the two source images are concatenated and sent to a convolutional layer and a Sigmoid activation function for automatically learn the weights. Therefore, the focus map $\boldsymbol{M}$ can be generated.

3.2 Loss Function

The binary cross entropy (BCE) loss function, consistent for each pixel, is widely used in binary classification tasks. However, in the multi-focus image fusion task, the pixels at the boundary between the focused and defocused regions are often classified incorrectly, which needs different weights compared to those of the pixels at other positions. To pay more attention to the border pixels, a weighted BCE (WBCE) loss function, with different weight assignement to each pixel by calculating the discrepancy between the central pixel point and the surrounding pixel points, defined as:

$$L_{wbce} = \frac{\sum_{i=1}^{H} \sum_{j=1}^{W} (1 + \gamma\alpha_{i,j}) BCE(P_{i,j}, G_{i,j})}{\sum_{i=1}^{H} \sum_{j=1}^{W} \gamma\alpha_{i,j}}, \tag{14}$$

where $P_{i,j}$ and $G_{i,j}$ are the values of the prediction mask and the true mask at pixel point (i,j), respectively. γ is a constant and is set to 5 in this paper. $\alpha_{i,j}$ is the weight assigned to pixel point (i,j), expressed as:

$$\alpha_{i,j} = \left| \frac{\sum_{m,n \in A_{i,j}} G_{m,n}}{|A_{i,j}|} - G_{i,j} \right|, \tag{15}$$

where $A_{i,j}$ is the area around the pixel. $|A_{i,j}|$ is the total number of pixels in the area, and (m,n) is the pixel within the area.

3.3　CCSR-Net-Based Multi-Focus Image Fusion

The schematic diagram of our proposed CCSR-Net-based image fusion method is shown in Fig. 2, consisting of three steps, namely initial decision map estimation, consistency verification, and fusion.

Initial Decision Map Estimation. The two source images are fed into the pre-trained network to obtain the focus map M. Each pixel in the focus map has a value between 0 and 1, indicating the focus attribute of the source image X. A higher value indicates a higher probability that the corresponding pixel belongs to the focused region in X, and vice versa. The initial decision map T is a binarization of the focus map M, shown as follows:

$$T(x,y) = \begin{cases} 1, & \text{if } M(x,y) > 0.5 \\ 0, & \text{otherwise} \end{cases}. \tag{16}$$

Consistency Verification. The derived initial decision map T may still contain some misclassified pixels. To further improve the quality of the fused image, a small region removal strategy in [15] is adopted to refine the initial decision map T. Specifically, we invert the regions in the initial decision map T that are smaller than the area threshold to obtain the final decision map D. In this paper, the area threshold is set to $0.01 \times H \times W$, where H and W are the height and the width of the source image, respectively.

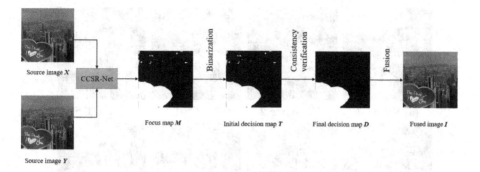

Fig. 2. Schematic diagram of the proposed CCSR-Net-based multi-focus image fusion method.

Fusion. The final decision map D is considered as a mask of the focused and defocused regions of the source image X, and denoted by 1 the all-1 matrix of the same shape as D. The fused image I is calculated by the following pixel-weighted averaging rule, expressed as:

$$I = X \cdot D + Y \cdot (1 - D). \tag{17}$$

4   Experiments

4.1   Experimental Settings

Training Details. 500 foreground images were selected from the datasets of [29, 30], 2000 images were chosen from COCO dataset [31], and 60 images were randomly selected as background images. To generate realistic training data with the defocus spread effect precisely modeled, we used a new α-matte boundary defocus model proposed in [32]. Our training dataset contains 30,000 multi-focus image pairs of size 512×512. The proposed method is implemented in Pytorch and trained on an NVIDIA TITAN RTX GPU. During the training process, the Adam optimizer is used, the initial learning rate is set to 0.0001, and the decay is 0.9 times of the original every 15 epochs.

Testing Data. The performance of our proposed CCSR-Net-based method is evaluated on two public datasets, namely "Lytro" and "MFFW". There are 20 pairs of multi-focus images in the "Lytro" dataset and 13 pairs of real-world multi-focus images collected on the Internet in "MFFW" dataset. The scenes in "MFFW" dataset are far more complicated and with a significant defocus spread effect, which brings challenge to the proposed fusion method.

Comparison Methods. The performance of the proposed method is compared with nine other typical multi-focus image fusion methods, including CSR [7], NSCT-SR [4], MADCNN [16], IFCNN [17], MFF-GAN [21], DRPL [18], ECNN [19], MFF-SSIM [20], and PSNN [33].

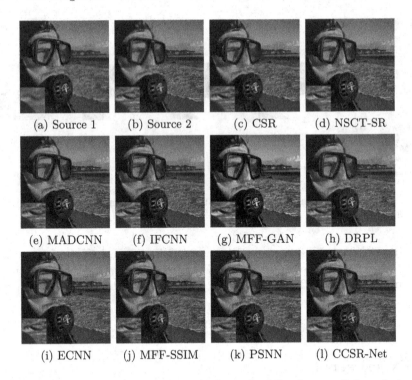

(a) Source 1   (b) Source 2   (c) CSR   (d) NSCT-SR

(e) MADCNN   (f) IFCNN   (g) MFF-GAN   (h) DRPL

(i) ECNN   (j) MFF-SSIM   (k) PSNN   (l) CCSR-Net

Fig. 3. Fusion results of different methods on the "diver" example from Lytro.

Evaluation Metrics. Six popular objective metrics are employed for performance evaluation, namely the normalized mutual information-based metric Q_{MI} [34], the nonlinear correlation information entropy Q_{NICE} [35], the image gradient-based metric Q_G [36], Yang's structural similarity-based metric Q_Y [37], Cvejie's structural similarity-based metric Q_C [38], and the human visual system (HVS)-based metric Q_{CB} [39]. These six metrics are all positive metrics, that is, metrics with high values imply better fusion performance.

Table 1. Objective assessment of different methods on "Lytro" dataset.

Methods	Q_{MI}	Q_{NICE}	Q_G	Q_Y	Q_C	Q_{CB}
CSR [7]	1.0003	0.8330	0.7353	0.9337	0.8796	0.7612
NSCT-SR [4]	1.1223	0.8410	0.7585	0.9781	0.9083	0.7940
MADCNN [16]	1.1052	0.8394	0.7564	0.9677	0.8988	0.7702
IFCNN [17]	0.9389	0.8298	0.7302	0.9528	0.8908	0.7294
MFF-GAN [21]	0.8141	0.8239	0.6549	0.8486	0.8045	0.6360
DRPL [18]	1.1869	0.8455	**0.7612**	0.9857	0.9082	0.8060
ECNN [19]	1.1854	0.8452	0.7573	0.9859	0.9071	0.8038
MFF-SSIM [20]	1.0591	0.8369	0.7568	0.9797	0.9048	0.7824
PSNN [33]	1.0155	0.8341	0.7437	0.9657	0.8974	0.7689
CCSR-Net	**1.1890**	**0.8456**	0.7587	**0.9880**	**0.9094**	**0.8067**

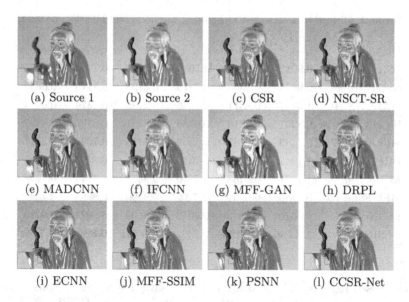

(a) Source 1   (b) Source 2   (c) CSR   (d) NSCT-SR

(e) MADCNN   (f) IFCNN   (g) MFF-GAN   (h) DRPL

(i) ECNN   (j) MFF-SSIM   (k) PSNN   (l) CCSR-Net

Fig. 4. Fusion results of different methods on the "statue" example from MFFW.

4.2   Comparison with State-of-the-Art Methods

Results on "Lytro" Dataset. Figure 3 dispalys the fusion results of different methods on the "diver" example from "Lytro" dataset. As shown in Fig. 3, the result of our method has clear boundary. Besides, with the aim of learning decision maps, the result of our method is not affected by chromatic aberration. In contrast, the results of other methods have different degrees of blurring near the boundary of the corresponding enlarged area, which may caused by the defocus spread effect (seen in Fig. 3(c), 3(d), 3(e), 3(f), 3(g), 3(h), 3(i), 3(j) and 3(k)). In addition, the result of MFF-GAN has an obvious chromatic aberration, probably due to the image color space conversion from RGB to YCbCr (seen in Fig. 3(g)). In contrast, our method can better distinguish the focus regions from complicated edges and achieve more competitive visual quality.

Table 1 shows objective assessment results of different method on "Lytro" dataset. The score of each metric in Table 1 are the average value of the scores derived from all examples from the testing data on the dataset. The best and the second best results are highlighted by bold and underline, respectively. It can be seen from Table 1 that our method achieved the first place for Q_{MI}, Q_{NICE}, Q_Y , Q_C and Q_{CB} , and the second place for Q_G, demonstrating the effectiveness of the proposed method.

Results on "MFFW" Dataset. Figure 4 dispalys the fusion results of different methods on the "statue" example from "MFFW" dataset. It can be seen from Fig. 4 that our proposed CCSR-Net-based method can well preserve the details of the source images, especially for the boundaries between the focused

Table 2. Objective assessment of different methods on "MFFW" dataset.

Methods	Q_{MI}	Q_{NICE}	Q_G	Q_Y	Q_C	Q_{CB}
CSR [7]	0.9012	0.8256	0.7069	0.8368	0.7760	0.6835
NSCT-SR [4]	0.9676	0.8300	0.7276	0.8798	0.8119	0.6844
MADCNN [16]	1.0037	0.8311	0.7266	0.9120	0.8374	0.6984
IFCNN [17]	0.8231	0.8219	0.6780	0.8731	0.8196	0.6427
MFF-GAN [21]	0.7383	0.8187	0.5950	0.7435	0.7092	0.5817
DRPL [18]	1.1342	0.8392	0.7315	0.9331	0.8479	0.7166
ECNN [19]	<u>1.1578</u>	<u>0.8408</u>	0.7326	0.9450	0.8541	0.7158
MFF-SSIM [20]	0.9985	0.8309	**0.7347**	<u>0.9590</u>	<u>0.8764</u>	<u>0.7344</u>
PSNN [33]	0.8670	0.8242	0.7076	0.8599	0.8021	0.6715
CCSR-Net	**1.1786**	**0.8426**	<u>0.7332</u>	**0.9781**	**0.8861**	**0.7451**

and defocused regions. All the other methods have deficiences to some extent. For instance, the results of four methods, CSR, NSCT-SR, DRPL, ECNN, all suffer from obvious artifacts (seen in Fig. 4(c), 4(d), 4(h), and 4(i)), while the rest five methods do not obviously suffer from artifacts. However, the edge diffusion with different degree of these five methods can be seen from the enlarged area (seen in Fig. 4(e), 4(f), 4(g), 4(j), and 4(k)).

Table 2 shows the objective assessment results of different methods on "MFFW" dataset. It can be proved again that our method achives either the first or the second place performance when compared to all the other methods, which also demonstrates the superior performance of our proposed method.

Table 3. Objective Performance of the proposed method with different values of T on "Lytro" dataset.

T	Q_{MI}	Q_{NICE}	Q_G	Q_Y	Q_C	Q_{CB}
1	1.1862	**0.8457**	0.7523	0.9861	0.9066	0.8014
2	<u>1.1878</u>	<u>0.8456</u>	<u>0.7582</u>	<u>0.9875</u>	<u>0.9091</u>	<u>0.8063</u>
3	**1.1890**	<u>0.8456</u>	**0.7587**	**0.9880**	**0.9094**	**0.8067**
4	1.1861	0.8453	0.7568	0.9868	0.9073	0.8055
5	1.1844	0.8455	0.7512	0.9848	0.9066	0.7975

4.3 Ablation Experiments

The value of iteration T impacts the performance of our proposed fusion method. Table 3 shows the objective performance with different values of T on "Lytro" dataset. It can be seen that the larger the value of iteration, the better the fusion performance. When the iteration value equals to 3, five metrics rank first and the other one metric ranks second, which is the best overall performance. Therefore, T is set to 3 to derive the sparse feature map.

5   Conclusion

In this paper, we propose an interpretable DL model for multi-focus image fusion to combine the advantages of DL-based and traditional methods. Specifically, an image fusion-oriented coupled CSR model is presented and its solving process is unfolded into a neural network named CCSR-Net via the deep unfolding technique. A new multi-focus image fusion based on the CCSR-Net is proposed. Experimental results show that the proposed method obtains the state-of-the-art performance in terms of both visual quality and objective assessment. In the future, we will focus on extending the proposed model to other image fusion tasks such as infrared and visible image fusion, multimodal medical image fusion, etc.

Acknowledgements. This work was supported by the National Natural Science Foundation of China under Grants 62176081, 62171176 and U23A20294

References

1. Liu, Y., Wang, L., Cheng, J., Li, C., Chen, X.: Multi-focus image fusion: a survey of the state of the art. Inf. Fusion **64**, 71–91 (2020)
2. Zhang, Q., Guo, B.: Multifocus image fusion using the nonsubsampled contourlet transform. Sig. Process. **89**(7), 1334–1346 (2009)
3. Li, S., Kang, X., Hu, J.: Image fusion with guided filtering. IEEE Trans. Image Process. **22**(7), 2864–2875 (2013)
4. Liu, Y., Liu, S., Wang, Z.: A general framework for image fusion based on multi-scale transform and sparse representation. Inf. Fusion **24**(1), 147–164 (2015)
5. Yang, B., Li, S.: Multifocus image fusion and restoration with sparse representation. IEEE Trans. Instrum. Meas. **59**(4), 884–892 (2010)
6. Ma, X., Hu, S., Liu, S., Fang, J., Xu, S.: Multi-focus image fusion based on joint sparse representation and optimum theory. Sig. Process. Image Commun. **78**, 125–134 (2019)
7. Liu, Y., Chen, X., Ward, R., Wang, Z.: Image fusion with convolutional sparse representation. IEEE Signal Process. Lett. **23**(12), 1882–1886 (2016)
8. Aslantas, V., Kurban, R.: Fusion of multi-focus images using differential evolution algorithm. Expert Syst. Appl. **37**(12), 8861–8870 (2010)
9. Bai, X., Zhang, Y., Zhou, F., Xue, B.: Quadtree-based multi-focus image fusion using a weighted focus-measure. Inf. Fusion **22**(1), 105–118 (2015)
10. Guo, D., Yan, J., Qu, X.: High quality multi-focus image fusion using self-similarity and depth information. Optics Commun. **338**(1), 138–144 (2015)
11. Nejati, M., Samavi, S., Shirani, S.: Multi-focus image fusion using dictionary-based sparse representation. Inf. Fusion **25**(1), 72–84 (2015)
12. Liu, Y., Liu, S., Wang, Z.: Multi-focus image fusion with dense sift. Inf. Fusion **23**(1), 139–155 (2015)

13. Bouzos, O., Andreadis, I., Mitianoudis, N.: Conditional random field model for robust multi-focus image fusion. IEEE Trans. Image Process. **28**(11), 5636–5648 (2019)
14. Zhang, Q., Liu, Y., Blum, R., Han, J., Tao, D.: Sparse representation based multi-sensor image fusion for multi-focus and multi-modality images: a review. Inf. Fusion **40**, 57–75 (2018)
15. Liu, Y., Chen, X., Peng, H., Wang, Z.: Multi-focus image fusion with a deep convolutional neural network. Inf. Fusion **36**, 191–207 (2017)
16. Lai, R., Li, Y., Guan, J., Xiong, A.: Multi-scale visual attention deep convolutional neural network for multi-focus image fusion. IEEE Access **7**, 114385–114399 (2019)
17. Zhang, Y., Liu, Y., Sun, P., Yan, H., Zhao, X., Zhang, L.: IFCNN: a general image fusion framework based on convolutional neural network. Inf. Fusion **54**, 99–118 (2020)
18. Li, J., Guo, X., Lu, G., Zhang, B., Xu, Y., Wu, F., Zhang, D.: DRPL: deep regression pair learning for multi-focus image fusion. IEEE Trans. Image Process. **29**, 4816–4831 (2020)
19. Amin-Naji, M., Aghagolzadeh, A., Ezoji, M.: Ensemble of CNN for multi-focus image fusion. Inf. Fusion **51**, 201–214 (2019)
20. Xu, S., et al.: Towards reducing severe defocus spread effects for multi-focus image fusion via an optimization based strategy. IEEE Trans. Comput. Imaging **6**, 1561–1570 (2020)
21. Zhang, H., Le, Z., Shao, Z., Xu, H., Ma, J.: MFF-GAN: an unsupervised generative adversarial network with adaptive and gradient joint constraints for multi-focus image fusion. Inf. Fusion **66**, 40–53 (2021)
22. Guo, X., Nie, R., Cao, J., Zhou, D., Mei, L., He, K.: Fusegan: learning to fuse multi-focus image via conditional generative adversarial network. IEEE Trans. Multimedia **21**(8), 1982–1996 (2019)
23. Wang, Y., Xu, S., Liu, J., Zhao, Z., Zhang, C., Zhang, J.: MFIF-GAN: a new generative adversarial network for multi-focus image fusion. Sig. Process. Image Commun. **96**, 116295 (2021)
24. Wang, X., Hua, Z., Li, J.: Multi-focus image fusion framework based on transformer and feedback mechanism. Ain Shams Eng. J. **14**(5), 101978 (2023)
25. Ma, J., Tang, L., Fan, F., Huang, J., Mei, X., Ma, Y.: Swinfusion: cross-domain long-range learning for general image fusion via swin transformer. IEEE/CAA J. Automatica Sinica **9**(7), 1200–1217 (2022)
26. Zhang, X.: Deep learning-based multi-focus image fusion: a survey and a comparative study. IEEE Trans. Pattern Anal. Mach. Intell. **44**(9), 4819–4838 (2022)
27. Gregor, K., LeCun, Y.: Learning fast approximations of sparse coding. In: Proceedings of the 27th International Conference on International Conference on Machine Learning, ICML 2010, pp. 399–406. Omnipress, Madison (2010)
28. Marivani, I., Tsiligianni, E., Cornelis, B., Deligiannis, N.: Multimodal deep unfolding for guided image super-resolution. IEEE Trans. Image Process. **29**, 8443–8456 (2020)
29. Xu, N., Price, B., Cohen, S., Huang, T.: Deep image matting (2017)
30. Rhemann, C., Rother, C., Wang, J., Gelautz, M., Kohli, P., Rott, P.: A perceptually motivated online benchmark for image matting. In: 2009 IEEE Conference on Computer Vision and Pattern Recognition, pp. 1826–1833 (2009)
31. Lin, T.-Y., et al.: Microsoft COCO: common objects in context. In: Fleet, D., Pajdla, T., Schiele, B., Tuytelaars, T. (eds.) ECCV 2014. LNCS, vol. 8693, pp. 740–755. Springer, Cham (2014). https://doi.org/10.1007/978-3-319-10602-1_48

32. Ma, H., Liao, Q., Zhang, J., Liu, S., Xue, J.H.: An -matte boundary defocus model-based cascaded network for multi-focus image fusion. IEEE Trans. Image Process. **29**, 8668–8679 (2020)

33. Jiang, L., Fan, H., Li, J., Tu, C.: Pseudo-siamese residual atrous pyramid network for multi-focus image fusion. IET Image Proc. **15**(13), 3304–3317 (2021)

34. Hossny, M., Nahavandi, S., Creighton, D.C.: Comments on 'information measure for performance of image fusion'. Electron. Lett. **44**, 1066–1067 (2008)

35. Wang, Q., Shen, Y., Zhang, J.Q.: A nonlinear correlation measure for multivariable data set. Physica D **200**(3), 287–295 (2005)

36. Xydeas, C., Petrovic, V.: Objective image fusion performance measure. Electron. Lett. **36**, 308–309 (2000)

37. Yang, C., Zhang, J.Q., Wang, X.R., Liu, X.: A novel similarity based quality metric for image fusion. Inf. Fusion **9**(2), 156–160 (2008)

38. Cvejic, N., Loza, A., Bull, D., Canagarajah, N.: A similarity metric for assessment of image fusion algorithms. Int. J. Signal Process. **2** (2006)

39. Chen, Y., Blum, R.S.: A new automated quality assessment algorithm for image fusion. Image Vis. Comput. **27**(10), 1421–1432 (2009)

FASONet: A Feature Alignment-Based SAR and Optical Image Fusion Network for Land Use Classification

Feng Deng[1,2], Meiyu Huang[1(✉)], Wei Bao[1], Nan Ji[1], and Xueshuang Xiang[1(✉)]

[1] Qian Xuesen Laboratory of Space Technology, China Academy of Space Technology, Beijing 100094, China
huangmeiyu2023@163.com, xiangxueshuang2023@163.com
[2] Xiangtan University, Xiangtan 411105, China
fengdeng@smail.xtu.edu.cn

Abstract. Land use classification using optical and Synthetic Aperture Radar (SAR) images is a crucial task in remote sensing image interpretation. Recently, deep multi-modal fusion models have significantly enhanced land use classification by integrating multi-source data. However, existing approaches solely rely on simple fusion methods to leverage the complementary information from each modality, disregarding the intermodal correlation during the feature extraction process, which leads to inadequate integration of the complementary information. In this paper, we propose FASONet, a novel multi-modal fusion network consisting of two key modules that tackle this challenge from different perspectives. Firstly, the feature alignment module (FAM) facilitates cross-modal learning by aligning high-level features from both modalities, thereby enhancing the feature representation for each modality. Secondly, we introduce the multi-modal squeeze and excitation fusion module (MSEM) to adaptively fuse discriminative features by weighting each modality and removing irrelevant parts. Our experimental results on the WHU-OPT-SAR dataset demonstrate the superiority of FASONet over other fusion-based methods, exhibiting a remarkable 5.1% improvement in MIoU compared to the state-of-the-art MCANet method.

Keywords: Land use classification · Multi-modal fusion · Feature alignment

1 Introduction

Land use classification has been widely used in disaster management, urban planning and environmental conservation [18]. Current classification algorithms mainly rely on single-modal remote sensing images, such as optical or SAR images. While optical images provide spectral information, they are susceptible to weather conditions and cloud cover, making classification challenging. SAR images are immune to weather conditions but suffer from coherent speckle

Q. Liu et al. (Eds.): PRCV 2023, LNCS 14434, pp. 298–310, 2024.
https://doi.org/10.1007/978-981-99-8549-4_25

noise, which significantly hinders their interpretation. [20]. Fortunately, recent technological advancements have made it possible to acquire multi-modal remote sensing data, and studies have shown that optical and SAR data can provide complementary information [22] [15]. However, integrating them directly into existing classification methods presents difficulties due to data heterogeneity. Recently, deep learning multi-modal fusion methods have been widely used in fusing optical and SAR images at the feature level to exploit their complementary information. In multi-modal feature fusion models, the main goal is to create a comprehensive and valuable fused feature for land use classification. Firstly, the extraction of discriminative unimodal features plays an important role, whose quality directly affects the performance of the fused features [14]. The deep convolutional neural network (DCNN) [11], as the powerful feature extractor, has been successfully applied in remote image classification [25] and segmentation [2]. However, most of these methods ignored the relationship between the features of each modality during feature extraction. Secondly, taking advantage of the information from both modalities is another issue. Most existing fusion methods use simple concatenation or different attention modules, which lead to redundant SAR/optical features or complex attention calculations during feature fusion.

To overcome the challenges mentioned above, this research proposes a novel multi-modal fusion network (FASONet) that operates on two main aspects.

Feature Extraction Process. To ensure semantic consistency while extracting features from optical and SAR images, it is crucial to address their visual disparities. Our solution is a cross-modal learning approach, where each mode learns from the other. The key is a feature alignment module (FAM) that measures the distance between high-level features from both modalities and subsequently incorporates this distance into the training loss function. By minimizing this distance, we further enhance the feature representation capabilities of optical and SAR images.

Feature Fusion Process. Rather than a straightforward concatenation or a summation fusion with equivalent weights for different modalities, we introduce a multi-modal squeeze-and-excitation fusion module (MSEM) to adaptively fuse the discriminative features by weighting each modality and removing irrelevant parts. This method is computationally efficient and can be integrated into various network architectures.

The main contributions of this research are as follows:

(1) We propose a novel multi-modal fusion network(FASONet), which comprehensively explores and utilizes the complementary information from multi-modal data in both feature extraction and fusion processes.
(2) Experimental results demonstrate that the FAM and MSEM proposed by us effectively improve the expression of single-modal features and the efficacy of fusion features, respectively.
(3) The proposed method improves the classification performance and provides new insights into the interpretation of multi-source remote sensing images.

The remainder of this paper is organized as follows. Section 2 gives the related work. Section 3 introduces the proposed method. Section 4 presents the experiment, including the experiment's setup and the experimental results. Finally, the conclusion is given in Sect. 5.

2 Related Work

Traditional land use classification methods encompass land mapping and remote mapping. While land mapping offers high accuracy, large-scale mapping requires substantial human resources, time, and financial investment [12]. Remote sensing images provide a cost-effective and convenient means to capture an overview of the study area. However, remote sensing mapping heavily relies on visual discrimination, which is prone to subjective judgments and can result in reduced accuracy.

Advances in computer technology and improvements in the resolution of remotely sensed images have facilitated the application of machine learning methods in land use classification, including artificial neural networks (ANN), support vector machines (SVM), random forests (RF) and cluster analysis [23]. For example, Taati [17] used SVM and maximum likelihood classifiers with TM images for land use classification. Gharaibeh improved [7] classification accuracy by combining ANN with cellular automata-Markov chain models. However, these classical machine learning models still produce relatively coarse and inaccurate results compared to land mapping. With the development of artificial intelligence techniques, deep network semantic segmentation models have emerged as a promising approach to land use classification, offering enhanced learning capabilities and comprehensive feature extraction. Notable models in this area include SegNet [1], which introduces maxpool indices for upsampling to eliminate the need to learn upsampling and preserve memory during inference; PSPNet [24], which proposes SPP to incorporate global contextual information; U-net [16], which incorporates skip connections to utilize semantic information at different levels; and Deeplabv3+ [3], which employs ASPP to capture rich spatial information. Notable achievements in land use classification utilizing these models include Dong [4], who achieved promising results using SegNet with high-resolution SAR images; Garg [6], who proposed an improved U-Net to enhance classification accuracy; and Zhang [21], who employed an enhanced Deeplabv3+ model for urban land use classification based on UAV-borne images. However, most of these methods use unimodal remote sensing (RS) imagery and are limited in their ability to leverage the advantages of multi-modal inputs. With the rapid development of RS technology, it is possible to obtain multi-modal RS data from the same area. Optical and SAR imagery, as two different Earth observation tools, can provide complementary information on the same land type for better land cover classification. Some deep learning methods used SAR and optical data to achieve multi-modal land use classification. For instance, Li [14] proposed a multi-modal bilinear fusion network (MBFNet) to fuse the optical and SAR features for land cover classification. Hosseinpour [8] introduced a multi-modal fusion network (CEGFNet) with a gate fusion module to extract complementary features from spectral and digital surface data. Li [15] proposed an end-to-end SAR and optical data fusion model(MCANet).

Building upon these advancements, we propose a novel method called SOFANet, which capitalizes on the valuable information from both modalities to achieve more accurate land use classification.

3   Method

We propose a novel method (FASONet) for land use classification by fusing optical and SAR images. As shown in Fig. 1, FASONet is built upon the classic encoder-decoder framework, with optical and SAR patches as the inputs to the network. The encoder module comprises a pseudo-siamese feature extraction module(PSEM), a multi-modal squeeze-excitation module(MSEM), and a low-high feature fusion module. The decoder module decodes the fused features and generates the final land use classification results. Additionally, we propose a feature alignment module(FAM) and design a feature alignment loss to align the high-level feature in the training process.

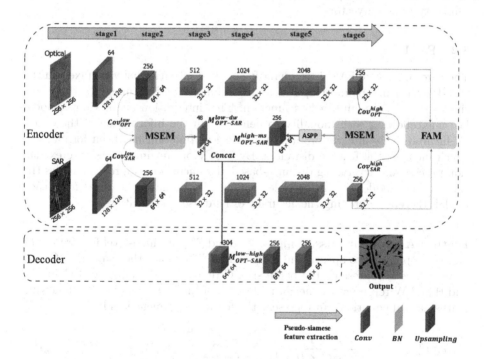

Fig. 1. Architecture of SOFANet

3.1   Network Structure

As shown in Fig. 1, optical and SAR images are input into PSEM. The convoluted Cov_{OPT}^{low} and Cov_{SAR}^{low} are fed into MSEM to produce $M_{OPT-SAR}^{low}$. The channel of the feature map is subsequently downsampled to obtain $M_{OPT-SAR}^{low-dw}$. On the other hand, $Cov_{OPT}^{low}/Cov_{SAR}^{low}$ are convoluted to generate Cov_{OPT}^{high} and Cov_{SAR}^{high} which are then entered into MSEM to produce $M_{OPT-SAR}^{high}$. The feature map is processed through ASPP and then upsampled to obtain $M_{OPT-SAR}^{high-sm}$. Finally, the high-level and low-level joint maps are stacked to create a low-high level joint map $M_{OPT-SAR}^{low-high}$, which is fed into the decoder via two convolutional layers. The outcome of semantic segmentation is obtained through upsampling.

Compare with MCANet [15]. We propose a novel module (FAM) to enhance the feature extraction capability of each modality. In addition, we have designed a new multi-modal fusion module (MSEM) to fuse the complementary information of the two modalities.

Compare with the Proposed Method in [13]. In the feature alignment module, instead of estimating the distance between distributions by the maximum mean discrepancy(MMD), which measures the distance between two probability distributions by comparing the difference in their expected values, we measure the distance between each pixel by Euclidean distance which is used to measure the similarity between vectors.

3.2 FAM

The core concept of FAM is to quantify the dissimilarities between pixels in the high-level feature pairs of optical and Sar images and to integrate these dissimilarity measures into a loss function. This integration enables cross-modal learning, allowing each modality to benefit from the information of the other. To facilitate this integration, a customized loss function has been formulated, which includes the feature distances. By jointly optimizing the feature extraction process and leveraging distance-based alignment, our approach enables the classification model to efficiently learn rich, discriminating features from each modal, thereby improving the accuracy of land use classification.

Feature Alignment Loss. Suppose f_{sar}^h and f_{opt}^h are high-level feature vector pair corresponding to optical and SAR images. They have the shape B × C × H × W, where B represents the batch size, C represents the number of channels, and H and W represent the height and width of the feature maps. The following mathematical equation can represent the feature alignment loss function:

$$L_{fa}(f_{sar}^h, f_{opt}^h) = \frac{1}{B \times C \times H \times W} \sum_{b=1}^{B}\sum_{c=1}^{C}\sum_{h=1}^{H}\sum_{w=1}^{W}(f_{sar,b,c,h,w}^h - f_{opt,b,c,h,w}^h)^2$$

(1)

where $f_{sar,b,c,h,w}^h$ represents the eigenvalues of the c channel, h-row, and w-column in the b-th sample, and $f_{opt,b,c,h,w}^h$ represents another eigenvalue of the c-channel, h-row, and w-column in the b-th sample. $\frac{1}{B \times C \times H \times W}$ is a normalization factor ensures that the loss value is independent of the size of the input data.

3.3   MSEM

Due to the heterogeneity of multi-modal data, the direct fusion of features leads to redundancy of features, which negatively affects the performance of the model. Combined with the attention mechanism, multi-modal features can be fused more efficiently. As shown in Fig. 2, the MSEM is primarily an improvement built upon the SE module [10], tailored to the characteristics of multi-modal data. Firstly, it concatenates the features from two modalities. However, considering that simple concatenation may introduce redundancy and affect classification performance, we introduce the attention mechanism SE module. The main objective of the SE module is to enhance the representation capability of feature maps by learning adaptive channel weights. Specifically, the SE module works by learning a weight coefficient for each channel, which is used to adjust the feature response of that channel. This weighting coefficient is learned based on statistics extracted from global features. As a result, it can adaptively weight each channel, reinforcing essential information while reducing the impact of less important details. Finally, the learned weights are applied to each modality, effectively reducing redundancy and achieving efficient fusion.

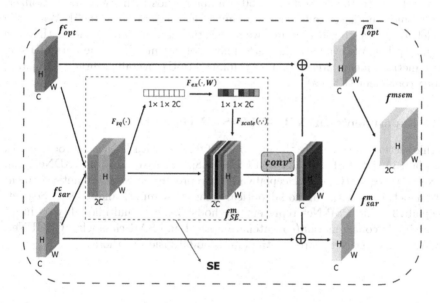

Fig. 2. Architecture of MSEM

3.4   Loss Function

In this paper, the loss function of SOFANet is divided into two components: the weighted cross entropy(WCE) loss (L_{wce}) and the feature alignment loss (L_{fa}). It can be expressed as:

$$Loss(p, \hat{p}) = L_{wce}(p, \hat{p}) + \beta L_{fa}(p, \hat{p}) \tag{2}$$

where p and $\hat{p}$ denote the ground-truth and prediction map, and the parameters β represent the weight coefficients of L_{fa}. We traverse the range of β from 2^{-10} to 2^{10} in binary and ultimately determine the optimal value of β to be 2^{-9}.

4   Experimental Results and Analysis

4.1   Experimental Setups

Data Description. The WHU-OPT-SAR dataset [15] comprises 100 high-resolution optical and SAR image pairs. To facilitate comparison with MCANet, we have cropped the dataset to 256×256 pixels, yielding 29,400 non-overlapping image blocks. The dataset is split into training, validation, and testing sets in a 6:2:2 ratio, with 17,640, 5,880, and 5,880 blocks, respectively.

Experimental Settings and Metrics. All experiments are conducted on a 64-bit Ubuntu 18.04 operating system. To ensure consistency, we used the same parameter settings for training in each set of experiments. Specifically, we set the batch size to 16 and performed 50 training rounds with the Adam optimizer. The momentum is set to 0.9, and the initial learning rate was 0.001. When the model error rate stops decreasing, we reduce the learning rate to one-tenth of the original value. We evaluate the performance of our model on the dataset using three metrics: mean intersection over union (MIoU), overall accuracy (OA), and kappa coefficient (Kappa).

4.2   Experiments on WHU-OPT-SAR Dataset

In this section, we compare the proposed SOFANet with U-Net [16], SegNet [1], Deeplabv3+ [3], PSP-Net [24], CEGFNet [8], MCANet [15] and ADNet [13]. U-Net(O), SegNet(O), and Deeplabv3+(O) represent classical semantic segmentation networks with unimodal optical images as input, and U-Net, SegNet, Deeplabv3+ and PSP-Net represent methods that use multimodality as input via a simple concatenation(concatenate optical and SAR channels). CEGFNet, MCANet and ADNet represent the multi-modal fusion methods.

Table 1. Land use classification results of different approaches.

method	IoU							OA	Kappa	MIoU
	Farmland	City	Village	Water	Forest	Road	Others			
U-Net(O)	59.1	69.9	37.3	76.1	84.3	20.3	0.8	79.6	72.5	49.7
U-Net	63.0	61.3	41.3	77.7	84.2	39.3	0.8	81.3	74.6	52.5
SegNet(O)	58.0	74.1	47.1	79.3	85.0	30.4	0.7	80.4	73.8	53.5
SegNet	61.9	70.9	46.4	79.9	84.1	31.5	0.3	82.3	75.9	53.6
Deeplabv3+(O)	63.7	92.4	40.1	74.3	83.5	46.9	0.4	83.6	77.5	57.3
Deeplabv3+	68.8	90.3	49.1	72.2	78.6	42.4	1.7	83.6	77.5	57.6
PSP-Net	67.3	87.5	41.8	**80.4**	84.9	42.2	**2.4**	84.7	79.1	58.1
CMGFNet	63.1	70.6	**49.2**	79.1	84.7	47.6	1.3	82.2	75.8	56.5
MCANet	68.7	93.4	46.3	80.2	**85.0**	36.2	0.8	84.8	79.3	58.1
ADNet	64.6	95.1	42.8	79.1	84.7	56.0	0.8	83.8	78.0	60.5
Ours	**69.3**	**96.9**	42.4	79.8	84.3	**69.0**	1.0	**85.2**	**79.8**	**63.2**

The results in Table 1 show that a simple concatenation method did not effectively use multi-modal information, which behaves almost the same as single-mode as input. Our method achieved the highest OA (85.2%) and MIoU (63.2%), outperforming other methods. Compared to MCANet, our method improved MIoU by 5.1%. Compared to ADNet, an improved approach to MCANet, we still have a 2.7% improvement in MIoU. This improvement is due to our FAM and MSEM. Figure 3 shows the visualisation results of the land use classification for five groups. The overall classification results of each method are similar, but SOFANet can discern the details of each ground object more accurately. As shown in (g), in the second and fifth groups, The results of MCANet show that many places, such as *farmland* and *forest*, are not successfully detected. As shown in (d), (e), (f) and (g), in the first and third groups, the results of U-Net, Deeplabv3+, CMGFNet and MCANet show large areas of confusion between *water* and *road*. As shown in (h), the above situation is improved in the SOFANet results. As the contours of water and roads are similar in SAR images, there is a clear difference in the appearance of *water* and *road* in optical images. We introduced FAM and MSEM to enhance the mutual learning of each modality while exploiting the multi-modal attention features to guide FASONet to better distinguish between *water* and *road*. Furthermore, as shown in Fig. 3, FASONet extracts less edge noise and more accurate edge details for each category than other methods.

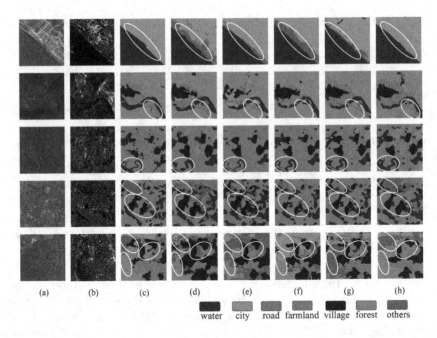

water   city   road   farmland   village   forest   others

Fig. 3. Comparison of the proposed method and four state-of-the-art models. (a) optical image. (b) SAR image. (c) GT. (d) U-Net. (e) Deeplabv3+. (f) CMGFNet. (g) MCANet. (h) Ours.

4.3   Ablation Study

We conduct a detailed ablation analysis in this section to demonstrate the importance and effectiveness of each module proposed in our method. All experiments are performed on the same dataset.

Table 2. Effectiveness of each module.

MSEM	FAM	OA	Kappa	MIoU
		84.3	78.6	57.3
√		85.8	80.5	61.5
	√	**86.1**	**80.7**	61.4
√	√	85.2	79.8	**63.2**

Table 3. Performance of each category under different modules.

module	Acc/IoU						
	Farmland	City	Village	Water	Forest	Road	Others
	(34.4%)	(4.7%)	(5.8%)	(14.2%)	(37.6%)	(1.0%)	(1.7%)
MSEM	68.4/74.5	**97.4/99.2**	41.7/74.2	79.0/**88.6**	**84.6/92.8**	58.3/89.4	0.9/5.1
FAM	**70.7/81.3**	87.9/88.1	**45.0**/73.8	**80.2**/87.5	83.8/91.0	61.4/87.5	0.9/5.1
MSEM+FAM	69.3/76.8	96.9/98.0	42.4/**83.4**	79.8/87.2	84.3/89.7	**69.0/89.0**	**1.0/10.7**

Effectiveness of Each Module. We test different combinations of modules on the test set to verify the effectiveness of the modules. "✓" represents the module used in this experiment. We replace all modules with the ordinary convolution and set this model as a baseline. The results in Table 2 show that adding FAM or MSEM can improve accuracy. The experimental results show a significant improvement in MIoU when both modules are used, with a 1.8% improvement compared to FAM but a decrease in OA and Kappa. This may be because the model focuses more on details and local features when using both FAM and MSEM, resulting in a decrease in OA for classes with larger proportions. However, the improvement in MIoU indicates that the model has enhanced the segmentation accuracy for classes with smaller proportions, which can be highly valuable for specific applications. Table 3 displays the Acc and IoU for different categories, and the values in () represent the percentage of each category. The results show that the accuracy of *farmland* and *forest* decreases when both modules are used, while there is a significant improvement for *road*, *others* and *village*.

Comparison of MSEM and Other Multi-modal Fusion Modules. Previous works have demonstrated the effectiveness of the attention mechanism and its applicability to various tasks. To further demonstrate the merit of our method, we choose several widely used modules CBAM [19], DANet [5], CA [9], and MCAM for comparison. M(0) represents no multi-modal fusion module. As shown in Table 4, MSEM achieves the best results in MIoU (61.5%) with a 6.4% improvement compared to M(0) and has a 1.6%to 3.4% improvement compared to other methods. The number of Params and MACs is significantly less than in other modules, which proves that our module is more efficient in calculations.

Effectiveness of FAM

Table 4. Effectiveness of MSEM and other fusion modules.

Method	Params(M)	MACs(G)	MIoU
M(0)	none	none	55.1
DANet	1.6	13.170	58.1
MCAM	0.23	1.879	58.1
CBAM	0.66	1.080	59.9
CA	0.18	1.091	59.9
MSEM(Ours)	**0.16**	**1.078**	**61.5**

Table 5. Effectiveness of the feature alignment position.

Method	OA	Kappa	MIoU
stage1	84.6	78.9	58.9
stage2	82.1	75.7	60.6
stage3	82.7	76.5	59.9
stage4	81.2	74.9	56.8
stage5	82.1	75.9	57.6
stage6(Ours)	**86.1**	**80.7**	**61.4**

Table 6. Effectiveness of FAM in single-modal classification.

	Method	OA	Kappa	MIoU
Optical	MCANet	81.3	74.6	49.6
	MCANet+FA	**81.4**	**74.8**	**51.1**
SAR	MCANet	74.6	64.9	43.4
	MCANet+FA	**78.2**	**69.7**	**43.8**

Table 7. Comparison of L_{fa} and other distance measurement.

Method	OA	Kappa	MIoU
MMD	81.0	74.6	58.3
CORR	82.5	76.4	60.2
COS	85.4	80.1	61.1
Ours	**86.1**	**80.7**	**61.4**

Effectiveness of the Feature Alignment Position. choosing which stage of features to perform feature alignment is vital. We select the last layer of the feature extraction module because the high-level features are rich in semantic information. Aligning these high-level features contributes to the ability to capture comprehensive information and improves land use classification. We conduct experiments on feature alignment at different stages. The results in Table 5 show that aligning high-level features (stage 6) has the best performance.

Effectiveness of FAM in Single-Modal Classification. In order to demonstrate the effectiveness of the FAM in enhancing the expression of single-mode features, we conduct an experiment using MCANet as the benchmark and adjusted it to use the extracted unimodal features for land use classification. Specifically, we have used the extracted optical features or SAR features for classification and compared their performances with and without the FAM. Our findings, as presented in Table 6, reveal that the optical features perform better than the SAR features for land use classification and that adding the FAM improves the single-modal classification.

Comparison of L_{fa} and Other Distance Measurement. In order to incorporate feature distances into the loss function, a suitable loss formulation should be designed. We choose several distance metrics, including Maximum mean discrepancy (MMD), Cosine similarity (COS), and Correlation coefficient (CORR) for comparison. In our alignment loss function, we insist on using the Euclidean distance concept and calculating the distance for each pixel in the feature. This distance calculation is added to the loss function during training. The effectiveness of this straightforward approach is unquestionable, as confirmed by the comparison results presented in Table 7.

5   Conclusion

In this article, we present a novel FASONet to efficiently fuse deep features of optical and SAR images for improving land cover classification. In FASONet, a feature alignment module (FAM) to enhance the expression of unimodal features and a multi-modal squeeze-and-excitation fusion module (MSEM) is proposed to combine the two modalities efficiently. Our experimental results on the dataset demonstrate that FASONet outperforms other fusion-based methods. Moreover, the efficacy of all our main elements is supported by extensive ablation studies.

References

1. Badrinarayanan, V., Kendall, A., Cipolla, R.: Segnet: a deep convolutional encoder-decoder architecture for image segmentation. IEEE Trans. Pattern Anal. Mach. Intell. **39**(12), 2481–2495 (2017)
2. Chen, L.C., Papandreou, G., Kokkinos, I., Murphy, K., Yuille, A.L.: Deeplab: semantic image segmentation with deep convolutional nets, atrous convolution, and fully connected crfs. IEEE Trans. Pattern Anal. Mach. Intell. **40**(4), 834–848 (2018)
3. Chen, L.-C., Zhu, Y., Papandreou, G., Schroff, F., Adam, H.: Encoder-decoder with atrous separable convolution for semantic image segmentation. In: Ferrari, V., Hebert, M., Sminchisescu, C., Weiss, Y. (eds.) ECCV 2018. LNCS, vol. 11211, pp. 833–851. Springer, Cham (2018). https://doi.org/10.1007/978-3-030-01234-2_49
4. Dong, Y., Li, F., Hong, W., Zhou, X., Ren, H.: Land cover semantic segmentation of port area with high resolution SAR images based on segnet. In: 2021 SAR in Big Data Era (BIGSARDATA), pp. 1–4. IEEE (2021)
5. Fu, J., et al.: Dual attention network for scene segmentation. In: Proceedings of the IEEE/CVF Conference on Computer Vision and Pattern Recognition, pp. 3146–3154 (2019)
6. Garg, L., Shukla, P., Singh, S.K., Bajpai, V., Yadav, U.: Land use land cover classification from satellite imagery using munet: a modified unet architecture. In: International Joint Conference on Computer Vision, Imaging and Computer Graphics Theory and Applications (2019)
7. Gharaibeh, A., Shaamala, A., Obeidat, R.M., Al-Kofahi, S.: Improving land-use change modeling by integrating ANN with cellular automata-markov chain model. Heliyon **6**(9), e05092 (2020)
8. Hosseinpour, H., Samadzadegan, F., Javan, F.D.: CMGFNET: a deep cross-modal gated fusion network for building extraction from very high-resolution remote sensing images. ISPRS J. Photogramm. Remote. Sens. **184**, 96–115 (2022)
9. Hou, Q., Zhou, D., Feng, J.: Coordinate attention for efficient mobile network design. In: Proceedings of the IEEE/CVF Conference on Computer Vision and Pattern Recognition, pp. 13713–13722 (2021)
10. Hu, J., Shen, L., Sun, G.: Squeeze-and-excitation networks. In: 2018 IEEE/CVF Conference on Computer Vision and Pattern Recognition (CVPR) (2018)
11. Kalchbrenner, N., Grefenstette, E., Blunsom, P.: A convolutional neural network for modelling sentences. Eprint Arxiv 1 (2014)
12. Langat, P.K., Kumar, L., Koech, R., Ghosh, M.K.: Monitoring of land use/land-cover dynamics using remote sensing: a case of tana river basin, Kenya. Geocarto Int. **36**(13), 1470–1488 (2021)
13. Li, W., et al.: Aligning semantic distribution in fusing optical and SAR images for land use classification. ISPRS J. Photogramm. Remote. Sens. **199**, 272–288 (2023)
14. Li, X., Sun, Y., Kuang, G.: Multimodal bilinear fusion network with second-order attention based channel selection for land cover classification. IEEE J. Sel. Topics Appl. Earth Obs. Remote Sens. **13**, 1011–1026 (2020)
15. Li, X., et al.: MCANET: a joint semantic segmentation framework of optical and SAR images for land use classification. Int. J. Appl. Earth Obs. Geoinf. **106**, 102638 (2022)

16. Ronneberger, O., Fischer, P., Brox, T.: U-Net: convolutional networks for biomedical image segmentation. In: Navab, N., Hornegger, J., Wells, W.M., Frangi, A.F. (eds.) MICCAI 2015. LNCS, vol. 9351, pp. 234–241. Springer, Cham (2015). https://doi.org/10.1007/978-3-319-24574-4_28

17. Taati, A., Sarmadian, F., Mousavi, A., Pour, C.T.H., Shahir, A.H.E.: Land use classification using support vector machine and maximum likelihood algorithms by landsat 5 tm images. Walailak J. Sci. Technol. (WJST) **12**(8), 681–687 (2015)

18. Talukdar, S., Singha, P., Mahato, S., Pal, S., Liou, Y.A., Rahman, A.: Land-use land-cover classification by machine learning classifiers for satellite observations-a review. Remote Sens. **12**(7), 1135 (2020)

19. Woo, S., Park, J., Lee, J.Y., Kweon, I.S.: CBAM: convolutional block attention module. In: Proceedings of the European Conference on Computer Vision (ECCV), pp. 3–19 (2018)

20. Xu, L., Zhang, H., Wang, C., Zhang, B., Liu, M.: Crop classification based on temporal information using sentinel-1 SAR time-series data. Remote Sens. **11**(1), 53 (2018)

21. Zhang, C., Li, M., Wei, D., Wu, B.: Enhanced deeplabv3+ for urban land use classification based on UAV-borne images. In: 2022 7th International Conference on Image, Vision and Computing (ICIVC), pp. 449–454. IEEE (2022)

22. Zhang, H., Wan, L., Wang, T., Lin, Y., Lin, H., Zheng, Z.: Impervious surface estimation from optical and polarimetric SAR data using small-patched deep convolutional networks: A comparative study. IEEE J. Sel. Topics Appl. Earth Obs. Remote Sens. **12**(7), 2374–2387 (2019)

23. Zhang, Y., et al.: MAAFEU-NET: a novel land use classification model based on mixed attention module and adjustable feature enhancement layer in remote sensing images. ISPRS Int. J. Geo-Inf. **12**(5), 206 (2023)

24. Zhao, H., Shi, J., Qi, X., Wang, X., Jia, J.: Pyramid scene parsing network. In: Proceedings of the IEEE Conference on Computer Vision and Pattern Recognition, pp. 2881–2890 (2017)

25. Zou, Q., Ni, L., Zhang, T., Wang, Q.: Deep learning based feature selection for remote sensing scene classification. IEEE Geosci. Remote Sens. Lett. **12**(11), 1–5 (2015)

De Novo Design of Target-Specific Ligands Using BERT-Pretrained Transformer

Yangkun Zheng[1], Fengqing Lu[1], Jiajun Zou[2], Haoyu Hua[1], Xiaoli Lu[3], and Xiaoping Min[1(✉)]

[1] Department of Computer Science and Technology, School of Informatics, Xiamen University of China, No. 422 Siming South Rd, Xiamen 361005, People's Republic of China
zhengyangkun@stu.xmu.edu.cn, fengqing.lu@foxmail.com, mxp@xmu.edu.cn

[2] Department of Artificial Intelligence, School of Informatics, Xiamen University of China, No. 422 Siming South Rd, Xiamen 361005, People's Republic of China
23020191153191@stu.xmu.edu.cn

[3] Information and Networking Center, Xiamen University of China, No. 422 Siming South Rd, Xiamen 361005, People's Republic of China
luxl@xmu.edu.cn

Abstract. The principal goal of drug design is to find ligand molecules that exhibit affinity to a given target protein. In recent years, deep generative methods have shown their promise in de novo drug design. However, most of these methods design molecules based on target-specific ligand datasets instead of targets' features and fail to design drugs against novel target proteins that barely have active ligand datasets. A fast and relatively accurate evaluation method is needed to evaluate algorithms capable of generating large numbers of molecules. In this work, we treat target-specific de novo drug design as a sequence-to-sequence generation task and propose a Transformer architecture that compensates for the lack of training data with a BERT pretraining approach to generate protein sequence-conditioned Target Ligand Molecules SMILES. First, we pre-train two self-attention blocks of Transformer on the large-scale amino acid sequence dataset and molecular SMILES dataset, respectively, to capture the feature representation of the target. Then we fine-tune the Transformer's encoder-decoder mutual attention block on the protein-ligand complex dataset to learn conditional generation using autoregressive supervised learning. The individual results do not demonstrate the effect of the generative algorithm, so we propose to evaluate the model by calculating the affinity distribution of the molecules. We also evaluate our method by designing ligands against three well-studied proteins. Furthermore, our model proposes molecules with binding affinities exceeding certain FDA-approved drugs in docking experiments.

Keywords: Molecular Generative Model · Deep Learning · Sequence-to-Sequence · Unsupervised Learning · Pretraining

Supported by The National Natural Science Foundation of China (Grant No. 62272399).

Q. Liu et al. (Eds.): PRCV 2023, LNCS 14434, pp. 311–322, 2024.
https://doi.org/10.1007/978-981-99-8549-4_26

1   Background

The ultimate goal of drug discovery and drug design is to seek novel chemical compounds that could bind to a predefined target protein and then modulate its biological functions in the desired way [26]. However, finding candidate ligands from the enormous and discrete chemical space is non-trivial.

The number of drug-like molecules is estimated to be 10^{60}, while the subset of molecules capable of binding a given target is much smaller [25,27]. Moreover, an imperceptible variation in a molecular structure might cause a drastic change in biochemical properties [32]. To reduce the time and cost of drug design, various computational methods, such as virtual screening [33], molecular docking [8], and quantitative structure-activity relationship (QSAR) models [6], have been developed to predict or characterize binding affinity of a small molecule to a target protein. However, these methods have inherent limitations while filtering on vast chemical libraries. A complete exploration of chemical space is computationally intractable, and they could not find novel compounds beyond existing libraries or optimize the molecular structure to improve input molecules' properties. The evaluation of the generative model is also a difficult point, because the quality of individual results is not enough to evaluate the quality of the model algorithm, and the generative algorithm can generate a large number of molecules, and it is difficult to measure all the molecules through docking experiments. The more accurate evaluation method evaluates a large number of molecules and then evaluates the advantages and disadvantages of the model algorithm more objectively as a whole.

In recent years, deep learning-based generative models are introduced to de novo drug design, which have the potential to overcome above-mentioned limitations. Molecular generative models could learn the underlying distribution of chemical space from given dataset, and then automatically sample this distribution to yield novel compounds with desired properties from scratch. These methods represent molecules by using simplified molecular input line entry system (SMILES) [37], a commonly used string-based representation of chemical structures, or molecular graph which preserves the topological relationship between atoms. Generative neural network like auto-regressive recurrent neural network (RNN) [11,18,23], variational autoencoder (VAE) [3,16,17,20,28] and generative adversarial network (GAN) [4,12,14] have been extensively utilized on molecular generation. Deep generative methods have achieved great success in designing molecules with desired physicochemical properties or biological activities toward a given target. However, a major of generative models are ligand-based, which merely learn chemical distribution from given molecular datasets and then produce molecules with similar structure or optimized desired properties. Hence ligand-based generative models are unavailable to design drug against novel target proteins or proteins with limited known ligand data because of their dependence on massive existing target-specific molecules.

More recently, some structure-based deep generative approaches [1,2,13,19, 22,29,38] are proposed to utilize structure information of target protein to generate molecules that could bind with given target. A part of these methods [1,19] represent target protein as graph and extract their signature with graph neural network. For instance, Aumentado-Armstrong et al [1] utilize two graph convolutional neural network to respectively extract features of protein binding site and ligand from a protein-ligand complex and then optimize compound's binding affinity by using gradient-based optimization on ligand's latent vector. Krishnan et al [19] extract protein's features by using graph attention network based VAE, extracts ligand's features by using sequence-based VAE, and then introduced conditional VAE and reinforcement learning methods to optimize given molecules' binding affinity according to ligand's features and protein's features. On the other hand, some works [22,29,38]characterize the three-dimensional structure information of the target protein and integrate it into conditional generative models. Skalic et al [29] develop named LiGANN , which generates the shape of ligands based on voxelized protein pocket with a conditional GAN and then decodes the generated ligand shapes into SMILES through a shape-captioning network. Besides, some works [2,13] utilize the amino acid sequences of target proteins as input and output candidate ligand's SMILES. Born et al [2] use a VAE to encode a protein sequence into latent vector and then decode it into target-specific molecule's SMILES with another VAE. Grechishnikova [13] consider target-specific de novo drug design as a translation from the amino acid sequence of protein to the SMILES string of ligand and apply the Transformer model [35] to this task.

Above-mentioned studies can be categorized as supervised learning methods that learn the mapping relationship between protein and ligand from protein-ligand complexes dataset. However, these models suffer from shortage of training data. For instance, PDBbind [36] only contains 19443 protein-ligand complexes crystal structures, while structure-based models must learn the complicated features extraction and mapping relation from limited data. Though BindingDB [10] contains 2513948 protein-ligand sequential pairs for sequence-based target-specific generative models, there are only 8839 unique protein targets in these pairs so it is still hard for existing approaches to learn amino acid sequences representation.

Fortunately, biological databases contain massive protein data and ligand data which are far more than protein-ligand complex data. In this work, we present a novel target-specific generative framework that utilizes protein data and ligand data to enhance model on feature extraction. Considering that there are a large number of amino acid sequence data in the protein database and SMILES sequence data in the ligand database, we deem the target-specific drug design as a sequence-to-sequence generation from amino acid sequence to SMILES string and introduce a Transformer on this task. Firstly, the self-attention modules of Transformer encoder and decoder are respectively pre-trained on amino acid sequence data and molecular SMILES data by using BERT's unsupervised learning methods [5]. Secondly, the encoder-decoder attention modules are finetuned on protein-ligand sequence data by using Transformer's supervised learning methods.

We chose to evaluate our model using computationally generated affinity distributions for molecules and protein targets, which are more robust to model algorithms than selecting a limited number of generated molecules for docking experiments. As valid validation, we use our model to design three molecules targeting three well-studied proteins: DRD2, 5HT2A, and JAK2. This approach designs molecules that are structurally similar to known ligands and predicted to be active by QSAR models. In addition, we find that some of the resulting molecules have binding affinities exceeding FDA-approved drugs through docking experiments.

2   Materials and Methods

We elaborate on the details of Transformer and BERT in the "Details of Methods" section of the supplementary file, and in this section we introduce how they are integrated into our generative framework, and finally describe our training process.

2.1   Network Framework

The three-dimensional structure of a protein has more details than the amino acid sequence. However, the measurement of 3D structure is more complicated than amino acid sequencing, which requires a lot of manpower and material resources. Generally, molecular generative models that take amino acid sequences as input can be trained with more data.

In this work, we consider the target-specific de novo drug design as a sequence-to-sequence generation task that would output the corresponding bioactive molecule SMILES sequences based on the amino acid sequence of the input target protein. To capture long-range dependencies between sequences, we adopt the Transformer as the targeted molecule generation network. As shown in Supplementary Fig. 1, the encoder and the decoder will respectively extract features containing context information from the input amino acid sequence and the partially generated SMILES sequence, then decoder utilizes the encoder-decoder attention module to calculate the correlation features between the amino acid sequence features and the SMILES sequence features. Finally, decoder outputs the probability distribution of the next SMILES character through the Feed Forward Module.

Since BERT is the encoder of the Transformer, we first train a BERT model on the BFD database [30], which contains 2.1 billion amino acid sequences. Then we initialize our Transformer encoder with the parameters of this BERT model so that encoder can learn to extract the context-related features of amino acid sequence in advance. The decoder of transformer also contains an embedding layer and a self-attention module to extract sequence features. Therefore, a BERT model is pretrained on the ChEMBL database [9], which contains 1.7 million molecules. Then the parameters of this BERT's embedding layer and self-attention module are used to initialize decoder's corresponding modules so that it can learn to extract the context-related features of SMILES sequence in advance. Finally, we freeze the parameters of word embedding layer and self-attention module in encoder and decoder, and then finetune the parameters of encoder-decoder attention module and Feed Forward of decoder with sequence-to-sequence generation on protein-ligand sequence database so that they can learn how to extract correlation information during generation. As shown in Fig. 1, the parameters of the green and orange modules are initialized by the BERT model trained on their respective large-scale databases, while the yellow part is the final training using the seq2seq method.

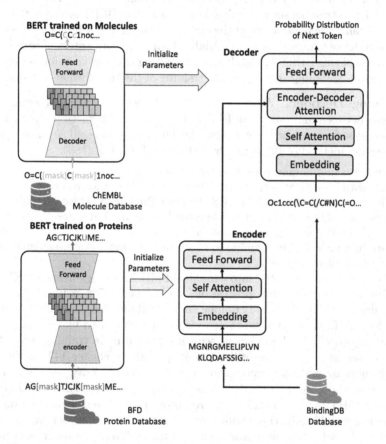

Fig. 1. Illustration of our network. The parameters of the green modules of the encoder are initialized by the BERT model trained on BFD database. The parameters of the orange modules of the decoder are initialized by the BERT model trained on ChEMBL database. The parameters of the yellow modules are trained with the seq2seq method on BindingDB database. (Color figure online)

2.2   Training Setting

Pretraining Encoder. The open-source library Pytorch-Transformers provides a pretrained model named ProtBert-BFD [7], which trains a BERT model on 2.1 billion amino acid sequences from the BFD database. Because of the limitation on computational conditions, this paper directly used the parameters of ProtBERT to initialize the encoder's parameters of our transformer.

Pretraining Decoder. 872462 molecules with biological activity (IC50, EC50) < 10um were collected from ChEMBL database and transformed into canonical SMILES string using RDKIT [21]. Because the grammar of SMILES is easier to be learned, this training set was utilized to train a small BERT model with only 50 epochs. Finally, the parameters of the embedding layer and self-attention module in the decoder of transformer were initialized with the parameters of corresponding modules of this BERT model. It is worth noting that the transformer encoder was not given any input during the Pretraining of the transformer decoder.

Finetuning Decoder. We selected 2412668 bioactive pairs of amino acid sequences between molecule from BindingDB database to finetune decoder with sequence-to-sequence generation tasks. The data were processed as follows, where the screening method of (2) comes from Kotsias P C et al [18]:

(1) Pairs with non-human proteins were removed.
(2) Pairs with biological activity (IC50, EC50) < 10 um were removed.
(3) Pairs without molecular SMILES string or PubChem ID were removed.
(4) Pairs with amino acid sequence lengths less than 50 and greater than 2050 or without Uniprot ID were removed.
(5) Pairs in which multiple amino acid sequences bind to one molecule at the same time were removed.
(6) Molecules were converted into canonical SMILES strings using RDKIT.
(7) After the above-mention procession, a total of 560704 pairs of data were obtained which include 1442 unique amino acid sequences and 295634 unique molecules. In order to avoid the wrong evaluation caused by the same amino acid sequence or its homologous sequence appearing in training set and test set at the same time, we utilized mmseqs2 [31] to cluster 1442 amino acid sequences by multi-sequence alignment. In order to verify the performance of our method, we reserved dopamine receptors D2 (DRD2), serotonin receptor 2A (5HT2A) and JAK2 kinase receptor in subsequent molecular docking experiments. Finally, these pairs were divided into training set, valid set and test set by 8:1:1, and we ensured that all the amino acid sequences clustered into one class are divided into the same set. 10% of the data is generally selected as the test set, and the size of the processed data set is shown in Table 1.

As mentioned above, we firstly initialized the parameters of the Transformer and freeze the parameters of the embedding layers and self-attention modules of both the encoder and the decoder. Secondly, we used the BindingDB training set to train the encoder-decoder attention module and Feed Forward module of the Transformer decoder for 30 epochs. After each epoch, we calculated the loss of the model on the valid set and saved the model with the minimum valid loss.

Table 1. The sizes of processed BindingDB dataset.

Datasets	Number of Amino Acid Sequences	Number of Protein-Ligand Complexes
Training Set	725	432,351
Valid Set	196	53,532
Test Set	520	53,751

3 Results and Discussion

3.1 Drug-Target Affinity Prediction

To verify the effectiveness of BERT's pretraining method and the potential of pretrained Transformer in designing target-specific ligands, DTA (Drug-Target Affinity) [24] prediction model was introduced to predict the affinity of generated molecules.

Training Details of DTA Model. DeepPurpose [15] is a DTA platform that provides training datasets, model training, and benchmark evaluation. We utilized DeepPurpose to quickly construct a DTA model, which uses convolutional neural network and recurrent neural network to respectively encode amino acid sequence and SMILES string and then predict the binding affinity between them using a multi-layer perceptron. After training for 50 epochs using the protein sequence-molecular sequence pair training set screened from the BindingDB dataset mentioned above, the MSE of the DTA model on the test set is 0.291, and the consistency index is 0.825.

Affinity Prediction. To demonstrate the effectiveness of BERT's pretraining method, we trained several networks as follows: (1) a Transformer without pre-training; (2) a Transformer with pretrained encoder, named Transformer-E; (3) a Transformer with pretrained decoder, named Transformer-D; (4) a Transformer with pretrained encoder and decoder, named Transformer-ED. Then we utilized each model to generate 100 molecules for each amino acid sequence in the test set and predicted their affinities with corresponding target with DTA model. The affinity here is the comprehensive score of IC50, EC50, Ki, Kd, etc. The higher the affinity score, the better the effect. Figure 2 shows the affinity distributions of molecules generated by these models.

Our analysis of affinity prediction and MOSES benchmarks is written in the Supplementary File, and we conclude that BERT's pre-training method has a certain enhancement effect on targeted drug design. And a BERT-pretrained Transformer can generate molecules similar to real bioactive molecules based on the amino acid sequence of a given target.

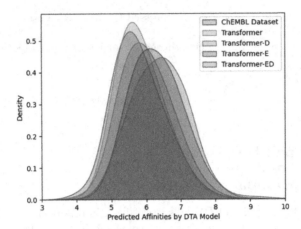

Fig. 2. Binding affinity of generated molecules predicted by DTA model. Each curve represents an affinity distribution of molecules generated toward amino acid sequences in test set by corresponding model.

3.2 Molecule Docking

In this section, we utilized AutoDock Vina [34] to dock the molecules generated by our method with given target proteins to analyze their binding ability. All encounter search parameters followed by AutoDock Vina's default.

Docking Settings and other docking results can be found in the "docking" section of the supplementary file.

Table 2. Minimum Binding Energy of selected molecules docking with 5HT2A

Source	Molecular Name	Minimun Binding Energy (Kcal/mol)
FDA-approved drug	Iloperidone	−9.8
FDA-approved drug	Lorpiprazole	−10.7
FDA-approved drug	Lumateperone	−10.3
Generated	Rank 1	−10.1
Generated	Rank 2	−10.8
Generated	Rank 3	−12.3
Generated	Rank 4	−11.8
Generated	Rank 5	−11.5

Docking Results. For 5HT2A, in addition to the five molecules screened by our methods, Iloperidone, Lorpiprazole and Lumateperone were selected from the DrugBank database as the benchmark. The SMILES strings of these molecules are shown in Table 2 of the Supplementary File. Table 2 shows the minimum

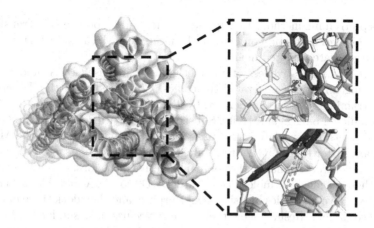

Fig. 3. Binding conformation of Lorpiprazole and Rank 3 molecule docking with 5HT2A. The blue-green molecule is Lorpiprazole, and the purple-red molecule is Rank 2 molecule, which binds in the same pocket as 5HT2A and has a similar binding conformation, and the yellow dotted line is a hydrogen bond. (Color figure online)

binding energy of the above-mentioned molecules docking with the target. In the docking results, all three drug bind to the same pocket of the target, and Lorpiprazole has the minimum binding energy. Among the generated molecules, the binding energy of Rank 2, Rank 3, Rank 4 and Rank5 is lower than that of Lorpiprazole, and the binding energy of Rank 3 is as low as −12.3 kcal/mol. As shown in Fig. 3, the blue-green molecule is Lorpiprazole, while the purple-red molecule is Rank 3, which binds to the same pocket as 5HT2A and has a similar binding conformation; both have hydrogen bonds to O, while Rank 3 has a total of three hydrogen bonds to the protein pocket and a higher binding fraction. The docking results for the other two targets are placed in the Supplementary File.

In conclusion, in the molecular docking experiments for DRD2, 5HT2A and JAK2, the molecules generated by our method and screened by DeepPurpose have similar docking binding conformations with FDA-approved drug, which bind to the same pocket of the target proteins, and some of the generated molecules have lower binding energy. This indicates that our method can generate candidate molecules with the amino acid sequence of the target protein as input.

4   Conclusion

In this work, we proposed a target-specific drug design method based on Transformer and BERT's pretraining. This approach considers the target-specific drug design task as a sequence-to-sequence generative task and represents protein with amino acid sequence while ligand with SMILES string. Then we introduced a Transformer model to learn their mapping relationship. Furthermore, we utilized

BERT's pretraining methods to help the encoder and decoder of this Transformer learning features extraction on large-scale databases, which enables Transformer to better generate targeted ligands

To verify the effectiveness of our method, we first used DTA model to predict binding affinity of generated molecules. Compared with the Transformer with pretraining, molecules generated by Transformer pretrained by BERT have higher predicted biological activity. Subsequently, we utilized our method to generate molecules toward DRD2, 5HT2A and JAK2, and calculated the MOSES benchmark. After the BERT's pretraining, the structures of generated molecules for each target were similar to that of real bioactive molecules. Finally, we utilized our method to generate molecules toward DRD2, 5HT2A and JAK2, screened the highly active molecules by using DeepPurpose, and then dock these molecules with these targets. Results show that some molecules have similar binding conformations and lower binding energy with existing FDA-approved drugs. This experiment demonstrates the potential of our method to generate candidate molecular drug toward given target proteins.

Our proposed approach shows great potential in target-specific de novo drug design and has the ability to design drugs for novel target proteins with limited active ligand datasets. Our method's use of only protein sequence information makes it applicable to novel target proteins without active ligand datasets, and our proposed evaluation method provides a comprehensive evaluation of the overall model performance.

References

1. Aumentado-Armstrong, T.: Latent molecular optimization for targeted therapeutic design. arXiv preprint arXiv:1809.02032 (2018)
2. Born, J., et al.: Data-driven molecular design for discovery and synthesis of novel ligands: a case study on SARS-COV-2. Mach. Learn. Sci. Technol. **2**(2), 025024 (2021)
3. Dai, H., Tian, Y., Dai, B., Skiena, S., Song, L.: Syntax-directed variational autoencoder for structured data. arXiv preprint arXiv:1802.08786 (2018)
4. De Cao, N., Kipf, T.: Molgan: an implicit generative model for small molecular graphs. arXiv preprint arXiv:1805.11973 (2018)
5. Devlin, J., Chang, M.W., Lee, K., Toutanova, K.: Bert: pre-training of deep bidirectional transformers for language understanding. arXiv preprint arXiv:1810.04805 (2018)
6. Dudek, A.Z., Arodz, T., Gálvez, J.: Computational methods in developing quantitative structure-activity relationships (QSAR): a review. Comb. Chem. High Throughput Screen. **9**(3), 213–228 (2006)
7. Elnaggar, A., et al.: Prottrans: toward understanding the language of life through self-supervised learning. IEEE Trans. Pattern Anal. Mach. Intell. **44**(10), 7112–7127 (2021)
8. Ferreira, L.G., Dos Santos, R.N., Oliva, G., Andricopulo, A.D.: Molecular docking and structure-based drug design strategies. Molecules **20**(7), 13384–13421 (2015)
9. Gaulton, A., et al.: Chembl: a large-scale bioactivity database for drug discovery. Nucleic Acids Res. **40**(D1), D1100–D1107 (2012)

10. Gilson, M.K., Liu, T., Baitaluk, M., Nicola, G., Hwang, L., Chong, J.: Bindingdb in 2015: a public database for medicinal chemistry, computational chemistry and systems pharmacology. Nucleic Acids Res. **44**(D1), D1045–D1053 (2016)
11. Goodfellow, I., Bengio, Y., Courville, A.: Deep Learning. MIT press, Cambridge (2016)
12. Goodfellow, I., et al.: Generative adversarial nets. Adv. Neural Inf. Process. Syst. (NIPS) **27**, 2672–2680 (2014)
13. Grechishnikova, D.: Transformer neural network for protein-specific de novo drug generation as a machine translation problem. Sci. Rep. **11**(1), 1–13 (2021)
14. Guimaraes, G.L., Sanchez-Lengeling, B., Outeiral, C., Farias, P.L.C., Aspuru-Guzik, A.: Objective-reinforced generative adversarial networks (organ) for sequence generation models. arXiv preprint arXiv:1705.10843 (2017)
15. Huang, K., Fu, T., Glass, L.M., Zitnik, M., Xiao, C., Sun, J.: Deeppurpose: a deep learning library for drug-target interaction prediction. Bioinformatics **36**(22–23), 5545–5547 (2020)
16. Jin, W., Barzilay, R., Jaakkola, T.: Junction tree variational autoencoder for molecular graph generation. In: International Conference on Machine Learning, pp. 2323–2332. PMLR (2018)
17. Kingma, D.P., Welling, M.: Auto-encoding variational bayes. arXiv preprint arXiv:1312.6114 (2013)
18. Kotsias, P.C., Arús-Pous, J., Chen, H., Engkvist, O., Tyrchan, C., Bjerrum, E.J.: Direct steering of de novo molecular generation with descriptor conditional recurrent neural networks. Nat. Mach. Intell. **2**(5), 254–265 (2020)
19. Krishnan, S.R., Bung, N., Vangala, S.R., Srinivasan, R., Bulusu, G., Roy, A.: De novo structure-based drug design using deep learning. J. Chem. Inf. Model. **62**(21), 5100–5109 (2021)
20. Kusner, M.J., Paige, B., Hernández-Lobato, J.M.: Grammar variational autoencoder. In: International Conference on Machine Learning, pp. 1945–1954. PMLR (2017)
21. Landrum, G., et al.: RDKIT: a software suite for cheminformatics, computational chemistry, and predictive modeling. Greg Landrum **8** (2013)
22. Masuda, T., Ragoza, M., Koes, D.R.: Generating 3D molecular structures conditional on a receptor binding site with deep generative models. arXiv preprint arXiv:2010.14442 (2020)
23. Méndez-Lucio, O., Baillif, B., Clevert, D.A., Rouquié, D., Wichard, J.: De novo generation of hit-like molecules from gene expression signatures using artificial intelligence. Nat. Commun. **11**(1), 10 (2020)
24. Öztürk, H., Özgür, A., Ozkirimli, E.: DeepDTA: deep drug-target binding affinity prediction. Bioinformatics **34**(17), i821–i829 (2018)
25. O'Hagan, S., Kell, D.B.: Analysing and navigating natural products space for generating small, diverse, but representative chemical libraries. Biotechnol. J. **13**(1), 1700503 (2018)
26. Ratti, E., Trist, D.: Continuing evolution of the drug discovery process in the pharmaceutical industry. Pure Appl. Chem. **73**(1), 67–75 (2001)
27. Reymond, J.L.: The chemical space project. Acc. Chem. Res. **48**(3), 722–730 (2015)
28. Simonovsky, M., Komodakis, N.: GraphVAE: towards generation of small graphs using variational autoencoders. In: Kůrková, V., Manolopoulos, Y., Hammer, B., Iliadis, L., Maglogiannis, I. (eds.) ICANN 2018. LNCS, vol. 11139, pp. 412–422. Springer, Cham (2018). https://doi.org/10.1007/978-3-030-01418-6_41

29. Skalic, M., Sabbadin, D., Sattarov, B., Sciabola, S., De Fabritiis, G.: From target to drug: generative modeling for the multimodal structure-based ligand design. Mol. Pharm. **16**(10), 4282–4291 (2019)
30. Steinegger, M., Mirdita, M., Söding, J.: Protein-level assembly increases protein sequence recovery from metagenomic samples manyfold. Nat. Methods **16**(7), 603–606 (2019)
31. Steinegger, M., Söding, J.: MMseqs2 enables sensitive protein sequence searching for the analysis of massive data sets. Nat. Biotechnol. **35**(11), 1026–1028 (2017)
32. Stumpfe, D., Dimova, D., Bajorath, J.: Composition and topology of activity cliff clusters formed by bioactive compounds. J. Chem. Inf. Model. **54**(2), 451–461 (2014)
33. Tanrikulu, Y., Krüger, B., Proschak, E.: The holistic integration of virtual screening in drug discovery. Drug Disc. Today **18**(7–8), 358–364 (2013)
34. Trott, O., Olson, A.J.: Autodock vina: improving the speed and accuracy of docking with a new scoring function, efficient optimization, and multithreading. J. Comput. Chem. **31**(2), 455–461 (2010)
35. Vaswani, A., et al.: Attention is all you need. Adv. Neural. Inf. Process. Syst. **30**, 1–11 (2017)
36. Wang, R., Fang, X., Lu, Y., Wang, S.: The PDBbind database: collection of binding affinities for protein- ligand complexes with known three-dimensional structures. J. Med. Chem. **47**(12), 2977–2980 (2004)
37. Weininger, D.: Smiles, a chemical language and information system. 1. introduction to methodology and encoding rules. J. Chem. Inf. Comput. Sci. **28**(1), 31–36 (1988)
38. Xu, M., Ran, T., Chen, H.: De novo molecule design through the molecular generative model conditioned by 3d information of protein binding sites. J. Chem. Inf. Model. **61**(7), 3240–3254 (2021)

CLIP for Lightweight Semantic Segmentation

Ke Jin and Wankou Yang[✉]

School of Automation, Southeast University, Nanjing 210096, China
jinke@seu.edu.cn

Abstract. The large-scale pretrained model CLIP, trained on 400 million image-text pairs, offers a promising paradigm for tackling vision tasks, albeit at the image level. Later works, such as DenseCLIP and LSeg, extend this paradigm to dense prediction, including semantic segmentation, and have achieved excellent results. However, the above methods either rely on CLIP-pretrained visual backbones or use none-pretrained but heavy backbones such as Swin, while falling ineffective when applied to lightweight backbones. The reason for this is that the lightweitht networks, feature extraction ability of which are relatively limited, meet difficulty embedding the image feature aligned with text embeddings perfectly. In this work, we present a new feature fusion module which tackles this problem and enables language-guided paradigm to be applied to lightweight networks. Specifically, the module is a parallel design of CNN and transformer with a two-way bridge in between, where CNN extracts spatial information and visual context of the feature map from the image encoder, and the transformer propagates text embeddings from the text encoder forward. The core of the module is the bidirectional fusion of visual and text feature across the bridge which prompts their proximity and alignment in embedding space. The module is model-agnostic, which can not only make language-guided lightweight semantic segmentation practical, but also fully exploit the pretrained knowledge of language priors and achieve better performance than previous SOTA work, such as DenseCLIP, whatever the vision backbone is. Extensive experiments have been conducted to demonstrate the superiority of our method.

Keywords: CLIP · semantic segmentation · lightweight

1 Introduction

In recent years, the CV community has made a lot of efforts to apply the achievements of NLP to the processing of visual tasks, and CLIP [1] is one of the most successful language-guided methods. In order to leverage a much broader source of supervision, namely the unlimited raw text on the Internet, and embrace the generality and usability that traditional supervised learning does not have, the authors of CLIP perform comparative learning training on 400 million image-text pairs to obtain a pair of image-text encoders. They demonstrate that the

Q. Liu et al. (Eds.): PRCV 2023, LNCS 14434, pp. 323–333, 2024.
https://doi.org/10.1007/978-981-99-8549-4_27

simple pretraining task of predicting which caption goes with which image is an efficient and scalable way to learn state-of-the-art image and text representations aligned in embedding space. After pretraining, natural language is used to reference learned visual concepts (or describe new ones), enabling zero-shot transfer of the model to downstream tasks such as OCR, action recognition in videos, geo-localization and object classification, which can basically be classified as image-level prediction tasks.

Shortly after CLIP was proposed, the problem of transferring the knowledge learned from image-text pairs to more complex dense prediction tasks, such as object detection and semantic segmentation, is visited by the community quickly [2,3]. The core of the problem is that, compared to image-level tasks such as object classification, pixel-level tasks not only require the ability to distinguish the concepts represented by the image, but also need to use spatial information to locate these concepts corresponding to pixels.

To solve the above-mentioned difficulty, scholars mostly convert the original image-text matching problem in CLIP to a pixel-text matching problem. Because they observed an interesting property of the feature map before the final pooling of CLIP's visual encoder, that it not only preserves spatial information, but also aligns with the text embeddings to some extent but not perfectly. Then they construct pixel-text score maps through inner product or other methods, to guide the learning of dense prediction models explicitly.

However, we notice that these methods rely on CLIP vision pre-training or heavy vision encoders trained on massive data, such Swin [4], and fall ineffective when applied to lightweight backbones. For an example, when using MobileNet [5] trained on ADE20K dataset [6] as a vision encoder, image feature and text feature can't be well aligned in embedding space, since the feature extraction ability of lightweight backbone is relatively limited compared to large models.

In this work, we propose a new feature fusion module which enables language-guided paradigm to be applied to lightweight network. The module takes inspiration from Mobile-Former [7] and is designed as a parallel architecture of CNN and transformer with a two-way bridge in between. On the one hand, the CNN takes the feature map from the vision encoder as input and stacks inverted bottleneck blocks. It leverages the efficient depthwise and pointwise convolution to extract spatial information and visual context of the feature map. On the other hand, the transformer takes the text embeddings from text encoder as input and stacks multi-head attention and feed-forward blocks.

The CNN and transformer in our module communicate through a two-way bridge to fuse the visual and text embedding and make them aligned by performing a lightweight cross attention we propose. The bridge, which is set at every bottleneck of CNN, feeds the image feature to text tokens in the transformer and introduces the text information to every pixel in the CNN reversely.

Extensive experiments demonstrate our method can greatly improve the segmentation performance with inference time slightly increasing which can be viewed as an acceptable compromise. The results of comparative experiments of multiple groups prove our following claims:

(1) The fusion module we propose can solve the difficulty that the language-guided paradigm cannot be well applied to lightweight visual backbones.
(2) The fusion module we propose is model-agnostic. It can fully exploit the pretrained knowledge of language priors and achieve better performance than previous SOTA work, DenseCLIP, even based on CLIP-pretrained models which DenseCLIP is designed for.

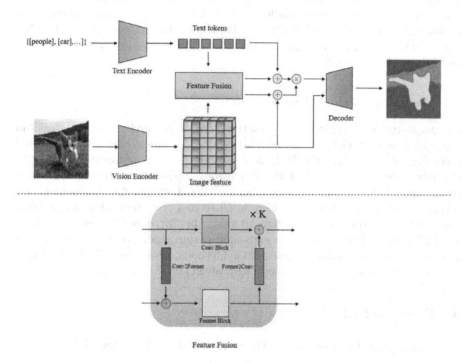

Fig. 1. The overall framework of our method and the pipeline of language-guided segmentation is demonstrated.

2 Related Work

2.1 Lightweight Visual Encoder

The deep neural network has changed the face of the field of visual recognition in a subversive way in the past ten years, but its high requirements for hardware computing performance have become a major obstacle to its application in production and life. Therefore, the research of lightweight network has become the focus of people's attention, which bridges the gap between academic research and industrial application.

MobileNetV2 [5] is based on an inverted residual structure where the shortcut connections are between the thin bottleneck layers and propose lightweight depthwise convolution to save computational cost. ShuffleNet [15] utilizes two proposed operations, pointwise group convolution and channel shuffle, to greatly reduce computation cost while maintaining accuracy. Combining the previous two works, Xception [13] proposes the depthwise separable convolution operation (a depthwise convolution followed by a pointwise convolution).

With the popularity of vision transformers, how to design lightweight transformers has also been visited frequently by the community. Swin [4] and following works propose window-based attention such that the receptive field is constrained to a pre-defined window size, which also inspires subsequent work to refine attention patterns [16,17]. Another track is to combine lightweight CNN and attention mechanism to form a hybrid architecture, such as MobileVit [18] and EfficientFormer [14].

2.2 Language-Guided Recognition

Language-driven recognition is an active area of research. Common tasks in this space include visual question answering [21], image captioning [20], and image-text retrieval [19]. CLIP [1] uses contrastive learning together with high-capacity language models and visual feature encoders to synthesize extremely robust models for zero-shot image classification. Later works [2,3] convert the original image-text matching problem in CLIP to a pixel-text matching problem and construct pixel-text score maps to guide the inference of dense prediction. Lseg [3] gets highly competitive zero-shot performance compared to existing zero- and few-shot semantic segmentation methods, while DenseCLIP [2] designs a more sophisticated framework and achieves SOTA.

3   Proposed Method

3.1   Language-Guided Semantic Segmentation Framework

The pipeline of the language-guided semantic segmentation framework is shown in Fig. 1. Before each segmentation, text prompts from the template "a photo of a [CLS]." with K class names are used as the input of the text encoder(CLIP-pretrained) to obtain text embeddings $T \in \mathbb{R}^{K \times C}$. After training, these embeddings are stored in memory for use in forward inference of the model, which can reduce the overhead brought by text encoder. On the other hand, the image encoder extract a language-compatible feature map $I \in \mathbb{R}^{H \times W \times C}$ from the input picture. Then T and I are both fed into the feature fusion module and realigned in the embedding space by this module.

$$[T', I'] = \text{Conv-Former}([T, I]) + [T, I] \tag{1}$$

After that, we correlate the visual and text embeddings by the inner product, creating a tensor of size $H \times W \times K$, defined as

$$S = I' \cdot T'^{T} \tag{2}$$

The tensor S is the score map we need and it characterizes the results of pixel-text matching. We can view the score map as segmentation results with a lower resolution and concatenate it to the last feature map to explicitly incorporate language priors, i.e., $X = [I', S] \in \mathbb{R}^{H \times W \times (C+K)}$. The modified feature map now can be directly used as usual in segmentation, followed by a popular decoder like semantic FPN.

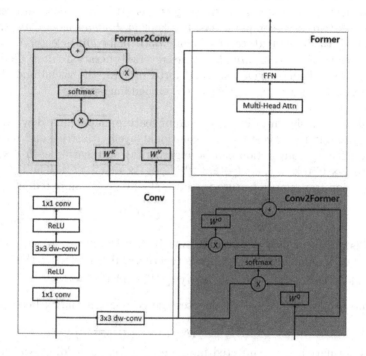

Fig. 2. Conv-Former consists of four blocks, namely *Conv, Former, Conv2Former* and *Former2Conv*, forming a parallel structure of CNN and transformer.

3.2 Structure of the Fusion Module

As shown in Fig. 1, the fusion module is formed by repeated stacking of the same component, which we named Conv-Former. Conv-Former consists of four blocks, namely *Conv, Former, Conv2Former* and *Former2Conv*, forming a parallel structure of CNN and transformer. The original image feature map $I \in \mathbb{R}^{H \times W \times C}$, is fed into the first *Conv*, while the first *Former* takes the corresponding text embedding $T \in \mathbb{R}^{K \times C}$ as input.

Efficient convolutional layers in the *Conv*, such as depthwise and pointwise convolution, extract the spatial information of the image feature map. *Former* leverage MultiHead-Attention operation to propagates the text embeddings forward. The bidirectional bridge, namely *Conv2Former* and *Former2Conv*, achieves information interaction and feature fusion between the two sides through a lightweight cross attention.

3.3 Design of Conv-Former

Bidirectional Bridge

Overall: Conv and *Former* communicate with each other through this two-way bridge. The two direction are realized by *Conv2Former* and *Former2Conv* respectively. A lightweight cross attention operation is proposed to model the process of feature fusion, where the projections (W^Q, W^K, W^V) are removed from *Conv* side to save computations, but kept at *Former* side. *Conv2Fomer* fuses image feature to text tokens before both embeddings enter *Conv* and *Former*, while *Former2Conv* does the reverse operation after *Conv* and *Former* output them. In this back and forth process, the visual and the text embeddings get closer to each other in the embedding space.

Conv2Former: As shown in Fig. 2, the input feature map is first down-sampled to reduce computational cost by a convolutional layer with stride and kernel size of 3. The effect of convolution can be regarded as aggregating each $3 \times 3 \times C$ patch into $1 \times C$ tokens, i.e., $I \in \mathbb{R}^{H \times W \times C} \longrightarrow \mathbf{x} \in \mathbb{R}^{\frac{H}{3} \frac{W}{3} \times C}$. The lightweight cross attention from image feature to text embeddings is computed as:

$$A_{\mathbf{x} \to I} = \text{MHSA}\left(IW^Q, \mathbf{x}, \mathbf{x}\right) \cdot W^O \in \mathbb{R}^{K \times C} \tag{3}$$

The W^Q is the query projection matrix and W^O is the output projection matrix which is used to combine multiple heads together. MHSA refers to the standard MultiHead-Attention function over query Q, K, and V as $softmax\left(\frac{QK^T}{\sqrt{d_k}}\right) \cdot V$. Specifically, $softmax\left(\frac{IW^Q(\mathbf{x})^T}{\sqrt{d_k}}\right)$ denotes the correlation matrix between each image patch and label token. Then $\left(softmax\left(\frac{IW^Q(\mathbf{x})^T}{\sqrt{d_k}}\right) \cdot V\right)_i$ is the visual context information to be injected into the ith label embedding, which is the weighted sum of each patch feature, with $\left(softmax\left(\frac{IW^Q(\mathbf{x})^T}{\sqrt{d_k}}\right)\right)_i$ serving as the weights.

Former2Conv: This block does the reverse operation of *ConvFormer*. Similarly, the cross attention from text embeddings to feature map is computed as:

$$A_{I \to \mathbf{x}} = \text{MHSA}\left(\mathbf{x}, IW^K, IW^V\right) \in \mathbb{R}^{H \times W \times C} \tag{4}$$

In this function, $\left(softmax\left(\frac{\mathbf{x}(IW^K)^T}{\sqrt{d_k}}\right) \cdot IW^V\right)_{ij}$ denotes the text context information to be injected into each pixel.

Conv and Former Block

Conv takes the feature map X as input and its output is taken as the input for *Former2Conv*. It uses a typical inverted bottleneck block and depthwise convolution in [5]. As shown in Fig. 3, the kernel size of depthwise convolution is 3 × 3 for all blocks.

Former is a standard transformer block including a Multi-Head Self-Attention (MHSA) and a feed-forward network (FFN). Expansion ratio 2 (instead of 4) is used in FFN. We follow [11] to use post layer normalization. *Former* is processed between *Conv2Former* and *Former2Conv*.

4 Experiments

To evaluate the effectiveness of our method, extensive experiments have been conducted on ADE20K and Cityscape [12]. Some representative vanilla methods and SOTA method DenseCLIP are taken as baseline. We hope to verify the following two points through experiments:

(1) The fusion module we propose can solve the difficulty that the language-guided paradigm cannot be well applied to lightweight visual backbones.
(2) The fusion module we propose is model-agnostic. It can fully exploit the pretrained knowledge of language priors and achieve better performance than DenseCLIP, even based on CLIP-pretrained models which DenseCLIP is designed for.

4.1 Set up and Implementation

Following common practice [8,9], we report the mIoU on the validation set of ADE20K and Cityscape. Since our method is designed for lightweight semantic segmentation, we also include the GFLOPs to evaluate the computation cost.

CLIP pretrained text encoder is used to generate text embeddings and We fix the text encoder during training to preserve the natural language knowledge. For fair comparisons, we take the Semantic FPN as the decoder.

The setting of the learning rate is special that the learning rate of the image encoder needs to be set to $\frac{1}{10}$ of other parts to preserve the knowledge in it, regardless of whether the backbone is CLIP-pretrained or ImageNet-pretrained. AdamW [10] is used instead of the vanilla SGD when the vision backbone is a transformer, following previous work.

4.2 For Lightweight Visual Backbone

We selected three representative lightweight visual backbone as the experimental objects, namely MobileNetV2, Xception [13], and EfficientFomer [14]. MobileNetV2 and Xception are both typical convolutional neutral network (CNN), while EfficientFormer combines lightweight CNN and attention mechanism to form a hybrid architecture. The experiment is conducted on the ADE20K, and the results are presented in Table 1.

From the experimental results, it can be seen that on the challenging ADE20K, DenseCLIP has a weak effect on performance improvement, and even get lower prediction accuracy than the vanilla method (only Semantic FPN) when it comes to MobileNetV2 and Xception. In contrast, our method achieve excellent performance that it is +7.1%, +4.8% and +3.8% mIoU higher than the vanilla method (only Semantic FPN) and +9.9%, +6.5% and +3.4% mIoU higher than DenseClip, with GFLOPs slightly increasing which can be viewed as an acceptable compromise.

DenseCLIP's poor performance proves the necessity of the feature fusion module, while excellent performance of our method verifies the effectiveness of the module.

Table 1. Semantic segmentation results on ADE20K: We report the results of three methods respectively. By comparing DenseCLIP and our method, the necessity and effectiveness of the feature fusion module we propose are justified.

Backbone	Method	mIoU	GFLOPs
MobileNetV2	Semantic FPN	25.1	32.7
	DenseCLIP + Semantic FPN	22.3	44.5
	Our Method + Semantic FPN	**32.2**	41.3
Xception	Semantic FPN	35.5	40.2
	DenseCLIP + Semantic FPN	33.8	51.6
	Our Method + Semantic FPN	**40.3**	47.5
EfficientFormer	Semantic FPN	42.1	48.9
	DenseCLIP + Semantic FPN	42.5	59.4
	Our Method + Semantic FPN	**45.9**	53.9

Table 2. Semantic segmentation results on **ADE20K** and **Cityscapes**.

Backbone	Method	Pretrained	mIoU(ADE20K)	mIoU(Cityscapes)
ResNet-50	Semantic FPN	ImageNet	38.6	74.5
	DenseCLIP + Semantic FPN	CLIP	43.5	75.9
	Our Method + Semantic FPN	CLIP	**44.9**	**76.3**
ResNet-101	Semantic FPN	ImageNet	40.4	75.8
	DenseCLIP + Semantic FPN	CLIP	45.1	77.1
	Our Method + Semantic FPN	CLIP	**46.7**	**77.5**
VIT-B	Semantic FPN	ImageNet	48.3	80.5
	DenseCLIP + Semantic FPN	CLIP	50.6	81.1
	Our Method + Semantic FPN	CLIP	**51.2**	**81.3**
Swin-T	UperNet	ImageNet	44.5	79.9
	DenseCLIP + UperNet	ImageNet	45.4	80.2
	Our Method + UperNet	ImageNet	**45.9**	**80.3**

4.3   For Any Visual Backbone

In the above, we claim that the fusion module we propose is not only effective for lightweight models and has outstanding generalization ability. Therefore, we selected a representative heavy model Swin-T for experiments. Moreover, we also selected three CLIP-pretrained visual backbones, which DenseCLIP is designed for, to verify the fact that our method can fully exploit the pretrained knowledge of language priors and achieve better performance than DenseCLIP.

Table 2 shows the results on ADE20K and Cityscapes respectively. Our Method achieves +1.4%, +1.6%, +0.6% and +0.5% higher mIoU than Dense-CLIP on ADE20K, and on Cityscapes the increase is +0.4%, +0.4%, +0.2% and +0.1%. This performance meets our expectations and verifies our claims.

4.4   Ablation Study

In this part, we want to further study the components of Conv-Fomer and demonstrate the effect of the lightweight cross attention operation. In addition, how many Conv-Formers need to be stacked repeatedly in the fusion module is also worth exploring. Therefore, we conduct ablation experiments for the above two problems.

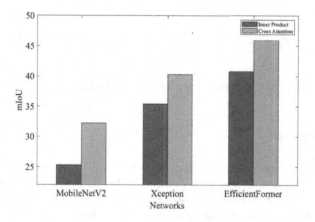

Fig. 3. The comparative experimental results on three lightweight networks.

Effect of the Cross Attention. In order to dispel doubts about whether cross-attention is really necessary, we replace it with an inner product operation, which can also calculate the similarity between two embeddings. Figure 3 shows the comparative experimental results on three lightweight networks, namely MobileNetV2, Xception and EfficientFomer. The performance gap exhibited in the figure illustrates the necessity of cross-attention.

Number of Stacked Conv-Former. In order to explore how many Conv-Formers are most suitable for stacking in the feature fusion module, we increase the number one by one and conduct experiments separately. This research is based on MobileNetV2, whose performance is most significantly improved by the feature fusion module. As shown in Fig. 4, stacking 6 times is the best choice, since when the number is less than 6, the performance is not optimal, and when the number is greater than 6, the performance tends to be saturated and wastes computing overhead.

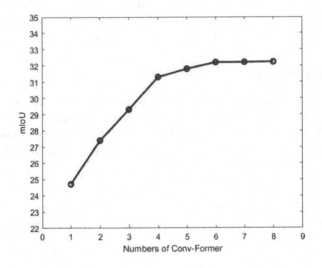

Fig. 4. Model performance curve with the number of blocks.

5   Conclusion

In this paper, we propose a feature fusion module to make language-guided lightweight semantic segmentation practical. The module is model–agnostic and achieve better performance than previous SOTA work. We conducted extensive experiments to demonstrate the superiority of our method.

References

1. Radford, A., et al.: Learning transferable visual models from natural language supervision. In: International Conference on Machine Learning (2021)
2. Rao, Y., et al.: DenseCLIP: language-guided dense prediction with context-aware prompting. In: 2022 IEEE/CVF Conference on Computer Vision and Pattern Recognition (CVPR), pp. 18061–18070 (2021)
3. Li, B., et al.: Language-driven semantic segmentation. ArXiv arxiv:2201.03546 (2022)

4. Liu, Z., et al.: Swin transformer: hierarchical vision transformer using shifted windows. In: 2021 IEEE/CVF International Conference on Computer Vision (ICCV), pp. 9992–10002 (2021)
5. Sandler, M., et al.: MobileNetV2: inverted residuals and linear bottlenecks. In: 2018 IEEE/CVF Conference on Computer Vision and Pattern Recognition, pp. 4510–4520 (2018)
6. Zhou, B., et al.: Semantic understanding of scenes through the ADE20K dataset. Int. J. Comput. Vision **127**, 302–321 (2016)
7. Chen, Y., et al.: Mobile-former: bridging MobileNet and transformer. In: 2022 IEEE/CVF Conference on Computer Vision and Pattern Recognition (CVPR), pp. 5260–5269 (2021)
8. Huang, Z., et al.: CCNet: criss-cross attention for semantic segmentation. In: 2019 IEEE/CVF International Conference on Computer Vision (ICCV), pp. 603–612 (2018)
9. Xiao, T., et al.: Unified perceptual parsing for scene understanding. ArXiv arxiv:1807.10221 (2018)
10. Loshchilov, I., Hutter, F.: Decoupled weight decay regularization. In: International Conference on Learning Representations (2017)
11. Vaswani, A., et al.: Attention is all you need. In: NIPS (2017)
12. Cordts, M., et al.: The cityscapes dataset for semantic urban scene understanding. In: 2016 IEEE Conference on Computer Vision and Pattern Recognition (CVPR), pp. 3213–3223 (2016)
13. Chollet, F.: Xception: deep learning with depthwise separable convolutions. In: 2017 IEEE Conference on Computer Vision and Pattern Recognition (CVPR), pp. 1800–1807 (2016)
14. Li, Y., et al.: Rethinking vision transformers for MobileNet size and speed. ArXiv arxiv:2212.08059 (2022)
15. Zhang, X., et al.: ShuffleNet: an extremely efficient convolutional neural network for mobile devices. In: 2018 IEEE/CVF Conference on Computer Vision and Pattern Recognition, pp. 6848–6856 (2017)
16. Chen, C.-F., et al.: CrossViT: cross-attention multi-scale vision transformer for image classification. In: 2021 IEEE/CVF International Conference on Computer Vision (ICCV), pp. 347–356 (2021)
17. Pan, Z., Cai, J., Zhuang, B.: Fast vision transformers with hilo attention. Adv. Neural. Inf. Process. Syst. **35**, 14541–14554 (2022)
18. Mehta, S., Rastegari, M.: MobileViT: light-weight, general-purpose, and mobile-friendly vision transformer. ArXiv arxiv:2110.02178 (2021)
19. Wang, Z., et al.: CAMP: cross-modal adaptive message passing for text-image retrieval. In: 2019 IEEE/CVF International Conference on Computer Vision (ICCV), pp. 5763–5772 (2019)
20. Xu, K., et al.: Show, attend and tell: neural image caption generation with visual attention. In: International Conference on Machine Learning (2015)
21. Agrawal, A., et al.: VQA: visual question answering. Int. J. Comput. Vision **123**, 4–31 (2015)

Teacher-Student Cross-Domain Object Detection Model Combining Style Transfer and Adversarial Learning

Lijun Wu, Zhe Cao, and Zhicong Chen[✉]

School of Advanced Manufacturing, Fuzhou University, Fujian, China
zhicong.chen@fzu.edu.cn

Abstract. Cross-domain object detection is challenging because object detection models are significantly susceptible to domain style. As a popular semi-supervised learning method, the teacher-student framework (pseudo labels from the teacher model supervise the student model) achieves significant accuracy gains in cross-domain object detection. However, it suffers from the domain shift and prone to generate low-quality pseudo labels, which limits the performance. To mitigate this problem, we propose a teacher-student framework that utilizes style transfer method, augmentation strategies, and adversarial learning to address domain shift. Specifically, we design a Fourier style transfer method to reduce the gap between source and target domains without altering the semantic information of the objects. Furthermore, we improve the data augmentation strategy, by weakly augmenting the images from the target domain, to avoid the teacher model biased to the source domain. Finally, we employ feature-level adversarial training in the student model which is trained based on images from all domains, allowing features derived from all domains to share similar distributions. This process ensures that the student model produces domain-invariant features. Our approach achieves state-of-the-art performance in several benchmark tests. For example, it achieved 51.6% and 49.9% mAP on Foggy Cityscapes and Clipart1K, respectively.

Keywords: Cross-Domain Object Detection · Unsupervised Domain Adaptation · Style Transfer · Adversarial Learning

1 Introduction

Convolutional neural networks have demonstrated remarkable object detection performance when trained on large-scale and high-quality annotated data [1–4]. However, when applied to unseen data, the detector suffers greatly from domain shifts such as changes in the weather, variations in the lighting, or image deterioration [5]. Unsupervised domain adaptive object detection (UDA-OD) techniques, that aim to transfer a model pre-trained based on a labeled source domain to an unlabeled target domain with different data distributions, have been proposed to address this problem.

© The Author(s), under exclusive license to Springer Nature Singapore Pte Ltd. 2024
Q. Liu et al. (Eds.): PRCV 2023, LNCS 14434, pp. 334–345, 2024.
https://doi.org/10.1007/978-981-99-8549-4_28

Various UDA-OD techniques have been proposed based on adversarial learning or self-training methods. Adversarial learning [6–8] aims to learn domain-invariant representation using domain classifiers and gradient reversal layers. In this way, the recognition accuracy can be improved. Very recently, self-training [6, 7, 9, 10] have been proposed to leverage teacher-student mutual learning to improve the performance on unlabeled target data progressively. Self-training typically involves a teacher model and a student model learning from each other, so as to generate pseudo labels for unlabeled images through studying the existing label knowledge. The self-training approach achieves great success in domain adaptive scenarios and has become today's most commonly used framework for cross-domain object detection. In addition to the optimization of the network structure, style transfer preprocessing can also be utilized to improve the performance of unsupervised domain adaptive object detection. Style transfer bridges the domain difference between the source and target domains by generating an intermediate domain. The most currently used style transfer is implemented based on Generative Adversarial Network (GAN) [11, 12]. It is demonstrated that the combination of GAN style transfer and the teacher-student self-training framework can significantly improve the system performance [9, 13].

However, the current teacher-student model and the most commonly used GAN-based style transfer still has some limitations. The significant domain differences between the source and target domain data can deeply influence the quality of the pseudo labels generated by the teacher model. The generated pseudo labels might bias toward the source domain if it is studied mainly based on the labeled source domain data. Fortunately, the target-like domain images generated by GAN, which transfer target style to source domain images and are visually similar to the target domain images, can mitigate the domain differences between the source and target domains. Nevertheless, the semantic information of the objects in the intermediate domain generated by GAN prone to be altered and therefore influence the system performance, as shown in Fig. 1.

To address the above problems, we propose a novel teacher-student self-training framework that employs three strategies to mitigate domain differences and improve the quality of the pseudo label. Firstly, inspired by [14], we propose a new Fourier style transfer method to generated the target-like images reduce the domain gap between the source and target domains through exchanging the low-frequency spectra of the source and target domains. In this way, the semantic information of the target images will not be altered and no additional training tasks is introduced. Secondly, we design an augmentation strategy which imports the target domain images into the teacher model after been weakly augmented, and to imports the images from target domain, source domain, and target-like domain into the student model after strongly, strongly and weakly augmented respectively. Such augmentation strategy hopes that the distribution of target domain data used to generate pseudo labels will not be disrupted as much as possible. Finally, we use adversarial learning to align the feature distribution between different data domains and learn the domain invariant features, and use mutual learning to help the teacher model produce high quality pseudo labels. We summarize the contributions of this paper as follows:

- Considering that the GAN-based [11] style transfer method may alter the semantic information of objects, we propose a Fourier style transfer method to exchanging the

336 L. Wu et al.

low-frequency spectra of the source and target domains through which the semantic information of objects are maintained.

- A novel framework is proposed which includes Fourier style transfer, augmentation strategies, mutual learning, and adversarial learning to address domain shift in cross-domain object detection.
- The proposed method provides state-of-the-art performance in several cross-domain benchmark tests, i.e., it reached 51.6% and 49.9% mAP on Foggy Cityscapes and Clipart1K, respectively.

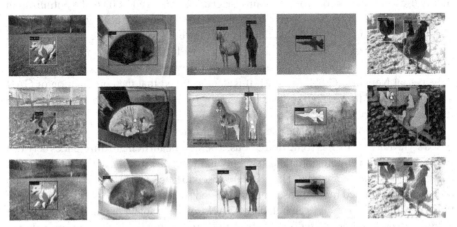

Fig. 1. The comparison between the GAN-based style transfer and Fourier-based style transfer. The first row is the original images, the second row is the target-like domain images generated by CycleGAN, and the third row is the target-like domain images generated by the Fourier-based style transfer module.

2 Related Works

Object detection has made significant progress in recent years [15, 16], and many excellent object detection models have emerged, such as YOLO and Faster RCNN. However, in many practical application scenarios, it is often difficult to obtain sufficient datasets, so unsupervised domain adaptive techniques are gradually introduced into the object detection domain [17, 18].

Unsupervised domain adaption aims to learn a model from additional labeled source domain to achieve satisfactory performance on the target domain. Compared to general vision tasks, the problem of object detection is more complicated since it has to predict the bounding box and class label for each object. Cross-domain object detection can be divided into two main categories: adversarial learning-based and self-training-based methods. Adversarial learning utilizes a gradient reverse layer (GRL) to map the feature across different domains [6–8, 19], and therefore improves recognition accuracy. Recently, researchers have explored methods to enable self-training on unmarked target domains. Basically, it extends teacher-student self-training methods from semi-supervised learning to domain adaptation [20]. Self-training methods [6, 7, 9, 10] can learn without labeling the target domain and usually involve the teacher model generating pseudo labels to update the student model. There has been a trend to use a combination of teacher-student self-training methods and adversarial learning [6, 7]. Adaptive Teacher (AT) [7] uses feature-level adversarial training in the student model to allow features from the source and target domains to share similar distributions, helping the model to produce domain-invariant features. However, the above approaches might suffer the same inherent issue in Mean Teacher (MT) [20] and influence the quality of pseudo labels generated for the target domain. In addition, style transfer that can bridge the domain gaps by generating an intermediate domain, is usually applied on the input side of the cross-domain object detection model [9, 13, 21]. It has been proposed to reduce the domain shift by augmenting the training samples with CycleGAN [11]. However, the images generated based on CycleGAN will alter object semantic information.

To address the above issues, we combine self-training strategy, adversarial learning method, and adds a style transfer module on the input side. It is worth to noting that, we make effort to improve the quality of pseudo-label problems and propose a style transfer method that will not alter the semantic information of objects.

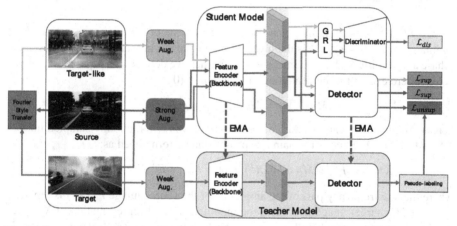

Fig. 2. Overview of our proposed framework. The proposed framework is trained via two learning streams: the Teacher-Student mutual learning stream and the adversarial learning stream.

3 Proposed Method

3.1 Fourier Style Transfer

In unsupervised domain adaptation (UDA), we are given a source dataset $D^s = \{(X_i^s, Y_i^s)\}_{i=1}^{N_s}$, where X^s is the source image, and $Y_i^s = \{B_i^s, C_i^s\}$ are the lables associated with X^s, where B_i^s denotes the bounding box annotations and C_i^s denotes corresponding class labels. Similarly, $D^t = \{X_i^t\}_{i=1}^{N_t}$ is the target dataset, where the ground truth object labels are absent. Generally, the cross-domain object detection network trained on D^s will have a performance drop when tested on D^t. Here, we propose Fourier Style Transfer generating intermediate domain to mitigate the domain gap between the two datasets.

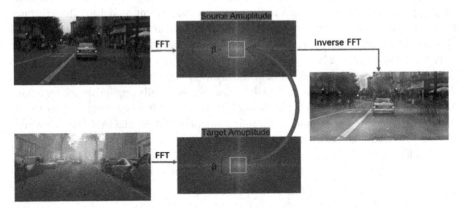

Fig. 3. Fourier Style Transfer: Mapping a source image to a target "style" without altering semantic information.

Let F^A, F^P be the amplitude and phase components of the Fourier transform F of an image. Accordingly, F^{-1} is the inverse Fourier transform that maps spectral signals (phase and amplitude) back to image space. Further, we denote with M_β a mask, whose value is zero except for the center region where $\beta \in (0, 1)$.

$$M_\beta(h, w) = 1_{(h,w) \in [-\beta H : \beta H, -\beta W : \beta W]} \tag{1}$$

here we assume the center of the image is $(0, 0)$. Given two randomly sampled images $X^s \in D^s, X^t \in D^t$, Fourier Domain Adaptation can be formalized as:

$$X^{s \to t} = F^{-1}\left(\left[M_\beta \circ F^A(X^t) + (1 - M_\beta) \circ F^A(X^s), F^P(X^s)\right]\right) \tag{2}$$

where the low frequency part of the amplitude of the source image $F^A(X^s)$ is replaced by that of the target image X^t. Then, the modified spectral representation of X^s, with its phase component unaltered, is mapped back to the image $X^{s \to t}$, whose content is the same as X^s, but will resemble the appearance of a sample from D^t. The process is illustrated in Fig. 3 where the mask M_β is shown in white.

Choice of β. After our tests, we found that the model has the best cross-domain detection performance when $\beta = 0.01$. Also, because the cross-domain object detection has a large difference in the dataset domains, we apply a smaller perturbation ($\delta = \pm 0.002$) to β to increase the model's generalization ability.

3.2 Augmentation Strategies

The cross-domain object detection model is ultimately tested and used on the target domain, so we input the target domain images into the teacher model after weak augmentation. This modification prevents the teacher model from being biased towards the source domain. For the target-like domain images generated by the Fourier style transfer module, we weakly-augmented it and input it into the student model. Weak augmentation does not destroy the domain style, so it can preserve the target domain style we have transferred [7]. We use strong augmentation for the source domain images, which will destroy the domain style of the source domain images and facilitate the detector to learn the invariant domain knowledge better [7, 10]. The whole augmentation strategies are shown in Fig. 2. A comparison of the augmentation strategies is shown in Table 1..

We use weak augmentation for the features input to the domain discriminator, which allows the discriminator to concentrate on optimizing domain differences rather than being influenced by strong augmentation.

The weak augmentation used in this paper includes randomly horizontal flipping and cropping. The strong augmentation includes: randomly color jittering, gray scaling, Gaussian blurring, and cutting patches.

Table 1. Comparison between augmentation strategies

Model	Source	Target	Target-like	Source-like
UMT [9]	Strong(Student)	Strong(Student)	Strong(Student)	Weak(Teacher)
Our	Strong(Student)	Waek(Teacher) Strong(Student)	Weak(Student)	\

3.3 Mutual Learning and Adversarial Learning

We train our model using two training streams which are the teacher-student mutual learning and the adversarial learning strategies.

Mutual Learning. We first use the labeled source and target-like images to optimize our model with the supervised loss $\mathcal{L}_{sup}$. Hence, the supervised loss for training and initializing the student model using the labeled images can be defined as:

$$\mathcal{L}_{sup1}(X^s, Y^s) = \mathcal{L}_{cls}^{rpn}(X^s, Y^s) + \mathcal{L}_{reg}^{rpn}(X^s, Y^s) + \mathcal{L}_{cls}^{roi}(X^s, Y^s) + \mathcal{L}_{reg}^{roi}(X^s, Y^s)$$
$$\mathcal{L}_{sup2}(X^{tl}, Y^s) = \mathcal{L}_{cls}^{rpn}(X^{tl}, Y^s) + \mathcal{L}_{reg}^{rpn}(X^{tl}, Y^s) + \mathcal{L}_{cls}^{roi}(X^{tl}, Y^s) + \mathcal{L}_{reg}^{roi}(X^{tl}, Y^s)$$
$$\mathcal{L}_{sup} = \mathcal{L}_{sup1} + \mathcal{L}_{sup2}$$

(3)

where X^s is the source domain image, Y^s is the corresponding label. X^{tl} is the target-like domain image, and the target-like domain image and the source domain image have the same label Y^s.

As the labels are unavailable in the target domain, we adopt the pseudo-labeling method to produce pseudo labels on the images from the target domain to train the student model. Hence, after obtaining the pseudo labels from the teacher model, we can update the student model with the loss:

$$\mathcal{L}_{unsup}\left(X^t, \widehat{C^i}\right) = \mathcal{L}_{cls}^{rpn}\left(X^t, \widehat{C^i}\right) + \mathcal{L}_{cls}^{roi}\left(X^t, \widehat{C^i}\right)$$

(4)

where $\widehat{C^i}$ denotes the pseudo labels produced by the teacher model on the target domain.

To produce high-quality pseudo labels, we apply Exponential Moving Average (EMA) to update the teacher model by temporally copying the weights of the student model. The update procedure can be defined as follows:

$$\alpha\theta_t + (1 - \alpha)\theta_s \rightarrow \theta_t \tag{5}$$

where θ_t and θ_s denote the network parameters of teacher and student, respectively. The weight smooth coefficient parameter α is set to 0.9996.

Adversarial Learning. In our model, only the source domain and the target-like domain have labels, and the labels are identical (all from the source domain), so both the teacher model and the student model are prone to bias towards the source domain during mutual learning. If the bias is not addressed at the domain feature level, the teacher model will always bias toward the source domain in mutual learning.

A domain discriminator D is placed after the feature encoder E on the student model to achieve adversarial learning. The discriminator aims to discriminate where the derived feature $E(X)$ is from. On the other hand, the feature encoder E is encouraged to produce features that confuse the discriminator D. Hence, such adversarial optimization objective function can be defined as the following:

$$\mathcal{L}_{dis} = \max_E \min_D \mathcal{L}oss \tag{6}$$

Fortunately, we can add a Gradient Reverse Layer (GRL) between the feature encoder and the discriminator to produce reverse gradient, simplifying the min-max optimization.

Total Loss. The total loss $\mathcal{L}$ for training our proposed model is summarized as follows:

$$L = \mathcal{L}_{sup} + \lambda_{unsup} \cdot \mathcal{L}_{unsup} + \lambda_{dis} \cdot \mathcal{L}_{dis} \tag{7}$$

where λ_{unsup} and λ_{dis} are the hyper-parameters used to control the weighting of the corresponding losses.

4 Experiments

We conduct our experiments on five public datasets, including Cityscapes [22], Foggy Cityscapes [23], PASCAL VOC [24], Clipart1k [25], and Watercolor2k [25].

4.1 Implementation Details

Following [7], we employ Faster RCNN as the base detection model in our model and implement it using Detectron2. Either of the network ResNet-101 or VGG16 pre-trained on ImageNet is used as the backbone. Following the implementation of Faster RCNN with ROI-alignment [7], we scale all images by resizing the shorter side of the image to 600 while maintaining the image ratios. Our learning rate during the entire training stage without applying any learning rate decay. For the hyperparameter, we set the $\beta = 0.01$, $\delta = \pm 0.002$, learning rate $r = 0.01$, $\lambda_{unsup} = 1.0$ and $\lambda_{dis} = 0.1$ for all the experiments for simplicity. Each experiment is conducted on Nvidia GPU V100 and implemented in PyTorch.

4.2 Experimental Settings and Evaluation

In this paper, we use average accuracy as the evaluation metric for each category in all experiments, using mean accuracy precision (mAP) as the evaluation metric for the model.

Real to Artistic Adaptation. In this setup, we use the PASCAL VOC [24] datasets as the source domain, and use the Clipart1k or Watercolor2k as the target domain, respectively. Following [7, 9], PASCAL VOC is combined from PASCAL VOC 2007 and 2012 with 16,551 images. Watercolor2k [25] is split in half into training sets and testing sets, each containing 1000 images. Clipart1k [25] is split in half into training sets and testing sets, each containing 500 images. The experiment uses ResNet-101 as the backbone.

Adverse Weather Adaptation. Cityscapes is used as the source domain in this setup, and Foggy Cityscapes is used as the target domain. Cityscapes [22] is collected by capturing images from outdoor street scenes in normal weather conditions from 50 cities with diverse scenes. It contains 2,975 images for training and 500 for validation. Foggy Cityscapes [23] has the same categories and labels as Cityscapes, and it simulates foggy weather conditions based on the images in Cityscapes. The experiment uses VGG16 as the backbone.

4.3 Results and Comparisons

In this section, we report the performance of our method and other state-of-the-art approaches in Table 2., Table 3. and Table 4..

Real to Artistic Adaptation. The results of Real to Artistic on Clipart1k are shown in Table 2. and on Watercolor2k in Table 3.. First, our model achieves state-of-the-art performance with 49.9% of mAP. We note that UMT [9] uses CycleGAN for style transfer at the image level, which reduces the domain gap visually. However, the altered semantic information of objects limits the performance of UMT. Since the MGA [26] does not give the mAP for each class, we use "–" instead. Similar results can be found for experiments conducted on PASCAL VOC → Watercolor2k.

Adverse Weather Adaptation. The experimental results are presented in Table 4.. First, our model achieves state-of-the-art performance by 51.6% mAP. DA-Detect [21] generates auxiliary domain images to simulate images of inclement weather by composing sunny source domain images and rainy images. However, there will be differences between the auxiliary domain-generated target-like domain images and the real target domain distribution. Our style transfer method can generate target-like domains close to the target domain distribution, which helps the model achieve state-of-the-art results.

4.4 Ablation Studies

We further conducted ablation studies for each significant component in Table 5..

Table 2. The results of cross-domain object detection on the Clipart1k test set for PASCAL VOC → Clipart1k adaptation. The average precision (AP, in %) on all classes is reported.

Method	TIA [27]	UMT [9]	MGA [26]	DAB [6]	AT [7]	Our
aero	**42.2**	39.6	–	25.2	33.8	41.9
bcycle	**66.0**	59.1	–	75.7	60.9	60.5
bird	36.9	32.4	–	31.8	38.6	**39.7**
boat	37.3	35.0	–	42.3	49.4	**51.2**
bottle	43.7	45.1	–	32.5	**52.4**	48.1
bus	**71.8**	61.9	–	70.8	53.9	60.2
car	49.7	48.4	–	**57.2**	56.7	49.6
cat	18.2	7.5	–	**18.3**	7.5	7.9
chair	44.9	46.0	–	42.2	52.8	**54.6**
cow	58.9	67.6	–	**73.7**	63.5	67.2
table	18.2	21.4	–	**42.5**	34.0	35.8
dog	29.1	**29.5**	–	25.7	25.0	26.6
horse	40.7	48.2	–	41.1	**62.2**	60.7
m-bike	**87.8**	75.9	–	65.9	72.1	79.4
person	67.4	70.5	–	**77.4**	77.2	70.3
plnt	49.7	56.7	–	**58.0**	57.7	53.4
sheep	27.4	25.9	–	**47.9**	27.2	30.1
sofa	27.8	28.9	–	33.7	**52.0**	47.9
train	**57.1**	39.4	–	52.5	55.7	56.9
tv	50.6	43.6	–	53.4	54.1	**56.0**
mAP	46.3	44.1	44.8	48.4	49.3	**49.9**

Table 3. The results of cross-domain object detection on the Watercolor2k test set for PASCAL VOC → Watercolor2k adaptation. The average precision (AP, in %) on all classes is reported.

Method	bike	bird	car	cat	dog	person	mAP
UMT [9]	88.2	55.3	51.7	39.8	**43.6**	69.9	58.1
MGA [26]	–	–	–	–	–	–	58.1
AT [7]	**93.6**	56.1	**58.9**	37.3	39.6	73.8	59.9
Our	91.1	**56.9**	55.8	**42.3**	41.7	**74.6**	**60.4**

Fourier Style Transfer (FST) Module. We remove this module from the model, which showed 1.3%, 0.9% and 1.1% performance drop in the three cross-domain detection

Table 4. The results and comparison on cross-domain object detection on the Foggy Cityscapes test set for Cityscapes → Foggy Cityscapes adaptation. The average precision (AP, in %) on all classes is reported.

Method	bus	bicycle	car	mcycle	person	rider	train	truck	mAP
DAB[6]	**57.1**	38.7	**68.9**	35.2	46.1	46.5	50.8	35.6	47.4
MGA[26]	50.7	42.8	60.6	38.3	43.9	49.6	39.0	29.6	44.3
DA-Detect[21]	51.2	39.1	54.3	31.6	36.5	46.7	48.7	30.3	42.3
TIA[27]	52.1	38.1	49.7	37.7	34.8	46.3	48.6	31.1	42.3
UMT[9]	56.5	37.3	48.6	30.4	33.0	46.7	46.8	34.1	41.7
AT[7]	56.3	51.9	64.2	38.5	45.5	**55.1**	**54.3**	35.0	50.9
Our	55.4	**54.4**	62.0	**44.1**	**51.7**	53.5	52.5	**39.2**	**51.6**

experiments, respectively. This indicates that the Fourier style transfer module can mitigate the inter-domain gap between two datasets. Meanwhile, we tested the performance of the GAN-based style transfer method, replacing the Fourier style transfer module with the CycleGAN style transfer module, which showed 3.4%, 2.3% and 2.2% performance drop in the three cross-domain detection experiments, respectively.

Augmentation Strategies. We remove augmentation strategies from the model, which showed 6.4%, 5.9% and 7.1% performance drop, respectively. This indicates that data augmentation strategies are essential for adaptive object detection in semi-supervised domains.

Adversarial Learning. We remove adversarial learning from the model, which showed 3.3%, 7.6% and 5.0% performance drop, respectively. This indicates that adversarial learning is effective for adaptive object detection.

Table 5. The ablation studies on our method.

Source Target	Cityscapes Foggy Cityscapes	PASCAL VOC Clipart1k	PASCAL VOC Watercolor2k
Our	51.6	49.9	60.4
Our w/o FST	50.3(−1.3)	49.0(−0.9)	59.3(−1.1)
Our w/o FST + CycleGAN	48.2(−3.4)	47.6(−2.3)	58.2(−2.2)
Our w/o Aug Our w/o λ_{dis}	45.2(−6.4) 48.3(−3.3)	44.0(−5.9) 42.3(−7.6)	53.3(−7.1) 55.4(−5.0)

5 Conclusion

In this paper, we proposed a novel framework to address the cross-domain object detection task. Through the designed Fourier style transfer module, we can generate a target-like domain images that do not alter the semantic information of the object and reduce domain shift at the image level. We also design an appropriate augmentation strategy that can relieve the bias of the teacher model toward the source domain. The model utilizes two training streams, mutual learning, and adversarial learning, to help the student model generate domain-invariant features and help the teacher model generate high-quality pseudo labels. Extensive experiments are conducted on multiple benchmark datasets, and the results clearly show that our model surpasses the existing state-of-the-art models.

References

1. Tong, J., Chen, T., Wang, Q., Yao, Y.: Few-shot object detection via understanding convolution and attention. In: Pattern Recognition and Computer Vision: 5th Chinese Conference, PRCV 2022, Shenzhen, China, 4–7 November 2022, Proceedings, Part I, pp. 674–687. Springer, Heidelberg (2022). https://doi.org/10.1007/978-3-031-18907-4_52
2. Zhou, K., Deng, K., Chen, P., Hu, Y.: An improved lightweight network based on MobileNetV3 for Palmprint recognition. In: Pattern Recognition and Computer Vision: 5th Chinese Conference, PRCV 2022, Shenzhen, China, 4–7 November 2022, Proceedings, Part I, pp. 749–761. Springer, Heidelberg (2022). https://doi.org/10.1007/978-3-031-18907-4_58
3. Gu, J., et al.: MSINet: twins contrastive search of multi-scale interaction for object ReID. In: Proceedings of the IEEE/CVF Conference on Computer Vision and Pattern Recognition, pp. 19243–19253 (2023)
4. Jin, L., et al.: Rethinking the person localization for single-stage multi-person pose estimation. IEEE Trans. Multimedia (2023)
5. Michaelis, C., et al.: Benchmarking robustness in object detection: Autonomous driving when winter is coming. arXiv preprint arXiv:1907.07484 (2019)
6. Liu, X., Zhang, B., Liu, N.: Cross-domain object detection by dual adaptive branch. IEEE Sens. J. **23**, 1199 (2023)
7. Li, Y.-J., et al.: Cross-domain adaptive teacher for object detection. In: Proceedings of the IEEE/CVF Conference on Computer Vision and Pattern Recognition, pp. 7581–7590 (2022)
8. Li, G., Ji, Z., Qu, X.: Stepwise domain adaptation (SDA) for object detection in autonomous vehicles using an adaptive CenterNet. IEEE Trans. Intell. Transport. Syst. **23**(10), 17729–17743 (2022)
9. Deng, J., Li, W., Chen, Y., Duan, L.: Unbiased mean teacher for cross-domain object detection. In: Proceedings of the IEEE/CVF Conference on Computer Vision and Pattern Recognition, pp. 4091–4101 (2021)
10. Chen, M., et al.: Learning domain adaptive object detection with probabilistic teacher. arXiv preprint arXiv:2206.06293 (2022)
11. Zhu, J.-Y., Park, T., Isola, P., Efros, A.A.: Unpaired image-to-image translation using cycle-consistent adversarial networks. In: Proceedings of the IEEE International Conference on Computer Vision, pp. 2223–2232 (2017)
12. Gong, R., Li, W., Chen, Y., Gool, L.V.: Dlow: domain flow for adaptation and generalization. In: Proceedings of the IEEE/CVF Conference on Computer Vision and Pattern Recognition, pp. 2477–2486 (2019)
13. Rodriguez, A.L., Mikolajczyk, K.: Domain adaptation for object detection via style consistency. arXiv preprint arXiv:1911.10033 (2019)

14. Yang, Y., Soatto, S.: FDA: fourier domain adaptation for semantic segmentation. In: Proceedings of the IEEE/CVF Conference on Computer Vision and Pattern Recognition, pp. 4085–4095 (2020)
15. Zou, Z., Chen, K., Shi, Z., Guo, Y., Ye, J.: Object detection in 20 years: a survey. Proc. IEEE (2023)
16. Ge, P., Ren, C.-X., Xu, X.-L., Yan, H.: Unsupervised domain adaptation via deep conditional adaptation network. Pattern Recogn. **134**, 109088 (2023)
17. Wang, M.,et al.: Reducing bi-level feature redundancy for unsupervised domain adaptation. Pattern Recogni. 109319 (2023)
18. Li, W., Li, L., Yang, H.: Progressive cross-domain knowledge distillation for efficient unsupervised domain adaptive object detection. Eng. Appl. Artif. Intell. **119**, 105774 (2023)
19. Tian, Q., Yang, H., Lu, Z., Liu, M.J.C., Engineering, E.: Unsupervised domain adaptation through adversarial enhancement and gradient discrepancy minimization. Comput. Electr. Eng. **105**, 108483 (2023)
20. Tarvainen, A., Valpola, H.: Mean teachers are better role models: weight-averaged consistency targets improve semi-supervised deep learning results. Adv. Neural Inf. Process. Syst. **30**, 1–10 (2017)
21. Li, J., Xu, R., Ma, J., Zou, Q., Ma, J., Yu, H.: Domain adaptive object detection for autonomous driving under foggy weather. In: Proceedings of the IEEE/CVF Winter Conference on Applications of Computer Vision, pp. 612–622 (2023)
22. Cordts, M., et al.: The cityscapes dataset for semantic urban scene understanding. In: Proceedings of the IEEE Conference on Computer Vision and Pattern Recognition, pp. 3213–3223 (2016)
23. Sakaridis, C., Dai, D., Van Gool, L.: Semantic foggy scene understanding with synthetic data. Int. J. Comput. Vision **126**, 973–992 (2018)
24. Everingham, M., Van Gool, L., Williams, C.K., Winn, J., Zisserman, A.: The pascal visual object classes (voc) challenge. Int. J. Comput. Vision **88**, 303–308 (2009)
25. Inoue, N., Furuta, R., Yamasaki, T., Aizawa, K.: Cross-domain weakly-supervised object detection through progressive domain adaptation. In: Proceedings of the IEEE Conference on Computer Vision and Pattern Recognition, pp. 5001–5009 (2018)
26. Zhou, W., Du, D., Zhang, L., Luo, T., Wu, Y.: Multi-granularity alignment domain adaptation for object detection. In: Proceedings of the IEEE/CVF Conference on Computer Vision and Pattern Recognition, pp. 9581–9590. (2022)
27. Zhao, L., Wang, L.: Task-specific inconsistency alignment for domain adaptive object detection. In: Proceedings of the IEEE/CVF Conference on Computer Vision and Pattern Recognition, pp. 14217–14226 (2022)

Computing 2D Skeleton via Generalized Electric Potential

Guangzhe Ma[1], Xiaoshan Wang[1], Xingchen Liu[1], Zhiyang Li[1(✉)] ⓘ,
and Zhaolin Wan[2]

[1] Dalian Maritime University, Dalian, China
lizy0205@dlmu.edu.cn
[2] Harbin Institute of Technology, Harbin, China

Abstract. As a fundamental step in visual shape analysis, finding a good shape representation has attracted a lot of interest. A skeleton, which not only retains the geometric characteristics of the shape but also has its complete topology, is regarded as a very efficient tool in visual shape analysis. Although a variety of skeleton computation methods have been proposed, the obtained skeletons still suffer from some limitations, i.e., being sensitive to noise or being inconsistent. In this study, we propose a method for computing the 2D skeleton based on a generalized electric field. Upon evenly placing unit positive charges on the shape contour, we find that the skeleton of the shape can be accurately extracted by a novel definition of electric potential and later pruned by the distribution of charge on the shape when the state of electrostatic equilibrium is reached. Extensive qualitative experiments are conducted on two public datasets, demonstrating that our method can produce a stable and concise skeleton and outperform several state-of-the-art methods.

Keywords: Skeleton extraction · Skeleton pruning · Generalized electric potential

1 Introduction

In computer vision, visual shape analysis has played an important role in the inferences about the properties of objects and scenes in the real world from 2D images, which commonly consists of many more particular tasks, i.e., shape representation, shape decomposition [8], shape recognition [17]. One of the fundamental steps is to find a high-quality representation of a given shape which is usually the silhouette of an object [16].

Generally, shape representations are designed based on the contour of the shape like shape signature [2], shape context [14], or the inside region of the shape, like moment representations [7], wavelet representations [10]. Contour-based representations are intuitive and concise but sensitive to variations or noise on the contour. By contrast, region-based representations are more robust but they have more information to handle.

Supported by the Natural Science Foundation of China (Nos. 62102059, 61672379, 61370198), the National Key R&D Program of China (No. 2021YFF0900503).

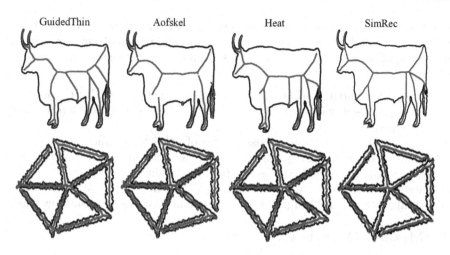

Fig. 1. Shape skeletons obtained by several classic skeleton computation algorithms.

Another popular choice is the skeletal or medial representation [15] that takes both advantages of the above two methods. The pioneering work is the medial axis of a shape, which was first proposed by Harry Blum, and later generalized to the skeleton [3]. The medial axis is defined as the locations where fire fronts meet when setting the contour of shape on fire and making the front advance inward at a constant speed. Due to the noise and shape variation, it is usually no need to find the exact location of medial points but rather to compute a stable medial axis called skeleton, a nearby location of medial points with the topology of the skeleton being stable.

Following the above grassfire analogy, various methods have been presented in computing the skeleton of shapes, which can be categorized into four categories: thinning, Voronoi diagram, distance map, and pruning methods. In Fig. 1, we choose 2 shapes and show the skeletons computed by 4 representative methods, which are selected from the above 4 categories, respectively. In the experimental section, we will give the details and more results of these methods. From Fig. 1, it can be seen that these methods still suffer from the following limitations.

On the one hand, the generated skeletons of a shape are inconsistent among the 4 methods. Take the cow for example. Although the shape is simple with no noise, there are not two very similar skeletons of the four generated skeletons. Actually, it is usually hard to provide the ground-truth skeleton of a shape when the shape is complex. On the other hand, these methods are still sensitive to noise, which causes the instability of the skeleton calculation. The man-made shapes in Fig. 1 have a certain amount of noise on the contour. These methods cannot avoid producing redundant branches on the skeleton, even for the skeleton pruning method of SimRec.

To address these issues, this paper proposes a 2D shape skeleton computation method based on a generalized electric potential. Our method belongs to the distance map methods. Moreover, unlike using classic fields like Euclidean distance field or heat field, we turn to the electric potential field and extend it to skeleton extraction and pruning. Our main contribution is summarized as follows.

- A skeleton computation method based on the generalized electric potential field is proposed, which can produce a stable and concise skeleton for noisy 2D shapes.
- The skeleton computation process consists of a skeleton extraction stage and a skeleton pruning stage, motivated by the physical facts of the electric potential field and the electric charge distribution on the shape.
- Various experiments are carried out on two public shape datasets, demonstrating that our method performs better than several state-of-arts methods.

2   Method Overview

2.1   Symbols

For a two-dimensional object O in an image, we assume its contour S is exacted and uniformly parameterized by an landmark sequence $P(n) = \{p_1, p_2, ..., p_n\}$, where $p_i = (x_i, y_i)$ is the i-th point, with x_i and y_i as its coordinates. $Q(m) = \{q_1, q_2, ..., q_m\}$ represents the set of points (pixels) inside the object O. $SP \subseteq Q(m)$ represents the skeleton of the shape S.

2.2   Overview

Figure 2 gives an overview of our skeleton computation methods. The skeleton computation consists of skeleton extraction and skeleton pruning. Given a two-dimensional object O shown in Fig. 2-(a), we obtain its contour S, the landmark sequence $P(n)$ and the inside point set $Q(m)$ as mentioned.

In the skeleton extraction phase, a unit positive charge is evenly placed at each point of $P(n)$, and the generalized potential at each point in $Q(m)$ can be obtained by the sum of the potentials of all the charges on $P(n)$, which is shown in Fig. 2-(b). The lighter the color, the smaller the generalized potential value is. It can be seen the skeleton of the shape is the ridge of the negative electric potential field. To this end, the skeleton is obtained and refined by the gradient of this potential field. See Fig. 2(c)–(f) for example.

In the skeleton pruning phase, we assume that the electric charge placed on $P(n)$ had reached the state of electrostatic equilibrium. For the skeleton with redundant branches like Fig. 2-(f), we calculate the generalized charge distribution on the shape shown in Fig. 2-(g). It can be seen that the charge has the characteristic of gathering at the sharp points and being sparse at the flat regions. We transform the skeleton into a graph and identify the ending points of the graph labeled by green-filled circles in Fig. 2-(h). These ending points are subsequently mapped on $P(n)$ by their largest inscribed circle. An observation is that the ending point of redundant branches is usually mapped at a low-charge region, which can be used to design a multi-round skeleton pruning approach. The pruned skeleton is shown in Fig. 2-(i).

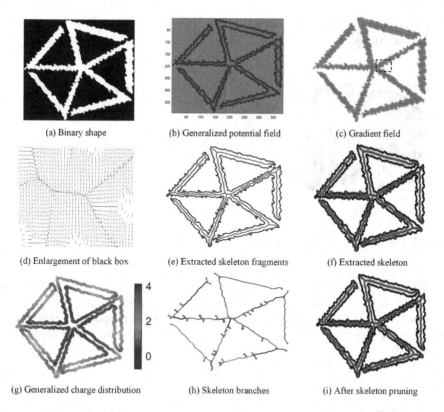

(a) Binary shape

(b) Generalized potential field

(c) Gradient field

(d) Enlargement of black box

(e) Extracted skeleton fragments

(f) Extracted skeleton

(g) Generalized charge distribution

(h) Skeleton branches

(i) After skeleton pruning

Fig. 2. The overview of our proposed skeleton computation method.

3   Skeleton Extraction by the Generalized Coulomb Field

3.1   Generalized Potential Model

By Coulomb's law, the electric potential at any observation point is the superposition of the electric potential of each charge in the scene. Since we have placed a unit positive electric charge on each point in $P(n)$, the electric potential at the point q_j in $Q(m)$ can be calculated by Eq. (1).

$$\varphi(q_j) = \frac{1}{4\pi\epsilon_0} \sum_{i=1}^{n} \frac{1}{||p_i - q_j||} \tag{1}$$

Here, ϵ_0 is the vacuum permittivity, and in practice, the constant term $\frac{1}{4\pi\epsilon_0}$ is omitted in the calculation since we are concerned with the local change of the potential in $Q(m)$. As q_j moves from the outline to the center of the shape, the value of $\varphi(q_j)$ will gradually decrease. Thus, like a distance field, the skeleton of the shape often corresponds to the ridges of the negative electric field. An obvious advantage over the distance field is that the electric potential field is the superposition of all the charges on $P(n)$ which will be more robust to noise and shape variations.

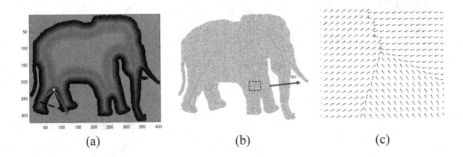

$$(a) \qquad\qquad (b) \qquad\qquad (c)$$

Fig. 3. Calculating the generalized potential field and the gradient field inside a shape.

However, directly using the traditional electric potential in Eq. (1), the obtained skeleton is often not accurate, and there are many redundant branches in practice. A generalized potential calculation model is proposed in this paper, as shown in Eq. (2). Here, we make two improvements against the traditional potential.

$$\varphi(q_j) = \frac{1}{4\pi\epsilon_0} \sum_{i=1}^{n} \frac{1}{id(p_i, q_j)^\alpha} \tag{2}$$

Firstly, we replace Euclidean distance with the inner distance [9]. $id(p_i, q_j)$ is defined as the length of the shortest path from p_i to q_j located entirely inside the shape. The motivation is explained as follows. A 2D shape is often composed of a main body and multiple branches, and the branches might do isometric (articulation) transformations, i.e., the walking of the elephant in Fig. 3-(a). The Euclidean distance between two points A, B will change at this time. However, the ideal shape skeleton should be an isometric transformation invariant, and the potential calculated by the Euclidean distance in Eq. (1) cannot be satisfactory. The inner distance, i.e., the length of the path colored in red in Fig. 3-(a), is a well-known isometric transformation invariant, which can be used to deal with the above problem.

Secondly, we introduce a distance control factor α. Actually, the location of a skeletal point is determined by only a part of the contours, not the entire contour. In Eq. (1), all charges on the contours will have an impact on an observation point, which affects the accuracy of the skeletal point indeed. Thus, a factor $\alpha > 1$ is introduced to decrease the influence of the charges far away.

3.2   Ridge Extraction of Generalized Electric Potential Surface

Figure 3-(a) shows the generalized potential field by Eq. (2), and the lighter color indicates the position with a smaller potential. It can be seen the skeleton of a shape is located on the ridge line of the negative electric potential surface. To illustrate this better, the gradients of the electric potential surface are calculated, which are shown in Fig. 3-(b). The unit gradient vector advances inward uniformly, and the ridge (skeletal) points are created, as the fronts meet. Motivated by this observation, the skeletal points can be detected by the Hessian matrix of the potential field.

$$H(x, y) = \begin{bmatrix} L_{xx}(x,y) & L_{xy}(x,y) \\ L_{yx}(x,y) & L_{yy}(x,y) \end{bmatrix} * I(x,y) \tag{3}$$

where $L_{xx}(x,y)$, $L_{xy}(x,y)$, $L_{yx}(x,y)$ and $L_{yy}(x,y)$ represent the second derivatives in the x and y directions, respectively. For each point in $Q(m)$, it can be judged whether it is a ridge point by calculating the eigenvalue of its Hessian matrix [5].

Meanwhile, for complex shapes, the skeletal points obtained above are usually skeleton fragments. It is necessary to repair the skeleton. As the gradient of an observation location points to the direction of the skeleton, this paper uses the direction of gradient to track the missing skeleton parts and connect these skeleton fragments to produce the final shape skeleton.

4   Skeleton Pruning by the Electric Charge Distribution

4.1   The Electric Charge Distribution on the Shape

After the charges on $P(n)$ are no longer in motion, the shape will reach the state of electric equilibrium. At this time, charges are gathered in the sharp regions and sparsely distributed in the flat or concave regions of the shape. Based on Coulomb's law and physical facts, the electric potential of each point of $P(n) = \{p_1, p_2, ..., p_n\}$ is equal, and can be calculated by Eq. (4).

$$AX = B \tag{4}$$

$$A = \begin{bmatrix} A_{11} & A_{12} & \cdots & A_{1n} & -1 \\ A_{21} & A_{22} & \cdots & A_{2n} & -1 \\ \vdots & \vdots & \ddots & \vdots & \vdots \\ A_{n1} & A_{n2} & \cdots & A_{nn} & -1 \\ 1 & 1 & 1 & 1 & 0 \end{bmatrix} \tag{5}$$

Where $A_{ij} = \frac{1}{id(p_i, p_j)^\beta}$, $X = [ed_1, ed_2, ..., ed_n, V]^T$, $B = [0, 0, ..., 0, n]^T$, ed_i represents the charge at the i-th point on the shape. V and n represent the potential and total amount of charge on the shape, respectively.

Like the generalized potential calculation model in Eq. (2), we introduce the inner distance and distance control factor β to improve the performance of the classic model. The difference is the optimal value of the distance control factor. In pruning extraction, the distance control factor α in Eq. (2) is used to decrease the influence of points far away from the observation point. In practice, a α larger than 1 is required, and the default value is 8. On the contrary, the control factor β in Eq. (4) is used to increase the influence of these points far away. A β smaller than 1 is required, and the default value is 0.1.

Algorithm 1. The algorithm of skeleton pruning.

Require: $SK = \{sk_1, sk_2, \cdots, sk_s\}$: original skeleton, $P(n) = \{p_1, p_2, \cdots, p_n\}$: shape,
  $ed = \{ed_1, ed_2, \cdots, ed_n\}$: electric charge distribution on $P(n)$, T: charge threshold.
Ensure: $SP = \{sp_1, sp_2, \cdots, sp_t\}$: pruned skeleton.
 1: Initialize Pruning = True
 2: Construct the skeleton graph $G =< SK, E >$, where E connects each point with
    its 8-neighbors.
 3: **while** Pruning **do**
 4:    Pruning = False
 5:    Calculate the endpoint set of G as EP
 6:    **for** each point ep in EP **do**
 7:       Find the point $p \in P$ that is nearest to the maximal inscribed circle of ep.
 8:       **if** the electric charge $ed(p) < T$ **then**
 9:          Delete ep, the edge connecting it and update G
10:          Pruning = True
11:       **end if**
12:    **end for**
13: **end while**
14: $SP =$ the vertex set of G

4.2   The Implementation of Skeleton Pruning

We prune the skeleton based on the following observation using the above charge distribution on the shape. A meaningful skeleton branch corresponds to a semantically clear part of the shape such as its limbs, head, and tail, while the redundant branch corresponds to the less prominent part of the shape where usually only a small amount of charge. The process of skeleton pruning is summarized in Algorithm 1.

The efficiency of pruning is mainly controlled by the threshold T. A detailed analysis of T is given in the following experimental section.

5   Experimental Results and Performance Analysis

We first test the parameters of our method, and then analyze the performance by the comparisons with several state-of-art methods. The skeleton computation experiments are conducted on shapes from two public shape datasets, MPEG-7 and S&V, with the codes implemented by MATLAB 2021a.

5.1   Parameters Analysis

The Parameters in Skeleton Extraction. To test the impact of α on the extracted skeleton, we select some representative shapes in the MPEG-7 database and conduct experiments under $\alpha = \{1, 2, 4, 6, 8, 12\}$, respectively. The experimental results are shown in Fig. 4. $\alpha = 1$ is the traditional electric potential field where many redundant and low-quality branches have been produced. Take the shapes of cock and pig in Fig. 4 for example. As the value of α increases,

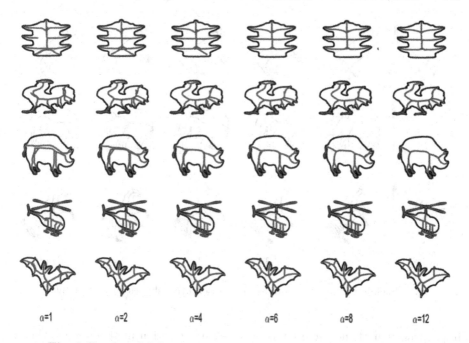

$\alpha=1$ $\alpha=2$ $\alpha=4$ $\alpha=6$ $\alpha=8$ $\alpha=12$

Fig. 4. Results of skeleton extraction for several shapes with varying α.

the redundant branches in the skeleton gradually disappear. It shows that larger α can reduce the influence of the charge far away from the observation point, and increase the accuracy of the skeleton.

In the experiment, we find that $\alpha = 8$ has achieved ideal results in most shapes, which is set as the default value.

The Parameters in Skeleton Pruning. The main parameters of our skeleton pruning include the distance control factor β and the charge threshold T. We fix $\beta = 0.1$ for all the experiments because the charge distribution on the shape reflects the geometric characteristics of the shape more accurately at this time. Next, we discuss the value of T, which is taken as the threshold to remove the redundant branches in Algorithm 1. As the average amount of electric charge is 1, a reasonable range of T is between 1 and 2. Figure 5 shows the pruning results when $T = \{1.1, 1.3, 1.5, 1.7, 1.9\}$ on the three representative shapes of cow, elephant, and horse. It can be seen the results are satisfactory at $T = 1.5$, which is chosen as the default value of T.

5.2   Comparisons with Existing Methods

To evaluate the performance of our method, we select 7 widely-used skeleton extraction methods for comparison: Thin, Voronoi, Dist, GuidedThin [6], Mapskel [4], Aofskel [11] and Heat [5]. Thin, Voronoi and Dist are the baseline

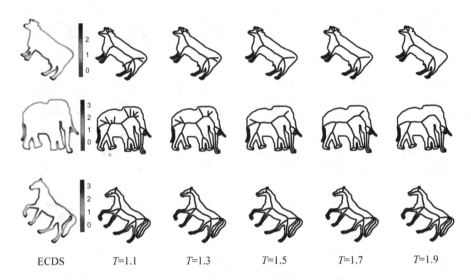

ECDS   T=1.1   T=1.3   T=1.5   T=1.7   T=1.9

Fig. 5. Skeleton pruning results for 3 representative shapes with different T.

implementation of the three kinds of skeleton extraction methods. GuideThin, Mapskel and Aofskel are the state-of-art approaches in these three kinds of methods. Heat is a recent method that computes the skeleton by the ridge lines of the heat surface. Moreover, we select 3 well-known skeleton pruning methods to make further comparisons: DCE [1], SimRec [13] and BPR [12]. Colu and ColuECDS are our methods, where Colu is achieved by the skeleton extraction process only.

Figure 6 gives the results of these methods on a camel shape. The camel has many meaningful parts with different scales, which is usually difficult for most methods to extract the skeleton accurately. From the results, we can see that except for the baseline implementation of Thin, Voronoi, and Dist, all the methods have produced good-looking skeletons. A challenging problem is the small bumps on the legs and humps of the camel. These methods haven't produced consistent results. Mapskel and our methods ColuECDS have produced the most satisfactory results since all the redundant skeleton segments caused by the mentioned bumps are removed.

To test the robustness of these methods to noise, we choose two representative noisy shapes from MPEG-7 databases. The results are shown in Fig. 7 and Fig. 8. In Fig. 7, the shape has a certain amount of noise on the contour, which poses a challenging problem for skeleton extraction. From the results, most methods suffer from this problem and have produced several redundant skeleton segments. Although the pruning methods of DCE [1], SimRec [13] and BPR [12] have pruned some of these segments, our method ColuECDS outperforms these 3 methods. More comparisons on another noisy device shape are given in Fig. 8. A similar conclusion can be drawn too.

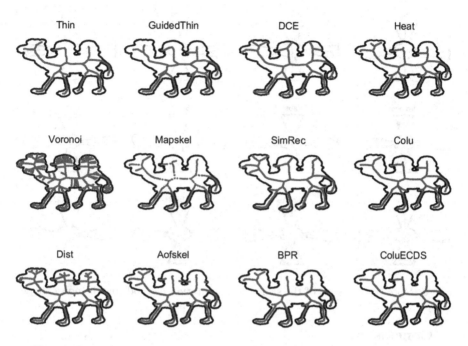

Fig. 6. Comparisons of the skeleton computation methods on a camel shape.

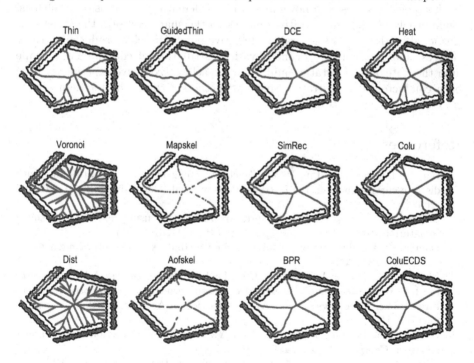

Fig. 7. Comparisons of the skeleton computation methods on a shape with noise.

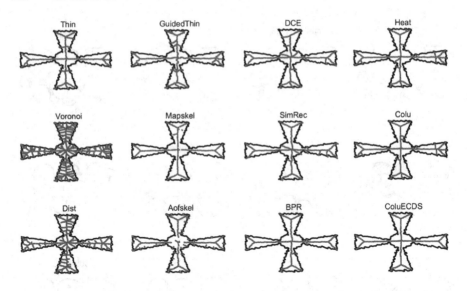

Fig. 8. More comparisons of the skeleton computation methods on the noisy shape.

6 Conclusion

In this paper, we present a novel method for calculating skeletons of 2D shapes. Upon evenly placing unit positive charges on the shape contour, the skeleton is extracted by the ridge of generalized negative potential fields produced by these charges, and subsequently pruned by the distribution of charge on the shape after the state of electrostatic equilibrium is reached. Extensive experiments have been conducted, demonstrating that our method performs better than several state-of-the-art competitors.

References

1. Bai, X., Latecki, L.J., Liu, W.Y.: Skeleton pruning by contour partitioning with discrete curve evolution. IEEE Trans. Pattern Anal. Mach. Intell. **29**(3), 449–462 (2007)
2. Barman, A., Dutta, P.: Facial expression recognition using distance and shape signature features. Pattern Recogn. Lett. **145**, 254–261 (2021)
3. Damon, J.: Rigidity properties of the Blum medial axis. J. Math. Imaging Vis. **63**(1), 120–129 (2021)
4. El-Gaaly, T., Froyen, V., Elgammal, A., Feldman, J., Singh, M.: A Bayesian approach to perceptual 3D object-part decomposition using skeleton-based representations. In: Proceedings of the AAAI Conference on Artificial Intelligence, vol. 29 (2015)
5. Gao, F., Wei, G., Xin, S., Gao, S., Zhou, Y.: 2D skeleton extraction based on heat equation. Comput. Graph. **74**, 99–108 (2018)
6. Iglesias-Cofán, S., Formella, A.: Guided thinning. Pattern Recogn. Lett. **128**, 176–182 (2019)

7. Joseph-Rivlin, M., Zvirin, A., Kimmel, R.: Momen(e)t: flavor the moments in learning to classify shapes. In: Proceedings of the IEEE/CVF International Conference on Computer Vision Workshops (2019)
8. Li, Z., Qu, W., Qi, H., Stojmenovic, M.: Near-convex decomposition of 2D shape using visibility range. Comput. Vis. Image Underst. **210**, 103243 (2021)
9. Ling, H., Jacobs, D.W.: Shape classification using the inner-distance. IEEE Trans. Pattern Anal. Mach. Intell. **29**(2), 286–299 (2007)
10. Osowski, S., et al.: Fourier and wavelet descriptors for shape recognition using neural networks - a comparative study. Pattern Recogn. **35**(9), 1949–1957 (2002)
11. Rezanejad, M., Siddiqi, K.: Flux graphs for 2D shape analysis. In: Dickinson, S., Pizlo, Z. (eds.) Shape Perception in Human and Computer Vision. Advances in Computer Vision and Pattern Recognition. Springer, London (2013). https://doi.org/10.1007/978-1-4471-5195-1_3
12. Shen, W., Bai, X., Hu, R., Wang, H., Latecki, L.J.: Skeleton growing and pruning with bending potential ratio. Pattern Recogn. **44**(2), 196–209 (2011)
13. Shen, W., Bai, X., Yang, X., Latecki, L.J.: Skeleton pruning as trade-off between skeleton simplicity and reconstruction error. Sci. China Inf. Sci. **56**, 1–14 (2013)
14. Yang, C., Fang, L., Fei, B., Yu, Q., Wei, H.: Multi-level contour combination features for shape recognition. Comput. Vis. Image Underst. **229**, 103650 (2023)
15. Yang, C., Indurkhya, B., See, J., Grzegorzek, M.: Towards automatic skeleton extraction with skeleton grafting. IEEE Trans. Vis. Comput. Graph. **27**(12), 4520–4532 (2020)
16. Yu, X., Gao, Y., Bennamoun, M., Xiong, S.: A lie algebra representation for efficient 2D shape classification. Pattern Recogn. **134**, 109078 (2023)
17. Zou, Z., Chen, K., Shi, Z., Guo, Y., Ye, J.: Object detection in 20 years: a survey. In: Proceedings of the IEEE (2023)

Illumination Insensitive Monocular Depth Estimation Based on Scene Object Attention and Depth Map Fusion

Jing Wen[1,2(✉)], Haojiang Ma[1,2], Jie Yang[1,2], and Songsong Zhang[1,2]

[1] Shanxi University, Taiyuan, China
`wjing@sxu.edu.cn`
[2] Key Laboratory of Computer Intelligence and Chinese Processing of Ministry of Education, Taiyuan, China

Abstract. Monocular depth estimation (MDE) is a crucial but challenging computer vision (CV) task which suffers from lighting sensitivity, blurring of neighboring depth edges, and object omissions. To address these problems, we propose an illumination insensitive monocular depth estimation method based on scene object attention and depth map fusion. Firstly, we design a low-light image selection algorithm, incorporated with the EnlightenGAN model, to improve the image quality of the training dataset and reduce the influence of lighting on depth estimation. Secondly, we develop a scene object attention mechanism (SOAM) to address the issue of incomplete depth information in natural scenes. Thirdly, we design a weighted depth map fusion (WDMF) module to fuse depth maps with various visual granularity and depth information, effectively resolving the problem of blurred depth map edges. Extensive experiments on the KITTI dataset demonstrate that our method effectively reduces the sensitivity of the depth estimation model to light and yields depth maps with more complete scene object contours.

Keywords: Monocular depth estimation · Scene object attention · Weighted depth map fusion · Image enhancement · Illumination insensitivity

1 Introduction

Monocular depth estimation plays an important role in autonomous driving [3, 22] and AR [1, 19]. Despite rapid progress in self-supervised depth estimation [7, 14, 21, 26], generating accurate depth maps remains challenged since the prior of ground truth is hard to acquire. We observe that some models are sensitive to lighting conditions, resulting in inaccurate depth predictions in low light environments. Traditional convolutional networks sometimes fail to effectively perceive global contextual information, leading to object omissions in the depth

Supported by organization x.

map. Moreover, the network does not fully leverage the different scale-specific details contained in depth maps, which results in blurry depth edges and details.

In this paper, we propose a low light image filtering algorithm to address the sensitivity of models to lighting conditions by improving the quality of images in the training dataset. Meanwhile, we present a depth estimation method based on scene object attention mechanism and weighted depth map fusion. We compute similarity vectors between arbitrary positions in the feature maps using convolutional networks to increase the networks receptive field and enhance the contextual information in the feature maps to resolves the issue of blurry depth edges in adjacent regions. Moreover, pixel-adaptive convolution is introduced for fusing features from the pre-trained semantic network to deal with the limitation of spatial weight sharing, since the limitation leads to content-agnostic convolutions. To summarize the main contributions of our work:

- We delicately design a low-light enhancement module to reduce the influence of lighting on depth estimation.
- We propose scene object attention mechanism to enhance the correlation between images, depth, and semantic information, enabling the model to resolve the problem of scene object omissions.
- Weighted depth map fusion is presented to fuse depth maps of different scales, thereby obtaining a depth map with rich contour information.

2   Related Work

Framework. Monodepth2 is a classic depth estimation method based on deep learning, proposed by Godard et al. [7], which solve the problem of occlusions and objects movement. Guizilini et al. [8] proposed a novel MDE model based on PackNet, which effectively reduces information loss when the encoder-decoder restores the original resolution.

Low Light Enhancement. Although there are some methods for low light enhancement, e.g. EnlightenGAN [11], those methods are not particularly designed for depth estimation. Therefore, more efforts should be made to address the impact of low light environments on depth estimation.

Attention Mechanism. Attention mechanism can improve the accuracy, robustness and scene awareness of depth estimation. Hu et al. [10] proposed a channel attention mechanism that assigns different weights to each channel, representing degrees of importance varying. Woo et al. [18] connected different attention mechanisms in a cascaded and parallel manner. Yang et al. [23] proposed a module based on non-local means and attention mechanisms to introduce a module that captures long-range dependencies between pixels in feature maps.

Multi-scale Fusion. Wu et al. [20] proposed a multi-level contextual and multi-modal fusion network, MCMFNet, for integrating multi-scale contextual feature mappings, and learning object edges from depth map. The low-resolution depth maps proposed in [25] lack spatial and object information in the scene, while the high-resolution depth maps contain rich scene and detail contents but lack depth semanteme.

Semantic Auxiliary Information. One of the most representative depth estimation methods that utilize semantic information as auxiliary guidance is proposed by Jiao et al. [12]. The auxiliary information can be divided into two categories: one uses a certain aspect of perceptual information alone and the other uses a combination of multiple scene perceptual information [15]. The key of guiding the depth estimation task with both types of auxiliary information lies in enhancing the models ability to integrate information, which is crucial for leveraging semantic auxiliary information.

3 Method

3.1 Problem Setup

The aim of self-supervised MDE tasks is to predict the depth from a RGB image without ground truth, which usually use the geometric constraints of stereo pairs or monocular videos as the supervised signals. Taking monocular video as input of training for instance, we input a single RGB image I_t to the depth network and get the corresponding depth map D_t. Then we predict the relative pose $T_{t \to t'}$, through the pose network, between target image I_t and source image $I_{t'}$ adjacent to I_t temporally. Finally, with the D_t, $T_{t \to t'}$ and camera intrinsics K, $I_{t' \to t}$ is synthesized.

The overall photometric reprojection error is calculated by combining it with the target image

$$L_p = \sum_{t'} pe(I_{t'}, I_{t' \to t}), \tag{1}$$

where pe represents the photometric reprojection error

$$pe(I_a, I_b) = \frac{\alpha}{2}(1 - SSIM(I_a, I_b)) + (1 - \alpha)\|I_a - I_b\|_1. \tag{2}$$

Considering the significance of edge information in images, we employ the edge-aware smoothness Loss L_s [6] to improve the edge prediction results:

$$L_s = |\alpha_x d_t^*|e^{-|\alpha_x I_t|} + |\alpha_y d_t^*|e^{-|\alpha_y I_t|}, \tag{3}$$

where d_t^* represents the mean-normalized inverse depth.

We refer to [8] to perform a removal operation on the pixels with photometric loss higher than the corresponding undistorted synthetic target image. This approach is used to mask out static pixels

$$M_p = \min_{I_{t'}} L_p(I_t, I_{t'}) > \min_{I_{t'}} L_p(I_t, I_{t' \to t}). \tag{4}$$

We employ the instantaneous velocity to constrain the translational component of the estimated camera pose [8]

$$L_v(\hat{t}_{t' \to t}, v) = \left| \|\hat{t}_{t' \to t}\| - |v| \Delta T_{t \to t'} \right|, \tag{5}$$

where $\hat{t}_{t' \to t}$ represents the predicted translational component of the pose by the pose estimation network, v denotes the instantaneous velocity, and $\Delta T_{t \to t'}$ represents the time difference between the target frame and the source frame.

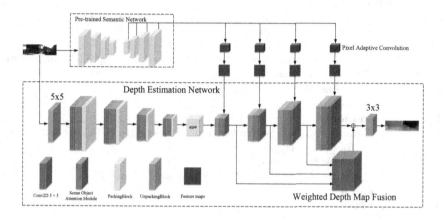

Fig. 1. Overview of Framework.

3.2 Proposed Architecture

The overview of our pipeline is illustrated in Fig. 1, the baseline is PackNet, and the pre-trained semantic segmentation network uses FCN [9]. As shown in the Fig. 1, we obtain semantic feature maps from the semantic network and extract them through pixel-adaptive convolution module to generate semantic-aware depth feature maps.

In the depth estimation network, the input image is first passed through encoder to obtain high-dimensional features. Then, the obtained feature maps are fed into the atrous spatial pyramid pooling layer. Afterwards, the processed feature maps are concatenated with the semantic-aware depth feature maps generated by the pixel-adaptive convolutional module and used as the input to the decoder. Finally, the depth maps at different scales are fed into the weighted depth map fusion module. Notably, we incorporate the scene object attention mechanism in both the encoder and decoder. The encoder module, EncBlock, is composed of three Conv2D 3×3 convolutional layers, the scene object attention module, and the PackingBlock module [8], connected in sequence. The decoder module primarily consists of DecBlock, which includes the UnpackingBlock module, the scene object attention module, and a Conv2D 3×3 layer.

Algorithm 1. Low-light image filtering algorithm

Input: Input dataset $I = \{i_1, i_2, ..., i_N\}$,height $H = \{h_1, h_2, ..., h_N\}$,width $W = \{w_1, w_2, ..., w_N\}$

Output: low-light dataset$\{R_1\}$

1: $win_h = 75$
2: $win_w = 75$
3: **for** $j = 1 \rightarrow N$ **do** //j represents the image index
4:      $hh \Leftarrow (win_h - 1)/2, ww \Leftarrow (win_w - 1)/2$
5:      **for** $i = hh \rightarrow h_j + hh$ **do**
6:          **for** $k = ww \rightarrow w_j + ww$ **do**
7:              $count \Leftarrow 0$
8:              **for** $r = -hh \rightarrow hh$ **do**
9:                  **for** $c = -ww \rightarrow ww$ **do**
10:                     **if** The pixel value at (r+i,c+k) is within the specified range **then**
11:                         $count =$count+1
12:                     **end if**
13:                 **end for**
14:             **end for**
15:             **if** $count \geq$80% of the number of pixels contained in the window **then**
16:                 $R_1 \Leftarrow R_1 \cup \{i_j\}$
17:                 **break** from line 5
18:             **else** Adjusting the stride for the next window movement
19:             **end if**
20:         **end for**
21:     **end for**
22: **end for**
23: **return** R_1 // outputs

3.3   Low Light Image Enhancement

To reduce the effect of light on depth estimation, we introduce the Enlighten-GAN to enhance the low-light images. The EnlightenGAN utilizes a U-Net as the generator. Since it is necessary to utilize both global and local information for image enhancement, the model employs a dual-discriminator setup comprising a global discriminator and a local discriminator. More important, to ensure that only low light images are enhanced, we propose a low light image filtering algorithm. The algorithm scans the input images with a 75 × 75 window, which examines whether the pixel values within the window fall within the range of low light pixel values every move. The image will be classified as low light, if the number of low light pixels is more than 80% of the total pixels within the window. The details of the low light image filtering algorithm are shown in Algorithm 1. To improve the efficiency, the low light image filtering algorithm could adaptively adjust the stride distance according to the previous execution result. Specifically, if low light pixels number is less than 20% of the total pixels within the scanning window, the stride for the next window movement will be increased.

3.4   Scene Object Attention Mechanism

To deal with the limitation of 2D and 3D convolutions in exploring global context, the scene object attention mechanism (SOAM) is proposed to learn similar semantic feature groups and their relationships in the context, which could make better use of global contextual information, such as lighting, posture, texture, deformation and occlusion. When using standard convolutions, the model could not obtain correct contextual information for predicting pixel depth because those information might be located in non-contiguous positions that the convolutional operations cannot reach. The SOAM is able to enhance the correlation of the pixel belonging to the same object, to ensure consistency and continuity in the depth of the same object, and improve the accuracy of depth estimation and reducing visual errors of object omissions. The structure of the SOAM is illustrated in Fig. 2. Inspired by the concepts of spatial attention and channel attention, we employed 1×1 convolutions to integrate and interact cross-channel information. Then matrix multiplication was applied to capture relationships between different spatial positions of the feature maps. Finally, these multiplied feature maps were added to the processed original feature maps, yielding an enhanced spatial-awareness feature map. This approach enables the network to perceive the dependencies between any two elements within the global context.

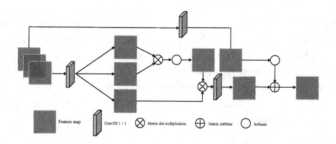

Fig. 2. Scene Object Attention Module.

3.5   Weighted Depth Map Fusion

To address the problem of blurry depth map edges, we design a weighted depth map fusion (WDMF) module. The feature map from lower-level networks possess complementary characteristics to the feature map from higher-level networks. By fusing feature map at different scales, we can obtain fused features with superior semantic representation and diverse spatial feature details, which helps the network model generate depth maps with rich contour information. The WDMF module is illustrated in Fig. 3. To enhance the spatial and channel dependency between features, we incorporate the CBAM module [10] into the weighted depth map fusion network model.

In addition, we insert the atrous spatial pyramid pooling module between encoder and decoder to refine the semantic scene, which helps the network enlarge the receptive field, and gather more local features.

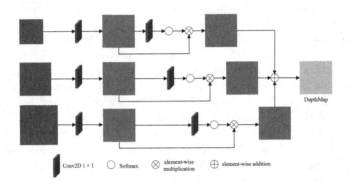

Fig. 3. Weighted Depth map Fusion Module.

3.6 Pixel Adaptive Convolution

The spatial weight sharing property of convolutions means all positions to share the same parameters, which is useful for the dense prediction tasks, e.g. semantic segmentation. However, it is not suitable for depth estimation, due to the optimal gradient direction of every pixel is supposed to be different. The property of spatial sharing usually leads to suboptimal results for the model [16]. To address the problem, inspired by bilateral filtering, we introduced the pixel-adaptive convolution [16] as follows,

$$v_i' = \sum_{j \in \Omega(i)} K(f_i, f_j) W[p_i, p_j] v_j + b \qquad (6)$$

where $f_i \in R^d$ denotes the d-dimensional feature space, which refers to the specific features of the semantic network, $v = (v_1, \dots, v_n)$ represents an image with n pixels, $v_i^{x'} \in R^c$ denotes the convolution output of each pixel, and c denotes the number of channels. $P = (x, y)^T$ is the coordinate information, $[p_i - p_j]$ is the offset of the pixel point in the image and $W_{k \times k}$ is used to complete the weight value of the convolution operation. $\Omega(i)$ denotes the $k \times k$ convolution window around that pixel, and $b \in R^c$ denotes the bias term. K is a kernel function.

4   Experimental Results and Analysis

4.1   Experimental Data Sets and Evaluation Metrics

We evaluate the MDE performance of our model on the KITTI dataset [5] using the data split of Eigen et al. [4]. We remove static frames, using 39810, 4424, and 697 images for training, validation, and test, respectively. There are two sets of metrics to evaluate the model, the first set is AbsRel, SqRel, RMSE, and RMSElog (the smaller the better), the second set is accuracy $\sigma_1, \sigma_2, \sigma_3$ (the larger the better).

4.2 Parameter Setting

Multiple experiments on the KITTI dataset are conducted to analyze the performance of the network model and compare it with previous models. The network model is implemented on PyTorch, and the model is trained on two Nvidia 1080Ti. Adam [13] optimizer with the default betas 0.9 and 0.999 is used. Initial depth and pose network learning rate are 2×10^{-4} and 5×10^{-4} respectively. As the experiment proceeds, the learning rate will decay by 2 times every 40 iterations, and the weight value α in SSIM is 0.85. The batch size is set to 2 and the maximum number of iterations for training is set to 100.

4.3 Analysis of Results

The comparison results on KITTI demonstrate accurate perception of depth-related contextual information and rich semantic details, effectively addressing the issues of sensitivity to illumination, object omissions, and blurring of depth map scene object edges. As shown in Table 1, our proposed method exhibits improvements across various metrics compared to previous approaches. The metrics of our model decrease by 0.016 in AbsRel and 0.451 in RMSE compared to the second-best performing model, while also increase by 0.040 in accuracy for the threshold $\sigma < 1.25$.

To demonstrate the performance advantages of our proposed method more clearly, we compare the visual results with different methods in [7,17], as shown in Fig. 4. Due to the low-light enhancement module, effective enhancement of dark regions is achieved while preserving texture information and avoiding excessive exposure artifacts. This allows the model to discriminate the depth of vehicles and background even in low light. Furthermore, The WDMF can capture clearer structures by fusing detail features from depth maps of different resolutions. The SOAM enables the model to make more comprehensive use of global contextual information, effectively addressing the issue of object omissions.

4.4 Analysis of Ablation Experiments

To further explain how the components of our method contribute to the overall performance, we conduct a comprehensive ablation study on the KITTI dataset, as shown in Table 2, to analyze and evaluate the individual components of our proposed method. We observe that the baseline model, without incorporating any of our proposed components, achieves the lowest performance. It is shown that the SOAM could not only improve quantitative performance metrics, but effectively deal with the issue of object omission in Fig. 4. Whereas, the WDMF

Table 1. Comparison of evaluation metrics of different methods on the dataset KITTI.

Algorithm	Dataset	AbsRel	SqRel	RMSE	RMSElog	$\sigma < 1.25$	$\sigma < 1.25^2$	$\sigma < 1.25^3$
Monodepth2 [7]	K	0.115	0.903	4.863	0.193	0.877	0.959	0.981
PackNet [8]	K	0.111	0.785	4.601	0.189	0.878	0.960	0.982
GeoNet [24]	K	0.149	1.106	5.567	0.226	0.796	0.935	0.975
DF-Net [27]	K	0.146	1.182	5.215	0.213	0.818	0.943	0.978
Struct2Depth [2]	K	0.141	1.026	5.291	0.215	0.816	0.945	0.979
Ours	K	0.095	0.819	4.150	0.168	0.918	0.967	0.982

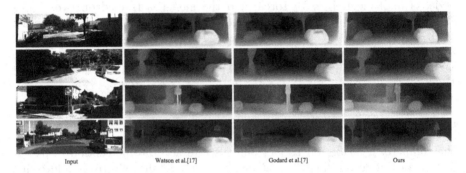

Input   Watson et al.[17]   Godard et al.[7]   Ours

Fig. 4. Visualization comparison results of different methods.

did not exhibit significant enhancement in quantitative results, but it effectively mitigated the issue of depth edge blurring in qualitative results (Fig. 4).

Table 2. We evaluate the performance of our scene object attention mechanism (S), weighted depth map fusion (W), and pixel-adaptive convolution (P) contributions.

	AbsRel	RMSE	RMSElog	$\sigma < 1.25$	$\sigma < 1.25^2$
PackNet	0.111	4.601	0.189	0.878	0.960
PackNet+S	0.110	4.566	0.185	0.879	0.961
PackNet+S+W	0.110	4.560	0.189	0.864	0.954
PackNet+S+W+P	0.095	4.150	0.168	0.918	0.967

The visual comparison results of our ablation experiments on the low light image enhancement module are shown in Fig. 5. From the first row, we can observe that the results with the module are more continuous, accurate, and better aligned with the actual depth of the car window. The second row demonstrates the comparison of depth map completeness in low light environments. The results obtained with the low light image enhancement exhibit more complete depth information, indicating its ability to reduce the model's sensitivity to illumination and perform better in complex environments.

4.5   Model Convergence and Execution Efficiency

During the model training process, the loss function value of our method gradually decreases to a stable level over the training iterations, without exhibiting phenomena such as loss stagnation or significant oscillations. This indicates favorable convergence behavior. To contrast the model's execution efficiency, we compare the testing time and prediction performance of different methods, with the RMSE selected as the prediction performance metric. The comparison of testing time and RMSE of various methods on the KITTI dataset is shown in Fig. 6. The computational time of our model is significantly lower than the models in [7,27], and although our method's testing time per image is slightly higher than the method in [8], we exhibit a much lower RMSE compared to other methods. Therefore, considering the trade-off between execution time and prediction error, our method demonstrates superior prediction performance.

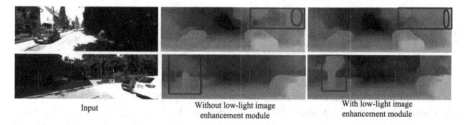

| Input | Without low-light image enhancement module | With low-light image enhancement module |

Fig. 5. Comparison results of predictive integrity of depth map in low light environment.

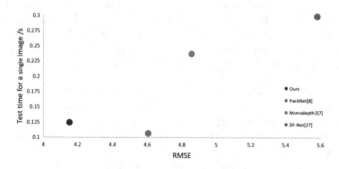

Fig. 6. Test time and RMSE comparison among different methods.

5   Conclusion

In this paper, we propose a novel method to address the challenges of object omissions, edge blurring, and illumination sensitivity in depth estimation models. The incorporated semantic information from the pre-trained network and the low light enhancement module can generate depth maps in high quality. We utilize the SOAM to enhance the correlation between images, depth, and scene semantics, thereby improving the contextual information and the correlation among pixels belonging to the same object. The WDMF and the dilated spatial pyramid pooling module address the problems of edge blurring and detail loss by incorporating multi-scale target information in the network, resulting in clearer edge contours of scene objects. In the future, we plan to incorporate multimodal information into our model, such as object categories and positions within images. By combining these semantic cues with depth estimation, a better understanding of the scene's geometric structure and improved accuracy in depth estimation can be achieved.

Acknowledgement. This research project is supported by Shanxi Scholarship Council of China (2022-008), 1331 Project.

References

1. Benavides, F.T., Ignatov, A., Timofte, R.: PhoneDepth: a dataset for monocular depth estimation on mobile devices. In: Proceedings of the IEEE/CVF Conference on Computer Vision and Pattern Recognition, pp. 3049–3056 (2022)
2. Casser, V., Pirk, S., Mahjourian, R., Angelova, A.: Depth prediction without the sensors: leveraging structure for unsupervised learning from monocular videos. In: Proceedings of the AAAI Conference on Artificial Intelligence, vol. 33, pp. 8001–8008 (2019)
3. Cheng, Z., et al.: Physical attack on monocular depth estimation with optimal adversarial patches. In: Avidan, S., Brostow, G., Cissé, M., Farinella, G.M., Hassner, T. (eds.) Computer Vision, ECCV 2022, Part XXXVIII. LNCS, vol. 13698, pp. 514–532. Springer, Cham (2022). https://doi.org/10.1007/978-3-031-19839-7_30
4. Eigen, D., Fergus, R.: Predicting depth, surface normals and semantic labels with a common multi-scale convolutional architecture. In: Proceedings of the IEEE International Conference on Computer Vision, pp. 2650–2658 (2015)
5. Geiger, A., Lenz, P., Stiller, C., Urtasun, R.: Vision meets robotics: the KITTI dataset. Int. J. Robot. Res. **32**(11), 1231–1237 (2013)
6. Godard, C., Mac Aodha, O., Brostow, G.J.: Unsupervised monocular depth estimation with left-right consistency. In: Proceedings of the IEEE Conference on Computer Vision and Pattern Recognition, pp. 270–279 (2017)
7. Godard, C., Mac Aodha, O., Firman, M., Brostow, G.J.: Digging into self-supervised monocular depth estimation. In: Proceedings of the IEEE/CVF International Conference on Computer Vision, pp. 3828–3838 (2019)
8. Guizilini, V., Ambrus, R., Pillai, S., Raventos, A., Gaidon, A.: 3D packing for self-supervised monocular depth estimation. In: Proceedings of the IEEE/CVF Conference on Computer Vision and Pattern Recognition, pp. 2485–2494 (2020)

9. He, K., Zhang, X., Ren, S., Sun, J.: Spatial pyramid pooling in deep convolutional networks for visual recognition. IEEE Trans. Pattern Anal. Mach. Intell. **37**(9), 1904–1916 (2015)

10. Hu, J., Shen, L., Sun, G.: Squeeze-and-excitation networks. In: Proceedings of the IEEE Conference on Computer Vision and Pattern Recognition, pp. 7132–7141 (2018)

11. Jiang, Y., et al.: EnlightenGAN: deep light enhancement without paired supervision. IEEE Trans. Image Process. **30**, 2340–2349 (2021)

12. Jiao, J., Cao, Y., Song, Y., Lau, R.: Look deeper into depth: monocular depth estimation with semantic booster and attention-driven loss. In: Ferrari, V., Hebert, M., Sminchisescu, C., Weiss, Y. (eds.) ECCV 2018. LNCS, vol. 11219, pp. 55–71. Springer, Cham (2018). https://doi.org/10.1007/978-3-030-01267-0_4

13. Kingma, D.P., Ba, J.: Adam: a method for stochastic optimization. arXiv preprint arXiv:1412.6980 (2014)

14. Lyu, X., et al.: HR-depth: high resolution self-supervised monocular depth estimation. In: Proceedings of the AAAI Conference on Artificial Intelligence, vol. 35, pp. 2294–2301 (2021)

15. Ranjan, A., et al.: Competitive collaboration: joint unsupervised learning of depth, camera motion, optical flow and motion segmentation. In: Proceedings of the IEEE/CVF Conference on Computer Vision and Pattern Recognition, pp. 12240–12249 (2019)

16. Su, H., Jampani, V., Sun, D., Gallo, O., Learned-Miller, E., Kautz, J.: Pixel-adaptive convolutional neural networks. In: Proceedings of the IEEE/CVF Conference on Computer Vision and Pattern Recognition, pp. 11166–11175 (2019)

17. Watson, J., Firman, M., Brostow, G.J., Turmukhambetov, D.: Self-supervised monocular depth hints. In: Proceedings of the IEEE/CVF International Conference on Computer Vision, pp. 2162–2171 (2019)

18. Woo, S., Park, J., Lee, J.-Y., Kweon, I.S.: CBAM: convolutional block attention module. In: Ferrari, V., Hebert, M., Sminchisescu, C., Weiss, Y. (eds.) ECCV 2018. LNCS, vol. 11211, pp. 3–19. Springer, Cham (2018). https://doi.org/10.1007/978-3-030-01234-2_1

19. Wu, C.Y., Wang, J., Hall, M., Neumann, U., Su, S.: Toward practical monocular indoor depth estimation. In: Proceedings of the IEEE/CVF Conference on Computer Vision and Pattern Recognition, pp. 3814–3824 (2022)

20. Wu, J., Zhou, W., Luo, T., Yu, L., Lei, J.: Multiscale multilevel context and multimodal fusion for RGB-D salient object detection. Sig. Process. **178**, 107766 (2021)

21. Xiong, M., Zhang, Z., Zhong, W., Ji, J., Liu, J., Xiong, H.: Self-supervised monocular depth and visual odometry learning with scale-consistent geometric constraints. In: Proceedings of the Twenty-Ninth International Conference on International Joint Conferences on Artificial Intelligence, pp. 963–969 (2021)

22. Xue, F., Zhuo, G., Huang, Z., Fu, W., Wu, Z., Ang, M.H.: Toward hierarchical self-supervised monocular absolute depth estimation for autonomous driving applications. In: 2020 IEEE/RSJ International Conference on Intelligent Robots and Systems (IROS), pp. 2330–2337. IEEE (2020)

23. Yang, Z., Wang, P., Xu, W., Zhao, L., Nevatia, R.: Unsupervised learning of geometry with edge-aware depth-normal consistency. arXiv preprint arXiv:1711.03665 (2017)

24. Yin, Z., Shi, J.: GeoNet: unsupervised learning of dense depth, optical flow and camera pose. In: Proceedings of the IEEE Conference on Computer Vision and Pattern Recognition, pp. 1983–1992 (2018)

25. Zhan, H., Garg, R., Weerasekera, C.S., Li, K., Agarwal, H., Reid, I.: Unsupervised learning of monocular depth estimation and visual odometry with deep feature reconstruction. In: Proceedings of the IEEE Conference on Computer Vision and Pattern Recognition, pp. 340–349 (2018)
26. Zhou, T., Brown, M., Snavely, N., Lowe, D.G.: Unsupervised learning of depth and ego-motion from video. In: Proceedings of the IEEE Conference on Computer Vision and Pattern Recognition, pp. 1851–1858 (2017)
27. Zou, Y., Luo, Z., Huang, J.-B.: DF-Net: unsupervised joint learning of depth and flow using cross-task consistency. In: Ferrari, V., Hebert, M., Sminchisescu, C., Weiss, Y. (eds.) ECCV 2018. LNCS, vol. 11209, pp. 38–55. Springer, Cham (2018). https://doi.org/10.1007/978-3-030-01228-1_3

A Few-Shot Medical Image Segmentation Network with Boundary Category Correction

Zeyu Xu, Xibin Jia[✉], Xiong Guo, Luo Wang, and Yiming Zheng

Beijing University of Technology, Beijing, China
jiaxibin@bjut.edu.cn

Abstract. Accurate medical image segmentation is the foundation of clinical imaging diagnosis and 3D image reconstruction. However, medical images often have low contrast between target objects, greatly affected by organ movement, and suffer from limited annotated samples. To address these issues, we propose a few-shot medical image segmentation network with boundary category correction named Boundary Category Correction Network (BCC-Net). Of overall medical few-shot learning framework, we first propose the Prior Mask Generation Module (PRGM) and Multi-scale Feature Fusion Module (MFFM). PRGM can better localize the query target, while MFFM can adaptively fuse the support set prototype, the prior mask and the query set features at different scales to solve the problem of the spatial inconsistency between the support set and the query set. To improve segmentation accuracy, we construct an additional base-learning branch, which, together with the meta-learning branch, forms the Boundary Category Correction Framework (BCCF). It corrects the boundary category of the meta-learning branch prediction mask by predicting the region of the base categories in the query set. Experiments are conducted on the mainstream ABD-MR and ABD-CT medical image segmentation public datasets. Comparative analysis and ablation experiments are performed with a variety of existing state-of-the-art few-shot segmentation methods. The results demonstrate that the effectiveness of the proposed method with significant enhance the segmentation performance on medical images.

Keywords: Medical image segmentation · Prior Mask · Feature Fusion · Boundary Category Correction

1 Introduction

In the field of medical image segmentation, acquiring annotated data is a challenging task. Therefore, few-shot learning [2,6,14,18] is a good choice. Few-shot learning enables rapid knowledge acquisition from limited data and leverages a small number of annotated samples to extract category-specific features. By providing a few unseen category samples as the support set, the model can perform segmentation on unseen samples in the query set. In the field of natural image segmentation, few-shot learning methods have shown significant performance improvements [9]. However, few-shot medical image segmentation is still

© The Author(s), under exclusive license to Springer Nature Singapore Pte Ltd. 2024
Q. Liu et al. (Eds.): PRCV 2023, LNCS 14434, pp. 371–382, 2024.
https://doi.org/10.1007/978-981-99-8549-4_31

in its stages of development [21], and further research is needed to effectively construct few-shot medical image segmentation models.

During the testing phase of few-shot medical image segmentation, researchers have found that when the model tends to favor known categories, it may incorrectly identify neighboring irrelevant category regions as known categories. At the same time, there is a spatial inconsistency issue when using the central slice of the support image block to segment all slices in the query image.

To address the issues mentioned above, we propose a Boundary Category Correction Network (BCC-Net) for few-shot medical image segmentation. The network consists of two parts: a meta-learning branch and a base-learning branch. The meta-learning branch utilizes the Prior Mask Generation Module (PRGM), which generates prior mask without training. To better localize the query target, it incorporates a hierarchical structure of adaptive multi-scale semantic feature fusion called the Multi-scale Feature Fusion Module (MFFM), which enables information propagation from fine-grained to coarse-grained features. By constructing an additional segmentation network, the base-learning branch predicts the regions of the base categories in the query image and corrects the boundary category in the predicted mask from the meta-learning branch. This facilitates segmentation of new objects. The proposed method is evaluated on the ABD-MR and ABD-CT datasets, and the results are assessed by the Dice coefficient as the evaluation metric. The experimental results show that our method achieves higher accuracy in few-shot semantic segmentation compared to several state-of-the-art methods. The main contributions of this paper are as follows:

1. We propose a novel approach comprising a Prior Mask Generation Module (PRGM) and a Multi-scale Feature Fusion Module (MFFM). PRGM is designed to enhance the accuracy of target localization in query images. Additionally, MFFM dynamically integrates features from the support set prototype, the prior mask and the query set at multi-scale, effectively addressing the issue of spatial inconsistency between the support set and the query set.
2. An additional base-learning branch was constructed to correct the boundary categories of the meta-learning branch prediction mask by predicting the regions of the base categories in the query image.
3. In this paper, the proposed BCC-Net is experimented on the mainstream ABD-MR and ABD-CT medical image segmentation public datasets, and the experimental results show that the proposed method achieves state-of-the-art research results.

2   Related Work

2.1   Few-Shot Segmentation in Natural Image Field

In the field of natural image few-shot segmentation, a lot of studies have been proposed. Shaban et al. [12] first propose a two-branch approach to train a few-shot image segmentation model. The parameter branch receives an annotated image input and generates a parameter vector. This vector, along with the new image, is then passed into the segmentation branch to produce a segmentation mask for the query set. Wang *et al.* [19] propose the PANet model, which uses a non-parametric metric learning approach to convert the segmentation task to a pixel-by-pixel classification, and also propose a prototype alignment regularisation method to predict the result of the support set from the query set result. Zhang *et al.* [22] propose the CANet model that uses a parameterized dense comparison module to perform point-to-point feature comparison of the query set features. It then refines the prediction results using an iterative optimization module.

2.2   Few-Shot Segmentation in Medical Image Field

In few-shot segmentation methods for medical images, Roy *et al.* [12] first propose a specialized few-shot learning architecture designed for medical image segmentation, they use squeeze and excitation module to fuse information from the support image to the query image to guide the segmentation branch. To overcome the issue of limited samples, Chen *et al.* [11] propose a self-supervised few-shot segmentation model. They utilized superpixels and corresponding pseudo-labels to eliminate the need for manual annotation. Furthermore, they utilize adaptive local prototype pooling to retain local features in the representations, leading to an enhancement in segmentation accuracy. Sun *et al.* [4] employ a global correlation module to capture the long-range dependencies between the support set and query set features, and utilize discriminative regularization to enhance the discriminative ability of foreground features. Shen *et al.* [13] proposed the PoissonSeg semi-supervised few-shot segmentation model. They utilize graph learning to connect labeled and unlabeled images and employed spatial consistency calibration and Poisson learning to improve the segmentation quality. Ye *et al.* [20] they propose a dynamic framework, efficient consistency regularization strategy and introduce a new pseudo-label generation strategy.

3   Method

3.1   Overview

In this section, we give the details of our proposed network (BCC-Net) for few-shot medical image segmentation. We first present a overall framework architecture in Fig. 1. BCC-Net consists of a meta-learning branch and a base-learning branch. In the meta-learning branch, we design a PRGM, which can better localize the query target. Meanwhile, MFFM is designed with rich semantic information using a top-to-bottom adaptive multi-scale semantic information feature fusion module to solve the spatial inconsistency problem between the support sets and query sets. In order to further improve the segmentation accuracy, we propose an additional base categories segmentation network named base-learning branch to correct for the boundary categories of the meta-learning branch prediction masks.

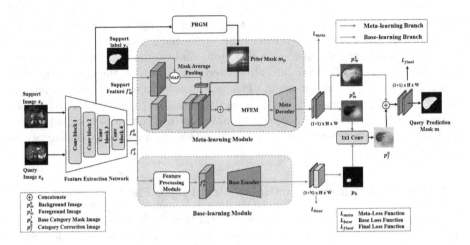

Fig. 1. Few-shot medical image segmentation network structure based on boundary category correction.

3.2 Prior Mask Generation Module

Based on [17], we propose a Prior Mask Generation Module (PRGM) to generate a rough segmentation prediction mask for the query set by using cosine similarity calculation of the high-level features of the query set and the support set, allowing our method to better localize the query target. By the support set features, the prior mask will locate key regions in the query set with high similarity to the support set, and can preserve the generalisation ability of the our method. The prior masks for CT images is shown in Fig. 2.

Specifically, the input query image is represented as I_Q, the input support image is denoted as I_S, the binary mask of the query set image is labeled as M_S, $\mathcal{F}$ represents the convolution feature extraction network, X_Q and X_S represent the high-level features of the query image and support image, respectively, which are represented as Eq. (1). $\odot$ represents the Hadamard product, which serves to remove the background part of the support image by setting to 0 and retain the foreground part of the support image by setting to 1.

$$X_Q = \mathcal{F}(I_Q), X_S = \mathcal{F}(I_Q) \odot M_S \tag{1}$$

The prior mask Y_Q is calculated from Eq. (2) to Eq. (5). First, the cosine similarity $\cos(x_q, x_s)$ between each pixel of X_Q and X_S is calculated, where q and s are one of the pixels of high-level features 1 to hw of query set and support set, respectively.

$$\cos(x_q, x_s) = \frac{x_q^T x_s}{\|x_q\| \|x_s\|}, q, s \in \{1, 2, \cdots, hw\} \tag{2}$$

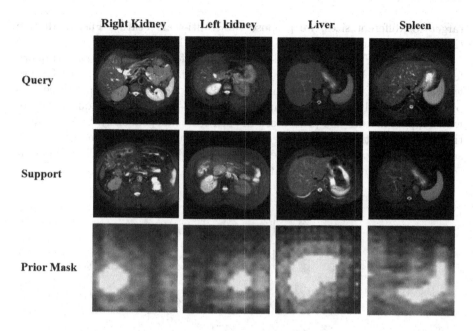

Fig. 2. Prior masks for CT images of four abdominal organs.

Then, the maximum similarity between all support pixels is taken as the response value c_q, calculated by Eq. (3):

$$c_q = \max_{s \in \{1,2,\cdots,hw\}} (\cos(x_q, x_s)) \tag{3}$$

Then we produce the prior mask Y_Q by reshaping C_Q to $h \times w$. We calculate Y_Q with a min-max normalization (Eq. (5)) to normalize the values between 0 and 1, ϵ is set to 10^{-7} in our experiments.

$$C_Q = [c_1, c_2, \cdots, c_{hw \times 1}] \in \mathbb{R}^{hw \times 1} \tag{4}$$

$$Y_Q = \frac{Y_Q - \min(Y_Q)}{\max(Y_Q) - \min(Y_Q) + \epsilon} \tag{5}$$

The key is to use the high-level features of the support and query set to generate prior masks, and then take the maximum value from a similarity matrix of size $hw \times hw$.

3.3   Multi-scale Feature Fusion Module

In few-shot medical image segmentation tasks, existing methods typically use global average pooling to extract category features. However, the area of query target may be larger or smaller than support samples, so global pooling on support images results in spatial information inconsistency. In order to handle targets of different sizes, we propose the Multi-scale Feature Fusion Module (MFFM) based on [17].

As shown in Fig. 3, MFFM takes the support feature, prior mask and query feature as input. It outputs the refined query feature from the support feature with enriched information. The enrichment process can be divided into three sub-processes. MFFM first projects input to different scales, interacts query feature with support feature and prior mask in each scale independently. Subsequently, the MFFM selectively passes essential information between the merged query-support features at different scales. Then, MFFM merges features in different scales to get the refined query feature.

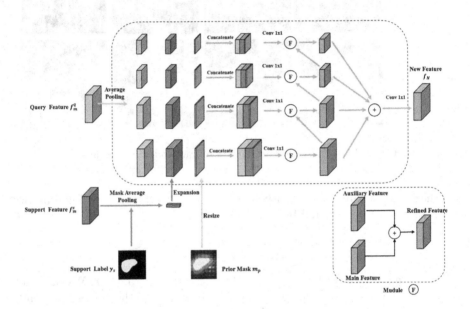

Fig. 3. Multi-scale Feature Fusion Module.

3.4   Boundary Category Correction Framework

In this paper, we propose the Boundary Category Correction Framework (BCCF) based on [8]. BCCF includes a meta-learning module and a base-learning module. The meta-learning module is used to identify new categories, the base-learning module is used to identify base categories, i.e. regions that do not need to be segmented, while providing highly reliable segmentation results and stable performance. The BCCF filter the coarse results output from the two learning modules in parallel to obtain accurate segmentation predictions. The two models are described in Fig. 1.

Meta-learning Module. The meta-learning module is shown in Fig. 1. In this module, the high-level features of the support set image and the query set image are extracted using the feature extractor, and then 1×1 convolution is applied to reduce the channel dimension and generate intermediate feature maps f_m^s and f_m^q, which are computed as Eq. (6).

$$f_m^s = \mathcal{F}_{1\times1}(\xi(x^s)) \in \mathbb{R}^{c\times h\times w}, f_m^q = \mathcal{F}_{1\times1}(\xi(x^q)) \in \mathbb{R}^{c\times h\times w} \tag{6}$$

The mask-averaged pooling is then applied to generate the target category-only feature vector v_s, computed as Eq. (7).

$$v_s = \mathcal{F}_{pool}(f_m^s \odot I(m^s)) \in \mathbb{R}^c \tag{7}$$

Then, the model activate the f_m^q target region under the guidance of v_s and generate the final prediction p_m by decoder network, shown as Eq. (8).

$$p_m = softmax(\mathcal{D}_m(\mathcal{F}_{guidance}(v_s, f_m^q)) \in \mathbb{R}^{2\times H\times W}) \tag{8}$$

In particular, $\mathcal{F}_{guidance}$ is an important step in few-shot image segmentation, which passes annotation information from the support branch to the query branch to provide specific segmentation information. $\mathcal{D}_m$ is the decoder of the meta-learning module, consisting of a 3×3 convolution and a 1×1 convolution. In this paper, a feature processing module is added before the decoder, and optional modules include MFFM [17], a pyramid pooling module PPM [17] and a void space pyramid pooling module ASPP [1].

Base-Learning Module. We propose the base-learning module to predict the regions of the base categories in the query set images, the model is shown in Fig. 1. In this paper, the PSPNet network [23] is used to aggregate the contextual information of different regions after the high-level features are extracted in order to fully exploit the ability of global contextual information and better improve the segmentation performance of the base categories.

The query set images are first fed into the feature extraction network to obtain the query set base categories features f_b^q, and the calculation process is shown as Eq. (9), where $\mathcal{F}_{conv}$ is a feature extractor that is not shared with the meta-learning branch.

$$f_b^q = \mathcal{F}_{conv}(x^q) \in \mathbb{R}^{c\times h\times w} \tag{9}$$

Then, the spatial scale of the intermediate features is gradually expanded by the feature processing module ξ and the final prediction p_b is generated by the decoder network $\mathcal{D}_b$, which is computed as Eq. (10).

$$p_b = softmax(\mathcal{D}_b(\xi(f_b^q))) \in \mathbb{R}^{(1+N_b)\times h\times w} \tag{10}$$

3.5 Loss Function

This paper adopts a two-stage training strategy, in which the base-learning branch is trained first and then the meta-learning branch is trained by fixing

Table 1. Segmentation results of different models on ABD-CT and ABD-MR datasets.

Method	Abdominal-CT					Abdominal-MRI				
	Spl[a]	Lkid[a]	Rkid[a]	Liver	Mean	Spl[a]	Lkid[a]	Rkid[a]	Liver	Mean
SE-Net	0.23	32.83	14.34	0.27	11.91	51.80	62.11	61.32	27.43	50.66
PANet	25.59	32.34	17.37	38.42	29.42	50.90	53.45	38.64	42.26	46.33
ALPNet	60.25	63.34	54.82	73.65	63.02	67.02	73.63	78.39	73.05	73.02
GCN-DE	56.53	**68.13**	**75.50**	46.77	61.73	60.63	76.07	83.03	49.47	67.30
PoissonSeg	52.33	50.11	47.02	58.74	52.05	52.85	50.58	53.57	61.03	54.51
YeNet	61.11	64.59	55.57	74.56	63.95	68.01	74.57	78.47	**73.84**	73.79
Ours	**67.57**	67.45	68.74	**77.49**	**70.31**	**73.31**	**80.98**	**84.55**	73.67	**78.13**

[a] Spl is Spleen. Lkid is Left Kidney. Rkid is Right Kidney

the parameters of the base-learning branch network. In the first stage of training, the loss function of the base-learning module is calculated between the prediction result p_b and the query set label m_b^q. The base-learning module uses CrossEntropy loss, and the calculation process is shown as Eq. (11), where N is the number of training samples for each batch.

On the meta-learning branch, Binary CrossEntropy loss is computed between the prediction result p_m and the query set label m^q to update parameters of the meta-learning module, and the loss function of the meta-learning module is computed as Eq. (12).

The final loss function for this network is computed between the predicted segmentation result p_i^q and the labelled segmentation mask m_i^q. $\mathcal{L}_{final}$ is based on binary crossentropy loss and defined as Eq. (13).

In the second stage of the training process of the model, the total loss L of the model consists of a combination of the weighted sum of the meta-loss $\mathcal{L}_{meta}$ and the final loss $\mathcal{L}_{final}$, defined as Eq. (14).

$$\mathcal{L}_{base} = \frac{1}{N}\sum_{i=1}^{N} CE(p_{m;i}, m_{b;i}^q) \tag{11}$$

$$\mathcal{L}_{meta} = \frac{1}{N}\sum_{i=1}^{N} BCE(p_{m;i}, m_i^q) \tag{12}$$

$$\mathcal{L}_{final} = \frac{1}{N}\sum_{i=1}^{N} CE(p_i^q, m_i^q) \tag{13}$$

$$\mathcal{L} = \mathcal{L}_{final} + \lambda\mathcal{L}_{meta} \tag{14}$$

4   Experiments

4.1   Datasets and Evaluation Metrics

The proposed method is evaluated on two abdominal datasets named ABD-CT [5] and ABD-MR [7]. ABD-CT is from MICCAI 2015 Multi-Atlas Abdomen Labeling Challenge. It contains thirty 3D abdominal CT of different patients. ABD-MR is from ISBI 2019 and contains twenty 3D T2-SPIR MRI scans. We employed five-fold cross-validation on the datasets and evaluate our model by Dice [11] loss.

4.2   Implementation Details

Drawing from the parameter settings of mainstream methods for natural image few-shot segmentation, such as SSL-ALPNet [11] and RP-Net [16]. The total number of effective slices from the CT and MR datasets used for training is less than 500. In the experiments, the base-learning branch parameters are as follows: the input image size is set to 256×256, the batch size is 8, the initial learning rate is 0.001, the SGD optimizer is employed, the training consists of 100 epochs, the learning rate is decayed by 0.98 every 2 rounds of training. Each slice of the support image is divided into 3 blocks, and the central slice of the block segments all the slices in the query block. To achieve higher segmentation performance, the feature extraction network utilizes ResNet101 [3] and pre-trained on the MS-COCO dataset [10]. Additionally, the experiments were conducted on an NVIDIA TITAN RTX graphics card. In meta-learning branch, the training consists of 200 epochs, each epoch contains 1000 iterations and the batch size is 1. The other parameters and settings are the same as in meta-learning branch.

4.3   Experiment Results

In this study, we conducted experiments with several state-of-the-art methods in few-shot segmentation field, including SE-Net [12], PANet [19], SSL-ALPNet [11], GCN-DE [15], PoissonSeg [13] and YeNet [20]. The results are shown in Table 1. The proposed method achieved an average Dice score of 70.31% and 78.13% on the two datasets, respectively. Specifically, the mean Dice of our method outperformed the seven competitive methods mentioned above by 58.4%, 40.89%, 7.29%, 8.58%, 18.26%, 8.36% on the ABD-CT dataset and 27.47%, 31.8%, 7.81%, 10.83%, 23.62%, 4.34% on the ABD-MRI dataset. Experiments show that BCC-Net achieve the state-of-the-art performance among all the methods.

Figure 4 demonstrate the performance of boundary segmentation. As Fig. 4 shows, the performance of the proposed method is better than SAM released by Facebook in 2023 on the liver area and similar in others. Liver and Right kidney are close to each other in the abdomen and are prone to boundary category segmentation out-of-place. It can be observed that the method we proposed can correct adjacent areas of the organ effectively.

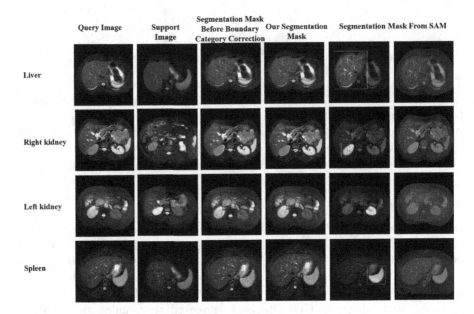

Fig. 4. Segmentation Results in MR Images.

Table 2. Quantified segmentation results for different modules.

	Spleen	Kidney L	Kidney R	Liver	Mean
Baseline	64.15	65.84	68.78	66.52	66.32
Baseline + PPM	67.20	69.98	72.88	69.63	69.92
Baseline + ASPP	64.53	69.79	71.63	68.24	68.55
Baseline + MFFM	67.81	71.32	73.96	69.45	70.64
BFP[a]	68.60	72.93	74.87	70.51	71.73
BFP[a] + BaseSeg	73.31	80.98	84.55	73.67	78.13

[a] BFP is equal to Baseline + MFFM + PRGM

4.4   Ablation Study

In order to further validate the effectiveness of each module we proposed in this
paper, we conduct ablation experiments on the ABD-MR dataset. Baseline is
the part of the boundary category correction network framework after removing
the base-learning module, PRGM and MFFM, keeping only the feature extraction part with the meta-learning network. BaseSeg represents the base-learning
branch. The specific results of ablation experiment are shown in Table 2. We
can see that the average Dice improved by 1.09% after using BFP, and using
BFP + BaseSeg is 6.4% higher compared to not using BaseSeg. This indicates
that PRGM, MFFM and BCCF is significantly effective in solving the semantic
segmentation problem of few-shot medical images.

5   Conclusion

In this paper we propose BCC-Net, a network exhibits good generalization performance for new medical category targets by utilizing knowledge from labels to guide the few-shot medical image segmentation task. The framework incorporates proposed PRGM and MFFM into the meta-learning branch. These modules can better localize the query targets and solve the spatial inconsistency problem between the support set and the query set. Moreover, we propose a base-learning branch to revise the boundary category predictions of the meta-learning branch. Experimental results show that our method can enhance the richness of the feature map and effectively correct the predicted boundaries of adjacent abdominal organs, improving the boundary segmentation accuracy of medical targets.

Funding. This work is supported by the National Natural Science Foundation of China (Nos. (82071876, 6217010009, 82372043, 82371904).

References

1. Chen, L.C., Papandreou, G., et al.: DeepLab: semantic image segmentation with deep convolutional nets, atrous convolution, and fully connected CRFs. IEEE Trans. Pattern Anal. Mach. Intell. **40**, 834–848 (2016)
2. Finn, C., Abbeel, P., Levine, S.: Model-agnostic meta-learning for fast adaptation of deep networks. arXiv arXiv:1703.03400 (2017)
3. He, K., et al.: Deep residual learning for image recognition. In: 2016 IEEE Conference on Computer Vision and Pattern Recognition (CVPR), pp. 770–778 (2015)
4. Heidari, M., Kazerouni, A., et al.: HiFormer: hierarchical multi-scale representations using transformers for medical image segmentation. In: 2023 IEEE/CVF Winter Conference on Applications of Computer Vision (WACV), pp. 6191–6201 (2022)
5. Kavur, A.E., Gezer, N.S., et al.: CHAOS challenge - combined (CT-MR) healthy abdominal organ segmentation. Med. Image Anal. **69**, 101950 (2020)
6. Koch, G.R.: Siamese neural networks for one-shot image recognition (2015)
7. Landman, B., Xu, Z., Igelsias, J., Styner, M., et al.: MICCAI multi-atlas labeling beyond the cranial vault-workshop and challenge. In: Proceedings of the MICCAI Multi-Atlas Labeling Beyond Cranial Vault-Workshop Challenge, vol. 5, p. 12 (2015)
8. Lang, C., Cheng, G., Tu, B., Han, J.: Learning what not to segment: a new perspective on few-shot segmentation. In: 2022 IEEE/CVF Conference on Computer Vision and Pattern Recognition (CVPR), pp. 8047–8057 (2022)
9. Li, G., Jampani, V., Sevilla-Lara, L., Sun, D., Kim, J., Kim, J.: Adaptive prototype learning and allocation for few-shot segmentation. In: 2021 IEEE/CVF Conference on Computer Vision and Pattern Recognition (CVPR), pp. 8330–8339 (2021)
10. Lin, T.-Y., et al.: Microsoft COCO: common objects in context. In: Fleet, D., Pajdla, T., Schiele, B., Tuytelaars, T. (eds.) ECCV 2014. LNCS, vol. 8693, pp. 740–755. Springer, Cham (2014). https://doi.org/10.1007/978-3-319-10602-1_48
11. Ouyang, C., Biffi, C., Chen, C., Kart, T., Qiu, H., Rueckert, D.: Self-supervision with superpixels: training few-shot medical image segmentation without annotation. In: Vedaldi, A., Bischof, H., Brox, T., Frahm, J.-M. (eds.) ECCV 2020. LNCS,

vol. 12374, pp. 762–780. Springer, Cham (2020). https://doi.org/10.1007/978-3-030-58526-6_45

12. Roy, A.G., Siddiqui, S., et al.: 'squeeze & excite' guided few-shot segmentation of volumetric images. Med. Image Anal. **59**, 101587 (2019)

13. Shen, X., Zhang, G., Lai, H., et al.: PoissonSeg: semi-supervised few-shot medical image segmentation via Poisson learning. In: 2021 IEEE International Conference on Bioinformatics and Biomedicine (BIBM), pp. 1513–1518 (2021)

14. Su, X., et al.: Amotivation, career engagement, and the moderating role of career adaptability of youth not in education, employment, or training (2020)

15. Sun, L., Li, C., Ding, X., Huang, Y., Wang, G., Yu, Y.: Few-shot medical image segmentation using a global correlation network with discriminative embedding. Comput. Biol. Med. **140**, 105067 (2020)

16. Tang, H., Liu, X., Sun, S., Yan, X., Xie, X.: Recurrent mask refinement for few-shot medical image segmentation. In: 2021 IEEE/CVF International Conference on Computer Vision (ICCV), pp. 3898–3908 (2021)

17. Tian, Z., Zhao, H., Shu, M., Yang, Z., Li, R., Jia, J.: Prior guided feature enrichment network for few-shot segmentation. IEEE Trans. Pattern Anal. Mach. Intell. **44**, 1050–1065 (2020)

18. Vinyals, O., Blundell, C., Lillicrap, T.P., Kavukcuoglu, K., Wierstra, D.: Matching networks for one shot learning. arXiv arXiv:1606.04080 (2016)

19. Wang, K., Liew, J.H., Zou, Y., Zhou, D., Feng, J.: PANet: few-shot image semantic segmentation with prototype alignment. In: 2019 IEEE/CVF International Conference on Computer Vision (ICCV), pp. 9196–9205 (2019)

20. Ye, Z., Zhang, W.: A dynamic few-shot learning framework for medical image stream mining based on self-training. EURASIP J. Adv. Sig. Process. **2023**, 1–19 (2023)

21. Yu, Q., Dang, K., Tajbakhsh, N., Terzopoulos, D., Ding, X.: A location-sensitive local prototype network for few-shot medical image segmentation. In: 2021 IEEE 18th International Symposium on Biomedical Imaging (ISBI), pp. 262–266 (2021)

22. Zhang, C., Lin, G., et al.: CANet: class-agnostic segmentation networks with iterative refinement and attentive few-shot learning. In: 2019 IEEE/CVF Conference on Computer Vision and Pattern Recognition (CVPR), pp. 5212–5221 (2019)

23. Zhao, H., Shi, J., et al.: Pyramid scene parsing network. In: 2017 IEEE Conference on Computer Vision and Pattern Recognition (CVPR), pp. 6230–6239 (2016)

Repdistiller: Knowledge Distillation Scaled by Re-parameterization for Crowd Counting

Tian Ni, Yuchen Cao, Xiaoyu Liang, and Haoji Hu[✉]

College of Information Science and Electronic Engineering, Zhejiang University,
Hangzhou, China
{nitian,Yuchen_Cao,22231080,haoji_hu}@zju.edu.cn

Abstract. Knowledge distillation (KD) is an important method to compress a large teacher model into a much smaller student model. However, the large capacity gap between the teacher and student models hinders the performance of KD in various tasks. In this paper, we propose Repdistiller, a knowledge distillation framework combined with structural re-parameterization to alleviate the capacity gap problem. Repdistiller makes the student model search for parallel branches during training, thus the capacity gap between the teacher and student models is decreased. After knowledge distillation, the searched branches are merged into the student network without causing any computation overhead for inference. Taking the crowd counting task as an example, Repdistiller achieves state-of-the-art performance on the ShanghaiTech and UCF-QNRF datasets, outperforming many well-established knowledge distillation methods.

Keywords: Crowd counting · Knowledge distillation · Structural re-parameterization

1 Introduction

Knowledge distillation (KD) has become a widely used method for deep neural network compression [3,10]. KD first constructs a set of teacher and student models. The teacher model is huge with a bigger amount of parameters, while the student model is lightweight with a smaller amount of parameters. The main idea of KD is to learn the output distributions of the teacher model and transfer the distribution knowledge from the teacher model to the student model. This helps the student model to imitate the teacher model, so as to learn the generalization ability of the teacher model.

However, having a more powerful and bigger teacher model in KD does not necessarily result in a better student model. When the capacity gap between the teacher and student is large, it is difficult for the student model to learn useful knowledge from the teacher model, which leads to accuracy degradation of the student model [20]. A common approach is to introduce several intermediate

Q. Liu et al. (Eds.): PRCV 2023, LNCS 14434, pp. 383–394, 2024.
https://doi.org/10.1007/978-981-99-8549-4_32

teacher assistant (TA) models [20]. Thus, knowledge is firstly transferred from the teacher to the TA models, then from the TA models to the student model.

Another recent trend to improve the performance of lightweight models is structural re-parameterization [8]. During training, structural re-parameterization injects several parallel branches to increase the capacity of the lightweight model. The performance would also improve correspondingly because the capacity of the model has been increased. All the injected branches are carefully designed, so that they can be merged into the lightweight model without causing any computation overhead during inference.

In this paper, we propose Repdistiller, a knowledge distillation framework combined with re-parameterization to alleviate the capacity gap problem. The capacity of the student model is firstly enlarged to decrease the gap between the teacher and student models. Then, knowledge of the teacher model is transferred to the enlarged student model. After distillation, the injected re-parameterization branches are merged into the student model to save computation.

Besides the proposed framework, the novelty of Repdistiller is embodied in other two aspects. Firstly, we implement network architecture search (NAS) to obtain the re-parameterization branches, compared with manually designing branches in the original re-parameterization process. The searched branches by NAS are more suitable for specific tasks, thus bringing further performance improvements. Another novelty is to apply gradient decoupling in the knowledge distillation process. By gradient decoupling, we disentangle the supervision of the teacher and the ground truth, making the KD process converge quickly with better performance.

We implemented RepDistiller on the crowd counting task. Many crowd counting models, such as CSRNet [16] and BL [19], adopt multi-branch structures, which are very complex in size and computation. In this paper, we compress the above models by RepDistiller, intensively decreasing the computation and storage, so that the compressed lightweight models can be applied in practical crowd counting scenarios.

The contributions of this paper are summarized as follows:

1. We introduce Repdistiller for the crowd counting task, which is one of the first attempts to combine knowledge distillation and re-parameterization to mitigate the capacity gap problem.
2. We implement NAS to obtain the re-parameterization branches, which further improves the performance compared with manually designing branches in original structural re-parameterization.
3. We apply gradient decoupling to distinguish the supervision of the teacher and the supervision of the ground truth in knowledge distillation, thus dynamically updating the re-parameterization branches of the student model.

2   Related Work

Knowledge distillation (KD) has become a widely used method for deep neural network compression [3, 10]. The capacity gap between the teacher and student models is a vital problem to hinder KD obtain good performance in application scenarios [20]. A common approach to solve this problem is to introduce several intermediate teacher assistant (TA) models [20] during the KD process. A recent trend is to apply KD on Transformer networks [17], combine KD with other network compression methods such as pruning and network architecture search [18], and theoretical analysis [22].

Structural re-parameterization means expanding the model structure horizontally or vertically, using complex networks with multiple branches during training, and merging them into a single branch for testing without further training, thus increasing speed. Structural re-parameterization techniques have shown great potential in improving model performance. Some re-parameterization methods have achieved surprising results on some network structures. The inclusion of a 1×1 convolutional layer and a residual connection in a 3×3 convolutional layer in RepVGG [8] resulted in a significant enhancement in the performance of VGG-like networks. DBB [7] provided a comprehensive overview of additional structural re-parameterization techniques and introduced the Diverse Branch Block, a versatile module that can be seamlessly incorporated into any convolutional network.

The task of crowd counting is to automatically estimate all crowd numbers in a surveillance scene. Crowd counting methods can generally be classified as those based on Direct Regression (DR) or based on Density Map Estimation (DME) [13]. The complex lighting conditions, severe crowd occlusion, viewing angle distortion, and different distributions of crowd density, make it difficult to generate high-quality crowd density maps [9]. Some methods like AG-FPN [5] try to solve such problems through attention blocks. By combining the high- and low-level features, high-quality density maps with accurate spatial locations can be adaptively generated.

3 Method

Figure 1 illustrates an overview of the Repdistiller framework for crowd counting. The proposed method consists of four steps. (1) **Model Selection.** Firstly, we select a lightweight student model, whose architecture is similar to the original teacher model, but the student model intensively reduces the block size of each component in the model. (2) **Re-parameterization.** The second step is to use NAS to search on the lightweight student model to obtain the corresponding re-parameterization structure. By doing this, the big capacity gap between the teacher and student models is alleviated, making preparation for the distillation process. (3) **Knowledge Distillation.** Then, we apply knowledge distillation to the teacher and student models, where the student model is a re-parameterized version produced by the second step. Because the capacity of the student model is increased through re-parameterization, the capacity gap between teacher and student models has been decreased. It is expected that the accuracy of the student model after KD should be increased correspondingly compared with the non-re-parameterized version. (4) **Merging Re-parameterization Branches.**

After knowledge distillation, the final step of Repdistiller is to merge all the re-parameterization branches into a single one. The merging step makes the student model produce the same inference results as the re-parameterized one. In addition, the computation load of the student model after the merging step is the same as the non-re-parameterized model. Thus, Repdistiller makes the student model both accurate and efficient by re-parameterization.

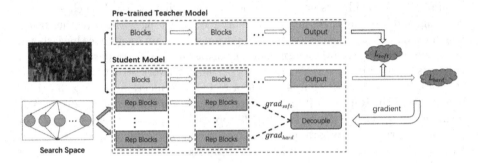

Fig. 1. The overall framework of Repdistiller.

3.1   Model Selection

We select two well-established networks for crowd counting, BL [19] and CSR-Net [16], as the teacher models. We compress the channel numbers of each teacher model to 1/4 of its original size for each block in the teacher model. Thus, we obtain two lightweight student models which are named as '1/4BL+Repdistiller' and '1/4CSRNet+Repdistiller', respectively.

Table 1 lists the parameters, FLOPs, GPU and CPU usage of the above four models on the ShanghaiTech and UCF-QNRF datasets. It can be seen that the student models intensively reduce the storage and computation compared with the teacher models. For example, on Shanghaitech Part_B, the '1/4CSR-Net+Repdistiller' takes 12.65 ms (7.8× speedup) on GPU and 3.09 s (11.62× speedup) on CPU to process a 768 × 1024 image, which is 7.8× faster on GPU and 11.62× faster on CPU than the original CSRNet model. These results further demonstrate the superiority of Repdistiller on inference efficiency.

3.2   Re-parameterization by Neural Architecture Search

The second step of Repdistiller is to increase the capacity of the student model by re-parameterization. Suppose that a student network consisting of N blocks, e.g., '1/4CSRNet+Repdistiller' consists of 16 blocks. All the blocks are convolutional layers with different kernel sizes. As shown in Fig. 1, the feature map F_n ($1 \le n \le N$) can be extracted from the previous block by

$$F_n = B_n(F_{n-1}), \quad F_0 = I. \tag{1}$$

Table 1. Comparison of inference efficiency. The unit of the parameter quantity is million (M). FLOPs represent the number of floating-point operations, and the unit is giga (G). GPU represents the inference time in milliseconds (ms) on an Nvidia GTX 1080 GPU. CPU represents the inference time in seconds (s) on an Intel(R) Core(TM) i5-8265UC CPU (1.6G).

Model	Params	ShanghaiTech Part_A(576 × 864)			ShanghaiTech Part_B(768 × 1024)			UCF-QNRF (2032 × 2912)		
		FLOPs	GPU	CPU	FLOPs	GPU	CPU	FLOPs	GPU	CPU
BL	21.50	205.32	47.89	14.71	324.46	70.18	3731	2441.23	595.72	195.58
1/4BL +Repdistiller	1.35	13.06	10.27	0.88	20.64	16.81	1.89	155.30	90.96	23.37
CSRNet	16.26	205.81	66.57	13.28	325.34	98.62	35.92	2447.91	823.84	173.46
1/4CSRNet +Repdistiller	1.02	13.09	8.88	1.53	20.69	12.65	3.09	155.69	106.08	26.34

Here B_n denotes the convolutional layer of the nth block, and I represents the input image.

To increase the capacity of the student model, Repdistiller adds C_n parallelly re-parameterization branches to each of the N blocks by network architecture search (NAS). Thus, Eq. (1) is rewritten as

$$F_n = B_n(F_{n-1}) + \sum_{c=1}^{C_n} \mathrm{RB}_{n,c}(F_{n-1}), \qquad (2)$$

where $\mathrm{RB}_{n,c}$ represents the cth re-parameterization branch of the nth block, which is searched from a search space S consisting of several operators.

We deliberately design 21 operators for the search space. The 21 operators can be divided into three categories – (1) Convolutional kernels of size 1×1. (2) Convolutional kernels of sizes $(3 \times 3) + (1 \times 1)^{0\sim3}$, which means that after 3×3 convolution, we further add several 1×1 convolutional kernels, and the number of 1×1 convolutional kernels could be 0, 1, 2 or 3. All convolutional kernels are separated by a batch normalization (BN) layer. Thus, there are 4 operators in this category. (3) Convolutional kernels of sizes $[(3 \times 1) + (1 \times 1)^{0\sim3}] \times [(1 \times 3) + (1 \times 1)^{0\sim3}]$. In this category, we use depthwise convolution [4] to reduce the parameters of the convolutional kernels, i.e., a (3×1) kernel is concatenated with a (1×3) kernel to replace a (3×3) kernel. For the (3×1) and (1×3) kernels, we further concatenate several (1×1) kernels, the same as category 2. All kernels are separated by a BN layer. Thus, there are 16 operators in this category. The above operators are specially designed for re-parameterization. For inference, all the operators can be merged into a 3×3 convolutional kernel without any computation overhead.

We implement ProxylessNAS [1] as the NAS search strategy. It assigns the weight parameters O and the architecture parameter α for each operator. Then, we recursively update O and α. When updating the weight parameters, the architecture parameters are frozen, and the operators are randomly activated based on the magnitude of the architecture parameter. Then, the weight parameters of the active operators are updated through backpropagation on the training set. When training the architecture parameters, the weight parameters are frozen, then the binary gates are reset and the architecture parameters on the validation set are updated. Finally, a compact architecture can be derived by pruning the operators with smaller architecture parameters.

Supposing that there are N parallelly aligned operators, their output is calculated by applying softmax to the N real-valued architecture parameters α_i:

$$m_{proxyLESS}(x) == \sum_{i=1}^{N} p_i O_i(x) = \sum_{i=1}^{N} p_i O_i(x) = \frac{\exp(\alpha_i)}{\sum_j \exp(\alpha_j)} O_i(x) \quad (3)$$

When updating the weight parameters, we change the output as a binary gate function:

$$m_{Binary}(x) = \sum_{i=1}^{N} g_i O_i(x), \quad (4)$$

where g_i is a binary gating parameter with the following form:

$$[g_1, g_2, ..., g_N] = \begin{cases} [1,0,...,0] & \text{with probability } p_1 \\ \quad ... \\ [0,0,...,1] & \text{with probability } p_N \end{cases} \quad (5)$$

Thus, only one operator is activated for each batch, and the weights of that operator are updated through backpropagation.

For training the architecture parameter, we freeze the weights and update α_i through backpropagation:

$$\frac{\partial L}{\partial \alpha_i} = \sum_{j=1}^{N} \frac{\partial L}{\partial p_j}\frac{\partial p_j}{\partial \alpha_i} = \sum_{j=1}^{N} \frac{\partial L}{\partial p_j}\frac{\partial\left(\frac{\exp(\alpha_j)}{\sum_k \exp(\alpha_k)}\right)}{\partial \alpha_i} = \sum_{j=1}^{N} \frac{\partial L}{\partial p_j} p_j(\delta_{ij} - p_i) \quad (6)$$

where $\delta_{ij} = 1$ if $i = j$ and $\delta_{ij} = 0$ if $i \neq j$.

After several iterations to recursively update O_i and α_i, we finally select the first C_n operators with the biggest α as the final re-parameterization network. In this paper, we select 5 operators ($C_n = 5$) for each block. By experiments, such selection well balances the accuracy and computation of the knowledge distillation process (Fig. 2).

3.3 Knowledge Distillation by Gradient Decoupling

After obtaining the student model by re-parameterization, we use knowledge distillation (KD) to transfer knowledge between the teacher and student models. The total loss for KD is a combination of the soft label loss $\mathcal{L}_{soft}$ (distillation task) and real label loss $\mathcal{L}_{hard}$ (original crowd counting task):

$$\mathcal{L}_{total} = (1-\alpha)\mathcal{L}_{soft} + \alpha\mathcal{L}_{hard} = (1-\alpha)||\mathcal{M}_s - \mathcal{M}_t||^2 + \alpha||\mathcal{M}_s - \mathcal{M}||^2, \quad (7)$$

where $\mathcal{M}$, $\mathcal{M}_s$ and $\mathcal{M}_t$ represent the ground truth density map, the density map of the student model and the density map of the teacher model, respectively. Here α is a weighting parameter to balance the two losses.

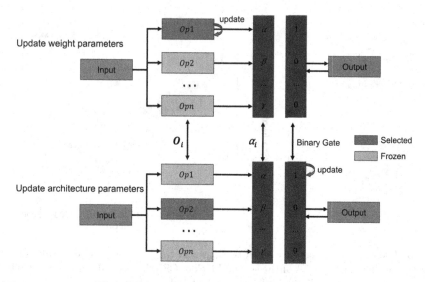

Fig. 2. The flowchart of the search strategy.

However, we find that L_{soft} and L_{hard} are intensively entangled together in many experiments. Figure 3 shows the gradient divergence effect of the two losses. During different iterations, the gradient directions of the two losses vary intensively, thus making the backpropagation of the total loss inefficient.

Inspired by gradient decoupling for multi-task learning [2], we implement gradient decoupling for knowledge distillation. For each iteration, we separately calculate the gradients of the two losses, i.e., $\frac{\partial L_{soft}}{\partial O_i(x)}$ and $\frac{\partial L_{hard}}{\partial O_i(x)}$. Then, the gradient of the total loss is modified by a winner-takes-all strategy:

$$\frac{\partial L_{total}}{\partial O_i(x)} = \begin{cases} \alpha \frac{\partial L_{soft}}{\partial O_i(x)}, & \text{if } \left|\frac{\partial L_{soft}}{\partial O_i(x)}\right| > \left|\frac{\partial L_{hard}}{\partial O_i(x)}\right| \\ (1-\alpha)\frac{\partial L_{hard}}{\partial O_i(x)}, & \text{if } \left|\frac{\partial L_{soft}}{\partial O_i(x)}\right| < \left|\frac{\partial L_{hard}}{\partial O_i(x)}\right| \end{cases} \tag{8}$$

Equation (8) means that in each iteration when we update the weighting parameters, we either use the supervision of the teacher model, or the supervision of the ground truth, but not both. By doing this, the loss entangle property is decoupled in order to make the training process convergent quickly.

After knowledge distillation, the final step is to merge all the re-parameterization branches into a single one. Since the operators are specially designed for re-parameterization, they can be merged into a 3×3 convolutional kernel without any computation overhead. Please refer to [8] for details of the merging process for re-parameterization.

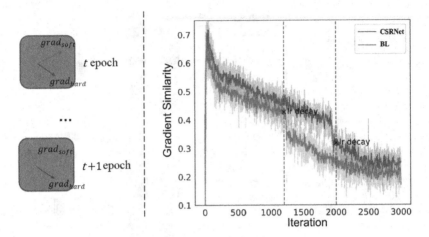

Fig. 3. The gradient divergence effect of L_{soft} and L_{hard} in knowledge distillation. We test the gradient similarities of the second block of CSRNet and BL on the ShaihaiTech Part B dataset. The gradient similarity is calculated as $\frac{1}{2}\left[cos(grad_{soft}, grad_{hard}) + 1\right]$, which lies in the range of $[0,1]$. It is seen that there are intensive gradient similarity changes during training.

4   Experiments

We choose two crowd counting datasets for our experiments – ShanghaiTech [25] and UCF-QNRF [11]. ShanghaiTech is one of the largest crowd counting datasets, which is composed of 1,198 images with 330,165 annotations. The dataset is divided into two parts according to different density distributions. UCF-QNRF is a benchmark dataset for crowd counting released in recent years. It consists of 1,535 challenging images with about 1.25 million annotations. The images in this dataset come with a wider variety of scenes and contain the most diverse set of viewpoints, densities, and lighting variations. As mentioned above, we choose CSRNet and BL as the teacher models, and the corresponding networks after reducing the number of channels as the student models.

4.1   Model Performance

Figure 4 shows the differences in crowd density maps produced by different methods. In the comparison, it can be seen that Repdistiller can present a quite clear density map without the dependency on the capacity of the network. Table 2 lists the experimental results on ShanghaiTech and UCF-QNRF datasets. It shows that efficiency gains can be obtained simply by reducing the number of channels, but the performance of the 1/4BL and 1/4CSRNet models without knowledge transfer is severely degraded. However, through the Repdistiller proposed in this paper, the lightweight model can achieve a large degree of performance recovery. For 1/4CSRNet, Repdistiller reduces the MAE and MSE indicators by 18.7 and 35.4 respectively on the ShanghaiTech Part_A dataset.

For 1/4BL, these two indicators are 26.4 and 44.5. This can intuitively prove that Repdistiller achieves good performance compensation. In fact, whether it is 1/4BL+Repdistiller or 1/4CSRNet+Repdistiller, the performance in terms of MAE and MSE can be very close to the corresponding original teacher network. For example, for 1/4BL+Repdistiller, the difference on ShanghaiTech Part_B is only 0.4 and 0.7. This excellent performance is attributed to the following reasons: (1) The Repdistiller designed in this paper fully compensates for the large capacity gap between the teacher and the student, and realizes the full transfer of knowledge; (2) The well-trained student model has fewer parameters and can effectively alleviate overfitting of crowd counting.

Table 2. Results on ShanghaiTech and UCF-QNRF datasets

Model	ShanghaiTech Part_A		ShanghaiTech Part_B		UCF-QNRF	
	MAE	MSE	MAE	MSE	MAE	MSE
BL	62.8	101.8	7.7	12.7	88.7	154.8
1/4BL	89.6	146.9	12.3	19.8	135.7	157.4
1/4BL+Repdistiller	63.2	102.4	8.1	13.4	90.8	157.1
CSRNet	68.2	115.0	10.6	16.0	109.0	194.0
1/4CSRNet	90.3	153.7	17.9	21.2	147.2	259.5
1/4CSRNet+Repdistiller	71.6	118.3	11.2	17.4	115.7	208.2

Fig. 4. The pictures are from the UCF-QNRF dataset. From left to right: the original images, the density maps estimated by MCNN [25], the density maps estimated by CSRNet, the density maps estimated by the method proposed in this paper, and the real density maps.

4.2 Comparisons with Other Compression Algorithms

Repditiller is a knowledge distillation method. To verify its superiority in compression, this paper also compares Repdistiller with some representative compression algorithms (BRECQ [15], QAT [12], L1Filter [14], PTD [21], FitNet [23], AT [24] and GID [6]). We compress CSRNet using these methods respectively and test the performance of the compressed models on Shanghaitech Part_A. It

Table 3. Performance comparison of different compression algorithms on Shanghaitech Part_A.

	Initial Network		Quantization		Pruning		Knowledge Distillation			
	CSRNet	1/4CSRNet	BRECQ	QAT	L1Filter	PTD	FitNet	AT	GID	Ous
MAE	68.2	90.3	82.5	75.4	84.3	78.4	87.4	75.6	73.2	**71.6**
MSE	115	153.7	138.9	135.2	145.4	129.4	141.9	128.4	123.3	**118.3**

can be seen in Table 3 that our method obtains the minimum MAE (71.6) and MSE (118.3). This illustrates the clear advantage of Repdistiller in compressing crowd counting models. In our opinion, the main reason is that Repdistiller is highly specific to crowd counting, which enables the student model to gain more knowledge from the teacher during the training phase.

4.3   Ablation Study

Table 4. Ablation experiment on 1/4CSRNet (Dataset: Shanghaitech Part_A).

	MAE	MSE
1/4CSRNet	90.3	153.7
RepVGG	80.6	138.2
DBB	75.5	130.7
NAS	72.9	123.4
Repdistiller	**71.6**	**118.3**

This paper further conducts experiments to evaluate the effect of different modules of Repdistiller. The ablation experiments are conducted on 1/4CSRNet, and the results are shown in Table 4. It can be seen that when the backbone network of 1/4CSRNet (VGG) is replaced by a structurally re-parameterized form (RepVGG and DBB), significant performance gains can be obtained. When using DBB, the MAE and MSE are reduced by 14.8 and 23.0. The performance improvement brought by the re-parameterization form does not come at the cost of reducing the inference speed, so it has a good landing value. In addition, when using the re-parameterized structure searched by Repdistiller, the MAE and MSE are reduced by 17.4 and 30.3 compared with 1/4CSRNet, and the performance is better than the fixed re-parameterized structure (RepVGG and DBB). This shows that compared to RepVGG and DBB, the search framework specially designed for crowd counting can find the most suitable re-parameterization structure. In addition, when using the full Repdistiller including the gradient decoupling module, 1/4CSRNet achieves the best performance (MAE 71.6, MSE 118.30).

5 Conclusion

In this paper, we propose a novel knowledge distillation framework incorporating structural re-parameterization methods for crowd counting. In addition, we decouple the gradients of the re-parameterization branches during training to help them focus on soft or hard labels. Our method improves the effectiveness of knowledge distillation and maintains a high speed in the inference stage. Extensive experimental results on ShanghaiTech and UCF-QNRF datasets show that Repdistiller can guarantee the model performance under the condition of greatly compressing the model. Compared with other compression methods, Repdistiller shows the best performance on the crowd counting task.

References

1. Cai, H., Zhu, L., Han, S.: ProxylessNAS: direct neural architecture search on target task and hardware. arXiv preprint arXiv:1812.00332 (2018)
2. Chen, Z., Badrinarayanan, V., Lee, C.Y., Rabinovich, A.: GradNorm: gradient normalization for adaptive loss balancing in deep multitask networks. In: Proceedings of the IEEE International Conference on Machine Learning, pp. 1–10 (2018)
3. Cheng, Y., Wang, D., Zhou, P., Zhang, T.: A survey of model compression and acceleration for deep neural networks. arXiv preprint arXiv:1710.09282 (2017)
4. Chollet, F.: Xception: deep learning with depthwise separable convolutions. In: Proceedings of the IEEE International Conference on Computer Vision and Pattern Recognition, pp. 1251–1258 (2016)
5. Chu, H., Tang, J., Hu, H.: Attention guided feature pyramid network for crowd counting. J. Vis. Commun. Image Represent. **80**, 103319 (2021)
6. Dai, X., et al.: General instance distillation for object detection. In: Proceedings of the IEEE/CVF Conference on Computer Vision and Pattern Recognition, pp. 7842–7851 (2021)
7. Ding, X., Zhang, X., Han, J., Ding, G.: Diverse branch block: building a convolution as an inception-like unit. In: Proceedings of the IEEE/CVF Conference on Computer Vision and Pattern Recognition, pp. 10886–10895 (2021)
8. Ding, X., Zhang, X., Ma, N., Han, J., Ding, G., Sun, J.: RepVGG: making VGG-style convnets great again. In: Proceedings of the IEEE/CVF Conference on Computer Vision and Pattern Recognition, pp. 13733–13742 (2021)
9. Gao, G., Gao, J., Liu, Q., Wang, Q., Wang, Y.: CNN-based density estimation and crowd counting: a survey. arXiv preprint arXiv:2003.12783 (2020)
10. Hinton, G., Vinyals, O., Dean, J.: Distilling the knowledge in a neural network. arXiv preprint arXiv:1503.02531 (2015)
11. Idrees, H., et al.: Composition loss for counting, density map estimation and localization in dense crowds. In: Ferrari, V., Hebert, M., Sminchisescu, C., Weiss, Y. (eds.) ECCV 2018. LNCS, vol. 11206, pp. 544–559. Springer, Cham (2018). https://doi.org/10.1007/978-3-030-01216-8_33
12. Jacob, B., et al.: Quantization and training of neural networks for efficient integer-arithmetic-only inference. In: Proceedings of the IEEE Conference on Computer Vision and Pattern Recognition, pp. 2704–2713 (2018)
13. Lempitsky, V., Zisserman, A.: Learning to count objects in images. In: Advances in Neural Information Processing Systems, vol. 23 (2010)

14. Li, H., Kadav, A., Durdanovic, I., Samet, H., Graf, H.P.: Pruning filters for efficient convnets. arXiv preprint arXiv:1608.08710 (2016)

15. Li, Y., et al.: BRECQ: pushing the limit of post-training quantization by block reconstruction. arXiv preprint arXiv:2102.05426 (2021)

16. Li, Y., Zhang, X., Chen, D.: CSRNet: dilated convolutional neural networks for understanding the highly congested scenes. In: Proceedings of the IEEE Conference on Computer Vision and Pattern Recognition, pp. 1091–1100 (2018)

17. Liu, Y., Cao, J., Hu, W., Ding, J., Li, L.: Cross-architecture knowledge distillation. In: Proceedings of the Asian Conference on Computer Vision, pp. 3396–3411 (2022)

18. Liu, Y., Cao, J., Li, B., Hu, W., Maybank, S.: Learning to explore distillability and sparsability: a joint framework for model compression. IEEE Trans. Pattern Anal. Mach. Intell. 45(3), 3378–3395 (2023)

19. Ma, Z., Wei, X., Hong, X., Gong, Y.: Bayesian loss for crowd count estimation with point supervision. In: Proceedings of the IEEE/CVF International Conference on Computer Vision, pp. 6142–6151 (2019)

20. Mirzadeh, S.I., Farajtabar, M., Li, A., Levine, N., Matsukawa, A., Ghasemzadeh, H.: Improved knowledge distillation via teacher assistant. In: Proceedings of the AAAI Conference on Artificial Intelligence, vol. 34, pp. 5191–5198 (2020)

21. Park, J., No, A.: prune your model before distill it. In: Avidan, S., Brostow, G., Cissé, M., Farinella, G.M., Hassner, T. (eds.) Computer Vision, ECCV 2022. LNCS, vol. 13671, pp. 120–136. Springer, Cham (2022). https://doi.org/10.1007/978-3-031-20083-0_8

22. Phuong, M., Lampert, C.H.: Towards understanding knowledge distillation. In: Proceedings of the International Conference on Machine Learning, pp. 1–10 (2019)

23. Romero, A., Ballas, N., Kahou, S.E., Chassang, A., Gatta, C., Bengio, Y.: FitNets: hints for thin deep nets. arXiv preprint arXiv:1412.6550 (2014)

24. Zagoruyko, S., Komodakis, N.: Paying more attention to attention: improving the performance of convolutional neural networks via attention transfer. arXiv preprint arXiv:1612.03928 (2016)

25. Zhang, Y., Zhou, D., Chen, S., Gao, S., Ma, Y.: Single-image crowd counting via multi-column convolutional neural network. In: Proceedings of the IEEE Conference on Computer Vision and Pattern Recognition, pp. 589–597 (2016)

Multi-depth Fusion Transformer and Batch Piecewise Loss for Visual Sentiment Analysis

Haochun Ou[1], Chunmei Qing[1,3(✉)], Jinglun Cen[1], and Xiangmin Xu[2,3]

[1] School of Electronic and Information Engineering, South China University of Technology, Guangzhou 510641, China
qchm@scut.edu.cn
[2] School of Future Technology, South China University of Technology, Guangzhou 510641, China
[3] Pazhou Lab, Guangzhou 510330, China

Abstract. Visual sentiment analysis is an important research direction in the field of affective computing, which is of great significance in user behavior prediction, visual scene construction, etc. In order to address the issue of low efficiency in high-level semantic perception of CNNs, in this paper, the Transformer architecture is introduced into the task of visual sentiment analysis, proposing the Multi-depth Fusion Transformer (MFT) model that allows nesting different architectures. Additionally, addressing the problem of imbalanced data samples and varying learning difficulties in visual sentiment analysis datasets, this paper proposes a batch piecewise loss suitable for visual sentiment binary classification tasks. Experimental results on multiple datasets demonstrate that the proposed method outperforms existing approaches. Visualization experiments comparing CNN and Transformer-based approaches demonstrate the effectiveness of the proposed method in semantic perception.

Keywords: Multi-depth Fusion Transformer · Batch Piecewise Loss · Sentiment analysis

1 Introduction

Recently, the Transformer-based model has achieved impressive results in visual tasks, surpassing many of the previous achievements of tasks dominated by CNN architectures. The use of attention mechanisms with CNN has achieved great results in visual sentiment analysis, greatly promoting visual sentiment analysis development. However, compared with Transformer, single-layer convolution in CNN cannot capture long-distance dependencies, and the receptive field is limited [1, 2]. The method of using deeper networks to obtain a larger receptive field in CNN leads to lower computational and representation space efficiency, making it difficult to capture semantic spatial correlation.

In addition, the study of Geirhos et al. [3] pointed out that CNN models trained on ImageNet tend to classify images by texture, which is different from human visual habits. A recent Princeton study suggests that Transformer is better at judging shapes

© The Author(s), under exclusive license to Springer Nature Singapore Pte Ltd. 2024
Q. Liu et al. (Eds.): PRCV 2023, LNCS 14434, pp. 395–406, 2024.
https://doi.org/10.1007/978-981-99-8549-4_33

and more similar to humans in semantic perception mechanism than CNN [4]. Models that use shape for classification are more robust. Under the same training conditions on the same dataset, Transformer is more inclined to classify images by shape and performs better than CNN [4]. Visual sentiment analysis has always been looking for a more suitable learning strategy or classification method. Transformer, which has a more similar perception mechanism to humans, is theoretically more suitable for the visual sentiment analysis task.

Therefore, this paper proposes a multi-depth fusion Transformer (MFT) model for the visual sentiment analysis task, which can address the issue of visual attention being selectively influenced by features at different levels. At the same time, a Batch Piece-wise Loss (BPLoss) is proposed to solve the problem of imbalanced datasets in visual sentiment tasks. During the training phase, the BPLoss is used to train the MFT model. The main contributions of this paper are summarized as follows:

- The Transformer model is introduced into the visual sentiment analysis task, and a suitable framework MFT is proposed according to the characteristics of the visual sentiment analysis task.
- A new loss function, BPLoss, is proposed, which considers the imbalance of positive and negative samples in the dataset and adopts different penalty strategies for the loss based on the proportion of samples in the current batch, thereby further improving the performance of the model.
- The effectiveness of the proposed MFT model and BPLoss is verified on 7 commonly used datasets, and the proposed method is visually compared with the representative CNN method on the EmotionROI and EMOd datasets to verify the proposed method in the effectiveness of semantic perception.

2 Related Work

2.1 Visual Sentiment Analysis

Research indicates that different regions in an image contribute differently to the expression of emotions. The emotional changes experienced by viewers while observing an image are often triggered by specific regions of the image [5–7]. Local region-based methods have achieved impressive performance gains on many visual sentiment datasets. Yang et al. [8] proposed WSCNet, which incorporates image local features into image sentiment analysis by weighting the final output features of the network through a cross-spatial pooling module. Similarly, Yang et al. [6] proposed the concept of emotion region, generated emotion score of local semantic information through sub-network, combined global and local information with CNN, and explicitly sensed local regions related to emotion in images only with single label data. Song et al. [9] integrated the concept of visual attention into image sentiment analysis, used a multi-layer neural network to model the significance distribution of each region of the image, and inferred the emotion category of the image in the region with the largest amount of information. Yadav et al. [10], based on NASNet, used the residual attention module to learn important areas related to emotion in images. Wu et al. [11] mentioned that the current work did not consider whether images without obvious local semantic objects are suitable for using local modules, therefore, they used target detection modules to determine whether to use local modules. Based on the work of Faster R-CNN [12], Rao et al. [7] further improved the model performance by combining local regions related to sentiment and features of different levels. Ou et al. [13] proposed MCPNet for object perception at different scales and multi-level emotion representation, introduced multi-scale adaptive context features into visual sentiment analysis tasks, and performed visual comparisons at different scales in the EmotionROI and EMOd. Xu et al. [27] proposed MDAN to disassemble the multi-level information, establishing affective semantic mapping to learn the level-wise discrimination, and classify emotions from both local and global.

2.2 Vision Transformer

Although MCPNet introduces the context features that can capture long-distance dependencies, it is only used in the output features of the backbone in the three stages, failing to fundamentally solve the problems of low semantic perception efficiency of CNN high-level and high cost of obtaining long-distance dependencies.

The core module of the Transformer is the multi-head self-attention module, which has the long-distance correlation feature to make it learn local correlation information from the shallow layer to the deep layer in the global perspective. At the same time, the multi-head self-attention mechanism ensures that the network can pay attention to various local dependencies. DETR [14], as a target detection framework, was the first model to successfully introduce Transformer into visual tasks. ViT [15] popularized Transformer to many computer vision tasks and achieved SOTA (State-of-the-art) in many fields such as object classification, object detection, semantic segmentation, and so on. Raghu et al. discussed the differences between traditional CNN and Transformer in their research and thought that compared with CNN. ViT can retain more spatial information than ResNet [16]. Therefore, ViT performs better in terms of performance. On the other hand, the proposed method by ST (Spatial Transformer) [17] integrates many successful experiences from CNNs into Transformers at a deeper level. Unlike ViT, which uniformly sizes features from different levels, the ST model structure resembles ResNet.

3 Multi-depth Fusion Transformer Framework Design

3.1 Overall Framework Process

The proposed MFT (Multi-Depth Fusion Transformer) architecture is shown in Fig. 1. The whole architecture can be divided into three parts: the Tokenizer module which maps the image data into sequence features, the multi-depth fusion encoder for learning sequence features, and the MLP module for classification. The frame takes the original image $I \in \mathbb{R}^{H \times W \times 3}$ as the input. Let the batch size of the input in the training stage be n, and $I_1, I_2, \cdots, I_n$ represents the image data of the current batch. A single image is used as an example to describe the execution process of MFT. The Transformer architecture can handle sequential data, i.e., two-dimensional matrices [num_token, token_dim]. Therefore, the image I is first input into the Tokenizer module to convert I into sequence features $X_p \in \mathbb{R}^{N \times d}$, where N represents the length of the sequence and d represents the number of channels of the features.

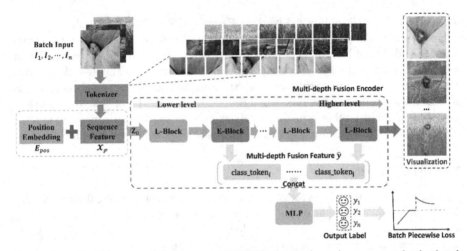

Fig. 1. Overall architecture of the proposed MFT. Firstly, the input images are tokenized and embedded into sequence features add with position embedding. Next, the multi-depth fusion encoder module employs a sequence of L-Blocks and E-Blocks to encapsulate various architectures for feature extraction purposes. Finally, the batch piecewise loss is proposed dealing with the imbalance of positive and negative samples in visual sentiment classification tasks. Specially, the model has another branch for visualization.

$$X_p = \text{Tokenizer}(I) \tag{1}$$

X_p can be entered directly into the Transformer model. However, since the relative position information between image blocks will be lost after the image is expanded into a sequence. To introduce position information into the sequence feature, MFT adds position-coding to the sequence feature X_p, just like most current studies [14, 17, 18]. By adding X_p and the learnable matrix $\mathbf{E}_{pos} \in \mathbb{R}^{N \times d}$ representing position information, the sequence feature $\mathbf{Z}_0$ of the input encoder can be obtained.

$$\mathbf{Z}_0 = X_p + \mathbf{E}_{pos} \tag{2}$$

$\mathbf{Z}_0$ represents the image sequence feature with location information. Since attention is affected by features at different levels, a multi-depth fusion encoder is used to learn the input $\mathbf{Z}_0$. The multi-depth fusion encoder is composed of multiple blocks, and different depth information is extracted from different blocks to obtain multi-depth fusion features $\hat{y}$.

$$\hat{y} = \text{Encoding}(\mathbf{Z}_0) \tag{3}$$

The classifier adopts the MLP module which is commonly used in the Transformer model. The sentiment classification result y of the image I is obtained by passing the inputs representing different depth information y through the MLP (Multi-Layer Perceptron).

$$y = \text{MLP}(\hat{y}) \tag{4}$$

To address the issue of imbalanced data samples and varying learning difficulties, we propose BPLoss (Batch-wise Segmented Loss) in the MFT architecture. In Fig. 1, $y_1, y_2, \cdots, y_n$ correspond to the classification result of the input images $I_1, I_2, \cdots, I_n$. The proportion of positive and negative samples in the batch data is obtained from $L_1, L_2, \cdots, L_n$ which correspond to the sentiment labels of the input images $I_1, I_2, \cdots, I_n$. In the training phase, different penalty strategies are adaptively adapted according to the proportion of positive and negative samples in the current batch to improve the performance of the model.

3.2 Tokenizer Module

The Transformer module requires the input of the sequence feature, so the input image needs to be converted first. The function of the Tokenizer module is to convert the input image I into sequence feature X_p. As shown in Fig. 2, Tokenizer has three steps in total. Firstly, the input image I is converted into a feature graph $X_p^{(1)} \in \mathbb{R}^{\frac{H}{r} \times \frac{W}{r} \times d}$ of a specified size through the linear projection network and then expanded to obtain the sequence feature $X_p^{(2)} \in \mathbb{R}^{\frac{H \times W}{r^2} \times d}$, where $\frac{H \times W}{r^2}$ is the sequence length. The third step is to add [class]token to the sequence of features according to the backbone.

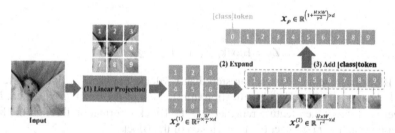

Fig. 2. An illustration of the Tokenizer module. The input image is linearly projected, then expanded, and finally the [class]token is added to the first position.

In this paper, two linear projection networks are used, namely, a single convolution layer and ResNet50. The step size and convolution kernel size adopted by a single convolution layer are both r, which is equivalent to dividing the image into $\frac{H \times W}{r^2}$ blocks, and then converting each image block into features. Therefore, each position of the obtained sequence feature corresponds to the information of an image block. Although CNN has low computational efficiency in high-level semantic information, it is very effective in low-level feature perception. The transformer is good at capturing semantic information in images. The visual sentiment induction of images is affected by both low-level features and high-level semantic features. Using ResNet50 in linear projection networks can be a combination of the advantages of CNN and Transformer. Subsequent experiments will prove that under the same encoder structure, using ResNet50 as a linear projection network is more effective than using a single convolutional layer.

3.3 Multi-depth Fusion Encoder

The multi-depth fusion encoder is the core of MFT and the main part of sequence feature learning. As shown in Fig. 3, each layer in the multi-depth fusion encoder is divided into two stages, one is L-Block (Learning block) and the other is E-Block (Extraction block). Each Block contains two main sub-modules: MSA (Multi-head self-attention) and MLP. Suppose the multi-depth fusion encoder has H layers of Block, let $\mathbf{Z}_{i-1}$ be the input of the current Block, and the output is $\mathbf{Z}_i$, the operation of a Block is as follows.

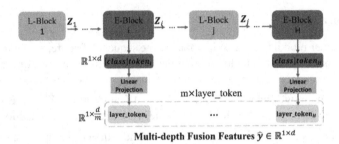

Fig. 3. Structure diagram of multi-depth fusion encoder.

$$\mathbf{Z}_i = \text{Block}_i(\mathbf{Z}_{i-1}), \, 1 \le i \le H \tag{5}$$

E-Block has one more extraction step than L-Block. For each E-Block, the feature [class]token representing the classification information of the Block is extracted. MFT will get the [class]token feature representing the Block information through the linear projection function $f(\cdot)$ to get the layer_token of the Block.

$$\text{layer\_token}_i = f([\text{class}]\text{token}_i) \tag{6}$$

$$\hat{\mathbf{y}} = [\text{layer\_token}_i, \text{layer\_token}_{i+1}, \cdots , \text{layer\_token}_H] \tag{7}$$

Here, $f(\cdot)$ uses a single fully connected layer, and $f(\cdot)$ of different depths is independent of each other. The depth selection of E-block is further discussed in the experimental section. Finally, MFT merges multiple level features of layer_token to obtain $\hat{\mathbf{y}}$. Here the size of the feature sent into the classifier by the MFT and its corresponding backbone is the same. Assuming that the size of the feature input to the classifier using the selected Transformer architecture is $\mathbb{R}^{1 \times D}$, then $\hat{\mathbf{y}} \in \mathbb{R}^{1 \times D}$. Assuming there are m E-Blocks and layer_token $\in \mathbb{R}^{1 \times \frac{D}{m}}$.

3.4 Batch Piecewise Loss

Most of the commonly used datasets of visual sentiment analysis are imbalance distributed. Inspired by Focal Loss [19], we proposed the BPLoss (Batch Piecewise Loss)

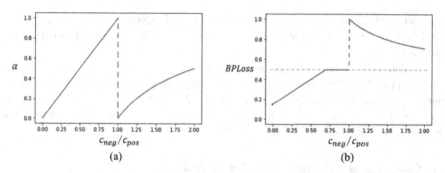

Fig. 4. α and BPLoss with c_{neg}/c_{pos} change value curve.

and calculated different Loss according to the number of positive and negative samples in each batch in the training stage.

$$Loss(x) = \begin{cases} (\alpha + \beta) \times L(x), & c_{pos} \geq c_{neg} \text{ and } \alpha = c_{neg}/c_{pos} \\ (L(x))^{\alpha}, & c_{pos} < c_{neg} \text{ and } \alpha = 1 - c_{pos}/c_{neg} \end{cases} \tag{8}$$

c_{neg} and c_{pos} are the number of negative and positive samples in batch respectively. α is the adaptive weight of BPLoss. When $c_{pos} \geq c_{neg}$, let $\alpha = c_{neg}/c_{pos}$, weaken the loss mainly uses the form $\alpha \times L(x)$. And in the case of $c_{pos} \geq c_{neg}$, the bias coefficient β is added. The main function of β is to increase the value of c_{neg}/c_{pos}, so as not to penalize loss too much. $\beta = 0.3$, here. And in order to prevent extreme cases (for example, c_{neg} and c_{pos} values are very close, resulting in $\alpha + \beta > 1$), BPLoss also adopted a partition measure, forcing $\alpha + \beta = 1$ when $\alpha + \beta > 1$. However, when the batch contains a large number of negative samples, the strategy is to increase the loss of the current batch and the nonlinear calculation of loss. Because $L(x) < 1$, so we use $(L(x))^{\alpha}$, and let $\alpha = 1 - c_{pos}/c_{neg}$.

As the sum of c_{neg} and c_{pos} is the size of the current batch and is a constant value n, the value curve of α is shown in Fig. 4(a). In this paper, the cross-entropy loss commonly used in visual sentiment analysis is adopted as $L(x)$, and its value is set to be 0.5. The curve of BPLoss with varying c_{neg}/c_{pos} values is shown in Fig. 4(b). Due to the bias coefficient $\beta = 0.3$, the black dot in Fig. 4(b) represents the minimum value of BPLoss (when c_{neg} is 0). The orange dotted line represents the value of BPLoss without penalty, which is $L(x)$. Due to β and partition measures, as shown in Fig. 4(b), when $0.7 < c_{neg}/c_{pos} < 1$, $\alpha + \beta = 1$, $BPLoss = L(x)$, the value curve of BPLoss will appear a segment parallel to the X-axis.

4 Experiments

This section uses three representative Transformer models as backbone, namely ViT [15], Hybrid [15], and ST [17], and pre-trained on ImageNet. When the backbone is ViT or Hybrid, r = 16; when the backbone is ST, r = 4. The encoder structures of ViT and Hybrid are the same, but the Tokenizer is different. ViT uses a single convolutional layer, and Hybrid uses ResNet50. The parameters of ResNet50 used in this paper are consistent with those in the paper [15]. ST and ViT have different encoders but use the same Tokenizer.

The datasets and basic settings used in this paper refer to the experimental settings of MCPNet [13]. As a form of data augmentation, images input into the MFT framework are randomly flipped horizontally and cropped to size 224 × 224 or 384 × 384 using center cropping to reduce overfitting. All experiments are modeled in the Pytorch framework and performed on a GeForce GTX1080Ti GPU.

4.1 Comparisons with State-of-the-Art Methods

For fair comparison, this paper adopts a 5-fold cross-validation strategy to evaluate current advanced methods, are shown in Table 1 and Table 2. Unlike the other methods, the proposed method is based on the Transformer structure as shown in Table 1 on five commonly used datasets. Compared with the methods based on CNN such as MCPNet, the proposed MFT is 3.1% higher than MCPNet method in the FI dataset and 2.96% higher than MCPNet in the EmotionROI dataset. In ArtPhoto, Abstract and IAPSsubset three small datasets, the performance of the proposed method is better than that of MCPNet. At the same time, we also compares on the two binary small datasets Twitter I, Twitter II in Table 2. The three subsets of Twitter I and the Twitter II also demonstrate the sophistication of this paper's approach. The three subsets of Twitter I represent different classification difficulties. However, MFT has a further improvement over MCPNet in these three datasets, and the classification accuracy score in Twitter I_5 is more than 90%.

Table 1. Classification results of different models on five different datasets

Model	Acc				
	FI	EmotionROI	ArtPhoto	Abstract	IAPSsubset
VGG-16 [20]	73.95	72.49	70.48	65.88	87.20
ResNet101 [21]	75.76	73.92	71.08	66.64	88.15
PCNN [22]	73.59	74.06	71.47	70.26	88.65
Rao(b) [23]	79.54	78.99	74.83	71.96	90.53
Zhu [24]	84.26	80.52	75.50	73.88	91.38
AR [6]	86.35	81.26	74.80	76.03	92.39
Rao(c) [7]	87.51	82.94	78.36	77.28	93.66
WSCNet [8]	86.74	82.15	80.38	77.84	94.61
PDANet [26]	87.25	82.27	80.27	78.24	95.18
MCPNet [13]	90.31	85.10	79.24	78.14	94.82
MDAN [27]	91.08	84.62	91.50	83.80	96.37
MFT + BPLoss	93.41	88.06	80.21	80.58	95.75

Table 2. Classification results of different models on Twitter I and Twitter II

Model	Acc			Twitter II
	Twitter I_5	Twitter I_4	Twitter I_3	
PCNN [22]	82.54	76.52	76.36	77.68
VGG-16 [20]	83.44	78.67	75.49	71.79
AR [6]	88.65	85.10	81.06	80.48
RADLNet [10]	89.10	83.20	81.30	81.20
GMEI&LRMSI [11]	89.50	86.97	81.65	80.97
MCPNet [13]	89.77	86.57	83.88	81.19
MFT + BPLoss	91.48	87.72	84.92	83.93

4.2 Ablation Experiments

The ablation experimental results of the combination of MFT and BPLoss are shown in Table 3. Simultaneous application of MFT and BPLoss has obvious improvement in the three backbones, among which there is a nearly one-point improvement in ViT-384, which proves the effectiveness of the method our proposed. Since the goal of this paper is to propose a general method, in addition to the selection of depth, other model parameters are not adjusted according to different frameworks, but unified parameters are used. Therefore, the proposed method will have different effects on different backbones and the BPLoss application has obvious improvement in classification performance in MFT, with an improvement of 0.37% in MFT.

Table 3. Module ablation experiment with different backbone on FI

Module		Backbone					
BPCLoss	MFT	ViT		Hybrid		ST	
		224	384	224	384	224	384
		91.40	91.78	91.59	92.58	92.06	93.04
$\checkmark^a$		91.56	91.80	91.79	92.54	91.98	93.01
$\checkmark$		91.70	92.17	91.99	92.78	92.12	93.15
	$\checkmark$	91.86	92.37	92.19	92.97	92.24	93.22
$\checkmark$	$\checkmark$	91.95	92.74	92.35	93.17	92.37	93.41

[a] indicates the use of BPLoss without β

4.3 Visualization

Here, MCPNet [13], with the best classification effect in the current visual sentiment analysis task, is selected as the representative model based on the CNN method for comparison. MCPNet is adopted to visualize the multi-cue sentiment features output by the model in the way of a class activation map in the visualization. The visualization method adopted by Transformer is different from that of CNN. The commonly used method is to combine multiple self-attention modules. Here, the Transformer visualization method proposed by Chefer et al. [25] is adopted. The input size of MCPNet is 448 × 448 and that of MFT is 224 × 224. Although the input size of MFT is half that of MCPNet, subsequent visualizations show that MFT performs well even at smaller input sizes. Both MCPNet and MFT can capture long-distance dependencies in images. However, MCP-Net only introduced a self-attentional module in the last three stages of Res-Net101, whereas the multi-head self-attentional module is the main module of MFT. Therefore, MFT has stronger global semantic awareness.

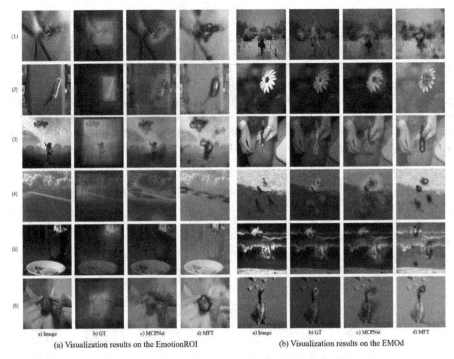

(a) Visualization results on the EmotionROI   (b) Visualization results on the EMOd

Fig. 5. Visual comparisons of GT, MCPNet and MFT.

Capturing the long-distance dependence is helpful to understand the semantic information of the image. As can be seen from Fig. 5, MFT shows excellent semantic awareness in Fig. 5(a-3), Fig. 5(a-4), Fig. 5(b-3), Fig. 5(b-5), etc. In Fig. 5(a-3), two main parts attract viewers: the balloon and the human. Compared with MPCNet, MFT can capture these two areas more clearly. Compared with other contents in Fig. 5(a-4), although the plane at the edge of the image is small, it can directly attract our attention. MPCNet focuses on the aircraft and the entire flight path, which does not fit the label well, but MFT can accurately focus on the aircraft. Figure 5(b-3) shows the scene of injection. In this example, MPCNet does not perform well, but MFT focuses on the needle area, which is similar to our human attention habit.

5   Conclusions

In this paper, we discuss the effect of Transformer architecture in visual sentiment analysis and propose MFT to combine different level features under long-distance dependence characteristics for analysis according to task characteristics. Meanwhile, aiming at the problem of unbalanced dataset of visual sentiment analysis, the BPLoss suitable for visual sentiment binary classification task is proposed. In the training stage, BPLoss adopts different punishment strategies according to the proportion of positive and negative samples in the current batch data to improve the performance of the model. Finally, the effectiveness of the proposed method in semantic perception is verified by visual comparison with the representative method based on CNN.

Acknowledgement. This work is partially supported by the following grants: National Natural Science Foundation of China (61972163, U1801262), Natural Science Foundation of Guangdong Province (2022A1515011555, 2023A1515012568), Guangdong Provincial Key Laboratory of Human Digital Twin (2022B1212010004) and Pazhou Lab, Guangzhou, 510330, China.

References

1. Wu, B., Xu, C., Dai, X., et al.: Visual transformers: Token-based image representation and processing for computer vision. arXiv preprint arXiv:2006.03677 (2020)
2. Zheng, S., Lu, J., Zhao, H., et al.: Rethinking semantic segmentation from a sequence-to-sequence perspective with transformers. In: Proceedings of the IEEE/CVF Conference on Computer Vision and Pattern Recognition, USA, pp. 6881–6890. IEEE (2021)
3. Geirhos, R., Rubisch, P., Michaelis, C., et al.: ImageNet-trained CNNs are biased towards texture; increasing shape bias improves accuracy and robustness. In: International Conference on Learning Representations (2018)
4. Tuli, S., Dasgupta, I., Grant, E., et al.: Are convolutional neural networks or transformers more like human vision? In: Proceedings of the Annual Meeting of the Cognitive Science Society, vol. 43 (2021)
5. You, Q., Jin, H., Luo, J.: Visual sentiment analysis by attending on local image regions. In: Proceedings of the Thirty-First AAAI Conference on Artificial Intelligence, pp. 231–237. AAAI Press, USA (2017)
6. Yang, J., She, D., Sun, M., et al.: Visual sentiment prediction based on automatic discovery of affective regions. IEEE Trans. Multimedia **20**(9), 2513–2525 (2018)
7. Rao, T., Li, X., Zhang, H., et al.: Multi-level region-based convolutional neural network for image emotion classification. Neurocomputing **333**(6), 429–439 (2019)
8. Yang, J., She, D., Lai, Y.K., et al.: Weakly supervised coupled networks for visual sentiment analysis. In: Proceedings of the IEEE Conference on Computer Vision and Pattern Recognition, pp. 7584–7592. IEEE, USA (2018)

9. Song, K., Yao, T., Ling, Q., et al.: Boosting image sentiment analysis with visual attention. Neurocomputing **312**, 218–228 (2018)

10. Yadav, A., Vishwakarma, D.K.: A deep learning architecture of RA-DLNet for visual sentiment analysis. Multimedia Syst. **26**(4), 431–451 (2020)

11. Wu, L., Qi, M., Jian, M., et al.: Visual sentiment analysis by combining global and local information. Neural. Process. Lett. **51**(3), 2063–2075 (2020)

12. Ren, S., He, K., Girshick, R., et al.: Faster R-CNN: towards real-time object detection with region proposal networks. IEEE Trans. Pattern Anal. Mach. Intell. **39**(6), 1137–1149 (2017)

13. Ou, H., Qing, C., Xu, X., et al.: Multi-level context pyramid network for visual sentiment analysis. Sensors **21**(6), 2136 (2021)

14. Carion, N., Massa, F., Synnaeve, G., Usunier, N., Kirillov, A., Zagoruyko, S.: End-to-end object detection with transformers. In: Vedaldi, A., Bischof, H., Brox, T., Frahm, J.-M. (eds.) ECCV 2020. LNCS, vol. 12346, pp. 213–229. Springer, Cham (2020). https://doi.org/10.1007/978-3-030-58452-8_13

15. Dosovitskiy, A., Beyer, L., Kolesnikov, A., et al.: An image is worth 16 × 16 words: transformers for image recognition at scale. In: Proceedings of the International Conference on Learning Representations, ICLR 2021 (2021)

16. Raghu, M., Unterthiner, T., Kornblith, S., et al.: Do vision transformers see like convolutional neural networks? In: Advances in Neural Information Processing Systems. MIT Press (2021)

17. Liu, Z., Lin, Y., Cao, Y., et al.: Swin transformer: hierarchical vision transformer using shifted windows. In: Proceedings of the IEEE/CVF International Conference on Computer Vision, pp. 10012–10022. IEEE, USA (2021)

18. Shaw, P., Uszkoreit, J., Vaswani, A.: Self-attention with relative position representations. In: Proceedings of the 2018 Conference of the North American Chapter of the Association for Computational Linguistics: Human Language Technologies, USA, pp. 464–468 (2018)

19. Lin, T.Y., Goyal, P., Girshick, R., et al.: Focal loss for dense object detection. In: Proceedings of the IEEE International Conference on Computer Vision, pp. 2980–2988. IEEE, USA (2017)

20. Simonyan, K., Zisserman, A.: Very deep convolutional networks for large-scale image recognition. In: Proceedings of the International Conference on Learning Representations, ICLR, USA (2015)

21. He, K., Zhang, X., Ren, S., et al.: Deep residual learning for image recognition. In: Proceedings of the IEEE Conference on Computer Vision and Pattern Recognition, pp. 770–778. IEEE, USA (2016)

22. You, Q., Luo, J., Jin, H., et al.: Robust image sentiment analysis using progressively trained and domain transferred deep networks. In: Proceedings of the Twenty-Ninth AAAI Conference on Artificial Intelligence, pp. 381–388. AAAI, USA (2015)

23. Rao, T., Li, X., Xu, M.: Learning multi-level deep representations for image emotion classification. Neural. Process. Lett. **51**(3), 2043–2061 (2020)

24. Zhu, X., Li, L., Zhang, W., et al.: Dependency exploitation: a unified CNN-RNN approach for visual emotion recognition. In: Proceedings of the Twenty-Sixth International Joint Conference on Artificial Intelligence, IJCAI 2017, Australia, pp. 3595–3601 (2017)

25. Chefer, H., Gur, S., Wolf, L.: Transformer interpretability beyond attention visualization. In: Proceedings of the IEEE/CVF Conference on Computer Vision and Pattern Recognition, pp. 782–791. IEEE, USA (2021)

26. Zhao, S., Jia, Z., Chen, H., et al.: PDANet: polarity-consistent deep attention network for fine-grained visual emotion regression. In: Proceedings of the 27th ACM International Conference on Multimedia, pp. 192–201 (2019)

27. Xu, L., Wang, Z., Wu, B., et al.: MDAN: multi-level dependent attention network for visual emotion analysis. In: Proceedings of the IEEE/CVF Conference on Computer Vision and Pattern Recognition, pp. 9479–9488 (2022)

Expanding the Horizons: Exploring Further Steps in Open-Vocabulary Segmentation

Xihua Wang[1], Lei Ji[2], Kun Yan[2,3], Yuchong Sun[1], and Ruihua Song[1(✉)]

[1] Gaoling School of Artificial Intelligence, Renmin University of China, Beijing, China
{xihuaw,ycsun}@ruc.edu.cn, songruihua_bloon@outlook.com
[2] Microsoft Research Asia, Beijing, China
leiji@microsoft.com
[3] SKLSDE Lab, Beihang University, Beijing, China
kunyan@buaa.edu.cn

Abstract. The open vocabulary segmentation (OVS) task has gained significant attention due to the challenges posed by both segmentation and open vocabulary classification, which involves recognizing arbitrary categories. Recent studies have leveraged pretrained Vision-Language models (VLMs) as a new paradigm for addressing this problem, leading to notable achievements. However, our analysis reveals that these methods are not yet fully satisfactory. In this paper, we empirically analyze the key challenges in four main categories: segmentation, dataset, reasoning and recognition. Surprisingly, we observe that the current research focus in OVS primarily revolves around recognition issues, while others remain relatively unexplored. Motivated by these findings, we propose preliminary approaches to address the top three identified issues by integrating advanced models and making adjustments to existing segmentation models. Experimental results demonstrate the promising performance gains achieved by our proposed methods on the OVS benchmark.

Keywords: Open-vocabulary · Segmentation · Vision-language models

1 Introduction

The open vocabulary segmentation (OVS) task involves two challenges: precise image segmentation and accurate recognition of segmented regions with open vocabulary labels. While traditional vision models have made notable advancements in closed-set scenarios for image segmentation, they encounter difficulties in recognizing unrestricted semantics [1,28,36]. Recently, there has been significant development in large-scale vision-language models (VLMs). These models, which are pretrained on massive image-text pairs, demonstrate a remarkable understanding of open semantics and possess strong zero-shot capabilities and robustness for image level tasks [9,22]. Furthermore, recent studies have shown

© The Author(s), under exclusive license to Springer Nature Singapore Pte Ltd. 2024
Q. Liu et al. (Eds.): PRCV 2023, LNCS 14434, pp. 407–419, 2024.
https://doi.org/10.1007/978-981-99-8549-4_34

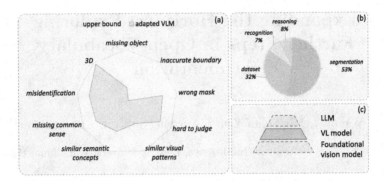

Fig. 1. Statistic of the 9 issues. (a) demonstrates the capability of existing model for each issue; (b) presents the frequency of 4 main issues; (c) shows collaborative integration of three foundational models has the potential to address the above issues and represents a promising future research direction.

that these VLMs can effectively transfer their image-level capabilities to region-level even pixel-level tasks [23,39]. This makes them efficient and promising solutions for open vocabulary segmentation tasks, as evidenced by their state-of-the-art performances on various benchmarks [14,34].

While VLM-based OVS methods have achieved superior performance on various OVS benchmarks, the upper limit of OVS still remains far from being fully realized, as depicted in Fig. 1. Recent works focus on improving the transfer ability of VLMs to OVS [7,14,30,32,34], but the performance bottleneck of VLM-based OVS methods remains unknown and has not been thoroughly discussed. In this paper, we offer a comprehensive analysis of the performance limitations encountered by the current state-of-the-art models when applied to OVS benchmarks. Furthermore, we present some preliminary attempts for these prominent limitations, to provide research insights for further exploring this problem.

At first, we conducted a comprehensive case analysis of the state-of-the-art solutions for OVS [14,30,34] and categorized the performance bottlenecks into the following types based on their frequency of occurrence: 1) segmentation issues (including missing elements, wrong mask, and inaccurate boundary), which require further improvement of the quality of image segmentation; 2) dataset issues (including similar visual patterns and concepts), which suggest that some annotations in the dataset are confusing; 3) reasoning issues, which result in non-common sense predictions; 4) recognition issues, which involve mis-classifying a segmentation into wrong object label or the requirement of 3D modeling.

We unexpectedly found that the current research focus in open vocabulary segmentation is mainly on *recognition issues* [14,30,34], which is only ranked fourth in terms of issue occurrence. However, there are three more frequent issues, namely segmentation issues, dataset issues, and reasoning issues. Recently, the progress of foundational vision models (FVMs), such as SAM [12], has provided more imagination space for the entire vision community. We found that combining new foundational vision models and making targeted adjustments to current OVS-adapted VL models for *segmentation issue* can bring stable and promising gains across the entire benchmark. Additionally, with the advancement of large language models (LLMs), their excellent common sense and reasoning abilities have been proven in various language and multi-modality tasks [15,21,26,37]. We found that the common sense reasoning ability of LLMs corresponds well to the *reasoning issues*. Therefore, we made targeted attempts to address common sense issues by combining LLMs with VLM-based OVS methods.

Our main contributions are three-fold: 1) We conduct a comprehensive analysis to show the performance bottlenecks of the current state-of-the-art models on the OVS benchmark. 2) We propose preliminary approaches to address the top 3 issues we found by integrating advanced models (LLMs and FVMs) and adjustments to current segmentation models. 3) Our methods show promising gains on the OVS benchmark.

2   Background

2.1   Open Vocabulary Semantic Segmentation

Different from traditional semantic segmentation, OVS poses a challenge by transitioning from closed sets to open sets, requiring models to handle open semantics effectively. Previous studies have explored approaches to facilitate open semantic understanding in traditional models by replacing labels with text embeddings [1,13,28,36]. These methods aimed to minimize reliance on additional annotated data and model design. Recent research has benefited from access to large-scale paired image-text data [9,22,25]. Many methods leverage weak supervision from image-text pairs to enable models to learn both segmentation and open semantic understanding capabilities. These approaches include methods with grounding loss and language supervision [8,31], omni-framework designs [35,41], Group-ViT based methods [16,18,29], and VLM-based methods [20].

Furthermore, an increasing number of recent studies directly utilize off-the-shelf VLMs pre-trained on large-scale image-text pairs [3,4,6,7,10,14,17,30,33,34]. Some works modify the attention weights in CLIP to focus on specific regions of an image [7,34], fine-tune CLIP to adapt to mask-centric images [14], add additional training modules to adapt to OVS [3,4,6,17], or utilize intermediate features from generative models to represent specific image regions [30]. By incorporating adaptive design with frozen or fine-tuned VLMs and training on a small amount of annotated segmentation data, these models achieve both strong semantic understanding and segmentation capabilities.

The challenge faced by VLM-based OVS methods lies in effectively transferring image-level semantic understanding capabilities to fine-grained and dense segmentation tasks. In our work, we provide an Issue Analysis perspective on this challenge. While most VLM-based OVS methods primarily focus on the Recognition Issue, which involves extracting accurate semantic embeddings for each mask area to unleash the capabilities of VLMs, there is insufficient attention given to other critical issues. Specifically, the Segmentation Issue aims to achieve higher segmentation quality, and the Reasoning Issue focuses on ensuring overall coherence and reasonableness of classification results. These limitations can be considered inherent to VLMs. By addressing these additional issues in our analysis, we aim to enhance the performance of VLM-based OVS methods.

2.2   Foundational Models Pretrained on Large-Scale Data

Foundational VLMs: As discussed above, VLMs [9,22] have demonstrated robust transfer capabilities and outstanding performance across various downstream tasks, including OVS. However, it is still worth noting that VLMs may have inherent limitations that prevent them from effectively addressing all the issues proposed in our work. **Foundational Language Models (LLMs):** LLMs have also gained attention due to their strong generalization and zero-shot abilities in language [26,37] and multimodal tasks [15,21]. However, the exploration of leveraging LLMs to enhance OVS is still limited. Considering the intricate nature of reasoning, we investigate the potential of leveraging existing LLMs' reasoning capabilities to address Reasoning Issue. **Foundational Vision Models (FVMs):** In recent times, there has been some exploration of foundational models of Computer Vision, particularly in the context of image segmentation [12,27]. Community efforts have led to significant advancements in training data, resulting in substantial improvements in the image segmentation capabilities of foundational models [2,11,12,38,42]. Nevertheless, the challenge of effectively combining these powerful FVMs with cross-modal understanding capabilities that possess the necessary strength for OVS remains unresolved.

The remarkable capabilities of foundational models instill confidence in the potential of an approach based on them for future research. Accordingly, this study explores the collaborative integration of these three FMs to address the aforementioned issues, thereby uncovering new possibilities for the OVS framework and pushing the performance boundaries of OVS benchmark further.

3   Issue Analysis in OVS

3.1   Analysis Settings

Issue Definition. We conducted a preliminary case study on OVS benchmark to examine the prediction results of state-of-the-art models. The objective of this study was to categorize the erroneous results into distinct and complete types. We identified 9 issue types across the models, which were further merged into 4 major categories: **segmentation issues, dataset issues, reasoning issues, recognition issues.** We illustrate the various issue types in details in Table 1. **Analysis Pipeline.** A random subset of each benchmark validation set was sampled for analysis. Given an input images and a state-of-the-art model, a semantic segmentation prediction result on input was generated. Each pixel in

Table 1. The detailed definition of the 4 categories and 9 issues.

Categories	Issues	Details
Segmentation	missing object	The object is not segmented out separately.
	wrong mask	The mask is not a valid object itself or the region is part of another object.
	inaccurate boundary	The segmentation boundary of the object is inaccurate, overlapped but not strictly aligned with groundtruth region, either too large or too small.
Dataset	similar visual pattern	The object predicted is visually similar to the groundtruth object. It is hard to judge which one is correct, e.g., hook or tower ring.
	similar concepts	The object predicted and the groundtruth is similar to each other semantically, e.g., chair v.s. armchair.
	hard to judge	Given the image, it is hard to judge whether the object is the predicted or the groundtruth label, both of which make sense.
Recognition	misidentification	The segmentation of the object is correct but misidentified as a wrong object.
	single-view 3D	To recognize the object accurately, extra single-view 3D data is required such as an extra depth channel.
Reasoning	missing common sense	The object should not appear in the current scene and the error is caused by lack of common sense reasoning capability.

the prediction is assigned a label from the open vocabulary, and each predicted label corresponds to a segmentation mask. Among the ground-truth (GT) masks of input image, we assume that the one with the maximum intersection over union with each predicted mask is its GT mask. Firstly, comparing each predicted mask to its GT mask, we identify any of the 9 specified issues for this predicted mask and increment the corresponding issue count accordingly. Multiple issues were allowed to be identified for each prediction. Secondly, if a label exists in the ground truth but is missed in the prediction results, we categorize it as "missing object issue" and increment the count of missed labels. Through this process, we obtain the issue counts for each state-of-the-art model on the benchmark dataset. To mitigate subjective factors, we invited 10 researchers to participate in this process, and the results are averaged to obtain the final issue statistics.

3.2 Datasets and Models

Datasets. We take four widely-used datasets in the OVS benchmark: ADE20k-150 [40], ADE20k-847 [40], Pascal-59 [19], and Pascal-459 [19]. **ADE20k-150, ADE20k-847** are large-scale scene understanding datasets with 20K training images and 2k validation images, consisting of 150/847 labels. **Pascal-59, Pascal-459** comprise 5k training and validation images, with 59/459 labels.
Models. We evaluate the recently VLM-based OVS methods that achieved outstanding performance: SAN [34], OVSeg [14], and ODISE [30]. Each model incorporates a VLM, such as CLIP [22] or a text-to-image generation model [24]. All

models were trained on the COCO dataset and evaluated on the ADE and Pascal datasets mentioned above.

3.3 Analysis Results

As shown in Fig. 1, we average the evaluation results of different models on different datasets. Surprisingly, most VLM-based OVS methods primarily focus on the recognition issues. However, our statistical analysis reveals that the three most frequently occurring issues are: segmentation, dataset and reasoning issues.

4 Preliminary Approaches for Bottleneck Issues

4.1 Preliminaries

We choose SAN [34] model with the superiors performance as our base model to verify our initial attempts. Given an image $i \in \mathbb{R}^{H \times W \times 3}$ and K categories, the SAN model generates Q mask proposals $\mathbf{M} \in \mathbb{R}^{Q \times H \times W}$ and computes similarity scores for each proposal with respect to the K categories $\mathbf{P} \in \mathbb{R}^{Q \times K}$. The semantic segmentation result $\mathbf{S} \in \mathbb{R}^{K \times H \times W}$ could be obtained by: $\mathbf{S} = \mathbf{P}^{\mathrm{T}} \times \mathbf{M}$. The standard semantic segmentation evaluation procedure involves applying the argmax operation to the first dimension of matrix $\mathbf{S}$, yielding the segmentation map where each pixel is assigned one of the K categories.

4.2 Refining Benchmark for Dataset Issue

According to the analysis in Sect. 3, it is evident that some annotations in the dataset may not be appropriate, leading to a significant portion of the model's errors on the benchmark. We specifically focus on a sub-Dataset-Issue: "similar concepts issue" and propose a preliminary solution to address a significantly determined subset of similar concepts, such as "bed - beds" and "step, stair - stairs, steps" obviously similar cases.

To create similar label sets, we employed a similarity-based approach by computing the similarity between phrases. Higher similarity scores indicate a closer relationship between phrases, and we grouped them accordingly. Each label l will be associated with a similar label set L, with the minimum set size 1, representing the label itself. In order to avoid using the same text encoder as the model, we utilized BERT [5] to calculate the phrase similarities and cosine similarity was employed to measure the similarity between phrases. Phrases with a similarity score greater than 80 were grouped together. Subsequently, we manually filtered the sets, removing those with significant semantic differences and retaining only those with highly similar semantics. With these label sets established, given the predicted label l_{pred} for each pixel and the ground truth label l_{gt}, we proceeded to make the following adjustments to the predicted label for each pixel:

$$l_{pred} = \begin{cases} l_{pred}, & \text{if } l_{pred} \notin L_{gt} \\ l_{gt}, & \text{if } l_{pred} \in L_{gt} \end{cases}. \tag{1}$$

4.3   Our Framework for Segmentation and Reasoning Issue

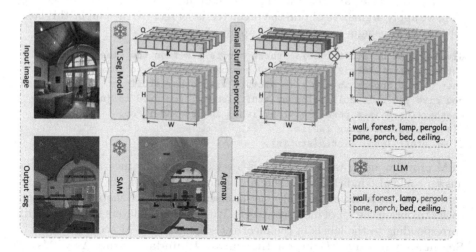

Fig. 2. Our framework consists of 3 components. Given an input image, the VL seg model generates matrices P and M. 1) the small stuff post-processing module corrects matrix P. 2) the corrected predictions are evaluated by LLM, which further adjusts matrix S. 3) the adjusted results are passed through SAM to generate the final segmentation output.

In this subsection, we primarily address the prevalent missing object and inaccurate boundary issues within Segmentation Issue and missing common sense issue within Reasoning Issue to propose a preliminary framework shown in Fig. 2.

Small Stuff Processing for Missing Object Issue. Regarding the missing object issue, our analysis revealed that a significant proportion of this issue is attributed to small stuffs, characterized by a relatively small GT mask ratio within the overall image. To tackle this, we implemented a post-processing step on matrix $\mathbf{P}$ to amplify small stuff's score, thereby enhancing their detectability through softmax operations. Specifically, for each layer $p \in \mathbb{R}^{1 \times K}$ in $\mathbf{P}$, the following adjustments were applied: $p = (1 - \lambda \cdot \frac{r_{area}^p}{Q}) \cdot p$, where r_{area}^p represents the ascending order rank of the area of p mask among all Q mask candidates, and Q signifies the layers of $\mathbf{P}$.

Large Language Models for Missing Common Sense Issue. Based on our observations, some predicted labels are clearly inconsistent with other labels and significantly deviate from the scene depicted in the image. For example, in an indoor scene with "beds, sofas, carpets, tables, and chairs", the roof of this small wooden house is mistakenly predicted as a "carport". This is a clear common-sense error that humans would not make due to their reasoning and common-sense abilities. When humans see an indoor scene, they would not associate it with a "carport", even though these two stuffs may look similar. These reasoning and understanding abilities between labels are lacking in VLMs but are crucial for this task. Considering the recent advancements in LLM [37], we introduce LLM (ChatGPT-turbo in our work) to address this common-sense issue.

Given an image i and the S matrix refined in the above Small Stuff Processing step, we apply argmax to obtain the predicted label set. We then input this label set into LLM using a question-based approach.

```
Given a list of stuff labels, these stuffs are from the same picture,
but some stuff labels maybe wrong, because according to common sense
they cannot appear in the same picture with other stuffs at the same time
now give you this collection of stuff labels, you need to judge
whether there is a possible wrong label in it, and if so,
please list the wrong labels.
Given stuff label list: {stuffs}.
```

LLM provides an answer set, which is a set of potentially incorrect labels (the answer could also indicate that there are no errors, i.e., an empty set). We repeat this process three times and take the intersection set U of the answer sets for each i to obtain the set of incorrect labels. We penalize the layers in the S matrix corresponding to the labels in the incorrect set and adjust M accordingly.

Segment-Anything-Model for Inaccurate Boundary Issue. As VLMs are pretrained at the image level and the added segmentation module is always lightweight, the generated quality of pixel-level segmentation is limited. Our analysis reveals that this limitation, known as the "Inaccurate Boundary Issue" also accounts for a significant portion. To address this, we introduce the Segment-Anything-Model (SAM) segmentation expert model to mitigate the problem by refining the VLM-based segmentation results. Given an input image i, we first utilize the adjusted M matrix from the previous two steps to obtain the final segmentation map S by applying argmax, assigning each pixel a label. Subsequently, we employ the SAM model on image i to generate a set of high-quality segmentation masks, denoted as m_i^{SAM}. We sort m_i^{SAM} in descending order based on their respective areas. For each SAM segmentation, we determine its label by cropping the corresponding semantic map S and selecting the label with the highest frequency. This process is repeated for subsequent SAM segmentation masks, with labels from smaller segmentation masks replacing previous ones.

5   Experiment Results

5.1   Dataset Refinement

We selected ADE20k-847 dataset from the four datasets with the highest number of semantic vocabulary and the greatest disparity in semantics compared to the training COCO dataset [34]. This dataset was used for preliminary dataset refinement to investigate the extent of differences in the performance of the same model on the dataset before and after annotation correction (paritially). This further illustrates the impact of dataset issue of OVS benchmark performance.

We employed the methodology described in Sect. 4 to refine a portion of the labels and showcased some of the corrected results in Table 3. The modified

Table 2. Comparison of our framework's base and large version with prior works. †
denotes the performance we achieved by reproducing the results with official code.

Method	VLM	Training Data	ADE-847	PC-459	ADE-150	PC-59
GroupViT	rand. init.	CC12M+YFCC	-	-	-	22.4
ODISE	SD-v1.3+CLIP-L	COCO	11.1	14.5	29.9	57.3
LSeg+	ALIGN EN-B7	COCO	3.8	7.8	18.0	46.5
OpenSeg	ALIGN EN-B7	COCO+Narr.	6.3	9.0	21.1	42.1
SimSeg	CLIP ViT-B/16	COCO	7.0	8.7	20.5	47.7
OVSeg	CLIP ViT-B/16	COCO	7.1	11.0	24.8	53.3
SAN	CLIP ViT-B/16	COCO	10.1	12.6	27.5	53.8
SAN†	CLIP ViT-B/16	COCO	10.2	16.7	27.5	54.1
Ours	CLIP ViT-B/16	-	**11.1**	**17.7**	**29.1**	**55.9**
MaskCLIP	CLIP ViT-L/14	COCO	8.2	10.0	23.7	45.9
SimSeg†	CLIP ViT-L/14	COCO	7.1	10.2	21.7	52.2
OVSeg	CLIP ViT-L/14	COCO	9.0	12.4	29.6	55.7
SAN	CLIP ViT-L/14	COCO	12.4	15.7	32.1	57.7
SAN†	CLIP ViT-L/14	COCO	12.8	20.8	31.9	57.5
Ours	CLIP ViT-L/14	-	**14.2**	**22.6**	**33.7**	**59.3**

dataset is denoted as ADE20k-847*. We compared the performance of the base
model and our model on both ADE20k-847 and ADE20k-847* as showed in
Table 4. Both the base model and the large model, with unchanged weights,
achieved performance gains on this simply corrected dataset. This observation
suggests that partial annotations in the dataset do have a certain impact on our
assessment of the model's capabilities on OVS benchmark.

5.2 Our Framework

We validated our method on the four datasets listed in Sect. 3, which are widely
used in OVS. For optimal performance, we employed the SAN model [34] as the
base model in our framework, specifically as our VL-seg module in framework.
Since the SAN model offers two versions based on different sizes of CLIP [22],
SAN-base and SAN-large, we conducted experiments with both sizes. The performance of the two models on the OVS benchmark is shown in Table 2. It is
noted that for fair comparison, all experiments in Table 2 were conducted using
the original dataset settings without any correction mentioned in this work.
Our framework achieved performance improvements across all settings. We also
conducted ablation experiments on the three sub-modules in our framework, as
shown in Table 5. Additionally, we present the cases of our framework in Fig. 3.

It is worth noting that our framework is training-free, requiring no additional
data for training. Besides, because of the powerful commonsense reasoning abilities of the LLM and the excellent pixel-level vision processing capabilities of

Table 3. Some cases of ADE20k-847 label correction (paritially) results.

label	similar labels
bed	beds
stairs, steps	step, stair
drawing	painting, picture
dvd	dvds; cd; video; cds; videos
ticket counter	ticket office; ticket booth

Table 4. The performance difference of the same model before and after label correction. ADE-847* denoted corrected ADE-847.

Method	ADE-847	ADE-847*
SAN-B/16	10.1	-
SAN-B/16†	10.24	10.63
Ours-B/16	11.11	11.57
SAN-L/14	12.4	-
SAN-L/14†	12.83	13.26
Ours-L/14	14.18	14.32

SAM, our proposed method achieves significant improvement on the existing benchmark. This further validates the issues discussed in Sect. 3 and demonstrates the potential of combining VLM-based segmentation models, LLMs and FVMs. We hope that our preliminary exploration will inspire further research on VLM+LLM+FVM models as illustrated in Fig. 1(c).

Table 5. Ablation experiments on the three sub-modules in our framework.

Method	SSP	LLM	SAM	ADE-847	PC-459	ADE-150	PC-59
SAN-B/16				10.1	12.6	27.5	53.8
SAN-B/16†				10.2	16.7	27.5	54.1
Ours-B/16	√			10.5	16.8	27.4	54.1
Ours-B/16		√		10.3	16.8	27.6	54.3
Ours-B/16	√	√		10.5	16.9	27.6	54.3
Ours-B/16	√	√	√	11.1	17.7	29.1	55.9
SAN-L/14				12.4	15.7	32.1	57.7
SAN-L/14†				12.8	20.8	31.9	57.5
Ours-L/14	√			13.1	20.9	31.7	57.3
Ours-L/14		√		12.9	21.4	32.0	57.7
Ours-L/14	√	√		13.2	21.4	32.0	57.7
Ours-L/14	√	√	√	14.2	22.6	33.7	59.3

6   Conclusion, Limitations and Future Work

This paper presents novel perspectives and initial attempts aimed at enhancing semantic understanding in computer vision. This work demonstrates some progress in this field, although it is evident that further analysis and refinement of proposed solutions hold potential. In particular, the current analysis lacks a direct correlation with mean Intersection over Union (mIoU) metric in terms

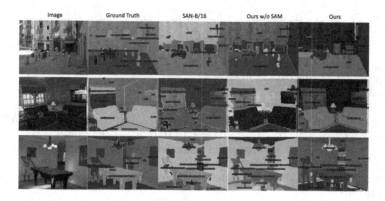

Fig. 3. Cases for SAN-B/16 and our model.

of error frequency. And the conclusions drawn from this analysis may not be applicable to all OVS frameworks, particularly those that do not base on VLMs. Furthermore, proposed approach can be considered preliminary. While this work has made initial corrections to ADE-847, more comprehensive refinements are reserved for future work and are not the primary focus of our this work. Lastly, it is important to note that the combination of LLMs and FVMs does not necessarily result in a multi-stage inference and training-free framework.

To push the boundaries of the field further, we propose the following directions for future research: **New benchmarks:** Develop benchmarks to evaluate semantic inclusion and foster open semantic understanding. **Unified framework:** Design elegant integrated structures, combining VLMs, LLMs and FVMs, to accommodate various tasks. **Semantic granularity:** Address the semantic granularity problem for accurate and efficient fine-grained semantic interpretation. **Expanding problem scope:** Incorporate a diverse range of vision problems, including 3D, segmentation-grounding, fine-grained semantic understanding in vision, and free form semantic understanding instead of open vocabulary.

Acknowledgements. This work was supported by the Fundamental Research Funds for the Central Universities, and the Research Funds of Renmin University of China (21XNLG28), National Natural Science Foundation of China (No. 62276268) and Kuaishou. We acknowledge the anonymous reviewers for their helpful comments.

References

1. Bucher, M., et al.: Zero-shot semantic segmentation. In: NeurIPS (2019)
2. Cen, J., et al.: Segment anything in 3D with NeRFs (2023)
3. Cha, J., et al.: Learning to generate text-grounded mask for open-world semantic segmentation from only image-text pairs. In: CVPR (2023)
4. Cho, S., et al.: CAT-Seg: cost aggregation for open-vocabulary semantic segmentation. CoRR (2023)

5. Devlin, J., et al.: BERT: pre-training of deep bidirectional transformers for language understanding. In: NAACL (2019)
6. Ding, J., et al.: Decoupling zero-shot semantic segmentation. In: CVPR (2022)
7. Ding, Z., et al.: Open-vocabulary panoptic segmentation with MaskCLIP. arXiv preprint arXiv:2208.08984 (2022)
8. Ghiasi, G., Gu, X., Cui, Y., Lin, T.Y.: Scaling open-vocabulary image segmentation with image-level labels. In: Avidan, S., Brostow, G., Cissé, M., Farinella, G.M., Hassner, T. (eds.) Computer Vision, ECCV 2022. LNCS, vol. 13696, pp. 540–557. Springer, Cham (2022). https://doi.org/10.1007/978-3-031-20059-5_31
9. Jia, C., et al.: Scaling up visual and vision-language representation learning with noisy text supervision. In: ICML (2021)
10. Karazija, L., et al.: Diffusion models for zero-shot open-vocabulary segmentation. CoRR (2023)
11. Ke, L., et al.: Segment anything in high quality (2023)
12. Kirillov, A., et al.: Segment anything. arXiv preprint arXiv:2304.02643 (2023)
13. Li, B., et al.: Language-driven semantic segmentation. In: ICLR (2022)
14. Liang, F., et al.: Open-vocabulary semantic segmentation with mask-adapted clip. In: CVPR (2023)
15. Liu, H., et al.: Visual instruction tuning. arXiv preprint arXiv:2304.08485 (2023)
16. Liu, Q., Wen, Y., Han, J., Xu, C., Xu, H., Liang, X.: Open-world semantic segmentation via contrasting and clustering vision-language embedding. In: Avidan, S., Brostow, G., Cissé, M., Farinella, G.M., Hassner, T. (eds.) Computer Vision, ECCV 2022. LNCS, vol. 13680, pp. 275–292. Springer, Cham (2022). https://doi.org/10.1007/978-3-031-20044-1_16
17. Lüddecke, T., et al.: Image segmentation using text and image prompts. In: CVPR (2022)
18. Luo, H., et al.: SegCLIP: patch aggregation with learnable centers for open-vocabulary semantic segmentation. In: ICML (2023)
19. Mottaghi, R., et al.: The role of context for object detection and semantic segmentation in the wild. In: CVPR (2014)
20. Mukhoti, J., et al.: Open vocabulary semantic segmentation with patch aligned contrastive learning. In: CVPR (2023)
21. OpenAI: GPT-4 technical report (2023)
22. Radford, A., et al.: Learning transferable visual models from natural language supervision. In: ICML (2021)
23. Rao, Y., et al.: DenseCLIP: language-guided dense prediction with context-aware prompting. In: CVPR (2022)
24. Rombach, R., et al.: High-resolution image synthesis with latent diffusion models. In: CVPR (2022)
25. Schuhmann, C., et al.: LAION-5B: an open large-scale dataset for training next generation image-text models. In: Advances in Neural Information Processing Systems (2022)
26. Touvron, H., et al.: LLaMA: open and efficient foundation language models (2023)
27. Wang, X., Zhang, X., Cao, Y., Wang, W., Shen, C., Huang, T.: SegGPT: segmenting everything in context. arXiv preprint arXiv:2304.03284 (2023)
28. Xian, Y., et al.: Semantic projection network for zero- and few-label semantic segmentation. In: CVPR (2019)
29. Xu, J., et al.: GroupViT: semantic segmentation emerges from text supervision. In: CVPR (2022)
30. Xu, J., et al.: Open-vocabulary panoptic segmentation with text-to-image diffusion models. In: CVPR (2023)

31. Xu, J., et al.: Learning open-vocabulary semantic segmentation models from natural language supervision. In: CVPR (2023)
32. Xu, M., et al.: A simple baseline for open-vocabulary semantic segmentation with pre-trained vision-language model. In: Avidan, S., Brostow, G., Cissé, M., Farinella, G.M., Hassner, T. (eds.) Computer Vision, ECCV 2022. LNCS, vol. 13689, pp. 736–753 . Springer, Cham (2022). https://doi.org/10.1007/978-3-031-19818-2_42
33. Xu, M., et al.: A simple baseline for open-vocabulary semantic segmentation with pre-trained vision-language model. In: Avidan, S., Brostow, G., Cissé, M., Farinella, G.M., Hassner, T. (eds.) Computer Vision, ECCV 2022. LNCS, vol. 13689, pp. 736–753. Springer, Cham (2022). https://doi.org/10.1007/978-3-031-19818-2_42
34. Xu, M., et al.: Side adapter network for open-vocabulary semantic segmentation. In: CVPR (2023)
35. Zhang, H., et al.: A simple framework for open-vocabulary segmentation and detection. arXiv preprint arXiv:2303.08131 (2023)
36. Zhao, H., et al.: Open vocabulary scene parsing. In: ICCV (2017)
37. Zhao, W.X., et al.: A survey of large language models (2023)
38. Zhao, X., et al.: Fast segment anything (2023)
39. Zhong, Y., et al.: RegionCLIP: region-based language-image pretraining. In: CVPR (2022)
40. Zhou, B., et al.: Scene parsing through ADE20K dataset. In: CVPR (2017)
41. Zou, X., et al.: Generalized decoding for pixel, image, and language. In: CVPR (2023)
42. Zou, X., et al.: Segment everything everywhere all at once (2023)

Exploring a Distillation with Embedded Prompts for Object Detection in Adverse Environments

Hao Fu[1], Long Ma[1], Jinyuan Liu[4], Xin Fan[2,3], and Risheng Liu[2,3]([✉])

[1] School of Software Technology, Dalian University of Technology, Dalian, China
[2] DUT-RU International School of Information Science and Engineering,
Dalian University of Technology, Dalian, China
`rsliu@dlut.edu.cn`
[3] Key Laboratory for Ubiquitous Network and Service Software of Liaoning Province,
Dalian, China
[4] School of Mechanical Engineering, Dalian University of Technology, Dalian, China

Abstract. Efficient and robust object detection in adverse environments is crucial and challenging for autonomous agents. The current mainstream approach is to use image enhancement or restoration as a means of image preprocessing to reduce the domain shift between adverse and regular scenes. However, these image-level methods cannot guide the model to capture the spatial and semantic information of object instances, resulting in only marginal performance improvements. To overcome this limitation, we explore a Prompts Embedded Distillation framework, called PED. Specifically, a spatial location prompt module is proposed to guide the model to learn the easily missed target position information. Considering the correlation between object instances in the scene, a semantic mask prompt module is proposed to constrain the global attention between instances, making each aggregated instance feature more discriminative. Naturally, we propose a teacher model with embedded cues and finally transfer the knowledge to the original student model through focal distillation. Extensive experimental results demonstrate the effectiveness and flexibility of our approach.

Keywords: Objection detection · Distillation · Underwater object detection · Foggy object detection

1 Introduction

In recent years, with the development of deep learning and autonomous mobile agents, object detection has not only achieved satisfactory results [3,11,23] on conventional datasets but has also been widely applied in real-world scenarios, such as autonomous driving cars and autonomous robots for performing various tasks. However, most existing detectors currently focus only on conventional

This work is partially supported by the National Natural Science Foundation of China (Nos. U22B2052).

scenarios, while ignoring the challenges that arise in common adverse scenarios such as underwater, and foggy conditions. These challenges are crucial for achieving robust perception and making correct path planning and decisions for autonomous mobile agents in the real world. Due to the significant domain shift, noise, color bias, and other degradation factors present in input images, the performance of conventional detectors is often greatly compromised when transferred to adverse scenarios. In underwater, and foggy conditions, the challenges of object detection under adverse conditions can be summarized as follows:

- Positional blur. Objects in adverse scenarios are often masked by noise from the background or other instances, which reduces the contrast between objects and the background or between objects themselves, causing blurred edge features and affecting the localization accuracy of subsequent object detection tasks, leading to missed detections.
- Feature degradation. Adverse scenarios involve complex degradation factors such as color distortion in underwater scenarios, and low visibility in foggy scenarios. These factors directly damage the texture details of the target, leading to damaged and indistinguishable semantic features, which in turn affect the classification accuracy of subsequent object detection tasks, leading to false positive detections.

To overcome the aforementioned issues, one approach is to use existing image enhancement methods [5,7,14–16,20,21] as a preprocessing step for degraded images to reduce the adverse effects of specific adverse scenarios or degradation factors, thereby reducing the domain shift between the source domain (conventional scenarios) and the target domain (adverse scenarios). However, this introduces complex image restoration networks for object detection, and it is almost impossible for such networks to have restoration effects on all images in all adverse scenarios. Another state-of-the-art approach is to use differentiable image preprocessing [8,18], which involves introducing differentiable white balance, gamma correction, sharpening, and dehazing operations and enhancing input images end-to-end through the guidance of the detection loss. This approach aims to mine more potential information from blurred objects and improve detection performance. However, this image-level approach cannot selectively restore or enhance features in target regions, leading to limited performance improvements for subsequent detection tasks.

To overcome these limitations, we propose an intuitive and efficient approach to convert as much label information as possible into guidance embeddings suitable for the model, to guide the model to mine potential information from instances in adverse scenarios while reducing the interference of background noise. We introduce a teacher model based on guidance embedding learning, called PED which accurately locates targets and extracts discriminative potential information from them through designed guidance embeddings. The knowledge learned by the teacher model is then transferred to the student model through focus distillation. To address the second challenge, we propose a semantic mask guidance module. Since the feature degradation of a single target instance may not support subsequent accurate classification tasks, we aim to establish relationships between target instances through attention mechanisms. By aggregating the features of strongly related instances, we can enhance the distinguishability of the semantic features of individual instances and increase the contrast with the surrounding background. We first use a multi-head self-attention mechanism to capture long-range dependencies between tokens and obtain an attention map. Then, we generate our semantic mask guidance embedding using the label information of object detection. This embedding ensures that the attention of background tokens to all other tokens and the attention of target tokens to background tokens are both 0, constraining global attention interaction between target tokens. This approach enriches the semantic features of each target token after aggregation. Afterwards, we embed the two proposed guidance modules into the detector to obtain our teacher model. Through focus distillation, we transfer the target potential information that the teacher model has mined based on the guidance embeddings to the student model, which is beneficial for subsequent object detection tasks.

In summary, our main contributions can be summarized as follows:

- We explore spatial and semantic guidance embeddings for object detection under adverse conditions and propose a teacher network based on guidance embeddings. The knowledge mined from guidance embeddings is transferred to the student model through focus distillation.
- We propose a spatial localization guidance module and a semantic mask guidance module to address the challenges of target position blurring and feature degradation in adverse environments, respectively.
- We conduct extensive quantitative and qualitative experiments on real underwater (URPC) and foggy (RTTS) datasets, demonstrating the effectiveness and flexibility of our approach.

2 Related Work

2.1 Object Detection

Object detection, as a fundamental task in the field of computer vision, has received widespread attention in recent years due to its applications in autonomous mobile agents. Currently, object detection methods can be roughly divided into two categories: two-stage detectors and one-stage detectors. However, the performance of both types of detectors is severely degraded under adverse conditions.

2.2 Enhancement for Detection

A direct method for object detection under adverse conditions is to use environment-specific enhancers [4,5,10,14,17,20,21,25] as a means of preprocessing degraded images, and then train and evaluate the detector on the enhanced images. Although many enhancement methods have achieved promising results in terms of visual quality and quantitative evaluation, only a few of them can bring slight improvements to downstream object detection tasks, while most enhancement methods lead to a decrease in detection performance.

In underwater environments, UWNet [21] proposes a compact convolutional neural network that maintains superior performance while saving computational resources. FGAN [7] uses conditional generative adversarial networks to achieve robust enhancement on paired or unpaired datasets and demonstrates its performance improvement on downstream underwater object detection and human pose estimation tasks. USUIR [5] constructs an unsupervised underwater image restoration method by exploiting the homogeneity between the original underwater image and the degraded image. TACL [16] proposes a twin adversarial contrastive learning method that achieves both visual-friendly and task-oriented enhancement.

In foggy environments, GridDehazeNet [19] considers the spatial structure of the image by dividing it into multiple grids and performing multi-scale feature extraction on each grid, thus preserving more image details while dehazing. MSBDN [4] proposes a bidirectional feature propagation strategy based on multi-scale feature extraction, which better utilizes the global and local context information of the hazy image. GCANet [1] proposes an inverse residual structure to prevent gradient vanishing and combines global contextual information with local feature information to improve the quality of image dehazing. FFANet [22] uses attention mechanisms to improve the feature fusion of multi-scale features, emphasizing information-rich features while suppressing irrelevant features that interfere with image dehazing.

2.3 Differentiable Image Processing

Another popular approach is to design traditional filters as differentiable image processing modules, and the hyperparameters of these modules are predicted by a small convolutional neural network guided by downstream detection losses, achieving task-specific image enhancement in an end-to-end manner. IA-YOLO [18] proposes a way of concatenating stacked DIP modules. GDIP performs various differentiable image processing operations in parallel and enhances images through a weighted combination of concurrent differentiable image processing operations. MGDIP [8] processes the degraded image multiple times with GDIP to obtain enhanced images more suitable for downstream detection tasks. IA-YOLO and GDIP have demonstrated the superiority of this DIP-based approach over existing domain adaptation methods, such as MS-DAYOLO [6], and multi-task learning methods, such as DSNet, for object detection in adverse scenarios.

3 Method

To endow existing general detectors with the ability to accurately perceive target locations and fully exploit target semantic information in adverse conditions, we

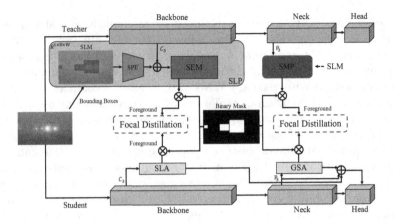

Fig. 1. The overall structure of our proposed PED. It mainly consists of a teacher model embedded with spatial location prompt and semantic mask prompt, and a naive student model. The teacher model transfers spatial and semantic knowledge to the student model through focal distillation.

propose a spatial location prompt module and a semantic mask prompt module. We also adopt a simplified focal distillation method to transfer knowledge learned from the teacher model to the student model, as shown in the Fig. 1. We take RetinaNet as an example to explain our method in detail.

3.1 Spatial Location Prompt

To address the problem of missing targets caused by feature loss, we propose a spatial location prompt module, called SLP, to guide the feature extraction network to mine the potential information of all targets in space by embedding target position information, even if some targets' features have been heavily degraded.

First, we generate a corresponding spatial location map (SLM) for each image based on the label information of the object detection:

$$SLM(i,j) = \begin{cases} 1, if \quad (i,j) \in BBoxes \\ 0, Others \end{cases} \tag{1}$$

where $BBoxes$ represents the bounding boxes, and i and j represent the horizontal and vertical coordinates of the feature map, respectively. If the feature point is inside the bounding boxes, then $SLM_{i,j} = 1$, otherwise it is 0.

To embed SLM as a prompt into the teacher network to guide it to extract the potential features of all targets in adverse scenarios during training, we propose a simple spatial positioning encoder (SPE), which consists of a convolutional layer with an input channel of 1, an output channel of 512, a stride of 1, and a convolution kernel size of 3 × 3. Using the SPE, we can map the SLM with a channel number of 1 to a spatial location prompt embedding with a channel number of 512.

Next, we aggregate the obtained spatial location prompt embedding with the C_3 feature layer of the teacher network and transmit it to the designed spatial enhancement module (SEM) for refining the target texture details and position features. Specifically, the SEM consists of two convolutional layers with a stride of 1 and a convolution kernel size of 3×3, and is interconnected with BatchNorm and ReLU activation layers in the middle.

The overall structure of the SLP is as follows:

$$C_3' = SEM(SPE(SLM) + C_3) \tag{2}$$

$$P_3' = P_3 + C_3' \tag{3}$$

Finally, the C_3' features refined based on spatial location information as a prompt are aggregated with the P_3 feature in the feature pyramid of the teacher network, and transmitted to the detection head for end-to-end training (Fig. 2).

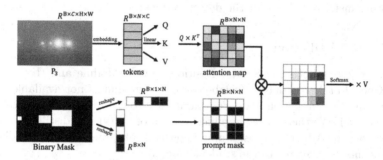

Fig. 2. The overall architecture of our proposed SMP. It uses the prompt mask to constrain the global self-attention between instance tokens, capturing the relationship between target instances. In the binary mask, white represents the presence of an object, while black represents the background. In the prompt mask, white areas represent values of 0, while black areas represent values of 1.

3.2 Semantic Mask Prompt

In adverse scenarios, the features of target instances are often subject to various complex degradation factors. Therefore, focusing only on the features obtained from the current target instance itself in the feature extraction process is often insufficient to support accurate regression and classification by the subsequent detection head, resulting in a decrease in detection accuracy. There is often complementary semantic information between multiple instances in the same image, and establishing relationships between multiple instances is used to improve the performance of object detection.

Here we propose a self-attention module based on semantic mask information prompt, which aims to constrain the attention map in the multi-head self-attention module through the target mask generated by label, and accurately establish the relationship between instances and instances, so that The aggregated instance feature semantic information is richer and more discriminative.

Specifically, the way we get the target mask according to the label information is the same as the way of SLM in the previous section. We use an embedding operation to convert the 2D feature map P_3 into a 1D tokens sequence. Then, we use a fully connected layer to map each token to its corresponding Q, K, and V feature vectors. We then use $(Q \times K^T / scale)$ to obtain the attention matrix between tokens, where the scale is $\sqrt{d_k}$. Next, we reshape $mask \in [B, H, W]$ into $mask_1 \in [B, N]$ and $mask_2 \in [B, 1, N]$. Then, we iterate over $mask_1$, and when $mask_1[i, j] = 0$, $Mask[i, j, :] = 0$, and when $mask_1[i, j] = 1$, $Mask[i, j, :] = mask_2[i, 0, :]$. Finally, we obtain the attention matrix based on the mask cue by computing $Attn \times Mask$, which can constrain the attention of tokens belonging to instances between other tokens belonging to the same instances, rather than aggregating features about background tokens, thereby reducing interference from background noise and aggregating strong semantic features from other instance tokens for the detection task.

3.3 Focal Distillation

Considering that both the Spatial Positioning Cue Module and the Semantic Mask Cue Module use detection label information that is not available during testing, we first embed the two cue modules into the detector to obtain our teacher model. We then use a simplified version of knowledge distillation [24] to transfer the learned spatial positioning knowledge of targets and highly discriminative semantic features to the student model.

Specifically, for the student model, in order to better utilize the knowledge transferred from the teacher model, we introduce a spatial location adaptation (SLA) layer and a global semantic aggregation (GSA) layer. The SLA contains only one convolution operation with a stride of 1 and a convolution kernel of 3×3. Its impact on the computation and parameter of the student model is minimal. The GSA consists of a multi-head self-attention module with a depth of 1 and 4 heads. The outputs of the two modules are aggregated with P_3 by element-wise addition and fed into the detection head for end-to-end training. The overall structure is very simple and efficient.

The focal distillation loss is as follows:

$$\mathcal{L}_{slp} = \alpha \cdot \sum (SLA(C_3^s) \cdot SLM, SLP(C_3^t \cdot SLM))^2 \qquad (4)$$

$$\mathcal{L}_{smp} = \beta \cdot \sum (GSA(P_3^s \cdot SLM), SMP(P_3^t \cdot SLM))^2 \qquad (5)$$

Where α and β are hyperparameters, we set them to 1 and 0.5, respectively. C_3^s and P_3^s are the feature layers of the student model, and C_3^t and P_3^t are the corresponding feature layers of the teacher model. Note, during the testing phase, the model completely removes the teacher model and retains only the student model along with two simple and efficient adaptive layers specifically designed for the student model.

3.4   Overall Loss

In summary, the total loss used to train the student model is as follows:

$$\mathcal{L}_{total} = \mathcal{L}_{cls} + \mathcal{L}_{reg} + \mathcal{L}_{slp} + \mathcal{L}_{smp} \tag{6}$$

Where $\mathcal{L}cls$ and $\mathcal{L}reg$ are the original classification loss and regression loss of the detector, respectively.

4   Experiments

All experiments are conducted using MMDetection [2]. During the training phase, all images are scaled to 448 448, with a batch size of 4. An AdamW optimizer with weight decay of 5×10^{-2} is employed with the training epoch of 12. The initial learning rate is set to 1×10^{-4} and decreased by a factor of 10 during epoch [8, 11]. We use PyTorch for the experiments and ran them on one 4090 GPU. All results are evaluated using the standard COCO Average Precision (AP) metrics [12].

Table 1. Comparing detection accuracy among our PED and other compared methods on the URPC2020 dataset. The best results are **bolded**.

Method		Holothurian	Echinus	Scallop	Starfish	mAP	AP_{50}	AP_{75}	mAR	AR_s	AR_m	AR_l
Baseline		30.1	39.5	27.0	41.2	34.4	68.5	30.7	47.1	27.0	46.2	48.5
Enhancer	UWNet	30.1	40.1	27.6	41.2	34.8	68.7	30.7	47.6	27.9	46.9	48.9
	FGAN	28.2	38.5	25.5	38.2	32.6	66.0	27.7	45.2	26.1	44.6	46.3
	USUIR	29.1	38.3	26.7	39.9	33.5	67.4	28.9	46.2	25.5	45.2	47.8
	TACL	28.0	37.5	26.1	38.5	32.5	65.7	27.9	45.5	23.6	43.9	47.4
Adverse	IA	28.0	38.7	25.5	39.7	33.0	66.5	28.1	45.8	23.4	44.9	47.5
	GDIP	28.8	38.7	25.3	40.7	33.4	67.1	28.6	46.4	25.4	46.0	47.7
	MGDIP	29.8	39.0	25.5	40.4	33.7	67.3	29.8	46.6	27.8	45.7	47.9
Ours		**32.1**	**41.8**	**29.7**	**43.6**	**36.8**	**71.4**	**33.1**	**49.7**	**29.1**	**48.9**	**51.0**

4.1   Underwater Scenes

In this section, we will demonstrate the performance of our proposed method on the real underwater dataset URPC2020.

Implementation Details. The URPC2020 dataset [13] consists of images taken in real underwater environments and includes a total of 6575 degraded images with annotations for four marine organisms: sea cucumbers, sea urchins, starfish, and scallops. For training our model on the URPC2020 dataset, we randomly selected 4602 images as our training set and used the remaining images as our test set. We used the classic RetinaNet detector as our baseline and embedded SLP and SMP into RetinaNet to obtain our teacher model, called PET-RetinaNet, while the student model is RetinaNet. In the comparative method, the enhanced model is cascaded in front of the detector as a means of image restoration.

Comparisons with State-of-the-Arts. Table 1 presents the quantitative results of our experiments. It is evident that our proposed method outperforms other methods in terms of both AP and AR. Specifically, our method achieves the best performance in terms of mAP, AP_{50}, and mAR, with improvements of 2.0%, 2.7%, and 2.1%, respectively, compared to the second-best model. Moreover, our method also shows a significant improvement in detecting small objects, which is a more challenging task. The AR_s of our method are 1.2% higher than those of the second-best model (Fig. 3).

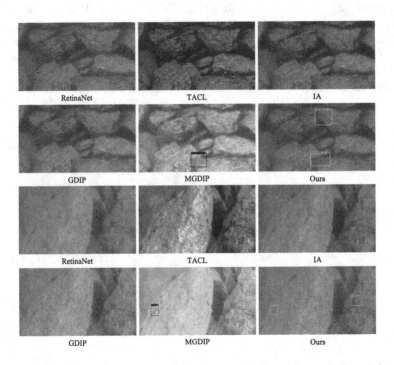

Fig. 3. Visual comparison with state-of-the-art methods on the URPC2020 dataset.

4.2 Haze Scenes

In this section, we will demonstrate the effectiveness of our method on the real-world foggy dataset RTTS, and also verify the flexibility of our method, which can adapt to various adverse environments simultaneously.

Implementation Details. The RTTS dataset [9] consists of 4322 images captured under real foggy conditions, containing various degradation factors in real foggy scenes and is very challenging. It mainly annotates five categories

Table 2. Comparing detection accuracy among our PED and other compared methods on the RTTS dataset. The best results are **bolded**.

Method		Bicycle	Bus	Car	Motorbike	Person	mAP	AP_{50}	AP_{75}	mAR	AR_s	AR_m	AR_l
Baseline		30.4	29.0	37.6	21.8	35.8	30.9	54.5	31.8	42.2	22.3	50.3	**63.5**
Enhancer	GridDehaze	29.7	28.8	37.6	22.0	35.8	30.8	54.3	30.1	41.6	21.1	50.6	62.9
	MSBDN	30.5	29.0	37.8	22.0	35.4	31.0	53.9	31.4	41.6	21.0	50.5	62.8
	GCANet	**32.6**	28.1	36.3	21.2	35.0	30.7	54.3	30.5	41.2	21.0	49.5	63.0
	FFANet	32.1	27.9	38.2	21.4	35.9	31.1	54.6	31.3	41.3	21.7	49.7	62.0
Adverse	IA	28.2	26.8	37.1	19.4	35.7	29.5	52.0	28.7	41.0	20.5	49.0	62.8
	GDIP	28.3	27.5	37.4	19.4	34.7	29.4	52.7	29.2	40.7	20.2	49.9	60.8
	MGDIP	30.5	27.0	37.0	20.5	34.7	29.9	53.2	29.9	40.7	19.7	49.1	62.8
Ours		31.0	**29.7**	**39.1**	**22.8**	**37.3**	**32.0**	**56.6**	**32.5**	**43.0**	**24.0**	50.8	63.5

of objects: person, car, bus, bicycle, and motorcycle. When using the RTTS dataset, we randomly selected 3889 images as our training set, and the remaining images were used for testing. Similar to the comparative experiments in underwater scenes, we also used RetinaNet as our baseline model and the student model, and embedded the SLP and SMP modules into the RetinaNet model to obtain our teacher model.

Comparisons with State-of-the-Arts. Table 2 provides the relevant quantitative results. It can be seen that our method outperforms previous models in all indicators except for bicycle, especially in the person category. In addition, compared to the second-ranked method, our method improves mAP and mAR by 0.9% and 0.8%, respectively, which also demonstrates the effectiveness of our method in adverse foggy environments.

4.3   Ablation Study

Table 3. An ablation study of two prompt modules (SLP and SMP) and an additional multi-head self-attention module (MHSA).

Model	SLP	MHSA	SMP	mAP	AP_{50}	AP_{75}	AP_s	AP_m	AP_l	mAR	AR_s	AR_m	AR_l
M1				34.4	68.5	30.7	10.9	33.5	36.6	47.1	27.0	46.2	48.5
M2	✓			36.2	70.6	32.6	12.6	34.7	38.8	49.1	28.3	48.1	50.6
M3		✓		35.7	70.4	31.5	12.0	34.7	37.9	48.4	28.8	47.9	49.4
M4		✓	✓	36.3	71.0	32.9	12.7	35.1	38.4	49.0	27.9	48.5	50.2
M5	✓	✓	✓	36.8	71.4	33.1	11.8	35.5	39.1	49.7	29.1	48.9	51.0

To analyze the effectiveness of each component proposed, we conducted a series of ablation experiments on the underwater dataset URPC. M1 serves as the baseline RetinaNet, and the teacher model is PET-RetinaNet. The specific objective results are shown in Table 3. The results of M2 demonstrate the importance of localization information for object detection under harsh conditions. Simply embedding SLP into the teacher model and distilling it can improve the final mAP by 1.8%, while also demonstrating the efficiency of SLP. M3 demonstrates that global interaction through attention mechanisms can indeed improve detection performance. Moreover, comparing the results of M3 and M4 shows that constraining this interaction as much as possible between instance tokens can better guide the model to establish interaction relationships between instances, thereby extracting more discriminative semantic features. Compared with other scenarios, our final model achieved the best performance, which also demonstrates the importance of mining the position information of targets and discriminative semantic features for object detection under harsh conditions.

References

1. Chen, D., et al.: Gated context aggregation network for image dehazing and deraining. In: 2019 IEEE Winter Conference on Applications of Computer Vision (WACV), pp. 1375–1383. IEEE (2019)
2. Chen, K., et al.: MMDetection: open mmlab detection toolbox and benchmark. arXiv preprint arXiv:1906.07155 (2019)
3. Chen, Q., Wang, Y., Yang, T., Zhang, X., Cheng, J., Sun, J.: You only look one-level feature. In: Proceedings of the IEEE/CVF Conference on Computer Vision and Pattern Recognition, pp. 13039–13048 (2021)
4. Dong, H., et al.: Multi-scale boosted dehazing network with dense feature fusion. In: Proceedings of the IEEE/CVF Conference on Computer Vision and Pattern Recognition, pp. 2157–2167 (2020)
5. Fu, Z., et al.: Unsupervised underwater image restoration: from a homology perspective. In: Proceedings of the AAAI Conference on Artificial Intelligence, vol. 36, pp. 643–651 (2022)
6. Hnewa, M., Radha, H.: Multiscale domain adaptive yolo for cross-domain object detection. In: 2021 IEEE International Conference on Image Processing (ICIP), pp. 3323–3327. IEEE (2021)
7. Islam, M.J., Xia, Y., Sattar, J.: Fast underwater image enhancement for improved visual perception. IEEE Robot. Autom. Lett. 5(2), 3227–3234 (2020)
8. Kalwar, S., Patel, D., Aanegola, A., Konda, K.R., Garg, S., Krishna, K.M.: GDIP: gated differentiable image processing for object-detection in adverse conditions. arXiv preprint arXiv:2209.14922 (2022)
9. Li, B., et al.: Benchmarking single-image dehazing and beyond. IEEE Trans. Image Process. 28(1), 492–505 (2018)
10. Lin, R., Liu, J., Liu, R., Fan, X.: Global structure-guided learning framework for underwater image enhancement. Vis. Comput. 1–16 (2021)
11. Lin, T.Y., Goyal, P., Girshick, R., He, K., Dollár, P.: Focal loss for dense object detection. In: Proceedings of the IEEE International Conference on Computer Vision, pp. 2980–2988 (2017)
12. Lin, T.-Y., et al.: Microsoft COCO: common objects in context. In: Fleet, D., Pajdla, T., Schiele, B., Tuytelaars, T. (eds.) ECCV 2014, Part V. LNCS, vol. 8693, pp. 740–755. Springer, Cham (2014). https://doi.org/10.1007/978-3-319-10602-1_48

13. Liu, C., et al.: A dataset and benchmark of underwater object detection for robot picking. In: 2021 IEEE International Conference on Multimedia and Expo Workshops (ICMEW), pp. 1–6. IEEE (2021)

14. Liu, J., et al.: Target-aware dual adversarial learning and a multi-scenario multi-modality benchmark to fuse infrared and visible for object detection. In: Proceedings of the IEEE/CVF Conference on Computer Vision and Pattern Recognition, pp. 5802–5811 (2022)

15. Liu, J., Wu, G., Luan, J., Jiang, Z., Liu, R., Fan, X.: Holoco: holistic and local contrastive learning network for multi-exposure image fusion. Inf. Fusion **95**, 237–249 (2023)

16. Liu, R., Jiang, Z., Yang, S., Fan, X.: Twin adversarial contrastive learning for underwater image enhancement and beyond. IEEE Trans. Image Process. **31**, 4922–4936 (2022)

17. Liu, R., Li, S., Liu, J., Ma, L., Fan, X., Luo, Z.: Learning Hadamard-product-propagation for image dehazing and beyond. IEEE Trans. Circuits Syst. Video Technol. **31**(4), 1366–1379 (2020)

18. Liu, W., Ren, G., Yu, R., Guo, S., Zhu, J., Zhang, L.: Image-adaptive yolo for object detection in adverse weather conditions. In: Proceedings of the AAAI Conference on Artificial Intelligence, vol. 36, pp. 1792–1800 (2022)

19. Liu, X., Ma, Y., Shi, Z., Chen, J.: Griddehazenet: attention-based multi-scale network for image dehazing. In: Proceedings of the IEEE/CVF International Conference on Computer Vision, pp. 7314–7323 (2019)

20. Ma, L., Ma, T., Liu, R., Fan, X., Luo, Z.: Toward fast, flexible, and robust low-light image enhancement. In: Proceedings of the IEEE/CVF Conference on Computer Vision and Pattern Recognition, pp. 5637–5646 (2022)

21. Naik, A., Swarnakar, A., Mittal, K.: Shallow-UWnet: compressed model for underwater image enhancement (student abstract). In: Proceedings of the AAAI Conference on Artificial Intelligence, vol. 35, pp. 15853–15854 (2021)

22. Qin, X., Wang, Z., Bai, Y., Xie, X., Jia, H.: FFA-net: feature fusion attention network for single image dehazing. In: Proceedings of the AAAI Conference on Artificial Intelligence, vol. 34, pp. 11908–11915 (2020)

23. Ren, S., He, K., Girshick, R., Sun, J.: Faster R-CNN: towards real-time object detection with region proposal networks. In: Advances in Neural Information Processing Systems, vol. 28 (2015)

24. Yang, Z., et al.: Focal and global knowledge distillation for detectors. In: Proceedings of the IEEE/CVF Conference on Computer Vision and Pattern Recognition, pp. 4643–4652 (2022)

25. Zhang, Z., Jiang, Z., Liu, J., Fan, X., Liu, R.: Waterflow: heuristic normalizing flow for underwater image enhancement and beyond. ACM MM (2023)

TEFNet: Target-Aware Enhanced Fusion Network for RGB-T Tracking

Panfeng Chen[1], Shengrong Gong[2(✉)], Wenhao Ying[2], Xin Du[3], and Shan Zhong[2(✉)]

[1] School of Information Engineering, Huzhou University, Huzhou 313000, China
[2] School of Computer Science and Engineering, Changshu Institute of Technology, Suzhou 215500, China
shrgong@suda.edu.cn, sunshine620@cslg.edu.cn
[3] School of Electronic and Information Engineering, Suzhou University of Science and Technology, Suzhou 215009, China

Abstract. RGB-T tracking leverages the fusion of visible (RGB) and thermal (T) modalities to achieve more robust object tracking. Existing popular RGB-T trackers often fail to fully leverage background information and complementary information from different modalities. To address these issues, we propose the target-aware enhanced fusion network (TEFNet). TEFNet concatenates the features of template and search regions from each modality and then utilizes self-attention operations to enhance the single-modality features for the target by discriminating it from the background. Additionally, a background elimination module is introduced to reduce the background regions. To further fuse the complementary information across different modalities, a dual-layer fusion module based on channel attention, self-attention, and bidirectional cross-attention is constructed. This module diminishes the feature information of the inferior modality, and amplifies the feature information of the dominant modality, effectively eliminating the adverse effects caused by modality differences. Experimental results on the LasHeR and VTUAV datasets demonstrate that our method outperforms other representative RGB-T tracking approaches, with significant improvements of 6.6% and 7.1% in MPR and MSR on the VTUAV dataset respectively.

Keyword: RGB-T tracking · Background elimination · Complementary information

1 Introduction

RGB-T tracking combines visible images (RGB) and thermal images (T) for object tracking, and is served as a cross-modal object tracking technique. The RGB image contains characteristics such as rich color and texture about the target but is sensitive to light conditions. While the T image is insensitive to light conditions due to its ability in reflecting the temperature distribution on the surface of the object, it fails to capture the detailed texture information of the object [1], as shown in Fig. 1(a). Therefore, precision and robustness for visual object tracking can be improved by leveraging the complementary characteristics of both modalities.

© The Author(s), under exclusive license to Springer Nature Singapore Pte Ltd. 2024
Q. Liu et al. (Eds.): PRCV 2023, LNCS 14434, pp. 432–443, 2024.
https://doi.org/10.1007/978-981-99-8549-4_36

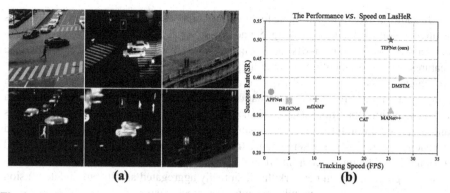

Fig. 1. (a) Comparison of RGB and T images. Top row is the RGB, bottom row is the T. (b) Comparison of tracking success and speed on the LasHeR dataset for different methods.

The extraction of single-modal feature for RGB-T affects the two-modal fusion significantly. Consequently, many works attempted to extract more accurate single-modal feature. An efficient approach is extracting single-modal features for both template and search region, and then estimating the similarity between them by cross-correlation [3–5]. Guo et al. [4] enhanced the extracted features from two modalities using channel attention and spatial attention modules, and obtained the final results through cross-correlation operations, ensuring both performance and computational efficiency. Zhang et al. [5] designed a complementarity-aware multi-modal feature fusion module to enhance the discriminability of cross-correlation features by reducing the modality differences between single-modal features. While cross-correlation linear operations exhibit high efficiency, they also result in the loss of semantic information. To tackle this issue, several studies introduced Transformers as an alternative approach for similarity measurement [6, 7], effectively addressing the problem of information loss. Feng et al. [6] first fused the extracted features from two modalities by concatenation and then utilize cross-attention to measure the similarity between the template and search regions. Zhu et al. [7] proposed a progressive fusion Transformer to integrate the features from both modalities and employ an encoder-decoder architecture for similarity measurement between the template and search regions.

However, the feature extraction and similarity measurement in the aforementioned methods are conducted in isolation, resulting in a lack of interaction between the template and search regions during the feature extraction process. This absence of interaction hinders the representation of discriminative features when the target undergoes significant deformation. Moreover, these approaches do not consider the utilization of background information. To address these issues, we propose TEFNet. Inspired by this work [8], we concatenate the template and search regions from different modalities, and input them into an encoder. We employ self-attention operations to enable interaction between the template and search regions, thereby better extracting discriminative features for the target. Furthermore, we introduce a background elimination module in the encoder to further mitigate the impact of background regions.

In addition to obtaining reliable features, the complementary fusion of two modalities is also crucial for RGB-T tracking. Some cross-modal fusion methods employed concatenation [9–11], element-wise addition [12, 13], or attention [14, 15] strategies to fuse features from two modalities. Mei et al. [14] utilized self-attention and cross-attention to achieve cross-modal global exploration and integrated global information through channel attention. Finally, the information from the two modalities was fused through a concatenation operation. Zhang et al. [16] combined the aforementioned strategies and proposed a combined approach using pixel-level fusion, feature-level fusion, and decision-level fusion to further enhance tracking performance. Some methods proposed attribute-based feature decoupling fusion [17–19]. Xiao et al. [19] proposed an attribute-based progressive fusion network that adaptively aggregated all attribute-related fusion features using self-attention and cross-attention mechanisms.

However, the aforementioned methods do not consider the impact of modality differences and cannot fully utilize complementary information, thereby hindering the feature representation capability of specific modalities in the fusion results. To address these issues, we propose a dual-layer fusion module. Firstly, this module utilizes a channel attention mechanism to compute reliable weights for the features of different modalities. These weighted features are then summed to obtain an initial fusion result. Subsequently, self-attention and bidirectional cross-attention operations are performed on the initial fusion result and the features of the two modalities, respectively, to enhance the feature representation capability in the fusion result. The experimental results demonstrate the effectiveness of the proposed method. As shown in Fig. 1(b), the success rate and speed comparison between TEFNet and other RGB-T tracking algorithms on the LasHeR dataset are presented.

The main contributions of our work can be summarised as follows:

- We propose an effective cross-modal tracking framework. This framework takes the concatenation of the two modalities as input to the encoder with a background elimination module. Self-attention is employed to accurately match the target and suppress background interference, enhancing the awareness of the target.
- We designed a dual-layer fusion module, which utilizes the preliminary fusion result as an intermediary for the interaction between the two modalities. This module effectively mitigates the influence of modality differences and fully leverages complementary information.
- Extensive experimental results demonstrate that our method surpasses representative approaches on the LasHeR [25] and VTUAV [16] datasets.

2   Methods

2.1   Overview

This section provides a detailed description of our proposed TEFNet. It consists of a target-aware feature extraction network, a dual-layer fusion module, and a target classification and regression network. The detailed architecture is depicted in Fig. 2. Firstly, we employ the patch embedding method to extract features from both RGB and T images, which are subsequently used as inputs to the feature extraction network. Secondly, the inputs undergo extraction and interaction using an encoder that incorporates a background elimination module. Thirdly, a dual-layer fusion module is employed to synergistically merge the extracted features from the search regions. Finally, the target is predicted using a classification and regression network.

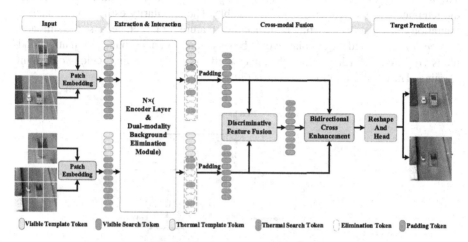

Fig. 2. The framework of TEFNet.

2.2 Target-Aware Feature Extraction Network

In the proposed method, we utilize Vision Transformer (ViT) [20] for feature extraction and similarity measurement between for different modalities. Firstly, the template and search images for each modality are cropped to patch embeddings with the same size. These embeddings are concatenated along the same dimension and then fed into the encoder, where self-attention is employed to facilitate intra-modality information interaction. The specific process is outlined as follows:

The first frame of the video is selected as the template image, while the subsequent frames of the video are used as the search images. By center-cropping around the target, we obtain the template $Z \in \mathbb{R}^{3 \times H_z \times W_z}$ and the search region $X \in \mathbb{R}^{3 \times H_x \times W_x}$. Next, the template image Z and search image X are cropped to patch embeddings. Assuming that the size of each patch is $P \times P$, then the number of patches in the template and search regions are $N_z = H_z W_z / P^2$ and $N_x = H_x W_x / P^2$, respectively. The patche sequence for the template and the search region can be obtained as $Z = \mathbb{R}^{N_z \times (3P^2)}$, $X = \mathbb{R}^{N_x \times (3P^2)}$. Afterward, the patch sequence for the template and the search region are projected onto the D-dimensional space by linear transformation as $Z_p = \mathbb{R}^{N_z \times D}$ and $X_p = \mathbb{R}^{N_x \times D}$, respectively. Through the above opearations, we can get the tokens of the templates for visible modality and for thermal modality, denoted as Z_p^v and Z_p^t, respectively. Similarly, the symbols X_p^v and X_p^t represent the tokens of the search regions for both modalities. All the tokens are then concatenated into a vector H:

$$H = [Z_p^v; X_p^v; Z_p^t; X_p^t] \tag{1}$$

436   P. Chen et al.

where the superscripts v and t represent the visible and thermal modalities, respectively. The token H is then coupled with position encoding and fed into transformer-based encoder, so the self-attention operation can be applied to facilitate infor-mation inter-action. The concatenation of Z_p^v and X_p^v, or Z_p^t and X_p^t, makes the interaction between the template and the search region be possible. The simularity estimation in such an interation can assist in extracting more accurate feature for target, because the similarity between the target and the template would be very small. A background elimination module is proposed to further reduce the adverse effects of background regions on feature learning. The structure of the encoder is illustrated in Fig. 3.

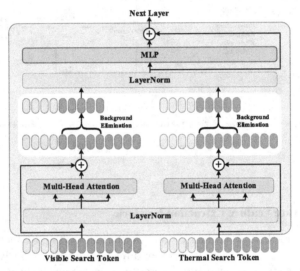

Fig. 3. Encoder structure with background elimination module.

The output of the single-modal multi-head self-attention can be written as *MSA*:

$$MSA = Softmax\left(\frac{[Q_zK_z^T, Q_zK_x^T, Q_xK_z^T, Q_xK_x^T]}{\sqrt{d_k}}\right) \bullet [V] \triangleq \begin{bmatrix} A_{zz} & A_{zx} \\ A_{xz} & A_{xx} \end{bmatrix} \bullet \begin{bmatrix} V_z \\ V_x \end{bmatrix}$$

(2)

where Q, K and V are the query matrix, key matrix and value matrix, respectively. The subscripts z and x indicate matrix entries belonging to the template and search regions, respectively. A_{zx}^v and A_{xz}^v represent the estimated similarity between the template and search region. After obtaining the output of multi-headed self-attention, the background elimination module is utilized to eliminate background tokens based on their attention weights. Higher weights indicate a higher likelihood of being associated with the target, while lower weights indicate background regions, as mentioned in [8]. However, we expanded it from a single-modal to a multi-modal setting. Specifically, each modality obtains its own set of discarded background tokens, and by taking their intersection, we obtain the common discarded background tokens. The final output of the encoder is denoted as R:

$$R \in \mathbb{R}^{(N_z^v + N_x^v + N_z^t + N_x^t) \times D} \tag{3}$$

After passing through all the encoders, the search region features of both modalities can be directly utilized for subsequent processing without requiring similarity measurement with the template.

2.3 Dual-Layer Fusion Module

Different modalities can mutually provide complementary information, thus harnessing this information can enhance the understanding of targets or scenes, ultimately improving tracker accuracy [2]. To better utilize this complementary information, we design a dual-layer fusion module comprising two stages. The first stage is called discriminative feature fusion, where the weights of different modality features are estimated by using channel attention mechanism, as shown in Fig. 4(a).

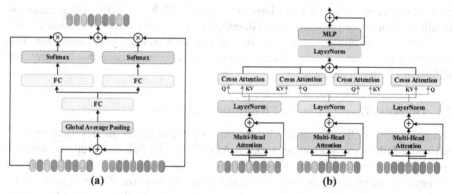

Fig. 4. (a) Discriminative feature fusion architecture. (b) Bidirectional cross-enhancement architecture.

Specifically, the features of the search regions for the two modalities, obtained from the feature extraction network, are represented as $R_v \in \mathbb{R}^{N_x^v \times D}$ and $R_t \in \mathbb{R}^{N_x^t \times D}$, respectively. We reshape their sizes as $R_x^v \in \mathbb{R}^{C \times H \times W}$ and $R_x^t \in \mathbb{R}^{C \times H \times W}$. The weights, calculated by adding R_x^v and R_x^t with an element-wise approach followed by global average pooling (GAP), fully connected layer (FC) and Softmax, are taken to weight the features. The channel wights for visible modalitiy is shown as Eq. (4).

$$W_x^v = Softmax(\text{FC}(GAP(R_x^v + R_x^t))) \tag{4}$$

Similarly, the channel weights W_x^t for thermal infrared can be obtained by the similar process, with the only diffference in FC. The weights W_x^v and W_x^t are multiplied with the original single-modal features and then are summed to generate the preliminary fusion result R_x^f:

$$R_x^f = R_x^v \times W_x^v + R_x^t \times W_x^t \tag{5}$$

It is worth noting that this module can discard the backgroud tokens by performing an intersection operation. The background tokens are assigned a value of zero after passing through the channel attention. Consequently, the intersection of these tokens represents the common discarded background tokens, which are set to zero, while the remaining parts are assigned values based on their weights.

The second stage is bidirectional cross-enhancement, and its structure is illustrated in Fig. 4(b). First, by enhancing the preliminary fusion result R_x^f through multi-headed self-attention, we can obtain:

$$\overline{R_x^f} = LN(R_x^f + MHA(R_x^f)) \tag{6}$$

The search region features of both modalities, R_x^v and R_x^t, are also experience to the same operation to obtain $\overline{R_x^v}$ and $\overline{R_x^t}$. The bidirectional cross-attention is utilized to enhance the representation capacity of different modalities in the fused features. Specifically, when $\overline{R_x^f}$ is taken as Q, $\overline{R_x^v}$ and $\overline{R_x^t}$ be regarded as K and V, while $\overline{R_x^v}$ and $\overline{R_x^t}$ are served as Q, $\overline{R_x^f}$ can be utilized as K and V. This will yield four results, which are then summed and fed into a Multi-Layer Perceptron (MLP) for further enhancement to obtain the final fusion result. Instead of directly applying cross-attention between the two modalities, we use the preliminary fusion result as an intermediary for interaction. This effectively addresses the issue of suppressing feature representation due to significant differences between modalities.

2.4 Objective Classification and Regression Networks

After obtaining the final fusion result, the features are reshaped into a 2D format and then fed into a fully convolutional network for prediction. The Full convolutional network consists of four stacked Conv-BN-ReLU layers. These layers are utilized to predict the target class, local offsets, and normalized bounding box size. The position with the highest classification score is considered the target location, and the normalized bounding box is calculated. Finally, the normalized bounding box coordinates are mapped back to the corresponding coordinates in the original image based on the image cropping ratio. During the training process, the total loss $\mathcal{L}$ is a weighted combination of multiple loss functions:

$$\mathcal{L} = L_{cls} + \lambda_{iou}L_{iou} + \lambda_{L_1}L_1 \tag{7}$$

where L_{cls} is the weighted focal loss [21] used for classification. For the prediction for bounding boxes, the L_1 loss and generalized IoU loss [22] are utilized for bounding box regression. The hyperparameters λ_{iou} and λ_{L_1} are set to 2 and 5, respectively.

3 Experiment

3.1 Experimental Details

We implemented our method using the PyTorch framework. The model training was performed on 4 Tesla P100 GPUs, and the inference speed was evaluated on an NVIDIA RTX3080 GPU. The input template image size was 128 and the search image size was

256. A vanilla ViT-Base [20] model pre-trained by MAE [23] was used as the the feature extraction network. We conducted training using the LasHeR and VTUAV datasets. The batch size was set to 16, and the initial learning rate for the backbone was set to 2×10^{-5}, while the remaining components were set to 2×10^{-4}. The training process consisted of 100 epochs, with 60,000 image pairs processed per epoch. The learning rate was reduced by a factor of 10 after 80 epochs. The model was trained using the AdamW optimizer [24]. The weight decay was set to 10^{-4}.

3.2 Public Dataset Evaluation

3.2.1 Evaluation on the LasHeR Dataset

LasHeR is the largest RGBT tracking dataset, comprising 1,224 video sequences. The dataset consists of 979 video sequences in the training set and 245 video sequences in the test set, amounting to a total of 730K image pairs. The dataset has 19 challenges such as fast motion, occlusion, and appearance similarity. In order to further assess the effectiveness of the proposed model, we compare its performance with eight RGB-T tracking algorithms on the LasHeR dataset, as presented in Table 1. The evaluation metrics used to measure performance include precision rate (PR), normalized precision rate (NPR), and success rate (SR).

Table 1. Comprehensive comparison on the LasHeR dataset. The best two results are shown in red and blue fonts.

Methods	Year	PR	NPR	SR	FPS
mfDiMP [10]	2019	44.7	39.5	34.3	10.3
CAT [17]	2020	45.0	39.5	31.4	20
FANet [26]	2021	44.1	38.4	30.9	19
MANet++ [12]	2021	46.7	40.4	31.4	25.4
DMCNet [11]	2022	49	43.1	35.5	2.3
APFNet [19]	2022	50.0	43.9	36.2	1.3
DRGCNet [15]	2023	48.3	42.3	33.8	4.9
DMSTM [14]	2023	54.1	50.3	40.0	27.6
TEFNet (Ours)	2023	63.0	59.1	50.1	25.5

The results demonstrate the effectiveness of our algorithm. We selected four video sequences from the Lasher test set and compared the tracking results with APFNet [19], MANet++ [12], and mfDiMP [10]. The visualized tracking results are shown in Fig. 5. TEFNet excels in accurately tracking and localizing small targets, out-of-view targets, and similar targets due to its effective utilization of background information, which enhances the discriminative nature of target features. Moreover, the bidirectional cross-enhancement fusion module achieves superior fusion results by effectively addressing modality differences. In contrast, the other three methods do not exploit background information. Specifically, APFnet [19] relies on attribute-based techniques, and its tracking accuracy notably diminishes when confronted with video attributes that lie beyond its training domain.

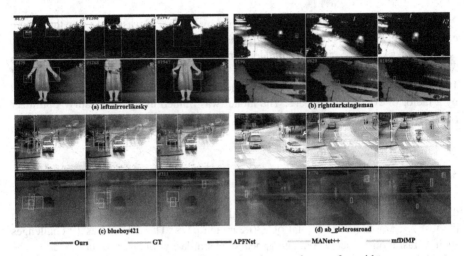

(a) leftmirrorlikesky (b) rightdarksingleman

(c) blueboy421 (d) ab_girlcrossroad

Ours ——— GT ——— APFNet ——— MANet++ ——— mfDiMP

Fig. 5. Comparison of our tracker with the other three trackers on four video sequences.

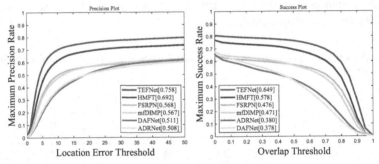

Fig. 6. Overall performance on the VTUAV test set. (a) Represents the maximum precision rate curve, (b) Represents the maximum success rate curve.

Evaluation on the VTUAV Dataset

The VTUAV dataset is a large-scale RGB-T dataset specifically designed for unmanned aerial vehicle (UAV) tracking. It comprises 500 video sequences categorized into short-term and long-term tracking scenarios, with an image resolution of 1920×1080. The dataset encompasses 10 challenging attributes. We merged the short-term and long-term tracking datasets for evaluation, As shown in Fig. 6. TEFNet was compared with five different RGB-T tracking algorithms (DAFNet [9], ADRNet [18], FSRPN [27], mfDiMP [10], HMFT [16]). The results indicate that TEFNet achieved a maximum accuracy of 75.8% and a maximum success rate of 64.9% on the dataset, surpassing HMFT with improvements of +6.6% and +7.1% respectively.

We visualize the tracking results in two videos, as shown in Fig. 7. The experimental results demonstrate that our method excels in capturing global information. It can promptly recognize the target when it goes out of the field of view and reappears within the view. Additionally, our method showcases stable tracking performance when encountering thermal crossover, effectively leveraging complementary information.

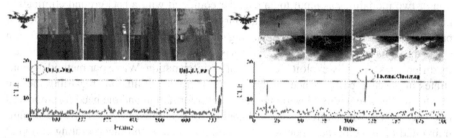

Fig. 7. Drone tracking for Visualization. Tracking targets are marked in red and CLE indicates central location error.

3.3 Ablation Experiments and Analysis

We pruned the Dual-layer Fusion Module by removing certain modules and modifying their structures for comparison with the original method. The first variant is TEFNet_dff, which removes the bidirectional cross-enhancement and retains only the discriminative featureusion module. The second variant, named TEFNet_QV, uses the preliminary fusion result as Q and the features from both modalities as K and V in the bidirectional cross-enhancement. The third variant, named TEFNet_VQ, swaps the roles, with the preliminary fusion result as K and V, and the features from both modalities as Q in the bidirectional cross-enhancement. The experimental results are shown in Table 2.

Table 2. Test results of different variants on the Lasher dataset

Methods	PR	NPR	SR	FPS
TEFNet	**63.0**	**59.1**	**50.1**	25.3
TEFNet_dff	54.5	50.8	43.2	34
TEFNet_QV	61.9	57.9	49.0	**28.4**
TEFNet_VQ	62.3	58.4	49.4	28.2

The experimental results demonstrate that the inclusion of the bidirectional cross-enhancement leads to significant improvements. It achieves an 8.5% increase in PR, an 8.3% increase in NPR, and a 6.9% increase in SR. These findings highlight the effectiveness of our bidirectional cross-enhancement in addressing the challenge of suppressed feature representation caused by significant modality differences. Moreover, the bidirectional cross-attention outperforms the unidirectional cross-modal attention, emphasizing the equal importance of both modality-specific feature representation and fused feature representation.

4 Conclusion

This paper presents an effective RGB-T tracking framework that utilizes the concatenation of two modalities as input to an encoder equipped with a background elimination module. This approach enables accurate target matching and effective suppression of background interference, thereby enhancing target perception. In the cross-modal fusion stage, direct interaction between modalities is avoided to eliminate the impact of modality differences and fully exploit complementary information. While our method demonstrates impressive performance on the dataset, there are limitations regarding the use of a fixed template without leveraging temporal information and the need for improved inference speed, which affects practical applications. Future directions involve streamlining and optimizing the model structure and parameters and incorporating temporal information by replacing the fixed template.

Acknowledgement. This work was supported by the National Natural Science Foundation of China (61972059, 62376041,42071438, 62102347), China Postdoctoral Science Foundation(2021M69236), Key Laboratory of Symbolic Computation and Knowledge Engineering of Ministry of Education, Jilin University (93K172021K01).

References

1. Tang, Z., et al.: A survey for deep RGBT tracking. arXiv preprint arXiv:2201.09296 (2022)
2. Zhang, P., Wang, D., Lu, H.: Multi-modal visual tracking: Review and experimental comparison. arXiv preprint arXiv:2012.04176 (2020)
3. Peng, J., Zhao, H., Hu, Z., et al.: Siamese infrared and visible light fusion network for RGB-T tracking. Int. J. Mach. Learn. Cybern. 1–13 (2023)
4. Guo, C., Xiao, L.: High speed and robust RGB-thermal tracking via dual attentive stream siamese network. In: IGARSS 2022–2022 IEEE International Geoscience and Remote Sensing Symposium, pp. 803–806. IEEE (2022)
5. Zhang, T., Liu, X., Zhang, Q., et al.: SiamCDA: complementarity-and distractor-aware RGB-T tracking based on Siamese network. IEEE Trans. Circuits Syst. Video Technol. **32**(3), 1403–1417 (2021)
6. Feng, M., Su, J.: Learning reliable modal weight with transformer for robust RGBT tracking. Knowl.-Based Syst..-Based Syst. **249**, 108945 (2022)
7. Zhu, Y., Li, C., Wang, X., et al.: RGBT tracking via progressive fusion transformer with dynamically guided learning. arXiv preprint arXiv:2303.14778 (2023)
8. Ye, B., Chang, H., Ma, B., et al.: Joint feature learning and relation modeling for tracking: A one-stream framework. In: Ferrari, V., Hebert, M., Sminchisescu, C., Weiss, Y. (eds.) ECCV 2018. LNCS, vol. 11206, pp. 544–559. Springer, Cham (2022). https://doi.org/10.1007/978-3-030-01216-8_33
9. Gao, Y., Li, C., Zhu, Y., et al.: Deep adaptive fusion network for high performance RGBT tracking. In: Proceedings of the IEEE/CVF International Conference on Computer Vision Workshops (2019)
10. Zhang, L., et al.: Multi-modal fusion for end-to-end RGB-T tracking. In: Proceedings of the IEEE/CVF International Conference on Computer Vision Workshops (2019)
11. Lu, A., Qian, C., Li, C., et al.: Duality-gated mutual condition network for RGBT tracking. IEEE Trans. Neural Netw. Learn. Syst. (2022)

12. Zhang, H., Zhang, L., Zhuo, L., et al.: Object tracking in RGB-T videos using modal-aware attention network and competitive learning. Sensors **20**(2), 393 (2020)

13. Cvejic, N., et al.: The effect of pixel-level fusion on object tracking in multi-sensor surveillance video. In: 2007 IEEE Conference on Computer Vision and Pattern Recognition, pp. 1–7. IEEE (2007)

14. Zhang, F., Peng, H., Yu, L., et al.: Dual-modality space-time memory network for RGBT tracking. IEEE Trans. Instrum. Meas. (2023)

15. Mei, J., Zhou, D., et al.: Differential reinforcement and global collaboration network for RGBT tracking. IEEE Sens. J. **23**(7), 7301–7311 (2023)

16. Zhang, P., Zhao, J., Wang, D., et al.: Visible-thermal UAV tracking: a large-scale benchmark and new baseline. In: Proceedings of the IEEE/CVF Conference on Computer Vision and Pattern Recognition, pp. 8886–8895 (2022)

17. Li, C., Liu, L., Lu, A., Ji, Q., Tang, J.: Challenge-aware RGBT tracking. In: Vedaldi, A., Bischof, H., Brox, T., Frahm, J.-M. (eds.) ECCV 2020. LNCS, vol. 12367, pp. 222–237. Springer, Cham (2020). https://doi.org/10.1007/978-3-030-58542-6_14

18. Zhang, P., et al.: Learning adaptive attribute-driven representation for real-time RGB-T tracking. Int. J. Comput. VisionComput. Vision **129**, 2714–2729 (2021)

19. Xiao, Y., Yang, M., Li, C., et al.: Attribute-based progressive fusion network for RGBT tracking. In: Proceedings of the AAAI Conference on Artificial Intelligence, pp. 2831–2838 (2022)

20. Dosovitskiy, A., et al.: An image is worth 16x16 words: transformers for image recognition at scale. arXiv preprint arXiv:2010.11929 (2020)

21. Law, H., Deng, J.: Cornernet: detecting objects as paired keypoints. In: Ferrari, V., Hebert, M., Sminchisescu, C., Weiss, Y. (eds.) Computer Vision – ECCV 2018. LNCS, vol. 11218, pp. 765–781. Springer, Cham (2018). https://doi.org/10.1007/978-3-030-01264-9_45

22. Rezatofighi, H., Tsoi, N., Gwak, J., et al.: Generalized intersection over union: a metric and a loss for bounding box regression. In: Proceedings of the IEEE/CVF Conference on Computer Vision and Pattern Recognition, pp. 658–666 (2019)

23. He, K., Chen, X., Xie, S., et al.: Masked autoencoders are scalable vision learners. In: Proceedings of the IEEE/CVF Conference on Computer Vision and Pattern Recognition, pp. 16000–16009 (2022)

24. Loshchilov, I., Hutter, F.: Decoupled weight decay regularization. arXiv preprint arXiv:1711.05101 (2017)

25. Li, C., Xue, W., Jia, Y., et al.: LasHeR: a large-scale high-diversity benchmark for RGBT tracking. IEEE Trans. Image Process. **31**, 392–404 (2021)

26. Zhu, Y., Li, C., Tang, J., Luo, B.: Quality-aware feature aggregation network for robust RGBT tracking. IEEE Trans. Intell. Veh. **6**(1), 121–130 (2020)

27. Kristan, M., Matas, J., Leonardis, A., et al.: The seventh visual object tracking vot2019 challenge results. In: Proceedings of the IEEE/CVF International Conference on Computer Vision Workshops (2019)

DARN: Crowd Counting Network Guided by Double Attention Refinement

Shuhan Chang[1], Shan Zhong[2]([✉]) [iD], Lifan Zhou[2], Xuanyu Zhou[1], and Shengrong Gong[1,2]([✉]) [iD]

[1] Northeast Petroleum University, Daqing 163318, China
[2] Changshu Institute of Technology, Suzhou 215500, China
sunshine620@cslg.edu.cn, shrgong@suda.edu.cn

Abstract. Although great progress has been made in crowd counting, accurate estimation of crowd numbers in high-density areas and full mitigation of the interference of background noise remain challenging. To address these issues, we propose a method called Double Attention Refinement Guided Counting Network (DARN). DARN introduces an attention-guided feature aggregation module that dynamically fuses features extracted from the Transformer backbone. By adaptively fusing features at different scales, this module can estimate the crowd for high-density areas by restoring the lost fine-grained information. Additionally, we propose a segmentation attention-guided refinement method with multiple stages. In this refinement process, crowd background noise is filtered by introducing segmentation attention maps as masks, resulting in a significant refinement of the foreground features. The introduction of multiple stages can further refine the features by utilizing fine-grained and global information. Extensive experiments were conducted on four challenging crowd counting datasets: ShanghaiTech A, UCF-QNRF, JHU-CROWD++, and NWPU-Crowd. The experimental results validate the effectiveness of the proposed method.

Keywords: Crowd counting · Double attention guided · Multi-stage refinement

1 Introduction

Crowd Counting employs machine learning algorithms to automatically analyze crowds in images or videos and provide estimations of their numbers. Crowd counting can effectively prevent potential hazards such as crowding and trampling and crowd overload, as well as analyze customer behavior and optimize resource allocation. Existing crowd counting methods can be divided into three categories: detection-based methods, regression-based methods, and deep learning-based methods.

Supported by the National Natural Science Foundation of China (61972059, 62376041, 42071438, 62102347), China Postdoctoral Science Foundation(2021M69236), Key Laboratory of Symbolic Computation and Knowledge Engineering of Ministry of Education, Jilin University (93K172021K01).

Q. Liu et al. (Eds.): PRCV 2023, LNCS 14434, pp. 444–456, 2024.
https://doi.org/10.1007/978-981-99-8549-4_37

Early crowd counting was conducted by detecting the number of pedestrians. Wu et al. [1] proposed silhouette oriented feature extraction methods to build body detectors. However, in complex scenes where pedestrians obscure each other, body features cannot be extracted accurately with a manual approach, hindering the further applications of detecting methods. To address the shortcomings of detection-based methods, subsequent studies have proposed regression-based counting methods to establish a mapping from manually extracted body features to image counts. For example, Idrees et al. [2] extracted different features for people from different angles and then weighted the number of people with the information represented by the different features. And Pham et al. [3] used the random forest to model the nonlinear mapping between image patches and density maps. Although regression-based methods can alleviate the reliance on detectors, they are still limited by hand-crafted features.

In recent years, with the development of deep learning, researchers have attempted to apply convolutional neural networks to crowd counting and use pseudo density maps as the primary regression target. This approach has been designed from several perspectives, including multi-scale structures [4], attention mechanisms [5], and auxiliary learning from related tasks [6], etc. Xu et al. [5] designed a multi-scale attention fusion block to cope with the problem of crowd scale variation by combining the idea of multi-scale information fusion and spatial attention filtering. Sindagi et al. [6] classified crowd counting as subtask-auxiliary learning, using cascaded CNN networks to jointly learn high-level prior and density map estimation. However, CNNs are not proficient in modeling global contexts.

Therefore, in recent years, self-attentive-based transformers have received increasing attention due to their ability to capture global features. For example, Liang et al. [7] used a pure Transformer model as the core of the network to accomplish the counting task. Chu et al. [8] proposed two Vision Transformer architectures, namely Twins-PCPVT and Twins-SVT, based on the Pyramid Vision Transformer architecture, which have shown excellent performance in image-level classification and dense detection. However, using only pure Transformer to generate features may result in generating features that focus mainly on the global context, thus ignoring fine-grained local information used for localization. To address this issue, some studies [9,10] combined the CNN and the Transformer. Combining the two models allows for a more comprehensive feature extraction while retaining the focus on fine-grained local information. To mitigate the negative impact of noisy annotations on the results, Dai et al. [11] proposed a model containing a convolutional head and a Transformer head that can supervise each other's noisy regions and collaboratively exploit different types of inductive biases.

Different from previous work, we propose a model called Double Attention Refinement Guided Counting Network, aimed at more accurately estimating the crowd numbers in high-density areas and completely alleviating interference from background noise. We combine Transformer and CNN to jointly extract features, so both global and fine-grained features can be captured. Afterward, we propose dual attention, namely, the global attention and the segmentation attention. The global attention can adaptively fuse features with different scales, while the multi-stage refinement method, guided by segmentation attention, facilitates the extraction of foreground region features with enhanced precision, effectively eliminating unwanted background features.

2   Proposed Method

2.1   Framework Overview

The network architecture of our proposed method is shown in Fig. 1. We take Twins-SVT consisting of four stages as our backbone. The stages in Twins-SVT are similar in structure and can generate feature maps at different scales. At the beginning of each stage, the input image or feature map is converted into fixed-size patches and then flattened into a series of vectors. By using Twins-SVT, both global and local features are extracted from the input image. Following Twins-SVT, the global attention-guided upsampling module (AUM) is constructed to fuse the global and local information by using attention-guided feature aggregation. Subsequently, the multi-stage refinement module is established to fuse the multi-scale features, where the feature generated in each stage is refined by the corresponding segmentation attention, to filter out background noise and highlight foreground features. Specifically, the ground truth segmentation maps generated from given point annotations are taken as the supervision information to refine the feature maps from each stage.

2.2   Attention-Guided Feature Aggregation

Although the Transformer is able to extract global features, higher-level feature maps still lack the fine-grained detail to be effectively reconstructed by simple upsampling. On the other hand, shallow-level feature maps contain rich detailed information to accurately locate individuals in the population. Therefore, in our approach, we propose an attention-guided feature aggregation module that leverages the advantages of both shallow and deep feature maps by fusing them together. This integration allows us to exploit the advantages of shallow feature maps in accurate individual localization while benefiting from the global contextual information provided by deep feature maps.

Traditionally, the output feature maps $X \in \mathbb{R}^{C \times H \times W}$ extracted from stages 3 and 4 of the backbone are expanded to 1/8 of the input size. Subsequently, they are fused with the feature maps generated from stage 2 using a layer-by-layer stacking method. However, traditional upsampling methods, such as nearest neighbor interpolation and bilinear interpolation, commonly with fixed interpolation kernels, can't make full use of the global context in the sub-pixel range. As a result, some important details in the feature map are neglected.

We propose an attention-guided upsampling module (AUM) for processing the feature maps $X \in \mathbb{R}^{C \times H \times W}$ output from stages 3 and 4 of the backbone network. As shown in Fig. 2, the feature map $F \in \mathbb{R}^{C \times H \times W}$ in our method is fed into two branches, namely branch 1 (upper part) and branch 2 (lower part). In branch 1, a global attention pooling operation is performed on the feature map F. Specifically, branch 1 consists of two parallel 1×1 convolution. The feature map F is convolved by 1×1 of N filters, and then the attention map $F_a \in \mathbb{R}^{HW \times N}$ is generated by the softmax function. At the same time, F is convolved with C filters by 1×1 and then reshaped to obtain the feature map F_1. Finally, F_1 and F_a are matrix multiplied to produce N semantic aggregation descriptors $S \in \mathbb{R}^{C \times N}$.

In branch 2, the feature map F is obtained by an ordinary bilinear interpolation operation, which is upsampled to k times to obtain the feature map $F_k \in \mathbb{R}^{C \times kH \times kW}$ ($k = 2, 4$). The feature map F_k is then convolved with N filters using a 1×1 convolution operation. After that, we normalize the resulting feature maps along the channel dimension, obtaining N-dimensional weight vectors F_v that represent the weights of each channel in the input features. Next, these channel weight vectors F_v are used to weight the semantic descriptor S, effectively distributing the global semantic information to each channel, and obtaining the feature map F_n with global semantic information.

Next, the feature map F_n is convolved by a 1×1 convolution and then fused with the feature map F_k obtained from the original upsampling, thus recovering the global context during the upsampling process and generating the global enhanced feature map F'.

2.3   Attention-Guided Multi-stage Refinement Method

Attention-Guided Refinement Module. The fusion features F, obtained from the attention-guided feature aggregation module, are fed into the multiple refinement modules to estimate the density map. The total number of modules is represented as T and each module comprises a Multi-scale Convolutional Block (MCB) and a Segmentation-Guided Attention Layer (SAL), as illustrated in Fig. 3. The MCB is composed of three dilated convolution layers and one conventional convolutional layer in parallel.

Inspired by the method of Wang et al. [12], we use the segmentation map as an attention map to regress the density map. This approach effectively emphasizes the contribution of foreground regions in estimating the density map while reducing the influence of background region features. To achieve this goal, we introduce the segmentation-guided attention layer (SAL), which generates the attention map by estimating the segmentation map. Additionally, the segmentation-guided attention layer is a convolutional layer. To restrict the output values to the range of 0–1, we apply a sigmoid layer to the output of the attention layer, resulting in the predicted segmentation attention map denoted as M_{att}.

The structure of the second stage is the same as that of the first stage. The input for the second stage is the concatenation of the feature map F^1_{seg} and the fusion feature F. By performing the same operations as in the first stage, we obtain the refined feature map F^2_{seg} in the second stage.

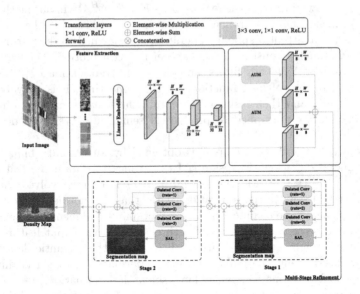

Fig. 1. The framework of Crowd Counting Networks Guided by Double Attention Refinement. First, the crowd images are transformed into vectors by Patch embedding and fed into the Transformer encoder. Next, the output feature maps from the second, third, and fourth stages are aggregated using the attention-guided feature aggregation module. The feature maps are refined using a segmentation attention-guided multi-stage refinement method, and ultimately, regression is performed using two simple convolutional layers.

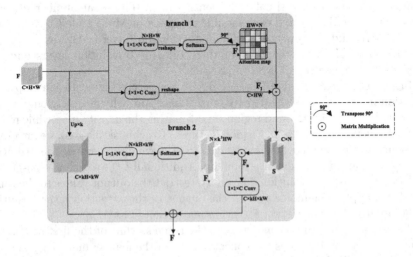

Fig. 2. Design details of the proposed attention-guided upsampling module.

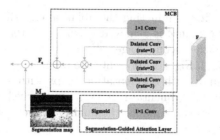

Fig. 3. Attention-Guided Refinement Module.

We conducted several ablation experiments to determine the optimal value of T. The best performance was achieved when the number of refinement stages was set to 2. Therefore, the refinement stage of our proposed network model consists of two refinement modules. The fusion features F serve as the input for both MCB and segmentation-guided attention layer, resulting in the generation of the feature F_c and the single-channel attention map M_{att}, respectively. Afterward, M_{att} as a mask is multiplied with F_c by element to element to obtain the feature map F_{seg}^1 for the first stage. Where F_{seg}^1 can be expressed by the following equation:

$$F_{seg}^1 = F_c \odot M_{att}, \tag{1}$$

where $\odot$ denotes the element-wise multiplication operation.

Generation of Segmentation Maps. To train our method, we generate ground truth segmentation maps M_{seg} for each stage of the refinement process by utilizing the given point annotations. The M_{seg} is served as the supervisory information for training the learned attention maps M_{att}. The generation of segmentation maps can be obtained by the following equation:

$$M_{seg}(x) = M(x) * O_s(x), \tag{2}$$

$$M(x) = \sum_{i=1}^{N} \delta(x - x_i), \tag{3}$$

where N represents the number of point annotations in the image, and $\delta(\cdot)$ denotes the delta function. $M(x)$ is a binary matrix with the same size as the image. The values of $M(x)$ are set to 1 at the point annotations and 0 otherwise. $O_s(x)$ represents an all-one matrix of size $s \times s$ centered at the position x. The segmentation map M_{seg} is obtained by convolving $M(x)$ with $O_s(x)$. From the resultant segmentation map M_{seg}, we can determine whether a pixel point belongs to the foreground or background based on the judgment value of 1 or 0.

The level of refinement and the detail in the generated segmentation map can be adjusted by setting a value for s. A smaller value of s corresponds to more refined segmentation maps, while a larger value yields smoother and more stable

segmentation maps. After conducting the ablation experiments, we determined
that the optimal performance is achieved when s is initialized as 17 in the first
stage and set to 13 in the second stage.

As depicted in Fig. 4, Fig. 4(a) presents the input image, while Figs. 4(b) and
4(c) display the ground truth segmentation maps with s values of 13 and 17,·
respectively. From the experimental results, it is observed that using different s
values in different stages can enhance the quality of the segmentation maps and
improve their robustness.

Fig. 4. Example input image and its corresponding ground truth segmentation map
with different s values.

2.4 Loss Function

The loss function consists of two parts, which can be written as:

$$\mathcal{L} = \mathcal{L}_{den} + \mathcal{L}_{seg}, \tag{4}$$

where $\mathcal{L}_{den}$ is the loss function used for the supervised density estimation task,
$\mathcal{L}_{seg}$ is used for the supervised segmentation task.

To optimize the segmentation attention map at each stage, we define the loss
function $\mathcal{L}_{seg}$ for the supervised segmentation task as the cross-entropy loss.

$$\mathcal{L}_{seg}(\Theta) = -\frac{1}{N}\sum_{i=1}^{N}\sum_{j,k}H_i(j,k), \tag{5}$$

where

$$H_i = M_i^{seg}log(\hat{M}_i^{seg}) + (1 - M_i^{seg})log(1 - \hat{M}_i^{seg}), \tag{6}$$

where $H(j,k)$ denotes an element in the matrix H. M_i^{seg} is the ground truth
segmentation map and $\hat{M}_i^{seg}$ refers to the predicted segmentation map.

In both refinement stages, we provide supervised signals. For each stage t,
we employ a specific loss function, denoted as $\mathcal{L}_{seg}^t$, tailored for the segmentation task. The overall segmentation loss function $\mathcal{L}_{seg}$ can be expressed in the
following form:

$$\mathcal{L}_{seg} = \alpha\sum_{t=1}^{2}\mathcal{L}_{seg}^t = \alpha\mathcal{L}_{seg}^1 + \alpha\mathcal{L}_{seg}^2, \tag{7}$$

where α is the weight for the two segmentation tasks. The optimal value of α
can be determined by grid search experiments.

3   Experiment

3.1   Implementation Details

Datasets. The proposed method is evaluated on the four largest crowd counting datasets: ShanghaiTech A [13], UCF-QNRF [14], JHU-CROWD++ [15], and NWPU-Crowd [16]. These datasets differ in terms of image resolution, number, crowd density, and color space.

Training Details. Random cropping and random horizontal flipping are utilized for data enhancement over all experiments. For the datasets UCF-QNRF, JHU-CROWD++, and NWPU-Crowd, a random crop size of 512 × 512 is adopted. Considering the relative low resolutions in images of ShanghaiTech A, the random crop size for such dataset is set to 256 × 256. The AdamW algorithm [17] is taken to optimize the proposed model, with the learning rate of 10^{-5}. By operating the grid search experiments, the loss balance parameters α is set to 0.6. All experiments are conducted using PyTorch on a single 16G Tesla P100 GPU.

Evaluation Metrics. We used two widely-used metrics to evaluate the performance of counting methods: Mean Absolute Error (MAE) and Mean Squared Error (MSE). The definitions of these metrics are shown below:

$$MAE = \frac{1}{N} \sum_{i=1}^{N} |y_i - \hat{y}_i|, \tag{8}$$

$$MSE = \sqrt{\frac{1}{N} \sum_{i=1}^{N} (y_i - \hat{y}_i)^2}, \tag{9}$$

where N represents the number of test images, and y_i and $\hat{y}_i$ denote the ground truth and predicted counts for the i-th test image, respectively.

3.2   Results and Analysis

We evaluated our model on the four datasets mentioned above and compared it with nine state-of-the-art methods. The quantitative results of counting accuracy are presented in Table 1. According to the results in Table 1, our DARN exhibits excellent accuracy on all four benchmark datasets. Specifically, on the UCF-QNRF dataset, our method achieved a new state-of-the-art performance, surpassing the previously top-performing methods GauNet [21] and CHS-Net [11]. On the challenging large-scale crowd counting dataset JHU-CROWD++, the MAE and MSE of our method reached 54.88 and 230.46, respectively, which are 2.9 and 14.64 lower than the previous optimal method, and the counting accuracy improved by 5% and 6%, respectively. On the NWPU-Crowd dataset, the MSE of our proposed method is also lower than that of the previous state-of-the-art method HDNet [22], and the counting accuracy is improved by 2.9%.

The visualization of our DARN is shown in Fig. 5.

3.3 Ablation Studies

To assess the effectiveness of the proposed DARN, we conducted extensive ablation experiments on two challenging datasets UCF-QNRF and JHU-CROWD++.

Fig. 5. Example input image and its corresponding ground truth segmentation map with different s values.

Table 1. Comparisons with the state of the arts on ShanghaiTech A, UCF-ONRF, JHU-Crowd++, and NWPU. The best performance is shown in bold and the second best is shown in underlined.

Method	UCF-QNRF		JHU-CROWD++		NWPU-Crowd		ShanghaiTech A	
	MAE	MSE	MAE	MSE	MAE	MSE	MAE	MSE
CMTL [6] (2017)	252.0	514.0	–	–	–	–	101.3	152.4
CSRNet [4] (2018)	–	–	85.9	309.2	121.3	387.8	68.2	115.0
CA-Net [18] (2019)	107.0	183.0	100.1	314.0	–	–	62.3	100.0
NoisyCC [19] (2020)	85.8	150.6	67.7	258.5	96.9	534.2	61.9	99.6
AutoScale [20] (2021)	87.5	147.8	65.9	264.8	94.1	388.2	60.5	100.4
GauNet [21] (2022)	<u>81.6</u>	153.7	58.2	<u>245.1</u>	–	–	<u>54.8</u>	**89.1**
HDNet [22] (2022)	83.2	148.3	62.9	276.2	**76.1**	<u>322.3</u>	**53.4**	89.9
MSANet [5] (2023)	83.39	145.19	<u>57.78</u>	253.33	–	–	–	–
CHS-Net [11] (2023)	83.4	<u>144.9</u>	–	–	–	–	59.3	97.8
DARN (Ours)	**79.68**	**133.22**	**54.88**	**230.46**	<u>76.2</u>	**312.91**	58.62	95.90

A. The Effect of Multi-stage Refinement. We conducted comparative experiments by stacking t refinement modules in the network, denoted as Net-t (t = {0, 1, 2, 3}), and the results are presented in Table 2. From the experimental findings, it can be observed that the proposed refinement method progressively improves counting accuracy. However, adding a third refinement module did not yield significant performance improvements. Instead, it increased computational time and even led to a decline in performance. Therefore, to strike a balance between accuracy and efficiency, our network ultimately adopts two refinement modules to accomplish the refinement of density maps.

Table 2. MAE, MSE of different configurations of our method on the UCF-QNRF and JHU-CROWD++ datasets.

Method	UCF-QNRF		JHU-CROWD++	
	MAE	MSE	MAE	MSE
Net-0	82.86	144.52	62.57	243.31
Net-1	81.16	140.42	59.33	240.83
Net-2	79.68	**133.22**	**54.88**	**230.46**
Net-3	**77.79**	138.07	57.48	239.3

Table 3. Performance comparison of applying Attention-guided Feature Aggregation Module on UCF-QNRF and JHU-CROWD++ datasets.

Method	UCF-QNRF		JHU-CROWD++	
	MAE	MSE	MAE	MSE
M_1	90.53	160.47	64.6	255.18
M_2	88.45	152.34	62.4	252.39
$M_2(EN)$	86.43	150.13	61.54	249.5
M_3	85.72	147.66	59.37	240.61
$M_3(EN)$	82.62	141.1	58.88	235.54
$M_1 + M_2 + M_3$	82.1	138.4	58.71	236.25
$DARN$	**79.68**	**133.22**	**54.88**	**230.46**

Furthermore, Fig. 6 showcases a comparison of density maps overlaid with different numbers of refinement modules. It is evident that as the refinement process proceeds, the quality of the density maps gradually improves, and the counting error diminishes accordingly. The segmentation attention maps effectively filter out background noise from the crowd, and the foreground features are further refined through the multi-stage approach. This substantiates the effectiveness of our proposed multi-stage refinement method.

Fig. 6. Attention-Guided Refinement Module.

B. The Effect of Attention-Guided Feature Aggregation Module. To assess the performance of Attention-guided Feature Aggregation Module, we conducted experiments on the features extracted at different stages and whether to use the attention-guided upsampling module, and the results are shown in Table 3. We denoted the output features of the three stages in the backbone as M_1, M_2, and M_3, representing shallow to deep levels. Additionally, we added "EN" after M_k when the attention-guided upsampling module was used in the upsampling process.

In the above experiments, we observed that using the attention-guided upsampling module can significantly improve the counting performance. Specifically, on the UCF-QNRF dataset, the MAE and MSE decreased by 3.6% and 4.4%, respectively, after adding the attention-guided upsampling module to feature M_3. In addition, from the feature aggregation experiments we find that the aggregated performance is significantly better than that of the single-feature solution. Overall, the introduction of the attention-guided upsampling module and the strategy of feature aggregation both play a positive role in improving the performance.

4   Conclusion

This paper aims to propose a high-performance crowd counting network to deal with the difficulty of counting dense crowd areas and background clutter interference counting problems in crowd counting. To this end, we introduce double attention, that is, propose a global attention-guided feature aggregation module and a segmentation attention-guided multi-stage refinement method. The former solves the problem that it is easy to lose fine-grained information in the process of crowd feature fusion, which makes it difficult to estimate the number of people in dense areas. The latter effectively filters out the background noise of the crowd and further refines the foreground features. At the same time, a multi-level supervision mechanism is used in the refinement stage to further distill features to recover the lost information. Through experiments on several major datasets, we demonstrate that the proposed double attention refinement guided counting network outperforms the current state-of-the-art on most of the major datasets used.

References

1. Wu, B., Nevatia, R.: Detection of multiple, partially occluded humans in a single image by bayesian combination of edgelet part detectors. In: Tenth IEEE International Conference on Computer Vision, pp. 90–97. IEEE (2005)
2. Idrees, H., Saleemi, I., Seibert, C., et al.: Multi-source multi-scale counting in extremely dense crowd images. In: Proceedings of the IEEE Conference on Computer Vision and Pattern Recognition, pp. 2547–2554. IEEE (2013)
3. Pham, V.Q., Kozakaya, T., Yamaguchi, O., et al.: Count forest: co-voting uncertain number of targets using random forest for crowd density estimation. In: Proceedings of the IEEE International Conference on Computer Vision, pp. 3253–3261. IEEE (2015)
4. Li, Y., Zhang, X., Chen, D.: CSRnet: dilated convolutional neural networks for understanding the highly congested scenes. In: Proceedings of the IEEE Conference on Computer Vision and Pattern Recognition, pp. 1091–1100. IEEE (2018)
5. Xu, Y., Liang, M., Gong, Z.: A crowd counting method based on multi-scale attention network. In: 2023 3rd International Conference on Neural Networks, Information and Communication Engineering, pp. 591–595. IEEE (2023)
6. Sindagi, V.A., Patel, V.M.: CNN-based cascaded multi-task learning of high-level prior and density estimation for crowd counting. In: 14th IEEE International Conference on Advanced Video and Signal Based Surveillance (AVSS), pp. 1–6. IEEE (2017)
7. Liang, D., Chen, X., Xu, W., et al.: Transcrowd: weakly-supervised crowd counting with transformers. SCIENCE CHINA Inf. Sci. **65**(6), 160104 (2022)
8. Chu, X., Tian, Z., Wang, Y., et al.: Twins: revisiting the design of spatial attention in vision transformers. In: Advances in Neural Information Processing SystemSL, vol. 34, pp. 9355–9366 (2021)
9. Lin, H., Ma, Z., Hong, X., et al.: Semi-supervised crowd counting via density agency. In: Proceedings of the 30th ACM International Conference on Multimedia, pp. 1416–1426. ACM (2022)
10. Lin, H., Ma, Z., Ji, R., et al.: Boosting crowd counting via multifaceted attention. In: Proceedings of the IEEE/CVF Conference on Computer Vision and Pattern Recognition, pp. 19628–19637. IEEE (2022)
11. Dai, M., Huang, Z., Gao, J., et al.: Cross-head supervision for crowd counting with noisy annotations. In: 2023 IEEE International Conference on Acoustics, Speech and Signal Processing (ICASSP), pp. 1–5. IEEE (2023)
12. Wang, Q., Breckon, T.P.: Crowd counting via segmentation guided attention networks and curriculum loss. IEEE Trans. Intell. Transp. Syst. **23**(9), 15233–15243 (2022)
13. Zhang, Y., Zhou, D., Chen, S., et al.: Single-image crowd counting via multi-column convolutional neural network. In: Proceedings of the IEEE Conference on Computer Vision and Pattern Recognition, pp. 589–597. IEEE (2016)
14. Idrees, H., et al.: Composition loss for counting, density map estimation and localization in dense crowds. In: Ferrari, V., Hebert, M., Sminchisescu, C., Weiss, Y. (eds.) ECCV 2018. LNCS, vol. 11206, pp. 544–559. Springer, Cham (2018). https://doi.org/10.1007/978-3-030-01216-8_33
15. Sindagi, V.A., Yasarla, R., Patel, V.M.: Jhu-crowd++: large-scale crowd counting dataset and a benchmark method. IEEE Trans. Pattern Anal. Mach. Intell. **44**(5), 2594–2609 (2020)

16. Wang, Q., Gao, J., Lin, W., et al.: NWPU-crowd: a large-scale benchmark for crowd counting and localization. IEEE Trans. Pattern Anal. Mach. Intell. **43**(6), 2141–2149 (2020)
17. Loshchilov, I., Hutter, F.: Decoupled Weight Decay Regularization. In: 7th International Conference on Learning Representations. ICLR (2019)
18. Liu, W., Salzmann, M., Fua, P.: Context-aware crowd counting. In: Conference on Computer Vision and Pattern Recognition, pp. 5099–5108. IEEE (2019)
19. Wan, J., Chan, A.: Modeling noisy annotations for crowd counting. In: Advances in Neural Information Processing Systems, vol. 33, pp. 3386–3396 (2020)
20. Xu, C., Liang, D., Xu, Y., et al.: Autoscale: learning to scale for crowd counting. Int. J. Comput. Vision **130**(2), 405–434 (2022)
21. Cheng, Z.Q., Dai, Q., Li, H., et al.: Rethinking spatial invariance of convolutional networks for object counting. In: Proceedings of the IEEE/CVF Conference on Computer Vision and Pattern Recognition, pp. 19638–19648. IEEE (2022)
22. Gu, C., Wang, C., Gao, B.B., et al.: HDNet: a hierarchically decoupled network for crowd counting. In: IEEE International Conference on Multimedia and Expo (ICME), pp. 1–6. IEEE (2022)

DFR-ECAPA: Diffusion Feature Refinement for Speaker Verification Based on ECAPA-TDNN

Ya Gao[1], Wei Song[1,2,3(✉)], Xiaobing Zhao[1,2,3], and Xiangchun Liu[1]

[1] School of Information Engineering, Minzu University of China, Beijing 100081, China
[2] National Lauguage Resource Monitoring and Research Center of Minority Languages, Minzu University of China, Beijing 100081, China
[3] Key Laboratory of Ethnic Language Intelligent Analysis and Security Governance of MOE, Minzu University of China, Beijing, China
songwei@muc.edu.cn

Abstract. Diffusion Probabilistic Models have gained significant recognition for their exceptional performance in generative image modeling. However, in the field of speech processing, a large number of diffusion-based studies focus on generative tasks such as speech synthesis and speech conversion, and few studies apply diffusion models to speaker verification. We investigated the integration of the diffusion model with the ECAPA-TDNN model. By constructing a dual-network branch architecture, the network further extracts and refines speaker embeddings under the guidance of the intermediate activations of the pre-trained DDPM. We put forward two methods for fusing network branch features, both of which demonstrated certain improvements. Furthermore, our proposed model also provides a new solution for semi-supervised cross-domain speaker verification. Experiments on Voxceleb and CN-Celeb show that DFR-ECAPA outperform origin ECAPA-TDNN by around 20%.

Keywords: Speaker verification · Diffusion model · Feature fusion

1 Introduction

Speaker verification (SV), as a biometric identification technology, is widely used in various fields such as financial transactions, security protection, and smart home systems due to its advantages of convenient data acquisition and non-contact verification.

The advancement of deep learning has led to the emergence of the d-vector [20] model and x-vector [18] model for speaker verification. These models leverage neural networks to extract speaker embeddings, effectively capturing speaker characteristics. Notably, their performance surpasses that of the traditional i-vector [3] model, marking a significant advancement in speaker verification technology. In recent years, the ResNet architecture, based on 2D-CNNs,

Supported by the Graduate Research and Practice Projects of Minzu University of China(SZKY2022085).

has gained popularity as a way of extracting speaker embeddings, which is commonly referred to as the r-vector [22]model. ECAPA-TDNN [5] is currently recognized as the prevailing model in speaker verification. It incorporates the one-dimensional convolutional neural network structure from the x-vector model and employs a frame-utterance feature extract method. To obtain multi-scale features, three 1-Dimensional SE Res2Blocks are employed in ECAPA-TDNN. Whether it is ResNet based model or ECAPA-TDNN, they are commonly trained using a classification objective to optimize the network. Different from the training phase, the testing phase is no longer for the purpose of classification. It only uses the intermediate output as speaker embedding, and measures the similarity between enroll utterance and test utterance through cosine similarity or PLDA algorithm.

In order to obtain better results, many architectures are proposed on the basis of ECAPA-TDNN. ECAPA CNN-TDNN [19] introduces 2D convolutional stem as the initial stage of feature processing, and therefore the benefits of TDNN (Time-Delay Neural Network) architecture and 2-D convolution will be integrated in this approach. MFA-TDNN [12] and SKA-TDNN [13] apply attention mechanisms to speaker verification tasks to facilitate context fusion in a data-driven manner and adaptively emphasize important areas for more effective information capture. PCF-ECAPA [23] proposed a progressive channel fusion strategy, which significantly improves the performance of TDNN models by extending the depth and adding branches to extract deep representations of features.

To extract speaker personality information more comprehensively, current researchers focus on optimizing from a single path by increasing network depth or integrating multi-scale features. Nevertheless, as the network depth increases and the parameter scale expands, training the model becomes more challenging, and in this situation achieving further performance improvements becomes increasingly difficult. This issue becomes especially pronounced when dealing with speaker verification tasks in short utterances. In the field of image processing, building multi-branch networks horizontally is another way to improve performance, which can achieve feature enhancement and refinement. For example, in Pix2PixHD [21], the global generator network branch and the local enhancer network branch are constructed, which implements the joint training of the generator by adding the features of the two network branches in element-wise. In the event image classification task, [11] uses three pre-trained networks as visual feature extractors for scene, object and human respectively, and fuses the output features of the fully connected layer of the three network branches for classification.

Denoising Diffusion Probabilistic Models (DDPM) [8] have attracted researchers for its ability to generate highly realistic and detailed results, which makes them the current preferred option among generation models. It realizes the modeling of data distribution by adding noise and denoising in the forward process and reverse process respectively. DDPM has demonstrated remarkable success in image inpainting and image super-resolution reconstruction [16], surpassing the performance of GANs (Generative Adversarial Networks), VAEs (Variational Autoencoders), and other generative models. Furthermore, DDPM has also made significant advancements in text processing [24] and video process-ing [9]. In the domain of audio processing, DDPM has attained state-of-the-art performance in areas such as speech synthesis [10], speech enhancement [17], and speech conversion [15]. [6] utilizes diffusion probabilistic-based multichan-nel speech enhancement as front-end to reconstruct the clean signal from the multichannel noisy input, improving far-field speaker verification performance. However, DDPM is not directly applied in speaker verification in this way. Recently, researchers have been exploring the application of DDPM in various domains beyond generation tasks. In [1], the intermediate activations of pre-trained DDPM noise learning network are extracted and utilized for training a specifically designed classifier. This approach leads to outstanding semantic segmentation results, indicating that DDPM can also function as representation learners. [4] further explores the applications of the pre-trained DDPM model in classification tasks and proposes a joint diffusion model that can simultaneously achieve excellent generation quality and accurate classification performance.

Inspired by these recent advances, we propose Diffusion Feature Refinement based on ECAPA-TDNN (DFR-ECAPA), a dual-branch speaker feature refine-ment method that combines diffusion models and ECAPA-TDNN. It aims to enhance the utilization of contextual information by leveraging separate net-work branches. We train the diffusion model by utilizing speech data without speaker labels first. Afterward, the intermediate activations of the pre-trained DDPM will be extracted and integrated with the ECAPA-TDNN for subsequent training. We further explored the two dual-branch featured fusion methods of DFR-ECAPA-ML and DFR-ECAPA-LL, and tested the performance of DFR-ECAPA under different duration utterances. At the same time, we explore the issue of cross-domain in speaker verification, and apply our model in the specific task of cross-lingual speaker verification. Under the guidance of the diffusion model, DFR-ECAPA performs approximately a 20% improvement compared to the version without the diffusion model. It shows that DFR-ECAPA has the capability to enhance target domain speaker verification accuracy in situations where the label of the target domain speaker is unavailable.

To sum up, our paper's contributions can be summarized as follows:

1. To the best of our knowledge, we are the first to apply a diffusion model directly to the task of speaker verification.

2. We further explored two dual-branch feature fusion approaches, which also has certain reference for other TDNN based models.

3. Experiments demonstrate that our proposed model also provides a new solution for semi-supervised cross-domain speaker verification.

2 Denoising Diffusion Probabilistic Models

In this section we will introduce the working mechanism of Denoising Diffusion Probabilistic Models (DDPM), including the forward diffusion process and the reverse denoising process. In the forward process, Gaussian noise is incrementally added to the original image x_0, causing gradual degradation of the image, which can be considered as a Markov chain process. The original data can be restored by constructing a neural network that predicts the noise in the reverse process.

Each step of the forward diffusion process is to add noise on the basis of the previous step. For the image x_{t-1}, the forward diffusion process can be expressed as:

$$q\left(\mathbf{x}_t \mid \mathbf{x}_{t-1}\right) = \mathcal{N}\left(\mathbf{x}_t; \sqrt{1-\beta_t}\mathbf{x}_{t-1}, \beta_t\mathbf{I}\right) \tag{1}$$

β_t is a predefined hyperparameter that takes a decimal value between 0 and 1. As t increases, β_t progressively increases, and $\mathbf{I}$ represents the identity matrix. At each step, a small quantity of Gaussian noise is added to the image. With increasing t, the final diffusion result x_T follows a standard Gaussian distribution, serving as the initial state for reverse process in the generation task. Typically, t is sampled following a specific schedule when training.

Based on the above formula, the noise image x_t at any given time t can be directly obtained from the original data x_0, without the need for iterative noise addition:

$$q\left(\mathbf{x}_t \mid \mathbf{x}_0\right) = \mathcal{N}\left(\mathbf{x}_t; \sqrt{\bar{\alpha}_t}\mathbf{x}_0, (1-\bar{\alpha}_t)\mathbf{I}\right) \tag{2}$$

And $\alpha_t = 1 - \beta_t$, $\bar{\alpha}_t = \prod_{i=1}^{t} \alpha_i$, by employing the reparametrization trick, x_t can be represented by x_0 and noise ϵ as:

$$x_t = \sqrt{\bar{\alpha}_t}x_0 + \sqrt{1-\bar{\alpha}_t}\epsilon, \quad \text{with } \epsilon \sim \mathcal{N}(0, \mathbf{I}) \tag{3}$$

If the noise image x_t at time t is given, the image x_{t-1} at the previous moment can be obtained through denoising. By starting with a random noise and iteratively sampling, a real sample can be generated. For simplicity, the reverse process is also considered as a Markov chain:

$$p_\theta\left(\mathbf{x}_{t-1} \mid \mathbf{x}_t\right) = \mathcal{N}\left(\mathbf{x}_{t-1}; \boldsymbol{\mu}_\theta\left(\mathbf{x}_t, t\right), \Sigma_\theta\left(\mathbf{x}_t, t\right)\right) \tag{4}$$

In practice, DDPMs typically transform the modeling objectives related to mean and variance mentioned above into the prediction of noise ϵ. They accomplish this by constructing a noise learning network $\epsilon_\theta(x_t, t)$ that aims to make the predicted noise as similar as the actual noise added during the forward process. This procedure is constrained by utilizing Mean Squared Error (MSE) loss:

$$L = \mathbb{E}_{x,\epsilon\sim\mathcal{N}(0,1),t}\left[\left\|\epsilon - \epsilon_\theta(x_t, t)\right\|_2^2\right] \tag{5}$$

3   Proposed Method

In our study, we introduce a strategy called Diffusion Feature Refinement which builds upon the ECAPA-TDNN model (DFR-ECAPA). This approach involves two main steps. Firstly, the data representations are extracted using the Diffusion model noise learning network. Secondly, these features are fused with the ECAPA-TDNN model for further training. In this section, we provide a separate introduction for each of these aspects.

3.1 Diffusion Model Representations

In [1], it performs thorough investigation into the image representation acquired by the diffusion model in the task of image segmentation. The noise learning network $\epsilon_\theta(x_t, t)$ utilizes the UNet structure, where the decoder section of UNet consists of 18 blocks. For the input image to be segmented, add noise at a specific time t, pass it through the pre-trained denoising network, extract the intermediate activation of the decoder, and further train it by the designed multi-layer perceptron (MLP) to obtain the predicted label. Experiments have proved that diffusion model has a certain ability to capture semantic information. In this paper, the proposed method employs audio filter-bank features as the target for the diffusion model to add noise and perform denoising. The Gaussian noise added during the forward process shares the same shapes as the network input. In the reverse process, we adopt the TDNN architecture, which has fewer parameters and is more appropriate for speech-related tasks, as a substitute for the commonly used UNet architecture in image processing. The structure of the noise learning network is illustrated in Fig. 1. The network input x_t represents the filter-bank features with added noise. The main component of the network consists of three SE-Res2Blocks. The size of the convolution kernel, the dilation rate, and the scale dimension follow the configuration of the three layers in ECAPA-TDNN. To ensure that the network output has the same shape as the input for calculating the loss, a TDNN block is added at the end of the network to reshape the features, and finally obtain the output feature $Y \in R^{F \times T}$, where $F = N \times 80$. When the network learns only noise and the variance is fixed, the value of N is 1. When both the noise and variance are learned, N is set to 2. We utilize the output from all SE-Res2Blocks as representations for the

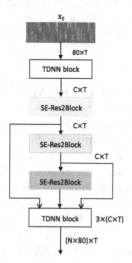

Fig. 1. Noise learning network in the diffusion model. Three SE-Res2Blocks follow the configuration corresponding to the three layers in ECAPA-TDNN.

diffusion model, and then combine them with the ECAPA-TDNN network to achieve feature refinement.

3.2   Dual Network Branch Feature Fusion

Regarding the fusion of the two network branch features of DDPM and ECAPA-TDNN, we explored two different ways of DFR-ECAPA-ML and DFR-ECAPA-LL.

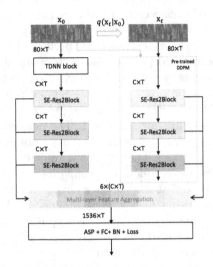

Fig. 2. Network structure of DFR-ECAPA-ML. The right branch is the pre-trained diffusion model. To simplify the illustration, two TDNN blocks are omitted, which are represented by black dotted lines. Use the noisy filter-bank features x_t as the input of the pre-trained diffusion model during training, and input the original features x_0 during inference.

Multi-Layer Feature Fusion (DFR-ECAPA-ML). The DFR-ECAPA-ML method incorporates concepts inspired by [1], where we consider the representations obtained from the diffusion model as a unified entity and process them uniformly, and its structure is shown in Fig. 2. To simplify the illustration, two TDNN blocks in diffusion model have been omitted.

Two attempts were conducted in this case. Firstly, we aim to combine ECAPA-TDNN and diffusion models as pre-training models, which means the parameters of MFA (Multi-layer Feature Aggregation), ASP (Attentive Statistic Pooling) and FC layers in ECAPA-TDNN are optimized while other weights fixed. We denote it as DFR-ECAPA-ML(A). Specifically, we extract the output features from 6 SE-ResBlocks in both ECAPA-TDNN and diffusion models, and then concatenat them along the frequency dimension. These features are then propagated to the following part and update the weight through loss backpropagation. Afterwards, we attempt to involve more parameters to learn the data representations extracted from the diffusion model, thus only keeping the weights of the neural network in the diffusion model fixed. It is named DFR-ECAPA-ML(B) to distinguish. In this way, the output features of the 6 SE-ResBlocks are still combined through frequency dimension at the MFA layer. We use the noisy filter-bank features x_t as the input of the pre-trained diffusion model during training, and input the original features x_0 during inference.

Layer2Layer Feature Fusion (DFR-ECAPA-LL). [1] also investigated the data representation capacity of the diffusion model across various time t and different decoder blocks. In terms of different blocks, shallower blocks provide more informative representations for smaller objects, whereas deeper blocks are more informative for larger objects.

Because of this, we consider the way of incorporating the features extracted from the pre-trained DDPM into the ECAPA-TDNN model in a gradual and decentralized manner. Consequently, we propose the Layer2Layer Feature Fusion approach, which is illustrated in Fig. 3. In this way, the features extracted from

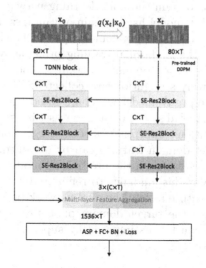

Fig. 3. Network structure of DFR-ECAPA-LL. The features extracted from each SE-ResBlock of the pre-trained DDPM will be fused into the same configuration layer in ECAPA-TDNN.

each SE-ResBlock of the pre-trained DDPM will be integrated into the same configuration layer in ECAPA-TDNN. Here we still adopt the way of features fusion along frequency dimension. To adapt the fusion method, we alter the input feature and input dimension of the first convolution kernel size of each SE-ResBlock in ECAPA-TDNN. The overall input of SE-ResBlock remains unchanged. Before entering the first convolutional layer, merge the features obtained from the pre-trained network with the original features. Modify the input dimension of the first convolutional layer to $2C$, while keeping the output dimension as C. In this context, C can be either 1024 or 512, which corresponds to its meaning in the original ECAPA-TDNN.

3.3   Cross-Domain Speaker Verification

Cross-domain is a significant challenge in the task of speaker verification. It refers to the mismatch between the training data and test data in terms of domain. Cross-domain challenges can arise from various factors, including language variations, background noise, channel disparities, and more. In real-world applications, obtaining speaker labels for the target domain is often challenging. To tackle cross-domain problems, semi-supervised learning methods have emerged as the primary solution, by training the model on the source domain and using unlabeled data of the target domain to realize domain adaptation. However, domain transfer is often time-consuming, and the effect is difficult to guarantee without corresponding labels.

The training of the diffusion model does not rely on paired data, only a subset of utterances is needed to train the noise learning network. This characteristic enables the DFR-ECAPA model have certain advantages in target domain feature learning in the absence of target domain speaker labels.

Instead of following the traditional training principle, we begin by utilizing unlabeled data from the target domain to train the noise learning network of DDPM, which is relatively straightforward collected. The intermediate activations obtained from the pre-trained DDPM are fused with the ECAPA-TDNN network, and then the labeled source domain dataset is used for further training. By learning the distribution characteristics of the target domain through the diffusion model, the target domain features can be obtained. The source domain data will extract features based on the pre-trained model, and new parameters will be determined under this guidance to refine and supplement the features. During the process of feature fusion, the dual network branch not only captures the features suitable for the target domain phonetic characteristics, but also leverages the powerful constraints imposed by supervised training of ECAPA-TDNN in the source domain.

4   Experiments

4.1   Dataset

VoxCeleb1&2 and CN-Celeb1 datasets were employed in this study, with CN-Celeb1 specifically utilized for cross-domain research purposes.

VoxCeleb is a widely used dataset in research on speaker verification. It consists of utterances from YouTube videos, covering thousands of speakers with diverse ages, genders, and accents. It is consisted of two parts: VoxCeleb1 [14] and VoxCeleb2 [2], and the speakers in these datasets are mutually exclusive. To train the network, we utilize the development set from the VoxCeleb2 dataset, which comprises a total of 5,994 speakers and 1,092,009 utterances of varying lengths. During model evaluation, we employ enroll-test pairs formed of the VoxCeleb1 test dataset (VoxCeleb1-O).

Despite VoxCeleb incorporating various languages, the majority of them are in English. To investigate the issue of cross-domain in speaker verification, we employed the CN-Celeb1 [7] dataset to simulate the scenario of cross-lingual speaker verification. CN-Celeb1 is a Chinese speech dataset that comprises 125,000 utterances from 1,000 Chinese celebrities. Among these, 800 speakers' utterances were utilized for training purposes, while the remaining formed test files. We utilized the development set of the VoxCeleb2 dataset, which includes speaker labels, along with the CN-Celeb1-T (training set), ignoring speaker labels, to train the model. For assessing the final performance, we employed the CN-Celeb1-E (evaluation set).

4.2   Training

The method is implemented with the Pytorch framework on one NVIDIA GeForce RTX 3090 GPU. During training, the audio input is randomly cropped for 2 s, and if the length is not enough, padding is performed. The model takes in 80-dimensional log mel-filterbanks obtained using a 25 ms hamming window with a 10 ms frame shift.

We set the hyperparameter T of the diffusion mode to 1000, and adopt the linear noise schedule. The model is trained with the hybrid loss objective, therefore, N in the network is set to 2, so that the network can learn noise and variance information at the same time. Adam optimizer is employed with a learning rate of $1e-4$. The batch size is set to 32, and we finally use DDPM pre-trained weights of 200k steps in the following training.

When we further train the DFR-ECAPA network, we add noise to the filterbank features using diffusion steps $t = 50$. The noisy features are then passed through the pre-trained DDPM to extract the speaker's diffusion features. The AAM-softmax objective function is applied with $m = 0.2$, $s = 30$. Model is trained with a batch size of 200 and optimized using the Adam optimizer with a weight decay of $2e-5$. The initial learning rate is set to $1e-4$. In the experiments in this paper, C takes the value of 1024.

4.3   Evaluation

We use equal error rate (EER) and minimum Detection Cost Function (minDCF) with $p_{target} = 0.01$, $C_{miss} = C_{fa} = 1$ to evaluate the final experimental results. The full utterances, as well as the 3 s segments and the 1.5 s segments were tested respectively.

4.4   Result

We tested the performance of the proposed DFR-ECAPA in different fusion methods and in cross-lingual speaker verification, and compared it with the original ECAPA-TDNN. Table 1 shows the performance of the corresponding evaluation indicators.

Table 1. DFR-ECAPA Performance on VoxCeleb1-O and CN-Celeb-E. ECAPA-TDNN is our reimplementation. *:Supervised training on Voxceleb2, directly applied to CN-Celeb. †:Pretrain the diffusion model using CN-Celeb without speaker labels.

Model	VoxCeleb1-O						CN-Celeb1-E	
	EER(%)			MinDCF			EER(%)	MinDCF
	Full	3.0 s	1.5 s	Full	3.0 s	1.5 s	Full	Full
ECAPA-TDNN	1.00	1.98	5.59	0.0755	**0.1349**	0.3403	17.23*	0.6097*
DDPM-Classifier	2.34	\	\	0.1675	\	\	\	\
DFR-ECAPA-ML(A)	0.98	1.98	5.42	0.0753	0.1387	0.3271	\	\
DFR-ECAPA-ML(B)	**0.96**	**1.88**	**5.18**	0.0785	0.1352	0.3378	13.64†	**0.5150**†
DFR-ECAPA-LL	0.97	1.90	5.38	**0.0735**	0.1414	**0.3200**	\	\

DDPM-Classifer is a network that we follow the idea of [1], retain the 2D architecture of the Unet denoising network, and make a simple modification to adapt to the speaker recognition task. The block takes {5,6,7,8,9,12}, t takes 50. Unlike semantic segmentation tasks, we do not need to do classification tasks at the pixel-level, so the method of extracting intermediate activations, upsampling, and retaining 8448 dimensional features for each pixel in the original paper is not applicable. Therefore, we used several convolution, statistical pooling and dropout operations instead. Experiments on VoxCeleb1-O show that this method can bring a certain accuracy rate, but its effect is far from the ideal level, therefore, we did not test on the other two durations which are more difficult. However, this proves that the intermediate activation of the diffusion model does capture the speaker's features to a certain extent, and it is feasible to use the diffusion model as a feature refiner.

First of all, we further refine the DFR-ECAPA-ML feature fusion method into two categories. In DFR-ECAPA-ML (A), we pre-trained the diffusion model and ECAPA-TDNN, while in DFR-ECAPA-ML (B) all parameters of ECAPA-TDNN are optimized in the second stage. The results show that the effect is better when more parameters are allowed to under the guidance of the diffusion feature. And the shorter the duration, the more the improvement. The EER of DFR-ECAPA-ML(B) has been improved by 4%, 5%, and 7% in the case of full utterances, the 3 s segments, and the 1.5 s segments, respectively. The performance of DFR-ECAPA-LL is between the above methods. We guess that if the extracted diffusion model features are integrated into the ECAPA model too early, it is difficult to judge what is useful information at the shallower layers.

Finally, we conduct experiments on the cross-domain problem, taking the cross-lingual case as an example. The original ECAPA-TDNN model is trained on paired data from Voxceleb and tested on CN-Celeb. The evaluation indicator EER is 17.23% and MinDCF is 0.6097. However, the ECAPA-TDNN obtained by training with labeled CN-Celeb data can achieve an EER of about 12.49%, which indicates that the domain mismatch problem does exist. Using our proposed DFR-ECAPA-ML(B) model and corresponding training methods, EER and MinDCF can obtain 13.64% and 0.515 respectively, and improve 20% and 15.5% respectively, approaching the recognition effect under supervised training. This shows that our model can indeed serve as a solution for semi-supervised cross-domain speaker verification in the absence of target domain labels.

5 Conclusion

Experiments demonstrate the value of DFR-ECAPA in applying diffusion models directly to speaker verification. The fusion of diffusion model representations indeed refines the features and enhances the accuracy of ECAPA-TDNN to some degree. Furthermore, our model exhibits competitiveness in semi-supervised cross-domain speaker verification, offering new idea to speaker verification with limited resources. The idea of DFR-ECA has certain exploration value in other TNND-based model. Currently, DDPMs are increasingly used in tasks other than generation tasks. Therefore, it can be seen that there are more ways to apply the idea of DDPMs algorithm to speaker verification.

References

1. Baranchuk, D., Rubachev, I., Voynov, A., Khrulkov, V., Babenko, A.: Label-efficient semantic segmentation with diffusion models. arXiv preprint arXiv:2112.03126 (2021)
2. Chung, J.S., Nagrani, A., Zisserman, A.: Voxceleb2: deep speaker recognition. arXiv preprint arXiv:1806.05622 (2018)
3. Dehak, N., Kenny, P.J., Dehak, R., Dumouchel, P., Ouellet, P.: Front-end factor analysis for speaker verification. IEEE Trans. Audio Speech Lang. Process. **19**(4), 788–798 (2010)
4. Deja, K., Trzcinski, T., Tomczak, J.M.: Learning data representations with joint diffusion models. arXiv preprint arXiv:2301.13622 (2023)
5. Desplanques, B., Thienpondt, J., Demuynck, K.: ECAPA-TDNN: emphasized channel attention, propagation and aggregation in TDNN based speaker verification. arXiv preprint arXiv:2005.07143 (2020)
6. Dowerah, S., Serizel, R., Jouvet, D., Mohammadamini, M., Matrouf, D.: Joint optimization of diffusion probabilistic-based multichannel speech enhancement with far-field speaker verification. In: 2022 IEEE Spoken Language Technology Workshop (SLT), pp. 428–435. IEEE (2023)
7. Fan, Y., et al.: Cn-celeb: a challenging Chinese speaker recognition dataset. In: ICASSP 2020–2020 IEEE International Conference on Acoustics, Speech and Signal Processing (ICASSP), pp. 7604–7608. IEEE (2020)
8. Ho, J., Jain, A., Abbeel, P.: Denoising diffusion probabilistic models. Adv. Neural. Inf. Process. Syst. **33**, 6840–6851 (2020)
9. Ho, J., Salimans, T., Gritsenko, A., Chan, W., Norouzi, M., Fleet, D.J.: Video diffusion models. arXiv preprint arXiv:2204.03458 (2022)

10. Huang, R., et al.: FastDiff: a fast conditional diffusion model for high-quality speech synthesis. arXiv preprint arXiv:2204.09934 (2022)
11. Li, P., Tang, H., Yu, J., Song, W.: LSTM and multiple CNNs based event image classification. Multimed. Tools Appl. **80**, 30743–30760 (2021)
12. Liu, T., Das, R.K., Lee, K.A., Li, H.: MFA: TDNN with multi-scale frequency-channel attention for text-independent speaker verification with short utterances. In: ICASSP 2022–2022 IEEE International Conference on Acoustics, Speech and Signal Processing (ICASSP), pp. 7517–7521. IEEE (2022)
13. Mun, S.H., Jung, J.w., Han, M.H., Kim, N.S.: Frequency and multi-scale selective kernel attention for speaker verification. In: 2022 IEEE Spoken Language Technology Workshop (SLT), pp. 548–554. IEEE (2023)
14. Nagrani, A., Chung, J.S., Zisserman, A.: VoxCeleb: a large-scale speaker identification dataset. arXiv preprint arXiv:1706.08612 (2017)
15. Popov, V., Vovk, I., Gogoryan, V., Sadekova, T., Kudinov, M., Wei, J.: Diffusion-based voice conversion with fast maximum likelihood sampling scheme (2022)
16. Rombach, R., Blattmann, A., Lorenz, D., Esser, P., Ommer, B.: High-resolution image synthesis with latent diffusion models. In: Proceedings of the IEEE/CVF Conference on Computer Vision and Pattern Recognition, pp. 10684–10695 (2022)
17. Sawata, R., et al.: Diffiner: A versatile diffusion-based generative refiner for speech enhancement (2023)
18. Snyder, D., Garcia-Romero, D., Sell, G., Povey, D., Khudanpur, S.: X-vectors: robust DNN embeddings for speaker recognition. In: 2018 IEEE International Conference on Acoustics, Speech and Signal Processing (ICASSP), pp. 5329–5333. IEEE (2018)
19. Thienpondt, J., Desplanques, B., Demuynck, K.: Integrating frequency translational invariance in TDNNs and frequency positional information in 2d Resnets to enhance speaker verification. arXiv preprint arXiv:2104.02370 (2021)
20. Variani, E., Lei, X., McDermott, E., Moreno, I.L., Gonzalez-Dominguez, J.: Deep neural networks for small footprint text-dependent speaker verification. In: 2014 IEEE International Conference on Acoustics, Speech and Signal Processing (ICASSP), pp. 4052–4056. IEEE (2014)
21. Wang, T.C., Liu, M.Y., Zhu, J.Y., Tao, A., Kautz, J., Catanzaro, B.: High-resolution image synthesis and semantic manipulation with conditional GANs. In: Proceedings of the IEEE Conference on Computer Vision and Pattern Recognition, pp. 8798–8807 (2018)
22. Zeinali, H., Wang, S., Silnova, A., Matějka, P., Plchot, O.: But system description to Voxceleb speaker recognition challenge 2019. arXiv preprint arXiv:1910.12592 (2019)
23. Zhao, Z., Li, Z., Wang, W., Zhang, P.: PCF: ECAPA-TDNN with progressive channel fusion for speaker verification. In: ICASSP 2023–2023 IEEE International Conference on Acoustics, Speech and Signal Processing (ICASSP), pp. 1–5. IEEE (2023)
24. Zhu, Z.,et al.: Exploring discrete diffusion models for image captioning. arXiv preprint arXiv:2211.11694 (2022)

Half Aggregation Transformer
for Exposure Correction

Ziwen Li, Jinpu Zhang, and Yuehuan Wang[✉]

National Key Laboratory of Science and Technology on Multispectral Information
Processing, School of Artificial Intelligence and Automation, Huazhong University of
Science and Technology, Wuhan 430074, China
yuehwang@hust.edu.cn

Abstract. Photos taken under poor illumination conditions often suffer from unsatisfactory visual effects. Recently, Transformer, avoiding the shortcomings of CNN models, has shown impressive performance on various computer vision tasks. However, directly leveraging Transformer for exposure correction is challenging. On the one hand, the global Transformer has an excessive computational burden and fails to preserve local feature details. On the other hand, the local Transformer fails to extract spatially varying light distributions. Both global illumination recovery and local detail enhancement are indispensable for exposure correction. In this paper, we propose a novel Half Aggregation Transformer (HAT) architecture for exposure correction with a key design of Half Aggregation Multi-head Self-Attention (HA-MSA). Specifically, our HA-MSA establishes inter- and intra-window token interactions via window aggregation and window splitting strategies to jointly capture both global dependencies and local contexts in a complementary manner. In addition, we customize an illumination guidance mechanism to explore illumination cues and boost the robustness of the network to handle complex illumination. Extensive experiments demonstrate that our method outperforms the state-of-the-art exposure correction methods qualitatively and quantitatively with cheaper computational costs.

Keywords: Exposure correction · transformer · illumination guidance

1 Introduction

When taking photographs, undesirable lighting conditions in a scene can introduce underexposure or overexposure, which is quite common and annoying. Poor lighting conditions can degrade image quality and lead to unsatisfactory results, so it is necessary to correct the exposure to improve the visual effect of the image. This technique has a wide range of applications, such as night photography and autonomous driving.

Traditional exposure correction methods include histogram equalization [37] and Retinex model-based methods [10,13] that use hand-crafted priors to adjust image brightness and contrast. However, these methods have limited robustness in the face of a complex scenes. Due to the powerful learning capability of

Q. Liu et al. (Eds.): PRCV 2023, LNCS 14434, pp. 469–481, 2024.
https://doi.org/10.1007/978-981-99-8549-4_39

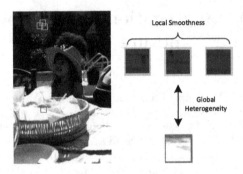

Fig. 1. Global heterogeneity and local smoothness exist in natural images. We explore them to design Half Aggregation Self-Attention to obtain global dependencies and local contexts at lower computational cost.

CNN, CNN-based methods have made significant progress for exposure correction. Existing CNN-based solutions either directly learn the pixel-level mapping relationship between degraded images to sharp images [1,12,14,32] or combine Retinex models to predict and adjust the illumination and reflection components [19,28,34], and these approaches have achieved impressive results. However, CNN-based models lack the ability to model long-range dependencies and the flexibility of content adaptation due to the locality constraints and weight sharing of convolution, which limits the performance of networks [9].

Transformer [22] captures global dependencies with its self-attention and avoids the drawbacks of CNN. Recently, after rapid progress in high-level vision tasks [3,36], Transformer has also been applied to low-level vision tasks [5,33]. These Transformer-based methods include the global Transformer [31] and the local Transformer [18,27]. The global Transformer computes self-attention for all tokens to establish long-range dependencies. However, this leads to the fact that its computational complexity is quadratic in spatial resolution, which is unaffordable for high-resolution features. Moreover, the global Transformer lacks the ability to retain local feature details and thus does not recover clear details well [6,11]. To address these issues, local Transformer restricts self-attention within local windows. However, local self-attention fails to extract spatially varying light distributions. The reason is that it prevents token interactions from non-local regions, leading to limited receptive fields and impairing the ability to model long-range dependencies. Even with shifting window strategies, global illumination cannot be modeled effectively [7]. Both global illumination recovery and local detail enhancement are crucial for image quality.

In order to capture both global dependencies and local contexts at cheap computational cost, we propose an novel self-attention mechanism for exposure correction. We are inspired by the local smoothing property possessed by global information. From a global perspective, there is global heterogeneity in the illumination distribution of natural images, which makes it necessary to exploit self-attention with global perception capabilities. From a local perspective, the presence of local smoothness in the global illumination implies similarity in local regions. Figure 1 depicts this phenomenon, where the white paper in the over-exposed region and the wall in the underexposed region have different patterns, but the illumination in the wall region are similar. Therefore, when calculating the self-similarity of non-local regions, local regions can share the same token representation to save computational cost. However, local regions sharing tokens will inevitably lose image detail information, so we introduce local window self-attention with low computational cost to capture the local context, thus ensuring complete recovery of global illumination and local details.

In this paper, we propose a novel Half Aggregation Transformer (HAT) for exposure correction. The key design of HAT that effectively provides complementary information with global and local properties lies in our Half Aggregation Multi-head Self-Attention (HA-MSA). Specifically, proposed HA-MSA contains Window Aggregation Multi-head Self-Attention (WA-MSA) and Window Split Multi-head Self-Attention (WS-MSA). WA-MSA aggregates individual windows into a token and computes self-attention among all tokens to establish token interactions across non-local regions, thus providing image-level receptive field and focusing on global illumination. WS-MSA restricts the self-attention within local windows, thus focusing on local details. Therefore, the proposed HA-MSA utilizes window aggregation and window splitting strategies to establish inter- and intra-window token interactions, thus learning to jointly extract global dependencies and local contexts in a complementary manner. In addition, to further improve the robustness for handling complex illumination, we customize an illumination guidance mechanism to explore illumination cues. We estimate a residual illumination map as the prior and designed a Gated Illumination Guidance Module (GIGM) to inject the illumination prior to further improve the performance of the network. Our proposed method is lightweight and surpasses state-of-the-art exposure correction methods. The main contributions of this work include

- We propose a novel self-attention mechanism called Half Aggregation Multi-head Self-Attention (HA-MSA), which can jointly capture global dependencies and local contexts to obtain complementary representations.
- We customized an illumination guidance mechanism to explore illumination cues to boost the robustness of the network to handle complex illumination.
- Extensive experiments on several datasets demonstrate that our method can outperform state-of-the-art methods with fewer parameters.

2   Related Work

2.1   Exposure Correction

Exposure correction has a long history of research. Traditional methods include histogram-based method [37] and Retinex-based methods [10,13,17,24]. CNN-based methods achieve impressive performance for exposure correction due to the powerful learning capability of CNNs. Some methods use CNNs to predict illumination and reflection components based on the Retinex model. For instance, RetinexNet [28] and KinD [34] predict and adjust the illumination and reflection of low-light images. RUAS [19] employs Retinex prior to automatically search for efficient architectures. URetinexNet [29] uses deep networks to unfold the optimization process of Retinex. Another type of approach leverages CNN to learn clear images directly [21,26,30,35]. DRBN [32] recursively decomposes and recomposes bands to improve the fidelity and perceptual quality. ZeroDCE [12] estimates higher-order curves to adjust images in a non-reference manner. ENC [14] designs exposure normalization and compensation to correct multiple exposure. FECNet [15] proposes a Fourier-based network with the interaction of frequency and space domains. However, CNN-based approaches lack the ability to model long-range dependencies and the flexibility of content adaptation, limiting the performance of exposure correction. Unlike previous exposure correction methods, we designed a lightweight self-attention mechanism to provide long-range dependency modeling capability and explored the guidance effect of illumination prior, and these designs yield state-of-the-art performance with fewer parameters.

2.2 Vision Transformer

Transformer [22] was originally proposed for natural language processing. Due to its excellent global modeling capability, it has been widely used in high-level vision tasks with superior performance, such as object detection [3]. In addition, Transformer has also been explored for low-level vision tasks [5,8,18,27,31,33]. Specifically, [31] proposes a signal-to-noise ratio-aware Transformer framework. [33] computes the self-attention along the channel dimension instead of the spatial dimension. SwinIR [18] and Uformer [27] use non-overlapping local windows and restrict self-attention within each window. [8] computes self-attention using rectangular windows and aggregates features of different windows to expand the receptive field. However, the global Transformer-based approach has an excessive computational burden while failing to preserve local feature details. And the local Transformer-based approach limits the receptive field of the network and cannot model long-range dependencies. In this work, we use window aggregation and splitting strategies to effectively and efficiently extract global dependencies and local contexts in a complementary manner.

3 Method

In this section, we first describe the framework of Half Aggregation Transformer (HAT). Then, we elaborate the Half Aggregation Multi-head Self-Attention (HA-MSA). Finally, we present the customized illumination guidance mechanism.

3.1 Overall Architecture

The overall architecture of proposed HAT is shown in Fig. 2(a). HAT is a three-layer symmetric U-shaped structure containing an encoder, a bottleneck and a

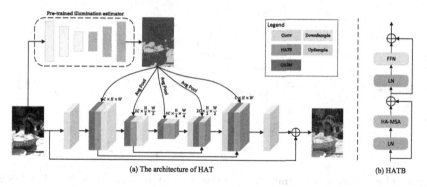

Fig. 2. (a) The architecture of our proposed HAT for exposure correction. (b) Illustration of HATB

decoder. To avoid the loss of shallow features due to downsampling, we adopt skip connections to integrate the features of the encoder and decoder. Given a degraded sRGB image $\mathbf{I} \in \mathbb{R}^{H \times W \times 3}$ as input, we first map the input image $\mathbf{I}$ to a low level feature $\mathbf{X}_0 \in \mathbb{R}^{H \times W \times C}$. Next, We then feed the features into the encoder, bottleneck and decoder in sequence, with each stage containing several HATBs. Finally, we remap the feature into the residual image $\mathbf{R}$ and adds it to the input image $\mathbf{I}$ to obtain a high quality reconstructed image: $\mathbf{I'} = \mathbf{I} + \mathbf{R}$. To further improve the robustness against complex illumination, we introduce a pre-trained illumination estimator to obtain the illumination prior, and use Gated Illumination Guidance Module (GIGM) to inject the prior at each stage of the network to modulate the feature map in a multi-scale manner.

As shown in Fig. 2(b), HATB is the basic module of the HAT, which contains a Half Aggregation Multi-head Self-Attention (HA-MSA) and a Feed-Forward Network (FFN). The calculation of HATB can be expressed as

$$\mathbf{F}_{mid} = \text{HA-MSA}(\text{LN}(\mathbf{F}_{in})) + \mathbf{F}_{in}, \tag{1}$$

$$\mathbf{F}_{out} = \text{FFN}(\text{LN}(\mathbf{F}_{mid})) + \mathbf{F}_{mid}, \tag{2}$$

where $\mathbf{F}_{in}$ denotes the input features, $\mathbf{F}_{mid}$ and $\mathbf{F}_{out}$ denote the output features of HA-MSA and HATB, respectively.

3.2 Half Aggregation Multi-head Self-Attention

Most previous exposure correction methods use CNNs, which prefer to extract local features and have limitations in modeling long-range dependencies. It is challenging to directly apply the existing Transformer. On the one hand, the global Transformer has an excessive computational burden and fails to preserve local feature details. On the other hand, the local Transformer limits the receptive field. To overcome the above drawbacks, we propose HA-MSA to jointly capture both global dependencies and local contexts at cheaper computational cost, which is more suitable for exposure correction.

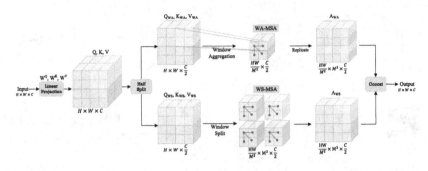

Fig. 3. Illustration of our Half Aggregation Multi-head Self-Attention (HA-MSA), which contains WA-MSA and WS-MSA.

As shown in Fig. 3, our HA-MSA consists of two components: Window Aggregation Multi-head Self-Attention (WA-MSA) and Window Split Multi-head Self-Attention (WS-MSA). Specifically, given a layer-normalized feature $\mathbf{X} \in \mathbb{R}^{H \times W \times C}$, we perform linear projection on $\mathbf{X}$ to obtain query Q, key K, value V, and then split along the channel. We can obtain two groups of features $Q_{WA}, K_{WA}, V_{WA} \in \mathbb{R}^{H \times W \times \frac{C}{2}}$ and $Q_{WS}, K_{WS}, V_{WS} \in \mathbb{R}^{H \times W \times \frac{C}{2}}$, which are used to calculate WA-MSA and WS-MSA, respectively.

In order to perceive the global illumination efficiently, WA-MSA employs the window aggregation strategy to compute the self-attention across non-local regions. First, we split the feature Q_{WA}, K_{WA}, V_{WA} into multiple windows by size of $M \times M$. Then, the individual window is averaged into a single independent token. This yields a total of $\frac{HW}{M^2}$ tokens from non-local regions. Subsequently, we compute the self-attention between all these aggregated tokens to provide a global receptive field and establish long-range dependencies. MA-MSA is formulated as

$$A_{WA} = \text{Softmax}(\frac{Q_{WA}K_{WA}^{T}}{\alpha})V_{WA}, \tag{3}$$

where $Q_{WA}, K_{WA}, V_{WA} \in \mathbb{R}^{\frac{HW}{M^2} \times \frac{C}{2}}$ denotes the tensor after reshaping and aggregation, and α denotes the normalization parameter.

To complement the local details, WS-MSA utilizes the window splitting strategy to compute the self-attention A_{WS} within the local window. We split features Q_{WS}, K_{WS}, V_{WS} into non-overlapping windows of size $M \times M$. Each token only interacts with other tokens within the same window.

After completing WA-MSA, we replicate A_{WA} to the original dimensions, then concatenate it with A_{WS} along the channel. As with other multi-head strategy, we divide the features into multiple heads to compute multiple attention maps in parallel and concatenate the output of all heads at the end.

Complexity. The computational complexity of our HA-MSA is $O(\text{HA-MSA}) = M^2HWC + \frac{(HW)^2}{M^4}C$. And the computational complexity of Global-MSA and Local-MSA is formulated as $O(\text{Global-MSA}) = 2(HW)^2C$ and $O(\text{Local-MSA}) =$

Fig. 4. Illustration of the residual illumination map, which can represent exposure patterns.

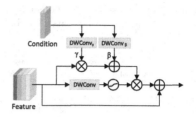

Fig. 5. Illustration of our GIGM, which explicitly injects illumination condition into the Transformer.

$2M^2HWC$. As an example, under the settings of H=W=256, C=16, and M=8, our HA-MSA requires only 0.6% of the computational cost compared to the Global-MSA. Compared to the limited receptive fields of Local-MSA, our HA-MSA is able to provide image-level receptive fields and establish global dependencies. Overall, HA-MSA can effectively and efficiently extract global dependencies and local contexts.

3.3 Illumination Guidance Mechanism

Degraded images often suffer from uneven distribution of illumination in spatial locations, and moreover different images or regions may have different patterns of overexposure and underexposure, which poses challenges to exposure correction. To address this problem, we customize an illumination guidance mechanism to explore illumination cues to further improve the robustness of the network to handle complex illumination and reconstruct well-lit images.

Illumination Estimation. We observe that the residual illumination map can accurately reflect the illumination variations. Let the illumination maps of the degraded and ground-truth be a and b, where $|a-b|$ can represent the overexposed region, and the underexposed region can be located by $|b-a|$, as shown in Fig. 4. That is, the exposure patterns and light variations can be distinguished and localized by the relative magnitude of the residual illumination map. Based on this observation, we try to use the residual illumination map as a guiding prior, expecting to perform different nonlinear transformations on regions with different illumination. We use the average of the three channels of the image to represent the light intensity. The illumination estimator uses a compacted UNet. Given a degraded image as input, the target of the illumination estimator is to predict a illumination map close to the reference. The illumination estimator will ignore color information and focus on the illumination distribution and texture structure of the image. We then combine the degraded illumination map to calculate the residual illumination map to provide guidance.

Multi-scale Illumination Guidance. We argue that there are discrepancies and mismatches between the image-level information of illumination cues and

Table 1. Quantitative comparisons of our method with the state-of-the-art methods on the MSEC and SICE datasets. The best performances are highlighted in bold.

Method	MSEC						SICE						Param
	Under		Over		Average		Under		Over		Average		
	PSNR	SSIM	PSNR	SSIM	PSNR	SSIM	PSNR	SSIM	PSNR	SSIM	PSNR	SSIM	
CLAHE [37]	16.77	0.6211	14.45	0.5842	15.38	0.5990	12.69	0.5037	10.21	0.4847	11.45	0.4942	/
RetinexNet [28]	12.13	0.6209	10.47	0.5953	11.14	0.6048	12.94	0.5171	12.87	0.5252	12.90	0.5212	0.84M
ZeroDCE [12]	14.55	0.5887	10.40	0.5142	12.06	0.5441	16.92	0.6330	7.11	0.4292	12.02	0.5311	0.079M
RUAS [19]	13.43	0.6807	6.39	0.4655	9.20	0.5515	16.63	0.5589	4.54	0.3196	10.59	0.4393	**0.003M**
DSLR [16]	13.14	0.5812	20.06	0.6826	15.91	0.6219	16.83	0.6133	7.99	0.4300	12.41	0.5217	0.39M
MSEC [1]	20.52	0.8129	19.79	0.8156	20.08	0.8145	19.62	0.6512	17.59	0.6560	18.58	0.6536	7.04M
DRBN [32]	19.74	0.8290	19.37	0.8321	19.52	0.8309	17.96	0.6767	17.33	0.6828	17.65	0.6798	0.58M
SID [4]	19.37	0.8103	18.83	0.8055	19.04	0.8074	19.51	0.6635	16.79	0.6444	18.15	0.6540	7.40M
LCDPNet [23]	22.24	0.8470	22.33	0.8630	22.29	0.8550	16.72	0.6346	17.53	0.5443	17.13	0.5895	0.28M
ENC-DRBN [14]	22.72	0.8544	22.11	0.8521	22.35	0.8530	21.77	0.7052	19.57	0.7267	20.67	0.7160	0.58M
ENC-SID [14]	22.59	0.8423	**22.36**	0.8519	22.45	0.8481	21.30	0.6645	19.63	0.6941	20.47	0.6793	7.45M
Ours	**23.39**	**0.8633**	22.22	**0.8629**	**22.73**	**0.8631**	**23.00**	**0.6942**	**21.16**	**0.7470**	**22.08**	**0.7206**	0.29M

(a) Input   (b) MSEC   (c) LCDPNet   (d) ENC-DRBN   (e) Ours   (f) GT

Fig. 6. Visual comparison on the SICE (top) and MSEC (bottom) dataset.

the feature-level information of the network, and directly fusing them leads to suboptimal results. Therefore, we design a Gated Illumination Guidance Module (GIGM), which explicitly injects the illumination prior into the Transformer. GIGM contains two designs of spatial feature transformation (SFT) and gating mechanism, see Fig. 5. Inspired by [25], we treat illumination as a condition, predict both scale and shift transformation matrices, and then flexibly modulate the feature maps. The gating mechanism uses a nonlinear activation function to generate weights and multiply them with features to control the information flow. Given an input X and a condition C, GIGM can be formulated as

$$SFT(X|C) = X \otimes W_\gamma(C) \oplus W_\beta(C), \tag{4}$$

$$Gate(X, C) = SFT(X|C) \otimes \sigma(W(X)), \tag{5}$$

$$\hat{X} = Gate(X, C) \oplus X, \tag{6}$$

where $\otimes$ and $\oplus$ denote element-wise multiplication and addition, σ denotes the GELU activation function, and W_γ and W_β denote the depth-wise convolution for generating the scale α and shift β. GIGM multi-scale injects illumination condition at each stage of the HAT. At each scale, we use an average pooling layer to adjust the resolution of the illumination map.

4   Experiment

4.1   Implementation Details

Our implementation is based on the Pytorch framework and is performed on one NVIDIA TITAN V GPU. We use Adam optimizer to train the model for 200 epochs with patch size of 384×384 and batch size of 4. The initial learning rate is $2e^{-4}$, and is gradually reduced to $1e^{-6}$ using cosine annealing strategy. Random cropping, flipping and rotation are used for data augmentation. To avoid out of memory when processing high resolution images, we set the window size of HA-MSA to 8×8. For the illumination estimator, we use only the L1 loss and then fix the weights. For HAT, We use the same loss function and balance weights as in the previous method [14].

We evaluated the performance of the proposed method on two multi-exposure datasets MSEC [1] and SICE [2]. The MSEC dataset is a multiple exposure dataset containing overexposed and underexposed images at five different exposure levels, containing 17675 images for training, 750 for validation, and 5905 for testing. For the SICE dataset, we treat the middle exposure level as the ground truth and the second and second-last exposure levels as the underexposed and overexposed images. The SICE dataset contains 1000 training images, 24 validation images, and 60 test images. The training and testing partitions of these datasets are consistent with related papers [1,14].

4.2   Comparisons with State-of-the-Art Methods

Quantitative Comparisons. To verify the superiority of our method, we compare it with state-of-the-art methods including CLAHE [37], RetinexNet [28], ZeroDCE [12], RUAS [19] , DSLR [16], MSEC [1], DRBN [32], SID [4], LCDP-Net [23], and ENC [14]. We use common PSNR and SSIM metrics for quantitative evaluation. Table 1 shows the exposure correction results on the MSEC and SICE datasets. Following the ENC [14], we average the results for overexposed and underexposed images, respectively. It can be seen that our method achieves the highest PSNR and SSIM with only 0.29M parameters compared to previous methods, which proves the superiority and efficiency of our method.

Visual Comparisons. We also provide visual results for qualitative comparison with the SOTA method, where the results for the MSEC and SICE datasets are shown in Fig. 6. It can be seen that our method can effectively cope with a variety of illumination conditions and produce pleasing results with suitable brightness, consistent color and fewer artifacts. These evidences demonstrate the consistent superiority of our method.

4.3   Ablation Study

To validate the effectiveness of the proposed method, we conducted ablation studies on the SICE dataset.

(a) Global-MSA   (b) Local-MSA   (c) Swin-MSA   (d) HA-MSA

Fig. 7. Visual comparison of self-attention mechanisms.

Table 2. Ablation study of self-attention mechanisms.

Setting	Global-MSA	Local-MSA	Swin-MSA	HA-MSA
PSNR	20.95	18.49	18.65	**21.56**
SSIM	0.6775	0.6793	0.6816	**0.7074**
Param	0.507M	0.239M	0.239M	0.239M

Self-Attention Mechanism. To validate the effectiveness of our HA-MSA, we compared it with different self-attention mechanisms, including Global-MSA [9], Local-MSA [20] and Swin-MSA [20]. For fairness, we use the same U-shaped architecture and remove the position encoding. Due to memory limitations, we downsample the feature maps of the Global-MSA. Table 2 presents the ablation results. Figure 7 displays the corrected results and the feature visualization of the self-attention. In general, PSNR focuses more on the global consistency and SSIM pays more attention to the local details. We can see that the Local-MSA has the lowest PSNR value and unreasonable illumination distribution in the image, which indicates that the Local-MSA ignores the global illumination, which severely degrades the performance of the network and the image quality. Swin-MSA slightly improves performance by shifting the window, but still cannot extract global information effectively. The Global-MSA recovers reasonable illumination, but it has the lowest SSIM value and the image details are blurred, so the Global-MSA lacks the ability to retain local details. In contrast, our HA-MSA achieves the highest PSNR and SSIM scores while the generated images and visualized features have consistent luminance distributions and clear texture details due to the ability of HA-MSA to extract global dependencies and local contexts, which proves the rationality and effectiveness of our design.

Illumination Guidance Mechanism. We evaluate the effectiveness of the illumination guidance mechanism. The *baseline* refers to the HAT without the illumination guidance mechanism. The *Retinex* uses Retinex-based illumination maps instead of residual illumination maps. The *w/ concat* denotes using concatenation to integrate illumination prior. The *w/ SFT* uses SFT to inject the illumination prior without the gating mechanism. As depicted in Table 3, our full

Table 3. Ablation study of illumination guidance mechanisms.

Setting	baseline	Retinex	w/ concat	w/ SFT	Ours
PSNR	21.56	21.53	21.66	22.01	**22.08**
SSIM	0.7074	0.7076	0.7107	0.7131	**0.7206**

model achieved the highest scores. Specifically, the comparison between *Retinex* and *Ours* demonstrates that the residual illumination map can better represent non-uniform light variations than the Retinex-based illumination map. Taking illumination as a condition to modulate Transformer features is a more appropriate choice than feature concatenation. Moreover the application of gating mechanism can further improve the performance, which demonstrates the effectiveness of illumination guidance.

5   Conclusion

In this paper, we present a powerful Half Aggregation Transformer model for exposure correction. We design HA-MSA to complementarily capture global dependencies and local context through window aggregation and splitting strategies at low computational cost. In addition, the illumination guidance mechanism is proposed to explore illumination cues to improve performance. Our proposed method is lightweight and extensive comparisons on several datasets demonstrate that the proposed method surpasses state-of-the-art methods.

References

1. Afifi, M., Derpanis, K.G., Ommer, B., Brown, M.S.: Learning multi-scale photo exposure correction. In: CVPR (2021)
2. Cai, J., Gu, S., Zhang, L.: Learning a deep single image contrast enhancer from multi-exposure images. TIP. **27**, 2049–2062 (2018)
3. Carion, N., Massa, F., Synnaeve, G., Usunier, N., Kirillov, A., Zagoruyko, S.: End-to-end object detection with transformers. In: Vedaldi, A., Bischof, H., Brox, T., Frahm, J.-M. (eds.) ECCV 2020. LNCS, vol. 12346, pp. 213–229. Springer, Cham (2020). https://doi.org/10.1007/978-3-030-58452-8_13
4. Chen, C., Chen, Q., Xu, J., Koltun, V.: Learning to see in the dark. In: CVPR (2018)
5. Chen, H., et al.: Pre-trained image processing transformer. In: CVPR (2021)
6. Chen, X., Li, H., Li, M., Pan, J.: Learning a sparse transformer network for effective image deraining. In: CVPR (2023)
7. Chen, X., Wang, X., Zhou, J., Dong, C.: Activating more pixels in image super-resolution transformer. In: CVPR (2023)
8. Chen, Z., Zhang, Y., Gu, J., Zhang, Y., Kong, L., Yuan, X.: Cross aggregation transformer for image restoration. In: NeurIPS (2022)
9. Dosovitskiy, A., et al.: An image is worth 16 × 16 words: transformers for image recognition at scale. In: ICLR (2021)

10. Fu, X., Zeng, D., Huang, Y., Zhang, X.P., Ding, X.: A weighted variational model for simultaneous reflectance and illumination estimation. In: CVPR (2016)

11. Guo, C.L., Yan, Q., Anwar, S., Cong, R., Ren, W., Li, C.: Image dehazing transformer with transmission-aware 3d position embedding. In: CVPR (2022)

12. Guo, C., et al.: Zero-reference deep curve estimation for low-light image enhancement. In: CVPR (2020)

13. Guo, X., Li, Y., Ling, H.: Lime: low-light image enhancement via illumination map estimation. TIP **26**, 982–993 (2017)

14. Huang, J., et al.: Exposure normalization and compensation for multiple-exposure correction. In: CVPR (2022)

15. Huang, J., et al.: Deep Fourier-based exposure correction network with spatial-frequency interaction. In: Avidan, S., Brostow, G., Cissé, M., Farinella, G.M., Hassner, T. (eds.) Computer Vision. ECCV 2022. LNCS, vol. 13679, pp. 163–180. Springer, Cham (2022). https://doi.org/10.1007/978-3-031-19800-7_10

16. Ignatov, A., Kobyshev, N., Timofte, R., Vanhoey, K., Van Gool, L.: DSLR-quality photos on mobile devices with deep convolutional networks. In: ICCV (2017)

17. Li, M., Liu, J., Yang, W., Sun, X., Guo, Z.: Structure-revealing low-light image enhancement via robust retinex model. TIP **27**, 2828–2841 (2018)

18. Liang, J., Cao, J., Sun, G., Zhang, K., Van Gool, L., Timofte, R.: SwinIR: image restoration using Swin transformer. In: ICCVW (2021)

19. Liu, R., Ma, L., Zhang, J., Fan, X., Luo, Z.: Retinex-inspired unrolling with cooperative prior architecture search for low-light image enhancement. In: CVPR (2021)

20. Liu, Z., et al.: Swin transformer: hierarchical vision transformer using shifted windows. In: ICCV (2021)

21. Ma, L., Ma, T., Liu, R., Fan, X., Luo, Z.: Toward fast, flexible, and robust low-light image enhancement. In: CVPR (2022)

22. Vaswani, A., et al.: Attention is all you need. In: NeurIPS (2017)

23. Wang, H., Xu, K., Lau, R.W.: Local color distributions prior for image enhancement. In: Avidan, S., Brostow, G., Cissé, M., Farinella, G.M., Hassner, T. (eds.) Computer Vision. ECCV 2022. LNCS, vol. 13678, pp. 343–359. Springer, Cham (2022). https://doi.org/10.1007/978-3-031-19797-0_20

24. Wang, S., Zheng, J., Hu, H.M., Li, B.: Naturalness preserved enhancement algorithm for non-uniform illumination images. TIP **22**, 3538–3548 (2013)

25. Wang, X., Yu, K., Dong, C., Loy, C.C.: Recovering realistic texture in image super-resolution by deep spatial feature transform. In: CVPR (2018)

26. Wang, Y., Wan, R., Yang, W., Li, H., Chau, L.P., Kot, A.: Low-light image enhancement with normalizing flow. In: AAAI, pp. 2604–2612 (2022)

27. Wang, Z., Cun, X., Bao, J., Zhou, W., Liu, J., Li, H.: UFormer: a general u-shaped transformer for image restoration. In: CVPR (2022)

28. Wei, C., Wang, W., Yang, W., Liu, J.: Deep retinex decomposition for low-light enhancement. In: BMVC (2018)

29. Wu, W., Weng, J., Zhang, P., Wang, X., Yang, W., Jiang, J.: Uretinex-net: Retinex-based deep unfolding network for low-light image enhancement. In: CVPR (2022)

30. Xu, K., Yang, X., Yin, B., Lau, R.W.: Learning to restore low-light images via decomposition-and-enhancement. In: CVPR (2020)

31. Xu, X., Wang, R., Fu, C.W., Jia, J.: Snr-aware low-light image enhancement. In: CVPR (2022)

32. Yang, W., Wang, S., Fang, Y., Wang, Y., Liu, J.: From fidelity to perceptual quality: a semi-supervised approach for low-light image enhancement. In: CVPR (2020)

33. Zamir, S.W., Arora, A., Khan, S., Hayat, M., Khan, F.S., Yang, M.H.: Restormer: efficient transformer for high-resolution image restoration. In: CVPR (2021)
34. Zhang, Y., Zhang, J., Guo, X.: Kindling the darkness: a practical low-light image enhancer. In: ICME (2019)
35. Zheng, C., Shi, D., Shi, W.: Adaptive unfolding total variation network for low-light image enhancement. In: CVPR, pp. 4439–4448 (2021)
36. Zheng, S., et al.: Rethinking semantic segmentation from a sequence-to-sequence perspective with transformers. In: CVPR (2021)
37. Zuiderveld, K.: Contrast limited adaptive histogram equalization. Graphics gems (1994)

Deformable Spatial-Temporal Attention for Lightweight Video Super-Resolution

Tong Xue, Xinyi Huang, and Dengshi Li[✉] [ID]

School of Artificial Intelligence, Jianghan University, Wuhan 430056, China
reallds@jhun.edu.cn

Abstract. Video super-resolution (VSR) aims to recover high-resolution video frames from their corresponding low-resolution video frames and their adjacent consecutive frames. Although some progress has been made, most existing methods typically use the spatial-temporal information of two adjacent reference frames to aid in enhancing the video frame super-resolution reconstruction effect. This makes it impossible for these methods to achieve satisfactory results. To solve this problem. We propose a deformable spatial-temporal attention (DSTA) module for video super-resolution. The deformable spatial-temporal attention module improves the reconstruction effect by aggregating favorable spatial-temporal information from multiple reference frames into the current frame. To speed up the model training, we select only the first s highly relevant feature points as the attention scheme. Experimental results show that our method with fewer network parameters has strong video super-resolution performance.

Keywords: Video super-resolution · Deformable spatial-temporal attention · Multiple reference frames

1 Introduction

The goal of video super resolution (VSR) is to predict realistic high resolution (HR) video from low resolution (LR) video. VSR differs significantly from single image super resolution (SISR) in that SISR only relies on intra-frame spatial correlation to recover the spatial resolution of a single image, whereas VSR uses inter-frame temporal coherence between adjacent frames to recover the spatial resolution of all video frames resolution of all video frames. The most straightforward VSR method is to feed continuous low resolution (LR) frames directly into the SISR model [7,13,27]. Although the spatial information of the video frames is exploited, they ignore the temporal correlation between the video frames.

An alternative method [11,19] is by estimating the optical flow between adjacent LR frames before reconstructing the high resolution (HR) frame, which is then used to align the LR frames. However, inaccurate optical flow can lead to unwanted artefacts in distorted frames. Therefore, recent approaches [4,5] have avoided artefacts due to optical flow estimation errors by using deformable convolution [6] to align adjacent frames at the level of the feature. Although deformable convolution can be used to capture features within a local area by expanding the convolution's field of perception. However, a large receptive field can impose a significant computational cost on the model. Currently, recurrent neural networks are a common approach in VSR tasks [3,10]. This approach can use internal memory states to handle long-term dependencies and improve the performance of VSR by aggregating the spatial information of two adjacent frames. In fact, different reference frames and different feature points have different effects on the final reconstruction of the VSR. The aggregation of unrelated feature information may cause ghosting and occlusion in the super-resolved video frames. Moreover, the recurrent network-based VSR model requires a large number of parameters and a long training time, which does not allow the model to be deployed in practical applications.

In order to solve the above problem, this paper proposes Deformable Spatial-Temporal Attention for Lightweight Video Super-Resolution (DSTALVSR). In DSTALVSR, we propose a Deformable Spatial-Temporal Attention (DSTA) mechanism. By marking the sampling points on the reference frame spatial-temporal features, the reference frame spatial-temporal features and the current frame spatial-temporal features are first fed into the optical flow estimation network. The offset between the current frame and the reference frame is obtained by the optical flow estimation network, and the generated offset and the sampled points are combined to filter out the features that are not relevant to the current frame by our deformable spatial-temporal attention mechanism. So that the spatial information of multiple reference frames is aggregated with the temporal information related to the position of the feature points in the current frame. In this way, the quality of HR reconstruction of the current frame can be improved. To further speed up the training and filter out irrelevant feature points, we select only highly relevant and deformable sampling points to synthesize new values of the target, avoiding model degradation due to irrelevant features. Before entering the DSTA module, we use a spatial-temporal extraction network with low computational complexity to extract the spatial-temporal features of the video to make our proposed DSTALVSR lightweight, as shown in Fig. 1, our model achieves lightweight and obtains the best performance at the same time. Our contributions are as follows: (1) We propose a deformable spatial-temporal attention mechanism that aggregates the spatial-temporal contextual information of multiple current reference system features. (2) We use a lightweight spatial-temporal extraction module to further extract the spatial-temporal features of the video. (3) We have conducted extensive experiments on benchmark datasets. The quantitative and qualitative evaluation results show that our proposed DSTALVSR outperforms existing methods while achieving lightweight features.

2   Related Work

2.1   Single Image Super-Resolution

SRCNN [7] was the first approach to use deep convolutional networks in a super-resolution task. This work inspired many CNN-based Single Image Super-

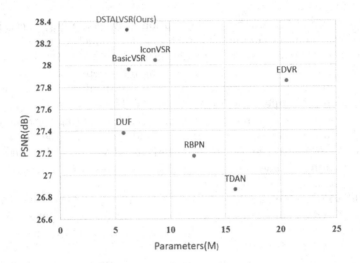

Fig. 1. Comparison of various VSR methods on the Vid4 dataset with an upsampling factor of 4. Blue and red dots represent other VSR methods, our proposed method, respectively. (Color figure online)

Resolution methods. For example, Kim et al. [12] proposed a residual learning strategy using a multilayer deep network, which showed significant improvement in super-resolution accuracy. Instead of applying traditional interpolation methods, [18] designed a subpixel convolutional network to efficiently upsample the low-resolution input. This operation reduces the computational complexity and enables a real-time network. Additional networks such as DBPN [8] and RDN [26] were proposed to further improve the performance of SISR using high quality large image datasets.

2.2   Video Super-Resolution

Spatial-temporal context information between LR frames plays an important role in the VSR task. Previous VSR methods can be classified as motion-compensated and non-motion-compensated based methods. Motion-compensated-based methods [11] first use classical algorithms to obtain optical flow and then construct a network for high-resolution image reconstruction. [2] integrate these two steps into a unified framework and train them in an end-to-end manner. Since accurate estimation of optical flow is a challenging task, for example, the method of [25] use 3D convolution to extract features directly from multiple frames and fuse spatial information of two adjacent frames to enhance the reconstruction of HR videos, which are usually computationally expensive despite their simplicity. In fact, they ignore the long-term spatial-temporal information in the video. In order to make better use of the long-term spatial-temporal information in videos, we propose DSTALVSR.

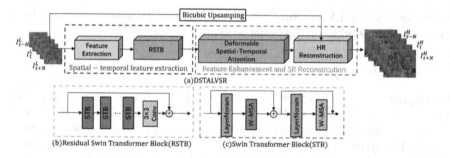

Fig. 2. (a) shows the structure of the model. It consists of two main components: the spatial-temporal feature extraction (STFE) subnet and the feature enhancement and SR reconstruction (FESR) subnet. Given an input LR frame, the STFE subnet first extracts the feature mapping F_i^{ST} and then feeds it to the FESR subnet along with the target LR frame I_t^L and bi-trivial interpolation to generate the output I_t^H. (b) shows the network results of RSTB, consisting of several STB modules and convolutional layers. (c) shows the specific structure of the Swin Transformer(STB).

3 Proposed Method

3.1 Network Architecture

The structure of the proposed network is shown in Fig. 2. Given 2n+1 consecutive LR frames $\{I_1^L,...., I_t^L,...,I_{2n+1}^L\}$, where $I_t^L \in R^{H \times W \times C}$ is the tth input frame. Our network goal is to generate the corresponding HR frames $\{I_1^H,...., I_t^H,...,I_{2n+1}^H\}$, where $I_t^H \in R^{4H \times 4W \times C}$. Our model consists of two main sub-networks, the spatial-temporal feature extraction network, and the spatial-temporal feature enhancement and super-resolution reconstruction network. Specifically, in the spatial-temporal feature extraction network we first use 3×3 convolutional layers in the feature extraction module to obtain shallow features $\{F_i^L\}_{i=1}^{2N+1}$ from the input 2N+1 LR video frames. Considering that these shallow features lack long-range spatial information due to the localized nature of the convolutional layer, this may lead to poor quality of the spatial-temporal feature enhancement and SR reconstruction modules. We would like to further extract these shallow features to obtain the spatial-temporal correlation between video frames. Inspired by the fact that Transformer [1,16] has the ability to model long-term dependencies. So we used the Residual Swin Transformer block (RSTB) [14] with low computational complexity proposed by Liang et al. to further extract the features of long-term spatial-temporal information $\{F_i^{ST}\}_{i=1}^{2N+1}$.

In order to enhance the final super-resolution reconstruction of the model, our proposed deformable spatial-temporal attention module needs to further enhance the spatial-temporal information of the frame feature sequence $\{F_i^{ST}\}_{i=1}^{2N+1}$. The 2N reference frame spatial-temporal feature sequences $\{F_{i,i\neq j}^{ST}\}_{i=1}^{2N}(i \in [1,2N], i \neq j)$ favorable spatial-temporal information are adaptively aggregated to the current frame feature F_j^{ST}. Finally, upsampling is performed through the pixelshuffle layer to complete the reconstruction of video frames from low to high resolution. The DSTA structure is shown in Fig. 3. A detailed description of each module is discussed in the following sections.

3.2 Spatial-Temporal Feature Extraction

Firstly, one 3 × 3 convolutional layer is used in the feature extraction module to obtain shallow features $\{F_i^L\}_{i=1}^{2N+1}$ from the input 2N+1 LR video frames. To further obtain the long-term spatial-temporal dependencies between video frames. We further extract the spatial-temporal features of video frames by using 5 Swin Transformer blocks. the structure of Swin Transformer blocks is shown in Fig. 2. The expressions are as follows

$$\{F_i^L\}_{i=1}^{2N+1} = H_{SF}\left(I_{LR}\right), \tag{1}$$

$$\{F_i^{ST}\}_{i=1}^{2N+1} = H_{RSTB}(\{F_i^L\}_{i=1}^{2N+1}) \tag{2}$$

where $H_{SF}()$ stands for 3 × 3 convolutional layers to extract shallow features of low-resolution video frames. $H_{RSTB}()$ stands for Residual Swin Transformer block [14].

3.3 Deformable Spatial-Temporal Attention

In Deformable Spatial-Temporal Attention, spatial-temporal information in the spatial-temporal features $\{F_{i,i\neq j}^{ST}\}_{i=1}^{2N}(i \in [1, 2N], i \neq j)$ of 2N reference frames is fused to the current frame spatial-temporal feature F_j^{ST}. To achieve deformable spatial-temporal attention for F_i^{ST} and F_j^{ST}. A set of sampling points is marked at F_i^{ST} and an offset Δp_θ is obtained through the offset generation network. The offset Δp_θ occurring between F_j^{ST} and $\{F_{i,i\neq j}^{ST}\}_{i=1}^{2N}(i \in [1, 2N], i \neq j)$ is obtained by optical flow estimation network $H_{spynet}(\cdot)$ [17]. F_j^{ST} and $\{F_{i,i\neq j}^{ST}\}_{i=1}^{2N}(i \in [1, 2N], i \neq j)$ are fed into the linear projection layer. According to the rules of the traditional attention mechanism [21]. F_j^{ST} generates the query vector $Q_j(p)$ and $\{F_{i,i\neq j}^{ST}\}_{i=1}^{2N}(i \in [1, 2N], i \neq j)$ generates the key vector $K_i(p_\theta)$ and the value $V_i(p_\theta)$

$$\Delta p_\theta = H_{spynet}(F_j^{ST} \cdot F_i^{ST}), \tag{3}$$

$$Q_j(p) = F_j^{ST} W_q, K_i(p_\theta) \ \ = F_i^{ST} W_k, V_i(p_\theta) = F_i^{ST} W_v \tag{4}$$

where p denotes the position of the query vector, p_θ the position of the deformable sampling point on F_i^{ST}. For ease of exposition, F_i^{ST} and $\{F_{i,i\neq j}^{ST}\}_{i=1}^{2N}(i \in [1, 2N], i \neq j)$ are equivalent in the formula. W_q, W_k, W_v all denote linear projection matrices. To reduce the computational complexity, by calculating the inner product between $Q_j(p)$ and $K_i(p_\theta)$ as the correlation values, we chose the first s points of the correlation values as the deformable sampling points. The expressions are as follows:

$$R_S = Q_j(p) \cdot K_i(p_\theta), \tag{5}$$

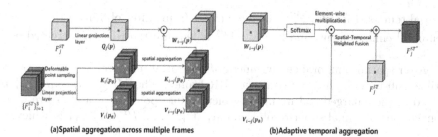

(a)Spatial aggregation across multiple frames   (b)Adaptive temporal aggregation

Fig. 3. Detailed process of Deformable spatial-Temporal Attention Mechanism (DSTA) module. DSTA is mainly divided into two processes: Spatial aggregation across multiple frames and Adaptive temporal aggregation.

where R_S denotes the value of correlation. The larger the value of R_S, the greater the relevance of the deformable sampling points. We use the Softmax function to calculate the spatial weight of the s deformable sample points to achieve the spatial aggregation of these deformable sample points. The spatially aggregated embedding feature vector is then obtained from the spatial weights of the deformed sampling points and the key vector $K_i(p_\theta)$ as follows:

$$\omega_s = \frac{exp\langle Q_j(p) \cdot K_i(p_\theta)\rangle}{\sum_{j=1}^{s} exp\langle Q_j(p) \cdot K_i(p_\theta)\rangle} \tag{6}$$

$$K_{i \to j}(p_\theta) = \omega_s \cdot K_i(p_\theta) \tag{7}$$

Where ω_s is the spatial weight of s deformable sampling points. $K_{i \to j}(p_\theta)$ is denoted as the spatially aggregated key vector. As with $K_{i \to j}(p_\theta)$, $V_{i \to j}(p_\theta)$ also implements spatial aggregation. Then by multiplying the query vector $Q_j(p)$ with the spatially aggregated $K_{i \to j}(p_\theta)$, we obtain the temporal weight $W_{i \to j}(p)$ for each reference frame. And aggregate these time weights via softmax function:

$$W_{i \to j}(p) = K_{i \to j}(p_\theta) \cdot Q_j(p) \tag{8}$$

$$\omega_t = \frac{exp\langle W_{i \to j}(p)\rangle}{\sum_{j=1}^{2N, j \neq i} exp\langle W_{i \to j}(p)\rangle} \tag{9}$$

where ω_t denotes the aggregated temporal attention weight. The spatial aggregated value vector $V_{i \to j}(p_\theta)$ and the aggregated temporal attention weight ω_t are weighted fused:

$$\omega_j^{st} = \sum_{j=1}^{2N, j \neq i} \omega_t \cdot V_{i \to j}(p) \tag{10}$$

where ω_j^{st} is denoted as spatial-temporal attention weight. The obtained spatial-temporal attention weight ω_j^{st} is multiplied by F_j^{ST} plus F_j^{ST} to obtain the spatial-temporal feature F_j^{ST*} of the current frame after obtaining the spatial-temporal enhancement, and finally F_j^{ST*} is fed into the HR reconstructed network.

After spatial-temporal enhancement of all spatial-temporal feature maps, we utilize sub-pixel layers to reconstruct HR frames. The $\{I_1^L,...., I_t^L,...,I_{2n+1}^L\}$ is added to the enlarged feature map by bicubic interpolation as a global residual connection. The final super-resolution output $\{I_1^H,...., I_t^H,...,I_{2n+1}^H\}$ is achieved.

4   EXPERIMENT

4.1   Experimental Settings

We conducted our experiments on the most commonly used VSR dataset, Vimeo-90K [23]

Training Dataset. Vimeo-90K [23] contains 64612 training samples. Each sample contains seven consecutive video frames of the same scene. The size of each frame is 448×256. We evaluated our model on three commonly used test datasets: Vid4 [15], UDM10 [24], and Vimeo-90K-T [23]. In our experiments, peak signal-to-noise ratio (PSNR) and structural similarity index (SSIM) are used as metrics.

Network Settings. We train our network using four NVIDIA GeForce GTX 2080Ti Gpus with a mini-batch size of three per GPU. The training of all datasets required 300 000 iterations. We use Adam as the optimizer and use a cosine learning rate decay policy with an initial value of 4e - 4. The input image is augmented by random cropping, flipping, and rotation. The clipping size is 64×64, which corresponds to an output of 256×256. We use SPyNet [17] for motion estimation to obtain the corresponding offsets. Charbonnier loss is adopted as our loss function, which is expressed as follows: $L\left(\hat{I}^H, I^H\right) = \sqrt{\left\|\hat{I}^H - I^H\right\|^2 + \epsilon^2}$, ϵ is set to 0.0001 in our experiments. Where $\hat{I}^H$ is the real HR video frame and I^H is the HR video frame estimated by our model.

4.2   Quantitative Comparison

To evaluate the performance. We compared the proposed with several state-of-the-art VSR methods: DUF [11], RBPN [9], EDVR [22], DUF [11], TDAN [20], BasicVSR [3], IconVSR [3]. For a fair comparison, we will use the same downsampling operation to generate VSR methods for LR frames for comparison.

We first analysed the quantitative comparison of the three test datasets. Table 1 gives the PSNR and SSIM values for the different methods tested on the Vid4, UMD10 and Vimeo-90K-T datasets. As some of these models are not yet publicly available, here we directly use the results shown in their publications. We observe that the proposed model (DSTALVSR) achieves the best performance in all three aspects. In particular, the PSNR value of Ours is 0.28 dB higher than the second best IconVSR method on the Vid4 dataset. DSTALVSR also achieves

Table 1. Quantitative comparison (PSNR/SSIM) on Vid4 [15], UDM10 [24] and Vimeo-90K-T [23] dataset for 4× video super-resolution. Red indicates the best and Blue indicates the second best performance

Methond	Vid4	UDM10	Vimeo-90K-T	Parameters(M)
Bicubic	21.80/0.5246	28.47/0.8253	31.30/0.8687	-
RBPN [9]	27.17/0.8205	38.66/0.9596	37.20/0.9458	12.2
DUF [11]	27.38/0.8329	38.48/0.9605	36.87/0.9447	5.8
EDVR [22]	27.85/0.8503	39.89/0.9686	37.81/0.9523	20.6
TDAN [20]	26.86/0.8140	38.19/0.9586	36.31/0.9376	16.2
BasicVSR [3]	27.96/0.8553	39.96/0.9694	37.53/0.9498	6.3
IconVSR [3]	28.04/0.8570	40.03/0.9694	37.84/0.9524	8.7
DSTALVSR(Ours)	28.32/0.8637	40.18/0.9702	38.02/0.9536	5.7

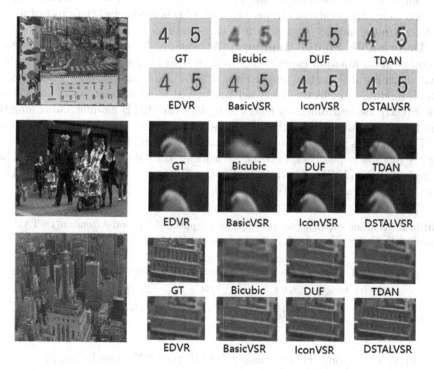

Fig. 4. Qualitative results for ×4 VSR on the Vid4 dataset.

PSNR gains of 0.15 dB and 0.18 dB on the UDM10 and Viemo90K-T datasets, respectively. In terms of model size, DSTALVSR can reduce the parameters by 34.4% compared to the second best IconVSR. Our model achieves the best results while being lightweight.

4.3 Qualitative Comparison

The visual comparison of our proposed DSTALVSR with other methods is shown in Fig. 4. In the first row, we can clearly see that the visual results achieved by DSTALVSR are almost identical to the real HR video frames. From the visual comparison in the third row, BasicVSR and IconVSR are the best results so far, and they still do not reconstruct more linear textures that match the visual characteristics. However, the architectural images reconstructed by the proposed DSTALVSR can show rectangular windows similar to the original images and their textures behave more closely to the real HR video frames.

4.4 Ablation Experiment

In this section, we perform an ablation study of the proposed deformable spatial-temporal attention mechanism and investigate the effect of using different deformation sampling points on the performance of this model and the effect of using different frame number aggregation in this module.

Deformable Spatial-Temporal Attention Mechanism. We conduct decomposition ablation experiments to investigate the effect of the deformable spatial-temporal Attention mechanism (DSTA), and also explore whether the RSTB module can effectively extract the spatial-temporal features of the video. We directly use convolutional layers to replace the RSTB module and eliminate the DSTA module in the network represented as our "Base" model. Only the RSTB module is put back into the network as a "Base+RSTB" model. Only the DSTA module is put back into the network as the "Base+DSTA" model. The complete DSTALVSR network is used as the "Base+RSTB+DSTA" model. The results are shown in Table 2.

Table 2. Ablation experiments of Deformable Spatial-Temporal Attention(DSTA) and Residual Swin Transformer block(RSTB) on the Vid4 dataset

Method	Base	Base+RSTB	Base+DSTA	Base+RSTB+DSTA
PSNR	26.37	26.93	27.58	28.32
SSIM	0.8013	0.8166	0.8411	0.8637

When adding only DSTA without RSTB, the PSNR can be improved from 26.37 to 27.58. This indicates that our proposed DSTA module is good at aggregating the spatial-temporal context information of multiple video frames. When RSTB is added for spatial-temporal feature extraction, the PSNR can be improved from 27.58 to 28.32. This indicates that RSTB can effectively extract the spatial-temporal features of the video, which shows that DSTA and RSTB can complement each other to play a good VSR effect.

Influence of Aggregating Different Frames Number During Inference. As shown in Table 3, we use different numbers of video frames for adaptive spatial-temporal information aggregation in the Vid4 dataset (33 frames).

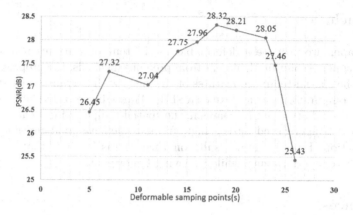

Fig. 5. Ablation study results for different deformable sampling points on the Vid4 dataset.

The performance is positively correlated with the number of aggregated frames. It proves that DSTA can effectively aggregate spatial-temporal information and perform long-term modeling. In fact, the performance benefit gradually decreases when the number of frames is greater than 20. This indicates that a small time interval cannot provide too much spatial-temporal information because the spatial-temporal information of adjacent frames is too similar.

Table 3. Ablation study results with different frame numbers on the Vid4[29] dataset.

#Frame	3	7	15	20	33
PSNR	28.11	28.24	28.28	28.31	28.32
SSIM	0.8591	0.8619	0.8629	0.8634	0.8637

The Effect of Different Deformable Sampling Points. We use different deformable sampling points in the video spatial-temporal features for spatial-temporal aggregation. As described in the main text, we use the inner product of the query vector and the key vector as the relevance score of the sampled points. As shown in Fig. 5, we achieved the best performance when collecting 18 deformable feature points in the spatial-temporal features of each reference frame. The reason is that choosing too many deformable sampling points tends to aggregate irrelevant features into the video, resulting in a degradation of our performance. With this ablation experiment, it is demonstrated that our DSTALVSR needs to manually test how many deformable sample points many times to achieve the best spatial-temporal aggregation. In fact, this is a limitation of DSTALVSR. One may need to retrain the model to intelligently sample the number of deformable points to improve its model, which we leave for future work.

5   Conclusion

In this paper, we propose a deformable spatial-temporal attention mechanism for video super-resolution. Different from previous methods, the proposed model extracts the spatial-temporal features of the video with a Transformer structure with low computational complexity (RSTB). Through the proposed deformable spatial-temporal attention mechanism, the spatial-temporal information of multiple frames is aggregated into a single frame to achieve good super-resolution reconstruction. Experimental results on three datasets show that our model achieves better performance while achieving lightweight.

References

1. Arnab, A., Dehghani, M., Heigold, G., Sun, C., Lučić, M., Schmid, C.: Vivit: a video vision transformer. In: Proceedings of the IEEE/CVF International Conference on Computer Vision, pp. 6836–6846 (2021)
2. Caballero, J., et al.: Real-time video super-resolution with spatio-temporal networks and motion compensation. In: Proceedings of the IEEE Conference on Computer Vision and Pattern Recognition, pp. 4778–4787 (2017)
3. Chan, K.C., Wang, X., Yu, K., Dong, C., Loy, C.C.: BasicVSR: the search for essential components in video super-resolution and beyond. In: Proceedings of the IEEE/CVF Conference on Computer Vision and Pattern Recognition, pp. 4947–4956 (2021)
4. Chan, K.C., Wang, X., Yu, K., Dong, C., Loy, C.C.: Understanding deformable alignment in video super-resolution. In: Proceedings of the AAAI Conference on Artificial Intelligence, vol. 35, pp. 973–981 (2021)
5. Chen, J., Tan, X., Shan, C., Liu, S., Chen, Z.: VESR-Net: the winning solution to Youku video enhancement and super-resolution challenge. arXiv e-prints, pp. arXiv-2003 (2020)
6. Dai, J., et al.: Deformable convolutional networks. In: Proceedings of the IEEE International Conference on Computer Vision, pp. 764–773 (2017)
7. Dong, C., Loy, C.C., He, K., Tang, X.: Image super-resolution using deep convolutional networks. IEEE Trans. Pattern Anal. Mach. Intell. **38**(2), 295–307 (2015)
8. Haris, M., Shakhnarovich, G., Ukita, N.: Deep back-projection networks for super-resolution. In: Proceedings of the IEEE Conference on Computer Vision and Pattern Recognition, pp. 1664–1673 (2018)
9. Haris, M., Shakhnarovich, G., Ukita, N.: Recurrent back-projection network for video super-resolution. In: Proceedings of the IEEE/CVF Conference on Computer Vision and Pattern Recognition, pp. 3897–3906 (2019)
10. Isobe, T., Jia, X., Gu, S., Li, S., Wang, S., Tian, Q.: Video super-resolution with recurrent structure-detail network. In: Vedaldi, A., Bischof, H., Brox, T., Frahm, J.-M. (eds.) ECCV 2020. LNCS, vol. 12357, pp. 645–660. Springer, Cham (2020). https://doi.org/10.1007/978-3-030-58610-2_38
11. Jo, Y., Oh, S.W., Kang, J., Kim, S.J.: Deep video super-resolution network using dynamic upsampling filters without explicit motion compensation. In: Proceedings of the IEEE Conference on Computer Vision and Pattern Recognition, pp. 3224–3232 (2018)

12. Kim, J., Lee, J.K., Lee, K.M.: Accurate image super-resolution using very deep convolutional networks. In: Proceedings of the IEEE Conference on Computer Vision and Pattern Recognition, pp. 1646–1654 (2016)
13. Li, Z., Yang, J., Liu, Z., Yang, X., Jeon, G., Wu, W.: Feedback network for image super-resolution. In: Proceedings of the IEEE/CVF Conference on Computer Vision and Pattern Recognition, pp. 3867–3876 (2019)
14. Liang, J., Cao, J., Sun, G., Zhang, K., Van Gool, L., Timofte, R.: SwinIR: image restoration using Swin transformer. In: Proceedings of the IEEE/CVF International Conference on Computer Vision, pp. 1833–1844 (2021)
15. Liu, C., Sun, D.: On Bayesian adaptive video super resolution. IEEE Trans. Pattern Anal. Mach. Intell. **36**(2), 346–360 (2013)
16. Liu, Z., et al.: Swin transformer: hierarchical vision transformer using shifted windows. In: Proceedings of the IEEE/CVF International Conference on Computer Vision, pp. 10012–10022 (2021)
17. Ranjan, A., Black, M.J.: Optical flow estimation using a spatial pyramid network. In: Proceedings of the IEEE Conference on Computer Vision and Pattern Recognition, pp. 4161–4170 (2017)
18. Sun, Y., Chen, J., Liu, Q., Liu, G.: Learning image compressed sensing with sub-pixel convolutional generative adversarial network. Pattern Recogn. **98**, 107051 (2020)
19. Tao, X., Gao, H., Liao, R., Wang, J., Jia, J.: Detail-revealing deep video super-resolution. In: Proceedings of the IEEE International Conference on Computer Vision, pp. 4472–4480 (2017)
20. Tian, Y., Zhang, Y., Fu, Y., Tdan, C.X.: Temporally-deformable alignment network for video super-resolution. In: 2020 IEEE CVF Conference on Computer Vision and Pattern Recognition (CVPR), pp. 3357–3366 (2020)
21. Vaswani, A., et al.: Attention is all you need. In: Advances in Neural Information Processing Systems, vol. 30 (2017)
22. Wang, X., Chan, K.C., Yu, K., Dong, C., Change Loy, C.: EDVR: video restoration with enhanced deformable convolutional networks. In: Proceedings of the IEEE/CVF Conference on Computer Vision and Pattern Recognition Workshops(2019)
23. Xue, T., Chen, B., Wu, J., Wei, D., Freeman, W.T.: Video enhancement with task-oriented flow. Int. J. Comput. Vision **127**, 1106–1125 (2019)
24. Yi, P., Wang, Z., Jiang, K., Jiang, J., Ma, J.: Progressive fusion video super-resolution network via exploiting non-local spatio-temporal correlations. In: Proceedings of the IEEE/CVF International Conference on Computer Vision, pp. 3106–3115 (2019)
25. Ying, X., Wang, L., Wang, Y., Sheng, W., An, W., Guo, Y.: Deformable 3d convolution for video super-resolution. IEEE Signal Process. Lett. **27**, 1500–1504 (2020)
26. Zhang, Y., Tian, Y., Kong, Y., Zhong, B., Fu, Y.: Residual dense network for image super-resolution. In: Proceedings of the IEEE Conference on Computer Vision and Pattern Recognition, pp. 2472–2481 (2018)
27. Zhou, Y., Wu, G., Fu, Y., Li, K., Liu, Y.: Cross-MPI: cross-scale stereo for image super-resolution using multiplane images. In: Proceedings of the IEEE/CVF Conference on Computer Vision and Pattern Recognition, pp. 14842–14851 (2021)

Author Index